Endorsements

Relating science, religion, mathematics and eugenics, this book challenges both scientists and theologians. It attempts to humanize science, empower religion, and make our common life flourishing. A critical, creative and inspiring book.

Professor Dr Kuruvilla Pandikattu SJ
Chair Professor of JRD Tata Foundation on Business Ethics
XLRI - Xavier School of Management, Jamshedpur, India.

In this wide-ranging study, Karim Hirji explores the linkages between religions (Hinduism, Buddhism, Christianity and Islam) and eugenics, science and mathematics. He uses an integrative approach that recognizes the interconnected, historical and dualistic nature of religion and of the other three entities.

Rich with illuminating case studies and illustrations, its discursive, reflective style draws the reader in on a fascinating journey of discovery. *Religion, Eugenics, Science and Mathematics* is a call for tolerance, equality and freedom of thought and of religious belief through the creation of a saner, gentler, more just global society, free of the toxic influences of neoliberalism.

Elizabeth Jones, MA
Teacher, Quaker, Former Co-editor of Christian Today, UK.

Under the premise that all humans are equal in dignity and have equal rights, this book critically explores the interplay between religion, science and mathematics on the theological and societal dimensions. Its call for principled unity among peoples of varied religious and cultural backgrounds is particularly important in the era of the global environmental crisis. The book is enriched by the author's personal experiences and the highlighting of the role of women in science and religion. The expose of the roles of esteemed scholars, academic institutions and theologians in the supremacist eugenics movement is an eye opener.

Professor Marjorie Mbilinyi
Professor of Education, University of Dar es Salaam (1968-2003)
Independent Researcher and Writer.

Like Hirji's earlier book, *Religion, Politics and Society,* this bulky book shines with brilliance and humanity. It emphatically demonstrates that the relationship between religions and the sciences is complex and interweaving. The pseudo-scientific doctrine of eugenics, the echoes of quantum theory in theology, the ambivalent perspectives of religions on mathematics and the standings of religions on climate change are among the many examples invoked to show both the conflict-ridden as well as complimentary relationship between science and religions.

Abdul Paliwala
Emeritus Professor of Law
University of Warwick

RELIGION, EUGENICS, SCIENCE AND MATHEMATICS

AN ETERNAL KNOT

Karim F Hirji

Daraja Press

Published by
Daraja Press
https://darajapress.com
Wakefield, Québec, Canada
&
Zand Press
https://zandgraphics.com
Nairobi, Kenya

ISBN 978-1-990263-22-4 (print)
ISBN 978-1-990263-73-6 (ebook)

Cover Design:
Joshua Folorunso, 4LR Creations, Nairobi, Kenya

Co-editors: Zarina Patel and Rosa Hirji

Library and Archives Canada Cataloguing in Publication

Title: Religion, eugenics, science and mathematics : an eternal knot / Karim F Hirji.
Names: Hirji, Karim F., author.
Description: Includes bibliographical references.
Identifiers: Canadiana 20230198953 | ISBN 9781990263224 (softcover)
Subjects: LCSH: Religion and science. | LCSH: Eugenics. | LCSH: Mathematics.
Classification: LCC BL240.3 .H57 2023 | DDC 201/.65—dc23

To:

Rosa Hirji, Farida Hirji and Zarina Patel
Indispensable Anchors

*The book of nature which we have to read
is written by the finger of God.*
Michael Faraday

*The more I study nature, the more I stand
amazed at the work of the Creator.*
Louis Pasteur

*No path leads from a knowledge of that
which is to that which should be.*
Albert Einstein

*I maintain that faith in this world
is perfectly possible
without faith in another world.*
Rosalind Franklin

CONTENTS

	Preface	xi
Chapter 01	Introduction	1
Chapter 02	Emergence	7
Chapter 03	Eugenics	35
Chapter 04	Science	101
Chapter 05	Mathematics	171
Chapter 06	Ruminations	241
Chapter 07	Finale	271
	Sources	301
	Credits	309
	References	313
	Author Profile	365

PREFACE

God has no religion.
The hands that serve
are holier than the lips that pray.
Mohandas K Gandhi

Science is not only compatible with spirituality;
it is a profound source of spirituality.
Carl Sagan

THIS BOOK explores the interplay between religion, and science and mathematics on both the theological and societal dimensions. Addressed in a preliminary fashion in my earlier book, *Religion, Politics and Society,* I now take a deeper look into the issue. Like its prequel, this book has an interdisciplinary, historical approach and primarily focuses on the four major world religions—Hinduism, Buddhism, Christianity and Islam—together with secularism. Some minor faith systems appear as well. The two books share the same foundational premise: All humans are equal in dignity and have equal rights. There are no chosen people; there is no chosen religion. We all are a part of the global human family.

Our world is engulfed within an existential social, economic, health and environmental crisis. Resolution of this crisis requires unity among peoples of varied backgrounds across the globe. Unity does not mean homogeneity; it means realizing that we are in the same planetary boat, acknowledging and respecting our differences, celebrating cultural and religious diversity, promoting science and interfaith dialogue, and joining our hands to confront our shared problems.

++++

In writing this book, I have been assisted by many. Foremost, I thank the co-editors Zarina Patel and Rosa Hirji whose meticulous editorial corrections and insightful comments considerably enhanced its quality. Abdul Paliwala and Elizabeth Jones deserve accolades for valuable comments and suggestions. Zahid Rajan of Zand Press and Firoze Manji of Daraja Press deserve much credit for their support and for expeditiously producing an elegantly designed book. And I thank Joshua Folorunso for generating an appealing cover design. As always, this book would not have seen the light of the day without the loving support from Farida Hirji, Rosa Hirji and Rafik Hirji.

This book uses US English spelling. The quotations at the beginning of each chapter do not necessarily reflect my views. They are meant to show the diversity of views on the subject. As they are available at many websites, no sources are given. The images used in the book are mostly in the public domain. Their sources appear in the Credits section.

Karim F Hirji
March 2023

CHAPTER 01: INTRODUCTION

Gravity explains the motions of the planets,
but it cannot explain who sets the planets in motion.
Isaac Newton

Science investigates; religion interprets.
Science gives man knowledge, which is power;
religion gives man wisdom, which is control.
Science deals mainly with facts;
religion deals mainly with values.
The two are not rivals.
Martin Luther King

ARE RELIGION and science mutually exclusive and opposed entities? Or are they complimentary entities? Or are they distinct but non-antagonistic entities? How does the relationship between them affect their roles in society? Do divine beings exist? Scientists and theologians, believers and secularists, express divergent opinions on such matters. Consider the views of the four pioneering scientists quoted at the outset.

Michael Faraday, the 19th century scientist who invented the electric motor and advanced electromagnetic and atomic theory, was a devout Christian. He saw the Bible as the primary moral guide for humans. But his religious beliefs did not affect his scientific work. When conducting experiments and formulating his theories, he studiously adhered to the methods of science. He integrated the two strands of his outlook by asserting that the laws of nature discovered by science were formulated by God.

Louis Pasteur made pioneering contributions to microbiology, epidemiology and vaccine development. A fervent Catholic and a meticulous scientist, he took it as his Christian duty to do research of benefit to people. Like Faraday, he did not interject his Christian beliefs into his laboratory. Nature was created by God. To discover its laws was to come closer to God and appreciate the majesty of His creation. He also expressed broad tolerance for other faith systems.

Albert Einstein, one of the greatest scientists of all time, did not go to church, did not believe in a personal God, and did not have high regard for the scriptures of any faith system. Yet, he held that science could not provide a moral guide for humans. Like Pasteur, he was entranced by the wondrous beauty of the Universe. But instead of theism, Einstein adopted a pantheist outlook that visualized nature as God and God as nature. He was a humanist, a socialist, and a spiritual but not religious scientist.

Rosalind Franklin, a leading expert in crystallography and virology, was a co-discoverer of the structure of the basic molecule of life, DNA. She also unraveled the structures of some viruses and coal. Though unfairly appreciated by fellow scientists during her life, she persevered in path breaking research until her death by cancer at the age of 37. She adhered to Jewish traditions at the social level but doubted the existence of a creator and wondered whether the creator was male or female. As a humanist, she rejected the notion of life after death. What mattered to her was her contribution to improve the lives of people.

Four eminent personages of science, dedicated to the vision of science as the fountain of knowledge and a tool for improving human welfare, yet far apart in their stand on religion. It is an indicator that the linkage of science to religion, and the manner in which that linkage affects their societal functions, is a complex issue in need of careful elucidation. (Note: The references to science in this book generally cover mathematics as well.)

1.1 SCHOOL CHEMISTRY

I learned in my secondary school chemistry class that the salt I sprinkle over my eggs at breakfast is composed of two elements, sodium and chlorine. The former is a volatile, soft metal, and the latter, a noxious gas. Ingesting a spoonful of sodium will burn your tongue and throat, and cause major injuries, if not death. And breathing in chlorine for more than a minute will inflame your respiratory system and maybe kill you. Yet, under right conditions, atoms of the metallic element combine with those of the gaseous element in a one-to-one ratio to form the inert white powdery salt that is an essential nutrient for humans and animals and has medicinal and industrial uses. It is also a food preservative. Found in abundance in sea water, it is extracted through a drying process.

Chemistry was a fascinating subject. I learned the properties of many of the hundred or so basic elements that exist in nature, how they are grouped in the Periodic Table and how they form simple and complex molecules. Organic chemistry was a venture into the molecular basis of life. Theory was combined with experiments. Testing the properties of chemicals and trying to determine their composition was a fascinating exercise.

School chemistry and physics were my first significant steps into science. Learning science was never an imposition. An array of wonderful teachers made it a memorable experience. There were two unspoken but key ingredients here, trust and truth. Our experiments only partly confirmed what we were learning; over 95% of the material in our books lay unverified by us. Yet, we did not doubt it. Why? Because we were told that eminent scientists had verified it through careful experiments. We trusted our teachers and assumed that they were fully conversant with the textbooks. If we pursued advanced level studies, we too would know the subjects in depth. Science was a community—students, teachers, professors, researchers and eminent scientists—of which we were a part.

Another aspect of school science, which was mentioned now and then, was that science was of benefit to humanity. Knowledge of the properties of common salt and its role in human metabolism, for example, was of use in the control of high blood pressure and treatment of dehydration.

1.2 RELIGIOUS STUDIES

I was born in Tanzania into the Ismaili sect of Shia Islam during the colonial era. Racial and religious boundaries were firm. My primary school, located in Lindi, a town in southern Tanzania, enrolled children from Asian families only. Instead of history of Africa, we were taught the history of India and a modicum of Hindu beliefs and rituals. The morning assembly began with a Hindu bhajan.

On Saturday mornings, I attended a religious studies class held at the Ismaili prayer house. The two-hour session taught us the beliefs and rituals of Ismailism. We were implored to have absolute faith in our spiritual leader (Imam), the Aga Khan, who, we were told, had divine powers. Our teacher was a gentle, patient young man. We memorized the prayers and hymns, learned the teachings of our Imam and were guided to be well behaved kids. These sessions were all the more enjoyable because most of my friends were also present.

My middle schooling was in Dar es Salaam. The school was run by the Aga Khan Education Board and the students were mostly Ismaili Asians. We had a religious studies class in each grade. The teacher, an Ismaili missionary, was a strict disciplinarian. He was hardly popular. His monotonous delivery generated intellectual stupor. We were supposed to learn the meaning and essence of the hymns and prayers, and the history of our faith, our Imams, Prophet Muhammad and Islam. But he hardly broached these topics. His words did

2

not remain with me for more than a few days. And we never got an introduction to the Quran, the principal holy book of Islam.

I did not attend a formal religious studies course thereafter, but the informal learning that had begun in early childhood continued. My parents, relatives and the missionaries who spoke in the prayer house taught me the different aspects of Ismailism. My grandmother was my best teacher. The hymns she recited to me remain in my mind to this day. I also read books and magazines relating to Ismailism and religion in general. I was enchanted with *The Memoirs of the Aga Khan* by Sir Sultan Mohamed Shah, the Aga Khan III, and *My Experiments with the Truth* by Mohandas K Gandhi.

My learning about our faith was, like my studies in science, characterized by trust and truth. The incompetence of my middle school teacher did not alienate me from our religion. I had absolute faith in our spiritual leader, and valued the hymns composed by Ismaili saints. They embodied spiritual truth. Religious education, daily prayer attendance and partaking in religious ceremonies firmly cemented me to the Ismaili community. It gave me a moral guide and showed the path to the liberation of my soul. It was an indispensable aspect of my life.

1.3 RELIGION VERSUS SCIENCE

Trust, truth, solidarity and utility underlay my early immersion in both science and religion. But I was hardly aware of it. It was much later that I came to reflect on the overlap. At that time, I had no inkling about the existence of a conflict between these two domains of my life. I implicitly accepted that they were two different aspects of life. Religious truth was spiritual truth; scientific truth was practical, worldly truth.

My horizon expanded after I joined the University of Dar es Salaam to study mathematics. It was 1968, an era of student radicalism. As elsewhere, activist students were placing academic subjects under critical scrutiny. Culture, religion and politics came under the microscope as well. Within a couple of months, I had joined the radical student group.

My formal studies and extensive independent reading in those days informed me that science and religion were similar and different. Both depended on trust. But religious trust was absolute. It was unshakable, with no room for doubt. To doubt was sinful. In science, trust was conditional; doubt was a virtue. Even the theories of eminent scientists were subject to critical inquiry. The truth of science was malleable if new evidence entailed a new theory. Later I realized that doubt also featured in the history and practice of religion while belief without evidence is also an aspect of science. The relationship between the two traditions is more complex than what appears at first sight. One is usually born into one's faith. But entry into the congregation of science depends on education, choice, ability and circumstances.

The vast compendium of science is essential for modern life. Electric power at home, for example, is a product of science. Religious belief is also useful, but in a different way. Religion gives emotional solace and hope in times of despair. It helps one cope with the stresses of daily life. The bonding between those in a religious community is qualitatively distinct from the bonding, for example, between the students, teachers, researchers and appliers of chemistry.

And as I was to learn later, there were aspects to the realities and histories of science and religion that singularly tarnished their image as humane, ethical, and noble pursuits. One cannot comprehend the natures of religion and science, and the linkage between them without exploring both of their aspects.

1.4 OBJECTIVES AND APPROACH

The objective of this book is to examine the relationship between religion and science (including mathematics) at the conceptual, societal and historical levels. In particular, it ponders the following questions:

Question 1: Are religion and science mutually exclusive, opposing entities?
Question 2: Do divine beings and divine realms exist?
Question 3: Are science and religion valid but different forms of truth?
Question 4: What are the societal roles of science and religion?
Question 5: Can science provide a tenable, exalted code of ethics?
Question 6: What are the futures of religion and science?
Question 7: Can religion and science cooperate in resolving the daunting, existential problems facing humanity today?

The queries are explored in relation to the four major faiths (Christianity, Islam, Hinduism, Buddhism), some minor religions, and secularism. A host of topics and case studies from science and religion together with personal biographies are employed to illustrate the main points. The doctrine of eugenics that was popular in the West in the first half of the 20th century forms a major case study.

++++

Nature and society are dynamic, interconnected entities with diverse components. They are integrated systems propelled by opposing tendencies. This book thereby follows an evidence-based approach whose three pillars are change (historical analysis), interconnectedness (systems analysis) and incorporation of opposing tendencies (dialectics). The Eternal (Endless) Knot symbol found in Hinduism, Jainism and Buddhism aptly symbolizes this approach.

The Eternal Knot

The Eternal Knot embodies the intertwining of truth, compassion and wisdom. The obverse sides of its strands represent falsehood, moral laxity and frivolity. Both types of facets exist within the diverse strands of religion and science. The Knot informs and cautions us that religion, science and how they relate to each other are complex, dynamic and convoluted matters that are inundated with extensive controversy in substantive and moral terms.

Religion and science have evolved under varied social formations, paleolithic and neolithic societies, feudalism, capitalism, imperialism, slavery, colonialism, neo-colonialism, neoliberalism and socialism. A basic grasp of these terms will facilitate the reading of this book.

This book is not based on primary research. It builds from secondary sources. And it shares the definition of religion used in RPS (2022):

> **Religion** is a system of beliefs, practices and symbols shared by a community that accepts the existence of divine being(s) and/or supernatural realms and has modes of worship, rituals, stories and rules of conduct (ethical norms) that are taken to be of divine origin.

Terms like faith system, belief system, faith, religious tradition and religious creed are used as equivalent terms for religion.

This book is a sequel to an earlier work, *Religion, Politics and Society* (Hirji 2022), where these questions were partially examined. Now we take a deeper dive into the subject. Though this book can be read on its own, the reader is advised to become familiar with the earlier work, which from here on will be referred to as RPS (2022).

The reader will benefit from a familiarity with terms encountered in religion-related contexts like Theism, Atheism, Agnosticism, Monotheism, Polytheism, Deism, Pantheism, Animism, Humanism, Skepticism, Freethinkers, Naturalism, Rationalism, Syncretism, New

Age Beliefs, Alternative Beliefs, Spiritualism, Fundamentalism, Evangelism, Irreligious, Non-Religious, Non-Theist, Non-Believer, Spiritual but Not Religious, Nothing in Particular, Ideology, Materialism, Idealism, and Realism. Readers not familiar with them should consult RPS (2022) or another source.

For now, we take science as a body of knowledge derived from an interacting process of observation, pattern identification, hypothesis formulation, and testing or prediction. Hypotheses that withstand extensive testing become theories, but are not cast in stone. Though the room for experimentation in the social sciences is smaller than in the natural sciences, their theories also depend on continued critical scrutiny.

This book shares a fundamental premise with RPS (2022): All humans are equal in dignity and all faith systems and secular creeds deserve equal respect and freedom to operate. Yet, that freedom does not include freedom to harm others. It favors open-minded discussion of religion, atheism, agnosticism and science with the proviso that it be a respectful dialogue that does not compromise on history and evidence.

The ultimate objective of this book is to promote peace and harmony between different faiths, secularism, science and cultural traditions so as to harness them in an endeavor to tackle the major social, economic, political and environmental problems facing the human race today.

One does not ask of one who suffers.
What is your country and what is your religion?
One merely says: You suffer, that is enough for me.
Louis Pasteur

CHAPTER 02: EMERGENCE

*The whole is more than
the sum of its parts.*
Aristotle

*I do not feel obliged to believe that the same God
who has endowed us with senses, reason, and intellect
has intended us to forego their use.*
Galileo Galilei

❈

All **PHENOMENA** in nature and society have constituent parts. A papaya tree has roots, branches, leaves, seeds, flowers and fruits, and it needs water, sunlight and nutrients. A college class has students, instructor, books and curriculum. Each part has a function. Some are essential, and some are not. How does a tree grow? What makes the learning process effective? Discovering the properties of a phenomenon is generally done in one of three ways:

Strict Reductionism: Investigate the properties of each part. The object, system or event is fully comprehended by combining these properties in an additive manner. Another name for reductionism is atomism.

Liberal Reductionism: Investigate each part and also the relationships between them. Knowledge of the object, system or event is incomplete without doing both.

Holism: Investigate each part, the relationships between the parts as well as the operation of the object, system or event as a whole. A totality generally has laws of being and change that cannot be gleaned from the properties of its parts. Holism is also called Systems Theory, Complexity Theory or Emergence Theory.

These disparate approaches to gaining knowledge occur in all natural and social science disciplines like medicine, alternative medicine, psychology, psychiatry and education together with religion and theology. The conceptual and practical discord between them arises in a variety of contexts. In the theological arena, holism and reductionism pertain to issues like the existence of God, soul, sin, morality, free will, consciousness, fate and divine retribution. In the worldly arena, they pertain to free will, consciousness, morality, genetics, life attainment, crime and punishment, and personal responsibility, among many other matters.

Levels

A natural or social system, or a system of ideas, has parts of diverse functionality. Each part has sub-parts, which in turn have sub-sub-parts, and so on. A human is a physical body, mind, abilities, knowledge, life history, and memories. The body has bones, muscles, blood vessels, nerves and heart, lung, brain, liver, stomach and other organs. These parts form systems like the digestive system, respiratory system, cardio-vascular system, nervous system and musculoskeletal system. Each constituent has tissues and cells. There are sub-cellular parts like nucleus, cytoplasm, membrane and mitochondria that are made from protein, fat and other molecules. And these molecules have atoms of different chemical elements.

A hierarchy of levels of components within components exists for all systems. But it is not a fixed scheme. A community is grouped according to economic status, culture, race or religion. For each grouping, sub-levels may be formed. Which grouping best explains how the community functions? Is a top-down (holistic) or a bottom-up (reductionist) approach better? For most systems, the experts remain divided on these questions.

2.1 MEDICINE AND HOLISM

The discord between reductionism and holism is particularly acute in health and medicine. Medical practice today is divided into two branches. The main branch has scientifically trained doctors and specialists—family medicine, pediatrics, cardiology, surgery, infectious diseases, endocrinology, neurology and scores of such areas. The other branch, called Complementary and Alternative Medicine (CAT), fields healers trained in health care modalities like acupuncture, massage therapy, herbalism, Ayurveda, Chinese medicine, meditation, yoga, homeopathy, folk medicine, reflexology, osteopathy, chiropractic and crystal therapy.

Medical doctors largely view CAT as quackery that mostly depends on the placebo effect, but may also harm the recipients. Yet, some of them employ CAT modalities like meditation and acupuncture in the treatment of mental ailments and pain. Much of CAT has not undergone the rigorous testing imposed by law on conventional therapies, especially medicinal drugs.

Advocates of CAT accuse mainstream doctors of treating the human body as a machine. Focusing on specific organs, a narrow drug and surgical approach is used to treat ill health. Natural remedies, nutrition and prevention are neglected. The adverse effects of what they prescribe are neglected. And so are the effects of the mind on the body. These shortcomings are attributed to the reductionist framework of scientific medicine.

The CAT practitioners claim they treat the human body as a whole. Instead of just relying on laboratory and clinical tests and dysfunction of an organ, they take the patient's concerns, lifestyle and environment into consideration and focus on prevention and natural, harmless remedies. Based on the tenet that the essence of life emanates from a life force energy, they argue that CAT is a holistic modality, both in theory and practice. In traditional Chinese medicine, the force is called Qi. In other contexts, it is elan vital, vital breath, or energy. Good health results when this force is in balance, and mental or physical pathology ensues when the balance is disturbed. The basic role of the health provider is to guide the afflicted to rebalance his or her life force. CAT holds that people should be empowered to take control of their own health.

++++

Take the case of diseases of the eye. An eye surgeon has intimate knowledge of the structure, parts and functions of eye and is well acquainted with the nerves, blood vessels, muscles and tissues surrounding the eye. She likely has a good grasp of the physiology and biochemistry of the eye and knows the varied ways eye diseases manifest. In theory, her purview starts from the eye, extends to the brain and the whole human body, and beyond to the environment.

Besides finely honing her surgical skills, she has to know about the relevant pharmaceuticals and supportive devices, the natural history and prevention of eye diseases, the role of genes in eye diseases, and effects of conditions like diabetes and hypertension on eye health. Ideally, she should also know that several major preventable eye ailments arise from adverse social and economic conditions, especially among the poor in the Global South nations.

In principle, nothing stands in the way of the eye specialist having a holistic perspective. Yet overspecialization is the norm today. Surgeons with poor skills at times practice. Unneeded eye operations are not uncommon. Some of the charges leveled by the CAT advocates are valid. Excessive use of anti-inflammatory and antibiotic has been documented as well.

The CAT modalities of eye care include use of biological substances (herbs, minerals, vitamins and antioxidant supplements, and a low fat diet rich in green vegetables), mind-body techniques (relaxation, prayer), physical therapies (yoga, cold compress, acupuncture, eye massage, frequent blinking) and lifestyle advice (no late night television or reading in poor light).

> *A peeled and grated fresh potato applied to the eyelids acts as an astringent and is said to have a healing effect.* (Astbury 2001).

Most CAT treatments lack evidence based on clinical trials. They derive from ancient traditions like folk medicine, Ayurveda and Chinese Medicine as well as common sense and the Internet. Some options are efficacious to a degree, some are harmful, and some are benign. Only a few have plausible, objective evidence of effectiveness. Thus, for some eye conditions,

> [The American Academy of Ophthalmology] *task force has acknowledged that acupuncture may be useful as an adjunctive therapy or as an acceptable alternative to conventional treatment.* (Astbury 2001).

While CAT practitioners spend more time with their patients and give more attention to diet and lifestyle issues, to call the current practice of various CAT modalities holistic is misleading. They utilize industrially prepared extracts from plants more than the seeds, leaves, flowers, and fruits of the herbal plants. CAT centers are run as profit making businesses just like private doctor practices. CAT practice is individual patient-oriented practice and is rarely involved in public health programs and nutrition drives. In the West, CAT practitioners cater to the affluent and the middle class, not the low-income and minority groups. In the Global South nations, traditional medicine is often accessed by the poor who are unable to afford the high cost conventional medical care. Free conventional care, when available, has major shortcomings including delays, poor quality of service, and corruption.

Present day CAT is a take-your-chance option. It is a profit driven venture with superficial trappings of holism. An open-minded ophthalmologist opines:

> *After thousands of years the human race has remained profoundly superstitious and prepared to try virtually any remedy when faced with a threat such as blindness. There is infinite scope for quacks and entrepreneurs and many harebrained schemes have fallen by the wayside but, nevertheless, a vast knowledge base has accumulated. Whether treatment is based on hard evidence, common sense, old wives' tales, or oriental wisdom, we should view it all with an open mind.* (Astbury 2001).

Similarly, the multiplicity of problems associated with mainstream medicine stem less from non-adherence to the principles of holism than from the ubiquity of the profit motive in the system, starting from the drug and device manufactures to pharmacies and hospitals. Financial goals including the rules of reimbursement of insurance firms influence the diagnosis and treatment given by doctors and surgeons. Another negative influence stems from drug companies. Sleek, misleading ads, the inducements (bribes) they offer to medical practitioners, and the lackluster oversight by the regulatory agencies factor into the provision of unneeded and harmful care. Other problems include the manner of funding of academic research, and medical training that neglects nutrition, the capacity of the body to heal itself, the placebo effect and public health. Geared towards specialization, it falls short on due cooperation among the specialists.

Both conventional and alternative medical practices are multi-billion dollar entities ensconced within the neoliberal order. Policies based on public health and prevention, equal access to good quality affordable health care of evidence-based treatment modalities often play a second fiddle. Money, not philosophy, is the key problem. Conventional medicine and CAT are afflicted by commercialization and unequal access to health care and reduction of personal interactions between patients and doctors. Home visits by family doctors, a fine aspect of past medical practice, are now a relic. That some doctors manage to enjoin conventional and complementary therapies signifies that an evidence-based, compassion driven integration of the two modalities is a distinct possibility.

We draw an important conclusion from this discussion: For assessing a form of practice or tradition, it is not sufficient to judge its philosophy only. The real assessment has to examine the practice as well. Reductionism and holism stand or fall not simply in terms of their philosophical essence but mainly on the basis of how they achieve the stated or desired aims in practice.

2.2 RELIGION AND HOLISM

The Paleolithic belief systems at the dawn of humanity were comprehensive, integrative outlooks. They blended knowledge of nature and spiritual values and ideas within a single package. But as the means of production advanced, these holistic doctrines gradually began to fracture. A separate body of knowledge, first in the form of techniques and technologies used by artisans, farmers, builders and others, and then as a nascent body of science ideas, emerged. Yet, for centuries, with humans remaining close to nature, science and faith traditions remained intertwined.

The inception of capitalist mode of production brought forth a tremendous growth in scientific knowledge. But science remained linked to religion. Major scientists attached religious significance to their discoveries. Chemists believed that life could not be produced from non-life and that the organic compounds found in living matter were a product of a God-given vital force acting on its elements and inorganic compounds.

Vitalist and spiritualist explanations are not testable. They are antithetical to the progress of science. As it began to stand on an independent experimental, observational and conceptual foundation, science dislodged such ideas from its framework. The library of science expanded astronomically. Both the natural and social sciences differentiated into specialized sub-disciplines. Each had its own lexicon, techniques and literature. The idea of science as a unitary system of thought weakened. While shedding vitalism, science reduced its attention to the interconnected features of natural and social phenomena. Solid walls were erected between specialties. Scientists identified as natural or social scientists in name only. They were astrophysicists, biochemists, cardiologists, paleontologists, and so on. As religions fractured within denominations, science fractured within specializations.

In this atmosphere, religion continued to adorn the mantle of a comprehensive vision that represents the ultimate truth. Only it gives a unitary identity to life, society and nature. Monotheistic religions like Christianity and Islam posit unity under the umbrella of a supreme creator while Buddhism holds that everything in the Universe is automatically interconnected. While science enhances material life and pursuit of secular goals, religion has an ethical, overarching perspective on birth, life, death and beyond. As science descended into mechanistic and atomistic thinking, religion unified the mind, body and spirit, and gave a sense of wholeness to life.

Human science fragments everything
in order to understand it,
kills everything in order to examine it.
Leo Tolstoy

Science was linked to reductionism; religion to holism. Holism became a dirty word in science, a relic of a superstitious age. For religion, reductionism became a heretical, unworthy

10

pursuit to be tolerated only to the extent it did some good in this life. It was a necessary evil. But scientists generally posited reductionism as a valid avenue for gaining insight into all phenomena under the sky and beyond. As a Nobel Prize winning scientist who studied the molecular basis of memory expressed it:

*In art, as in science, reductionism does not trivialize our perception
- of color, light, and perspective - but allows us to see
each of these components in a new way.*
Eric Kandel

We explore the holism versus reductionism schism from the vantage points of the four major religions.

Hinduism

Hinduism has a myriad of traditions, each with its special god or goddess. At the apex stands Brahman, the supreme deity of a triumvirate form. The traditions are unified through acceptance of the Vedas as the authoritative repository of divine wisdom, the idea that the Universe undergoes cycles of creation and destruction, and the doctrines of karma (fate) and dharma (duty). Life is affected by karma, but it also provides one an opportunity to transcend life's limitations by abiding by one's dharma. Life is not a purely deterministic process.

Hindu texts contain detailed guidance on religious rituals and on personal and social life. Ayurvedic medicine and yoga are associated with Hinduism. Hindu thinkers project their faith as an all-encompassing holistic philosophy. Swami Prabhupada derided reductionist science for placing matter over mind and denying the existence of atman (soul). Fathoming the essence of life is beyond the purview of science. Thus, he posed a challenge:

Go in your lab and put some chemicals together and produce life, and then you can come and tell me that life comes from matter. Swami Prabhupada (Egnor and Gallagher 2022a).

In practice, Hinduism adopts a pragmatic stance on reductionism. One of the Swami's followers declared that while it has limitations, it is also a useful tool:

[The] *reductionist world view is really good at a lot of things. Like if you get smashed up on the motorway, they're really good at putting you back together because musculoskeletal stuff is really mechanical and engineering principles. Reductionism works well for that kind of thing, but they really fail at looking at the bigger picture.* Arjuna Gallagher (Egnor and Gallagher 2022a).

A UK based Hindu organization, the Hindu Swayamsevak Sangh (HSS-UK), promotes social interconnectedness and tolerance, and stands against divisiveness based on race, religion, gender, nationality, and individualism. Its official stand is that unlike reductionist science and narrow ideologies, Hinduism recognizes the inherent unity in the diversity of life. It proclaims that compassion and universal family-hood ensuing from that recognition are the essence of dharma. Religion and science are not foes, but science has to function within its utilitarian domain while religion renders meaning and a moral code to humans and quenches their thirst for emotional equipoise. With its centuries of amiable disposition towards science, Hinduism is deemed an eminently suitable umbrella philosophy.

> *Hegel put forward the principle of thesis, anti-thesis and synthesis; Karl Marx used this principle as a basis and presented his analysis of history and economics; Darwin considered the principle of survival of the fittest as the sole basis of life; but we in this country [India] saw the basic unity of all life.* Deendayal Upadhyaya (HSS-UK 2022).

In recent times, the claim that Hinduism is a holistic, compassionate religion has come under serious doubt. The political space in India is now under the control of the exclusionary, supremacist Hindutva doctrine. It has not only generated deadly religion-based divisions within the nation but has also fomented an uneasy tension between science and Hinduism. The ruling neoliberal politicians have no compunction in using modern science and technology in a drive to make India an economic and military powerhouse. Yet, their ultra-nationalist outlook makes them proclaim that the Vedas contain many fundamental ideas of modern (Western) science like notions of energy and genetics. They proclaim that key results in mathematics like the idea of zero and the Pythagoras Theorem were discovered by Hindu mathematicians long before mathematicians in other places discovered them.

Some claims have a modicum of the truth. But, in general, they emanate from distortions of the history of science that legitimize the Hindutva agenda. That agenda seeks to make India a pure, homogenized Hindu nation, and suppress all other faith visions. Leading Hindu swamis are allying themselves with the hard-line politicians, and some gurus peddling supposedly traditional holistic therapies are reaping millions. School texts are being changed to reflect the often-flawed Hindutva claims. The new Hinduism bears little resemblance to the inclusive, compassionate faith of its enlightened sages, including Mahatma Gandhi. It is not holistic in content or practice.

Buddhism

Prominent scientists and scholars have presented Buddhism as the religion most amiable with science. Its pantheistic aroma, the absence of a creator, personal god, and the lack of rigid rules and rituals have garnered favorable ratings from eminent personalities like Albert Einstein and Bertrand Russell.

Buddhism has three main philosophical tenets: interconnectedness, duality and impermanence. Nothing exists by itself, in an isolated corner. There is no separate self or soul; it is a segment of the cosmic being. Every phenomenon is composed of opposing entities. The positive coexists with the negative. There is no good without bad, no creation without destruction, no light without dark. Everything is in a state of flux. Nature and life flow in the river of time, driven by interactions between opposing tendencies. Neither good nor bad are fixed; each has the potential to be transmuted into its opposite.

Existence is an organic whole; all events and processes are interlinked. Body and mind are integrally connected. Alan W Watts was a prominent proponent of this outlook. With advanced training in theology and Zen Buddhism, he at first became an Episcopal priest but later turned to college teaching and writing notable books on spirituality, Eastern religions and philosophy. His talks blended caustic satire and humor with insightful interpretations of Buddhism, Hinduism and Taoism. In the 1960s, radio stations across the US played his speeches. He formulated the holistic nature of Buddhism in a distinctly quirky fashion:

> *There is no such thing as a single, solitary event. The only possible single event is all events whatsoever. That could be regarded as the only possible atom; the only possible single thing is everything.*
> Alan W Watts

> [The] *prevalent sensation of oneself as a separate ego enclosed in a bag of skin is a hallucination which accords neither with Western science nor with the experimental philosophy-religions of the East.* Alan W Watts (BT 2019).

Emphasizing the integral nature of life, Buddhism holds that the fate of humans depends on the fate of the global ecosystem and vice versa. Basically, it is an environmentally friendly religion. The health modalities linked to Buddhism like meditation, mindfulness, acupuncture, acupressure and herbalism are deemed holistic approaches because they link the mind and the body. Some medical doctors integrate their practice with Buddhist principles and health practices.

The ubiquity of suffering (dukkha) is a primary Buddhist concern. Desire is the key cause of dukkha, and the Noble Eightfold Path is the means for liberation from dukkha. The internal and external, the mental and physical, must be in a state of balance. Aligning life in ways that are conducive towards that balance, like practicing mindfulness, is a viable conduit for liberation. Buddhism is thus considered a quintessential holistic philosophy.

> [The] *more firmly that holistic mindfulness is established, the more the individual effortlessly inclines to acting in an honest, harmless, and modest way.* (Amaro 2015).

Holistic mindfulness involves attaining internal repose and composure and living a modest life, cultivating a deep sense of connectedness with life and humanity, and conduct based on compassion.

> *The development of a more holistic mindfulness would not only support the growth of self-compassion and thereby an individual's own well-being, but it would also lead towards cultivating compassion for others.* (Amaro 2015).

Yet, mindfulness training today is a multi-billion-dollar practice that has gone far beyond its Eastern roots. Adopting varied elaborate formats, it is applied in health care, psychology, education, business, and military affairs (RPS 2022). Many scientific studies have documented the efficacy of meditation and mindfulness-based interventions for reducing stress and treating mental ailments. They may be used by themselves or as adjuncts to conventional therapies,

In the commercial environment of today, mindfulness and meditation have deviated from the traditional ethical goals. The Buddhist canon advocates them for personal liberation and cultivating compassion for humanity and life. Now the second goal has taken the back seat. These practices are now harnessed towards self-centered goals, corporate profiteering, neoliberal policies and militarism.

Buddhism has become intertwined with violence, but in a fractured manner. Buddhist monks in Myanmar have for long stood against military rule using non-violent civil disobedience tactics. Yet, they have also been in the forefront of the genocidal pogroms against the Muslim Rohingya people that have consumed thousands of lives and exiled over 750,000.

Many modern Buddhist practices are not consistent with the holistic teachings of the Buddha. Periodically attending retreats and mindfulness training enjoined with a consumerist lifestyle or promoting divisive politics is a reductionist, not holistic posture. It is holistic only if practiced in line with noble personal goals, compassion and social justice. Globally, the reductionist strand has been gaining the upper hand. A sizable segment of the Buddhists of today has allied itself with self-centered consumerism, social divisiveness and hate-filled, violent politics.

However, Buddhist leaders like the Dalai Lama continue to stress the integral nature of all life. Recognizing the interconnectedness of the global biosphere, and centrality of universal compassion, they advocate international unity in dealing with major problems like climate change and outbreaks of pandemics. But the wide practical gap between the leadership and the rank-and-file weakens the claim of holism.

Christianity

Christianity embodies several strands of holism, each with its own distinct vision. Transcendentalism critiques reductionism for accepting only two dimensions of reality, matter and mind, and ignoring the spiritual dimension.

> [Reductionism] *is self-contradictory, and transcendentalism is self-evident once we admit data from our three most valued and distinctively human powers, namely our power to think anything true, to choose anything good, and to appreciate anything beautiful.* (Kreeft 2008).

Christian evangelical circles warn that reductionism has polluted their creed. Instead of viewing Christian precepts as a whole, projecting the omnipotence of God, and seeing the love of God in broadest terms, churches focus on rituals and particularities like the cross, conversion and heaven. While the specifics are of value, they lose their import when broad spiritual truths are ignored.

> *We know that God is at work on his people through the full journey of their lives, from the earliest glimmers of awareness to the ups and downs of the spiritual life, but we emphasize the hinge of all spiritual experience: conversion.* (Taylor 2010).

Some devout evangelicals admonish their churches for excessive involvement in business affairs. That practice has swerved the churches away from their primary mission and reinforced reductionist or materialist tendencies. But there is a major paradox here: The mega televangelist churches dominating the religious arena are mega business empires. Faith is reduced to financial donations. Some strands of Christian holism emphasize augmentation of conversion and ministry with varied forms of social engagement.

Got Questions Ministries is a non-denominational Protestant evangelical group devoted to glorifying Jesus Christ and educating people about spirituality from a Christian perspective. Its experts opine that by reducing complexity to simplicity, reductionism is a flawed, anti-Biblical idea. The features of a complex organ like the brain are explained in terms of electrical signals and chemical reactions, but the social and spiritual aspects are ignored. Morality is reduced to a social contract, love to neurological or chemical reactions, and spiritual awareness to a placid mental state. Reductionism cannot explain Biblical notions like the three-in-one nature of God.

> *Your religious beliefs are nothing but the sum of human evolution, cultural mythology, and your own psychological make-up. …. Spiritually speaking, reductionism is frequently arrogance masquerading as analysis.* (GQ 2022a).

Food for the Hungry is an evangelical charity operating in twenty nations like South Sudan, Syria and the Philippines and serving poor and minority peoples. Holding everyone spiritually worthy, it does not discriminate by race, creed or nationality, respects local cultures and depends largely on local staff for field work. Its projects cover clean water, health care, food and education. Local leaders are consulted and involved in these projects.

The work of Food for the Hungry is inspired by the Biblical injunction to assist compassionately and holistically people in need and pain. Recognizing the three-in-one nature of God, and that humans were created in God's image, it fosters a holistic practice because the faith of Jesus Christ '*is holistic in every way*'.

God loves people, not just souls. It is a misguided theology that elevates the spiritual over the material and conceives of faith and ministry in primarily spiritual terms, just as it is wrong to elevate the material over the spiritual. It is also a misguided theology that separates the church from the hurting world, which needs it so desperately. In fact, when separated from the hurting world, the church and each of us cannot be what God calls us to be. (FH 2020).

Crossway, a Christian project, aims to address a major problem of modern society—stress and burnout—in a holistic manner. Today many people suffer from symptoms of burnout arising from financial and other pressures at work and home. Unable to function in a stable, healthy manner, they resort to harmful habits and cause further problems for themselves and their families. Pastor David Murray and the leaders of Crossway promote holistic prevention and resolution of stress and burnout.

> *Whether we look at the condition, the causes, or the cures for burnout, the evidence is clear: our bodies, minds, souls, emotions, consciences, relations, and vocations are all interconnected. We cannot neglect one part and expect the others to flourish.* (Murray 2017).

Stressing joint attention to the physical, mental and spiritual aspects of the malady, it advocates a gospel-based approach. Bible reading and prayers are deemed essential for any stress reduction program.

> *We cannot overwork our bodies and minds, for example, and expect to thrive in our spirituality and our relationships. Neither can we expect to neglect the soul and remain balanced and healthy in other parts of our lives.* (Murray 2017).

Christians for Social Action, a Christian NGO operating in the US and a few nations of the Global South, undertakes faith-based policy analysis and projects that promote economic, racial and environmental justice. Reflecting its holistic platform, it enjoins deep devotion to Jesus Christ with infusion of hope, peace and reconciliation through a wholesome embrace of the Gospel, cooperation across hitherto divided groups, and mobilization for social action.

Writing on its website, Christians for Social Action member Ronald Sider and his associates elaborate the notion of holistic ministry. God's salvation is all embracing. The Christian ministry should not just focus on specific concerns but also deal with personal, family, community, national and global concerns.

> *Holistic ministry* [is] *reaching your community with the whole gospel for the whole person through whole churches.* (Sider et al. 2019).

Standing on the pillars of evangelism and social action, Christians should engage with the spectrum of human problems but not ignore the mission to gain disciples of Jesus, the Savior. They should go beyond palliative, short-term programs and foster sustained spiritual and material elevation in this world and the next.

Word Made Flesh is a US-based but globally operating nondenominational Christian charity working among poor, distressed and abused groups like street children, sexually abused, trafficked women and children, and people caught up in civil wars. Seeking to protect human dignity and reconciliation in a holistic fashion, its key objective is '*to infuse hope, empower others and amplify the voices of those who are often not heard*'. Christian values and precepts form a guide for create sustainable solutions. To assist the disadvantaged, cooperation with local churches and Christians is mandatory.

> *Our purpose is for the redemption of the whole person toward the*
> *redemption of society. ... Our mission is to pray and work for signs of new*
> *creation, for shalom, for justice and for environments that enhance human*
> *flourishing.* (https://wordmadeflesh.org/about).

Christopher L Heuertz, then its executive director, elaborates the holistic vision of Word Made Flesh in *Christianity Today*, a major conservative outlet. Taking the existence of thousands of Christian denominations as a sign of the proclivity of Christians towards reductionism, he calls for a return to the essence of the teachings of Jesus Christ.

> *Whatever the issue—including issues no less comprehensive than church,*
> *gospel, or world—Christians are a divided people. Yet Christ shunned such*
> *ecclesial, theological, and human reductionism and division by maintaining*
> *a simple center based in love and reflected in unity.* (Heuertz 2009).

He faults Christians involved in serving the poor for operating in an elitist, fly in-fly out fashion, and glorifying poverty and loss of dignity. Conventional outreach is only a start.

> *Thoughtful, caring Christians must base their reconstruction of holism on a*
> *clearer vision of the church, the gospel, and the world with love as the only*
> *true indicator of integrity.* (Heuertz 2009).

The Salvation Army is perhaps the most well-known nondenominational Christian organization operating in about two-thirds of the nations across the world. It assists people affected by natural disasters, conflict and disease outbreaks as well as those enduring chronic problems like poverty, lack of access to clean water and education. It is also involved with promotion of social justice and women's rights.

Writing in a New Zealand based outlet of the organization, Nathan Holt assesses the holistic character of the Army's mission. Noting that it is a Western dominated group, he underlines the segmented nature of Western culture.

> *Everything in our culture in the West is segregated and compartmentalized.*
> *You go to school for your mind. You go to the gym or doctor for your body.*
> *You go to church for your spirit. You go home for community. You go to a*
> *therapist for your emotions. For every need you literally go to a different*
> *place. This is how we work in the West.* (Holt 2017).

Under the influence of this culture, the work of the Army has been fragmented. In particular, fulfillment of spiritual need has taken lesser import than fulfillment of material needs. Spirituality transcends all needs. To forget that is to dehumanize the children of God. The true mission thus is to primarily serve the spiritual goal while attending to material and emotional needs.

> *We are holistic people living among holistic people, all in search of a*
> *fullness of life through Jesus Christ—mind, body, spirit, family.* (Holt 2017).

Several Christian groups and experts engaged in nursing, public health and medicine advocate complementing conventional health and medical work with alternative medical therapies. The Bible, they say, favors nature-based preventive and curative practices and substances. Health has multiple components—physical, mental, social and spiritual. None can be ignored. Biblical wisdom like resting on the seventh day is good for heart health. They however caution that some of the practices that carry the banner of alternative medicine are harmful practices that veer towards heretical or demonic ideas.

++++

Different Christian groups have different visions of reductionism, and advocate varied strategies to enhance holism. Many equate materialism with reductionism, but none directly critiques neoliberalism, the major primer of consumerism, divisiveness and selfishness. And they mostly ignore the harmful role of Western nations and corporations in the Global South. They have grand descriptions for the projects they undertake; but when examined, apart from the religious aspect, the projects hardly differ from the conventional dependency generating Western-funded NGO projects. Their claim of holism does not pass critical scrutiny. The addition of the Bible onto a flood relief project does not make it holistic; it makes it sectarian.

A major complaint from the promoters of holistic Christianity is that their budget for evangelical work and spreading the gospel usually does not meet the need. Some estimates indicate that Christians in the US and Europe contribute about five times as much as for poverty alleviation and philanthropic work than to the basic mission of the Church. That imbalance weakens the holistic nature of their projects. It is another paradox: The purveyors of holistic Christianity are not that pleased that rank and file Christians value compassion more than inducing people to join the church.

Islam

Unlike Christianity, an explicit philosophical conflict between reductionism and holism has not taken center-stage in Islam. On the few occasions that Muslim theologians have addressed reductionism, they have expressed similar views. Their general point of departure is that Islam enjoins the believer to lead a holistic, non-compartmentalized life. Action is worship and worship is action; all need to be infused with Islamic values.

> *Indeed, my prayer, my sacrifice, my living, and my dying are for Allah, the Lord of the all that exists.* Quran 6:162 (IB 2018).

Allah is one, indivisible, omnipotent and most merciful. The integral nature of Allah behooves Muslims to integrate their thoughts, emotions and actions in personal, work and social dimensions in ways guided by Allah. More than prayer, faith is a way of life. Compassion and avoidance of immoral deeds are forms of prayer. What one does ought to express love for Allah. Islam is thereby a holistic faith.

Yet, Muslim scholars do not express absolute aversion to reductionism. They concede that the expansion of science requires specialization, and the complexity of modern life necessitates individuals to focus on certain actions and conduct. Rather than denying reductionist science, they embrace it since what it has found reflects the magnificence of Allah. Reductionist thought and practice are a start, not the end. True science is holistic just as true faith is holistic. Both integrate the material, mental and spiritual. There are sublime truths about life that a purely reductionist science cannot fathom.

In ascribing to holism, Islamic scholars draw inspiration from Al-Ghazali, the 11th century theologian, philosopher and mystic. He deployed a holistic scheme to classify knowledge and presented a model of pedagogy that was holistic in content and practice. The curriculum he promoted integrated religious studies with the sciences and subjects like language, law, literature and the arts.

> [The] *concept of al Ghazali curriculum has similar characteristics to the concept of holistic education which is characterized by intellectual, emotional, physical, and spiritual developments.* (Barni and Mahdany 2017).

17

Al-Ghazali also contributed to the then nascent field of psychology by linking the mind with the body and soul.

> [This] *interdisciplinary approach to understanding mental illnesses helped Muslim scholars conclude that their causes were multi-factorial: they postulated that biology, heritable factors (today known as genetics), environment, and spirituality could all be implicated. It was for this reason that Muslim scholars did not attribute mental illness to simply a weakness of faith. As such, their treatment regimens were also varied, and they did not prescribe prayer alone to combat mental illnesses. Along with the pre-modern medications, talk therapies and other forms of well being previously discussed, they also gave spiritual remedies in line with their understanding of holistic well being.* (Awaad et al. 2021).

By reducing them to collections of physical components, reductionism denies the humanness of humans and is not consonant with the Islamic rendition of humans as beings with body, mind and spirit. Overspecialized approaches miss the forest for the trees, and cannot comprehend that a human being is essentially driven by the soul, the seat of morality. Religious disbelief will endanger human morality and induce irresponsible conduct. An influential Muslim theologian opines that in the Islamic and Christian frameworks:

> [humans] *have a uniquely human part that is layered on top of the ape part and that controls it. The uniquely human part has self-consciousness, free will and inviolable dignity. There is nothing wrong with the biological and evolutionary study of humans, but there is something wrong with suggesting that that is all there is to humans. We believe that humans can transcend their physical limits and overcome the inner ape's instincts in order to do what is better, more just and more admirable.* (Marc 2018).

Islamic scholars tend to adopt holistic views on education, environmental science and stewardship, and societal analysis in which the teachings of the Quran and love for Allah occupy the central place. Some Muslim theologians consider Sufism, the mystical tradition within Islam, as the quintessential form of holistic thought and practice.

However, in appraising holism and reductionism, most Islamic scholars fail to notice the gap between theory and practice. They do not attend to neoliberalism, the principal promoter of morally deviant reductionism. They rarely examine the socio-economic policies of Muslim majority nations, and fail to interrogate the pro-capitalist, individualist practices of Muslims across the world. Thereby, they are unable to fathom the extent to which reductionist, materialist tendencies have penetrated the practice of their religion. In that respect, their holism is as flawed as that of the Christian denominations noted above.

++++

All the four major religions—Buddhism, Hinduism, Christianity and Islam—claim the mantle of holism by emphasizing the unity of body, mind and spirit. The interconnectedness of humans and of humanity with the Universe under a universal force (God, Allah, Brahman, Heaven) is an expression of holism. Some religious denominations tolerate other faith traditions under the rationale that all emanate from the same supreme being. Practical forms of religious holism link help for the poor and the marginalized to devotional activities and propagation of their faith. All faith traditions declare reductionism a materialist, anti-spiritual philosophy. But, owing to its practical utility, they grudgingly accept the results of modern reductionist science. But it comes with the proviso that science cannot perceive the whole truth; only religion can.

The meaning attached to the term holism varies for different religions. The importance attached to evangelism and secular assistance work varies, the nature of the assistance work differs, and the degree of tolerance towards other faiths also differs. Hardly any religious tradition espousing holism extends its purview onto structural socio-economic factors and critiques capitalism, neoliberalism, corporate globalization and militarism. If it is done, it is in the mildest of terms. Their high flowing rhetoric rarely ventures into a comprehensive interrogation of the human condition. Charity and reform, not fundamental change, inform their circumscribed holism.

2.3 MARXISM AND HOLISM

Marxism is a philosophy of society and nature that is combined with a program for social change. A brief overview of Marxism is provided in Chapter 9 of RPS (2022). For an extended, readable exposition, see the series of eight articles by P Thompson (Thompson 2011a, 2011b, 2011c, 2011d, 2011e, 2011f, 2011g, 2011h) at www.guardian.co.uk.

Karl Marx and Friedrich Engels were strongly influenced by two German philosophers, GWF Hegel and Ludwig Feuerbach. But they had quite distinct approaches to phenomenology. Hegel saw evolution of ideas through a dialectical process as the prime generator and director of history. Feuerbach placed primacy on material forces. For Hegel, realization of the Spirit climaxed history but the atheist Feuerbach dismissed that as idealist speculation. Valuing its holistic character, Marx and Engels utilized the Hegelian dialectic that was based on the unity and conflict of opposites. But they discarded the primacy accorded to ideas and spirit. And they adopted the import ascribed to material forces by Feuerbach, but dispensed with his mechanistic analysis.

The philosophy emerging from that blend is called Dialectical Materialism. It posits the interdependence and conflict between economic classes as the main driver of history. Ideas, in large part, reflect the social structure. Prevalent ideas generally promote the interests of the dominant class. Yet, at critical junctures, the constellation of ideas may gain a significant independent momentum. When economic conflict and disparity become acute, ideas that challenge the *status quo* gather greater prominence.

Marx and Engels were the major pioneers of the interdisciplinary approach to the study of human society. Visualizing human society as an integral entity, they disparaged compartmentalization of social science into distinct disciplines—political science, economics, sociology, social psychology, history, jurisprudence, and anthropology—as if there was little or no linkage between them. For Marx and Engels, the main task of social analysis was to accurately depict social reality and foster fundamental change in the capitalist order. There is no purely neutral social science; it either serves the oppressed or the oppressor class.

Recognizing that studying one factor at a time while keeping the others methodologically fixed often yields vital information, Marxist analysis employs information obtained by reductionist investigations. Yet, it holds that in itself, such information cannot disclose the social reality. Social systems have laws of operation and change can be gleaned only by observing the system as a whole. Marxism is a holistic (systemic) philosophy.

The purely reductionist approach either rationalizes the *status quo* or, at best, promotes minor reforms within the system. To change the system at its roots, a class-based holistic analysis is indispensable. Reductionist thinking in biology, sociology and psychology has justified racism, inequality, marginalization of minorities and slavery. Thus, reductionist genetics diverts the blame for poor educational achievement from social conditions onto personal factors. Socio-biology unjustifiably extrapolates laws relevant to the biological domain to the societal domain and interprets behaviors as purely biologically based behaviors. Complex behaviors are coded in genes. Criminality is mainly caused by genes.

Yet, some variants of Marxism degenerated into dogmatic theories that were based on superfluous data and flawed analysis. In Soviet Russia under Stalin, Lysenko's 'anti-reductionist' theories dismissed key genetic findings and led to disastrous results in the agriculture sector. Marxism cannot provide a 'holy book' for science.

> *As far as the knowledge content of science is concerned, Marxism of itself offers no especially privileged insights into the workings of nature - that is the job of science and scientists. But a dialectical methodology is an essential complement to reductionism.* (Clarke 2017).

Marxists are often accused of 'class-determinism' and ignoring societal practices like racism, patriarchy, xenophobia, ethnic marginalization, nationalism and religious discrimination. These divisions are subordinated to class divisions. Resolving class divisions is a perquisite resolving the other contradictions.

This type of thinking arises from a mechanistic interpretation of Marxism. Marxian dialectics takes the dynamic interaction between economics and culture and politics into account. The ruling class benefits by fomenting divisions among social classes. A divided people are less inclined to challenge its supremacy. The oppressed are pitted against each other: white against black, Hindu against Muslim, men against women. Cultural divisions become internalized over time and assume a logic of their own. Poor whites despise black people; it becomes a 'natural' thing to do. Dialectical Marxism confronts both modes of oppression contemporaneously. Racism within trade unions must be fought at the same time as fighting for better wages and work conditions.

Marxist holism differs from religious holism in two ways. The latter augments reductionism with spirituality, while Marxism resorts to class-based dialectical systems analysis. While recognizing the varied effects of religion in promoting and opposing the *status quo*, Marxism does not attribute those effects to forces beyond nature. Instead of the holy scriptures, it takes recourse to comprehensive and empirical social analysis to explain social change and phenomena like religion.

2.4 EMERGENCE

Holism occurs in two forms, spiritual and secular. Monotheistic holism posits the oneness of the Supreme Being and interconnectedness of His creation as the basis of the unity of body, mind and spirit. Ultimately, the spirit governs and gives the moral and emotional basis for human acts. Humans are discrete beings. They are responsible for their actions, and yet are a part of the universal whole under the purview of God. Polytheistic holism derives from the interconnectedness of the multiple divine beings.

> [No] *reaction to the wonder revealed by modern scientific efforts is more appropriate than a feeling of religious awe; and that that no worldview can match the scope of religion in its ability to reconcile scientific, emotional, and metaphysical strands into one holistic system.* (Wattles 2002).

Spiritual holism is a remnant of feudalistic modes of thought. Reductionism in its mature form is a child of capitalism. But capitalism has also spawned a secular form of holism that dispenses with divine entities yet asserts that systems of sufficient complexity have properties that cannot be gleaned from those of its components. A full explication of the operation of the system requires a study of its components, their interconnections, and of the system as a whole. Secular holism dispenses with effusive notions like vitalism, soul, or a universal life force (Qi, logos) that governs and guides all that exists.

The stupendous growth of science, much of it reductionist, and its extensive uses dealt a major blow to spiritual holism. The latter was slowly banished from the worldly domain into its own shell of beliefs and ideas. As the applications of science and technology seeped into all the domains of life, science became a required part of school curricula. Even the religious generally came to terms with reductionist science, often implicitly. But they continued to hold that their faith provided a unified perspective on life and the Universe. An expert in genetics could thus believe that genes were created by God for a specific purpose.

As science peered into complex systems like the human body, the brain and ecosystems, some scientists felt that the pure reductionist approach overlooked key things. Not succumbing to the avalanche of reductionist ideas, they began to develop a secular, scientific form of holism. But until recently, their vision was a tiny, mostly ignored vision within the natural and social sciences.

The global crises of capitalism in the 1930s, the expansion of the socialist world, and the rise of anti-colonial movements placed the capitalist tenets of individualism, competition and freedom from state control on the defensive. These tenets had ruled economic theories and policies and academic fields like sociology, psychology, anthropology and biology. The rise of welfare states in the West and further expansion of the socialist world after WW II gave credence to cooperation and state regulation. Instead of reliance on free market capitalism, UN development agencies and the World Bank promoted mixed economies and policies based on a multidisciplinary, multi-sectoral approach. The gains of the working-class movements reinforced the drive to limit the power of capital. It favored the flourishing of holistic perspectives in the various disciplines.

Holistic outlooks and social policies were dealt a major blow with the demise of the socialist world in the early 1990s and the import attained by neoliberal ideas in the aftermath. Free market, privatization, liberalization, individual initiative, deregulation and removal of controls on capital flow became guiding ideas for policy makers, development experts and academicians. Most fields in the social sciences fell in line with the neoliberal mode of thinking. Academic funds were channeled to specialized research. Targeted studies flowered while the interconnected perspectives played the second fiddle.

Almost within a decade, the neoliberal reductionist worldview once more fell into disrepute. Wholesale privatization generated serious crises in the arenas of health, education, water, power and transportation services in the Global South. Many privatized entities had to revert to public control. The Asian economic crises of the mid-1990s and the global financial meltdown of 2008 brought socialistic and Marxist ideas further into the public domain. Corporate moguls, business gurus and politicians espoused holistic management practices to cater for the emotional needs of employees and civil servants. Mindfulness training became popular in the world of money making and the corridors of state power.

Today, reductionism retains its dominant position within the natural and social sciences, and policy domains. Yet, holistic or system analysis has gained a firm but minor foothold in the academy. Alternative medicine, with its superfluous claim to be holistic, flourishes. The key thing is that in fields ranging from genetics, neurology, biology, medicine, to psychology, sociology, public policy and management, a new word has entered the lexicon: Emergence.

++++

Emergence Theory is the modern term for a field of philosophic and scientific modes of inquiry that encompasses Complexity Theory, Systems Theory, self-organizing systems, dynamic systems and secular holism. Stated simply, its basic tenets are:

Tenet 1: All phenomena or systems are composed of a hierarchy of levels or sub-systems.
Tenet 2: The constituents (parts) of any sub-system are more complex than those of the one immediately below it.
Tenet 3: A particular level (sub-system) is composed not just of its parts but also of the relationships between them.
Tenet 4: When the transition from one level to the next is a substantive one, the properties of the upper level cannot be predicted from the properties of the parts of the lower level and of the interactions between them.
Tenet 5: The critical (emergent) properties of the system can only be found from observing the operation of the upper-level system in its own terms.

Systems analysis (emergence, holism) obviates reductionism. It posits that a system's behavior is determined not just by the laws governing its parts but also by new laws emerging from the interactions between them. Complex collective entities have emergent laws that

21

operate at the system level. In psychology, for example, human behavior is explained at three levels:

The biological level, the lowest level, focuses on genes, neurotransmitters, hormones, gender and the like.

The personal level, the next highest, focuses on individual characteristics like upbringing, family life and history, education and economic status.

The community level, the uppermost level, focuses on life and conditions in the community in which the individual grew up and lives, and the relationship of that community to the rest of the society. It deals with the economic status and culture of the community, educational, health and general facilities, and special social, economic aspects of the community.

Consider a critical question: How can behaviors that cause harm to the individual or others be treated and prevented?

Biological (strict) reductionism starts and ends at the lowest level. **Genetic reductionism (determinism)** attributes complex human behavior to genetic factors, if not wholly then for the most part. Criminality and poor educational attainment are attributed to genes. For biological determinists, little can be done. The propensity of persons with 'bad' genes to engage in harm causing behavior is ingrained. But its actualization may be reduced by ameliorating the changeable risk factors. To protect future generations, the person with 'bad' genes may be isolated or prevented from breeding. While hardly anyone now advocates it, that policy has the potential to lead to eugenic extermination programs, as happened in Nazi Germany. This point is the focus of the next chapter.

Psychological (liberal) reductionism (determinism) starts at the personal level and considers biological factors as adjuvant factors. Early intervention for at-risk persons like counseling, tracking, special tutoring and financial assistance are needed. Those engaging in harmful conduct should be held to account and due penalties have to imposed. Appropriate punishment has a deterrent effect.

Emergence or Systems Theory starts at the societal level but entertains the effects of biological and psychological factors as well. Information gathered from genetic and psychological investigations informs collective level interventions. Poor educational attainment needs funding to improve schools, teacher training, school meals, books and other things. Improving job prospects induces students to attend to their studies. The systems approach may propose moderate to major reformist measures. But, in the final analysis, it promotes the replacement of the unequal social system by an egalitarian system.

Assessing the pros and cons of these approaches for mental disorders, a psychologist concludes:

> *Reductionism can overlook other causes behind behavior and is in danger of over-simplifying human behavior. Holism* [emergence] *makes it difficult to prioritize and use only one or two factors as a basis for therapy.* (Harper 2019).

Emergence Theory (Systems Theory) is now an established academic and research discipline with scores of related journals and academic departments. It has made major inroads in fields like neurology, genetics and ecology. The Santa Fe Institute in the US is a well-known research body in the West devoted to the study and applications of complexity (systems) theory.

Cases of emergent systems include protein molecules, cells, beehives, ant colonies, human organs like the brain, educational institutions, political parties, commercial and public bodies, ecosystems and human society. Emergence Theory posits an interactive two-way mode of causation between system levels—top-down and bottom-up. Examining a collapsed building, for example, a sociologist considers the profit seeking drive of the contractor; a civil engineer considers flaws in design and erection of beams and floors; a material scientist looks at the chemical and physical properties of cement and steel bars; and a physicist nails it down to atoms and molecules. Not all these assessments are relevant. Some are purely theoretical. If you seek safe construction, you will hardly pay attention to the physicist. Prevention generally requires attending to higher level assessments.

In principle, Emergence Theory reiterates the secular holistic perspective of Aristotle: The truth is the whole, implying that the whole is more than the sum of its parts. And it posits the interdisciplinary approach for the study of nature and human society. Scientist Peter U Tse discerningly observes:

> *Reality has no arbitrary professional boundaries.* (Tse 2015).

Marxism is a quintessential mode of Emergence Theory. It regards the economic factors and class conflict as the primary, long-term factors driving social change. Yet, they operate in a dynamic relationship with other material and ideological factors.

Many natural and social scientists remain in the reductionist mode. Not at ease with emergence related ideas, they accuse the proponents of Emergence Theory of surreptitiously injecting the discredited notion of vitalism into science. Causality runs in one direction, from the parts to the whole. Top-down causality, as in system level laws, is an unscientific proposition in their view.

These are not just academic disputes. For mental disorders, for example, the reductionist approach favors drug-based therapies while the holistic approach (Gestalt psychology, humanistic psychology) treats the person as a whole. It favors therapies like cognitive behavioral therapy, long term counseling and family-based interventions. Drug therapies treat the symptoms but not the underlying causes of mental conditions. They control acute symptoms but long term usage of psychoactive drugs may lead to addiction with serious withdrawal symptoms and induce major adverse effects. The cure becomes worse than the disease.

Reductionism is criticized for extrapolating from studies in the laboratory or highly controlled settings to the world of daunting complexities. Application of the one or two causative variables found is suspect in practice. The identification of risk factors focuses on statistical, not practical, significance. Based often on non-representative samples, the external validity of the controlled experiments is suspect. The experimental setting is too distinct from reality for its findings to be of value. And interactions between the potentially causative factors are usually ignored.

On the other hand, Emergence Theory is charged with ignoring quantitative analysis. Its proponents are accused of relying on vague, qualitative arguments. Importantly, they cannot prioritize the causative factors. Such charges seem credible as many spiritual traditions and pseudo-scientific practices in alternative or complimentary medicine fly the banner of holism. Making vague inconsistent claims is a hallmark of such therapeutic approaches. Major and minor religious traditions vaunt holism as their essential tenet. As a proponent of religious holism declared:

> [The] *most critical aspect of religion is the constitution of a vision and value of holism. Religion is the prime conveyor of values of holism (of whatever scale) in a world continuously fragmenting and reworking through politics and economics.* (Handelman 2015).

Some champions of religious holism have little aversion to astounding claims for the upper-level laws governing the cosmos. It only strengthens the case of the scientists wedded to reductionism.

I do not think that magic flying reindeer are refuted in the same way by the laws off empirical physics, any more than any other kind of magic is. It is not logically impossible that some entities perform acts which defy physical laws, if those entities are not merely physical entities. (Kreeft 2008).

Yet, Emergence Theory (Systems Theory, Complexity Theory, Dynamic Systems Theory) is a sound scientific discipline based on experimental, observational and evidence and philosophic rationale. Thousands of journal articles and many books deal with the theory and application of Emergence Theory. A minority, but a significant one, of physicists, chemists, biologists, environmentalists, economists, sociologists, psychologists, neurologists, development experts and agronomists doubt the validity and value of micro studies and interventions, and propose emergence type of alternatives. They utilize quantitative methods and simulation studies under novel experimental designs. The accusations of hand-waving and superficial methodology no longer applies to Emergence Theory.

On the religious front, theological holism is often linked with conservative social and political agendas. Yet, some theologians link spiritual holism with secular holism in, for example, dealing with global environmental concerns. In their view, protecting the environment is a religious act undertaken to safeguard God's creation. While many religious philanthropic groups uphold the banner of holism. their holism lacks comprehensive inclusion of key social and economic factors.

The prudent, science-based approach is to be aware of the advantages and limitations of both reductionism and Emergence Theory. They are complimentary approaches for unraveling the laws of nature and society, and undertaking actions to initiate minor and major change. The usual practical concern is over-reliance on reductionism and paying little attention to system level analysis and measures to deal with critical ethical, scientific and human issues.

We must ... go beyond reductionism
to a holistic recognition that biology and culture
interpenetrate in an inextricable manner.
Stephen Jay Gould

An Example

A driver on a freeway suddenly swerves his vehicle. A cascade of collisions ensues. Was it a random or caused event? What or who caused it? Who bears the responsibility?

Many causes are adduced: The driver says that a bird struck the windscreen, making him lose his focus. The police claim he was driving over the speed limit. A laboratory test indicates that his blood alcohol level was too high. Colleagues note that he had mishaps on the job. Worries about personal affairs and finances may have made him lose focus. A psychologist says that childhood trauma has made him an accident-prone person. A statistical investigator faults the recently raised freeway speed limit. A journalist blames the automakers' lobby which has stymied state investment in safer modes of public transport. An urban planner blames the location of residential, commercial and office structures that daily places millions of cars on the freeways and raises the probability of accidents. His lawyer says that he has a gene associated with impulsivity. Genetic factors beyond his control made him react the way he did on the freeway.

Who is right? What are the effect sizes of the different causative factors? Which factor bears the major share of the blame? No factor can be absolved in the causation of the event. But they are not isolated factors; they interact with one another. The key challenge is to build a model that integrates them in a cohesive manner and which has in-built flexibility for variability (randomness and free will). Admitting multi-factorial, multiple levels of causation, Emergence Theory enables such an integrative model of analysis.

++++

The philosophy of social and natural sciences continues to reverberate with strong contentions between reductionism and Emergence Theory. Science writer Eric Siegel is a hardcore reductionist. With examples from particle physics, he declares that reductionism has been so fruitful in so many the fields that in fields where its utility has not been established, it should be the fallback doctrine. The null hypothesis is to be used unless rejected. Emergence Theory is a speculative and approximate expedient choice, nothing more.

> [The] *formation of literally everything in our Universe, from atomic nuclei to atoms to simple molecules to complex molecules to life to intelligence to consciousness and beyond, can all be understood as something that emerges directly from the fundamental laws underpinning reality, with no additional laws, forces, or interactions required. ... As far as we can tell, the Universe is truly 100% reductionist in nature. Our ignorance about why certain emergent phenomena exist, and how they behave is no excuse for magical thinking.* (Siegel 2022).

Paul Humphreys, a philosopher, presents a similar vision:

> *The world is nothing but spatiotemporal arrangements of fundamental physical objects and properties. You and I, rocks and galaxies, toads and scrambled eggs are just processes, the successive states of which are spatial arrangements of elementary physical objects. These elementary physical objects, arranged in different configurations, account for all the astonishing variety that we encounter in our day-to-day lives.* (Frank 2021).

Though it counts eminent scientists like Stephen Hawking and Francis Crick in its ranks, hardcore reductionism is opposed by a growing number of scientists. They assert that a sufficiently complex system has properties that cannot, even in principle, be ascertained from the properties of its elemental parts and their interactions. These qualitatively new properties have to be unraveled by studying how the system operates as a whole. The computations required for reductionist analysis of a system with say a hundred or more parts needs a computer that exceeds the size of Universe. Apart from trivial cases, explication of systems through their parts and their interactions is an untenable and infeasible task. Total reliance on reductionism is a faith derived, not a scientific, philosophy.

Robert B Laughlin, a winner of the Nobel prize in physics, observes that nature is replete with spontaneous emergence of organized phenomena. The laws of the parts do not suffice to deduce the laws of the new organized collective. The latter are emergent laws. The basic laws of physics are emergent laws as well. Hence there are no ultimate particles or ultimate laws from which everything else is derivable. That flawed perception has been created by confusing mathematical formulation as the be-all and end-all of science. Though immensely helpful, mathematics is often misapplied, creating an image of pattern and predictability when there is none. Often, a qualitative approach is better.

Organized entities are to be examined in their totality. The dialectic between the parts and the whole is to be investigated as well. Laughlin boldly asserts that such a vision is increasingly permeating all areas of natural and social science to an extent that '*science has moved from the Age of Reductionism to the Age of Emergence*'. (Laughlin 2005, page 208).

Sabine Hossenfelder, a theoretical physicist and musician, also holds that excessive reliance on mathematical formulations in physics and other fields has generated unverifiable theories. Equations have obviated meaningful images and explanation. The pursuit of abstruse, metaphysical queries has to be done bearing the limitations of science in mind. Better say we do not know than produce equations of low predictive power that only a few can understand.

Anthony Leggett, also a Nobel awardee, shares the views of Laughlin:

I claim then that the important advances in macroscopic physics come essentially in the construction of models at an intermediate or macroscopic level, and that these are logically (and psychologically) independent of microscopic physics.

Anthony Leggett

Reductionism provides a myopic insight into natural and social phenomena. It overlooks that evolutionary, cumulative processes over time produce complex systems with novel modes of operation. As philosophers Brigitte Falkenburg and Margaret Morrison surmise:

A phenomenon is emergent if it cannot be reduced to, explained or predicted from its constituent parts... emergent phenomena arise out of lower-level entities, but they cannot be reduced to, explained nor predicted from their micro-level base. From an emergentist view, over the course of the Universe's history, new entities and even new laws governing those entities have appeared. (Frank 2021).

Reductionist methods have produced extensive insights in all natural and social science disciplines and spawned many useful substances and devices. But that is not the end of science. It has major theoretical and practical limitations that can be surmounted only through a systems level approach. Optimal science needs to blend reductionism with Emergence Theory. But thus far few scientists agree. A majority supports a hierarchy of authority of scientific disciplines that reflects a reductionist vision of reality. One partly humorous rendition of the hierarchy is:

Sociologists defer to Psychologists. Psychologists defer to Neurologists. Neurologists defer to Biologists. Biologists defer to Chemists. Chemists defer to Physicists. Physicists defer to Mathematicians. Mathematicians defer to God. (Frank 2021b).

Reductionism accuses Emergence Theory of bringing miracles into science. Yet, in its own way, it appears to interject divinities into science.

2.5 GENETICS

The molecule deoxyribonucleic acid (DNA) forms the basis of life. Inherited by a child from its parent, it is a long sequence of nucleic acid molecules (bases) arranged in a pair-wise fashion. Its four nitrogen containing bases are: Adenine (A), Thymine (T), Cytosine (C) and Guanine (G). The two linked strands are arranged so that A in one strand occurs with T in the opposite strand, and G occurs opposite C.

```
A   T   G   G   A   C   C   T   A   G   .   .   .   .   .

I   I   I   I   I   I   I   I   I   I   I   I   I   I

T   A   C   C   T   G   G   A   T   C   .   .   .   .   .
```

The human genome is a DNA string with about 3.1 billion nucleic molecules arranged in 23 coiled chromosomes. A gene is a contiguous segment of the DNA that begins from a particular location. Genes code for proteins from which cells, organs and functional molecules are built. Humans have about 21,000 genes of varied length. Some parts of the

DNA code for starting and stopping points of a gene but the exact role of a large portion, known as junk DNA, is a mystery. Nearly 99% of the DNA is identical for all humans.

The cell, a complex structure enclosed within a permeable membrane, is the basic form of life. Housed in the cell nucleus, the DNA manufactures protein molecules with the help of cellular organelles and molecules and reproduces the cell through cell division.

The process of generating new mammalian life starts with the unfolding of the strands DNA from the sperm and egg cells. One strand from the sperm cell joins one strand of the egg cell to make an embryonic cell. The embryo undergoes repeated divisions to generate specialized cells that generate organs and tissues like bones, muscles, heart, lungs, kidney, skin, veins and arteries. After a period that varies between animal species, a viable offspring issues from the womb.

The DNA has the same components and form in bacteria, plants and animals. What differentiates them is the sequencing of the nucleic acids and their genes. Yet, the differences are not as large as we may think. Humans and chimpanzees share around 98.7% of their genes; seventy percent or so of the genes in zebra fish and humans are identical; and between-human gene variation is just about 0.1%. But since variations in gene locations can number in the billions, it suffices to explain why one person acquired Type I diabetes as a child and another did not, why one person can readily digest dairy food, but in another, it induces abdominal discomfort. Mutations (small changes) within a gene or a set of genes affect whether a particular trait is expressed or not. When the gene encoding the oxygen-carrying protein in the red blood cell, hemoglobin, is mutated, you get sickle cell anemia.

The entire human genome was sequenced in the year 2000. Greeted with euphoria by the scientific community and the media, it was as if the ultimate frontier in biology had been reached:

> *Today we are learning the language in which God created life.* US President
> Bill Clinton (Carey 2012).

Gene technology has advanced rapidly. One's genetic risk for several diseases can be evaluated now by specialized firms for less than $1000. Complementing the long term familial and twin follow-up and cross-sectional studies of heredity, a literal avalanche of genetic studies has transpired. Volumes of data have been collected. And fantastic headline-grabbing claims have ensued.

Genetics is applied in agriculture, health and medicine, criminology, paternity determination and social policy. Medicinal drug development and individualized therapies are informed by genetic data. Genetically modified crops—corn, soy, canola, papaya—are on the market. Animals are cloned. Genomics is a multi-billion-dollar business. And genetics features prominently in the modern debates, secular and religious, on reductionism, holism, determinism, responsibility and free will.

2.6 THE NERVOUS SYSTEM

The nervous system performs perceptual and control functions in the body. Its elemental constituent is the neuron, a cell that transmits electro-chemical signals from one part of the body to another. There are more than 7 trillion neurons in the human body, of which about 100 billion are in the brain. Neurons vary in length from less than a millimeter to around a meter.

The nervous system has two main components, the central nervous system (CNS) and the peripheral nervous system (PNS). The CNS consists of the brain, the spinal cord and specialized nerves (bundles of neurons) emerging from the brain while the PNS has nerves emerging from the spinal cord that separate as they approach the target organ. Neurons, individually or serially, send electro-chemical signals from and to the brain. Other cells in the nervous system and blood vessels protect the nerves and facilitate their functioning.

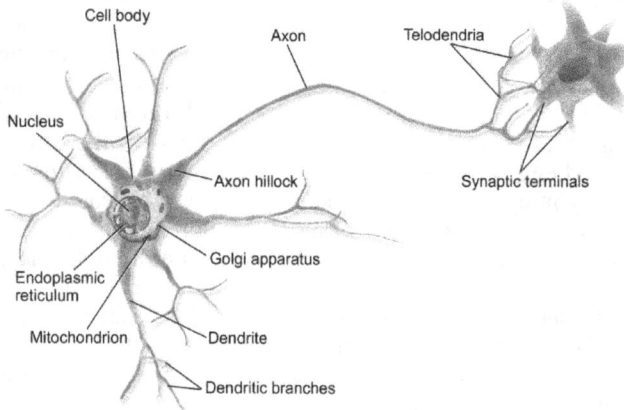

Neuron

Receiving sensory and other stimuli from within the body and the surroundings, the brain sends off signals to control bodily functions and actions. There are five principal sensory stimuli—sight, touch, smell, hearing and taste. Some signals control or indicate the state of the organs and processes in the body. Some signals propel purposeful acts (walking, deciding what to eat) and some activate automatic impulses (breathing, closure of eyelids if an insect flies too close by). The nerves of the PNS fall in two categories: the somatic nervous system and the autonomic nervous system. The former control conscious acts and the latter, involuntary functions of the body organs. Thinking, recall, emotions, feelings, motor skills, body temperature and other functions within the body are under the purview of the brain and entail their own patterns of neuronal activation.

With the brain as the master, the nervous system is the most complex organ in the human body. Sophisticated technology has enabled us to know a great deal about it and its components. For example, the feelings and control of hunger and thirst relate to locations with hypothalamus; and emotions like anger, fear, love and hate are connected to the limbic system. While major strides are being made, the scientific bases of two major phenomena linked to the brain are hardly understood. These are mind and consciousness.

++++

Intuitively, we know mind and consciousness. You are aware of your feelings, thoughts, intentions, actions and what is around you. You feel hungry or thirsty. You hope it will be a sunny day. You observe the actions of people around you. The constellation of mental entities gives you a unique identity. They signify that you have a mind. You are conscious.

Blindfolded subjects with hands and feet tied, and subject to unvarying sound and deprived of all sensory stimuli, hallucinate and become wildly angry and are not able to concentrate. Feeling as if they are in an unreal world, they 'lose their mind'. Though inextricably tied to our senses, our mind operates in a separate dimension. A person does not lose her sense of being just because she is blind. She may have a well-rounded personality and develop an acute sense of hearing.

The mind is not simply a configuration of fixed forms of electro-chemical signals. The brain center for love, if it exists, may be located with precision. But does that tell us much more than that if it is damaged, the person will not be able to feel or express love? But what of the nature, object, intensity of his love? No center in the brain harbors such things. They are systems level entities dependent on, but not reducible to, neurons and biochemical reactions. And most likely, love and other emotions primarily are system level features as well.

Suppose the part of the brain related to memory is damaged. She then loses her memories and is unable to form memories. Now assume that the techniques of surgery are

28

advanced enough to restore her ability to record and recall events. Will that restore her past memories? Surely not. They are gone. Human memory is not like computer memory. The latter reflects fixity. It can have copies, and can be stored, restored and deleted will. Human memory is not as pliable and cannot be stored on an external device.

There are about one quadrillion (10^{15}) inter-neuronal connections (synaptic junctions) in the brain. Our thoughts, feelings, memories, sense perceptions and much more arise from a dynamic constellation of activation of synapses. They are a whole-brain phenomena. The brain is the physical substrate upon which the mental superstructure of the mind rests. You can touch, see or dissect the brain. None of that applies to the mind.

Our mind gives us our sense of morality, our feelings right and wrong. It gives us a conscience. It also makes us aware (conscious) of ourselves and our social and natural environment. It is fundamental to our sense of self-worth and identity.

Body, brain, mind and consciousness feature centrally in the discussion of science and religion. The foundation laid here will facilitate our discourse on this point below and in the subsequent chapters.

2.7 FREE WILL

The notions of holism and reductionism relate to the concepts of free will and determinism. In basic terms, if a person can make autonomous choices from a range of alternatives, then she has free will.

> *Free will* is the capacity of agents to choose between different possible courses of action unimpeded. (Wikipedia 2022 – Free Will).

If the choices are decided, in part or full, by forces beyond her control, then she has little or no free will. The degree to which she has freedom to choose and act affects the extent of responsibility and accountability she has for her actions. The code of conduct (ethics, morality and law) that constrains personal behavior has to allow for the extent to which a person can act independently.

Strict reductionism accords with strict determinism and excludes independent causality at the level of the whole. Human choices arise from the biology and chemistry of the brain and other organs. Matter determines mind. There is no free will. Only a rare scientist adopts such a hard deterministic stand.

Free will is a complex systemic property, not reducible to a constellation of elemental entities. It involves biology, psychology, sociology and philosophy. We make a preliminary foray via a dive into a bee colony.

Honey Bees

The queen bee of a honeybee hive has hundreds of male bees and thousands of female bees as her subjects. Her stature depends on the fact that she is the only female with the biological ability to mate, produce off spring and perpetuate the colony. She emerges from a few young bees groomed with special nutrition. The queen bee mates with 10 to 12 male bees from her own and other colonies during the several flights she makes out of the hive. Storing their sperm cells in special sacs, she combines them with her eggs over time to begin a complex process that eventually generates thousands of male and female baby bees.

Male bees (drones), who have 16 chromosomes in their genomes, do no work. Their sole role is to mate, with their own queen or one from another hive. Once a drone mates, he is killed by the mating queen. When food supplies run low, the existent drones are starved by the worker bees and thrown out of the hive.

Female bees have 32 chromosomes in their genomes. As they are not sexually mature, they cannot procreate. They perform the multiplicity of tasks required to keep the hive functional. These include foraging (searching for and bringing back pollen, nectar, water and plant sap), hive construction and repair, cleaning, guarding and ventilating the hive, honey processing, and nursing the brood and the queen. Age is a key factor affecting what a bee does. Female bees have a life span of up to 7 weeks. In the first two to three weeks, they nurse the brood, serve the queen bee, process honey or clean the hive. For the rest of their lives, they forage and perform hive protection tasks. Allocation of the type of work done by the workers is also affected environmental and colony health related factors. The female bees in a hive have the same mother but possibly a different father. Hence they relate to one another as sisters and half-sisters.

Bees use a complex communication system to ensure that the hive operates smoothly, is well supplied with necessities, and is protected from predators. The pattern of activities within the hive also depends on the season. The bee hive is a complex community whose survival requires finely coordinated activities among the tens of thousands of multi-functional bees.

Reductionist biologists posit that the role of a bee in the hive and the nature of the hive are fully fixed by genes or neural factors. One study of the variations in the flight paths of foraging bees revealed that in terms of learning methods, choice and multiplicity of locations and number of trips, different bees appeared to be making their own choices. It was as if they had 'unique personalities'. But a detailed study showed that the flight path variations were fully explicable by the variations in the biochemistry of their brains.

> *The hapless bumble bee fumbling around the English countryside is simply a victim of a particular surge of proteins and hormones released into her tiny brain. She has no choice but to cooperate. Whether that results in wide-arcs of adventurous flower-seeking or careful and measured foraging is beyond the bee's willful control.* (Miksha 2016).

Yet, the overall evidence indicates that while there are many genetic correlates of bee characteristics and behavior, the variations in their behavior cannot be fully ascribed to genes or brain chemistry.

> *There is no special gene that controls whether a bee grows up to be a queen or a worker; their jobs in the hive are not determined by genetic makeup. .... Furthermore, there are no specific genes predisposing the workers to the different tasks they do.* (BS 2021).

Seasonal changes in the environment, availability of food sources, presence of predators and pathogens, the number of bees in the hive, the health of the queen and other factors have a significant effect on the pattern of activities in the hive. On a given day, how many bees will forage, how many will nurse, and what type of cargo will the foragers bring back depend on factors other than genes. Random factors also operate. At what exact time point will a nurse turn into a forager? Which drone will mate and which will not be able to mate during its life? Among the young female bees being specially groomed, how will the queen be selected? And by whom? The detailed history of the hive depends in a major way upon the interactions between genes and the environment as well as on random factors.

The honey bee is a complex biological organism and the beehive is a complex social system. Their natures are significantly dependent upon the environment. Thus, changes in the environment can turn certain genes on or off, and promote or inhibit certain activities. The beehive is an emergent system with a mode of operation and rules that are not fully explicable by genes.

Do bees have free will? Can a forager bee rest one day and not venture out? Can it decide to change its role? Unless it is sick, probably not. If the hive is attacked by wasps, can a defender bee sit back as others engage the attackers? Probably not. Gene-based instincts and the logic of the hive will drive it to play its role in defending the hive. Objectively, bees are socially responsible, and thus, moral beings. They live for themselves and for the good of the hive. If bees have free will, it is largely circumscribed free will.

++++

Evidence from many studies shows that free will is an evolutionary, adaptive entity. Many animals exhibit variable behaviors that seem to be willed behaviors. In humans, free will is associated with following or violating an established moral, ethical or legal code or a ritual. Some primatologists claim that animals—birds, dogs, cats, rodents, elephants, dolphins— exhibit elemental moral behavior. Frans de Waal, an authority in the field, argues that morality has evolved and is primed in genes. Observations indicate that elephants have elaborate grave site rituals for dead elephants; rodents desist from actions that would harm other rodents; birds behave in a violently jealous manner towards their mates; dolphins assist other sea creatures in trouble; dogs appear to feel sad if the owner is angry; chimpanzees punish members of their band who stray from the rules of conduct. They seem to exhibit emotions that have a moral component and appear able to differentiate right from wrong.

Animals with larger brains behave in empathetic, mutually supportive, peace-making and consoling ways towards others and do so even when no immediate rewards accrue to themselves. Their behavior resembles that of human infants.

> *Human morality is a deep-seated, natural trait grown from the social nature that natural selection has produced.* Frans de Waal (Rios 2007).

Human morality, though, is clothed in elaborate, explicit declarations of desirable and unwanted conduct. It is packaged with explicit rewards and punishments, now or beyond. Though rooted in brain biology, it encompasses a qualitatively wider range of emotions, values and behaviors than found in animals. Human morality is primarily a product of the long history of the development of human society.

Animals show behaviors such as teaching, decision making, multi-word vocabularies and playfulness that bring them close to humans. Wherever we draw the line, the key point is that morality and will are not unique to humans. They evolved over millions of years and are, to a degree, genetically based. Morality is both inherited and learned.

Ascribing morality to animals counters the opinion of Rene Descartes who regarded animals as biological machines lacking emotions and thought. It also contradicts the position that animal behavior is just competitive survival-based behavior and that no moral aspects exist. Animals lacking language and codes of morality just show biologically based behaviors onto which a moral component is being artificially transposed. Studies show that the truth lies somewhere in between. It is as erroneous to completely deny free will among animals as it is to exaggerate it. If bees are self-aware and have will, it is but in an elemental way. If they display morality, it is of a rudimentary nature. The difference between humans and animals is the relative importance of inheritance and learning, the complexity of the moral code and the range variations of will observed. Human morality lies miles above animal morality.

That animals have moral tendencies is not consistent with the idea that morality has a divine origin. It accords a biological basis to religion. But that does not imply that religion had no role in the development of codes of morality. Until recently, religion was deeply intertwined with prevailing social norms and moral values. Only in the past two hundred years have secular codes of morality triumphed over religious codes and come to dominate individual conduct. You do not steal not just because of religious strictures but because it is against the law of the land, because it will violate your personal beliefs and education, because you consider yourself a role model, because you fear getting caught, or for other reasons.

The varied expressions of free will and moral or immoral behavior patterns in humans needs to be explored through an approach that transcends theology and reductionist genetics and neuroscience. It has to recognize human behavior is affected not just by genes and the brain, but also by emergent laws that operate at the collective level. Human societies are emergent systems, much more complex than a beehive. With more intricate brains, humans are aware of their existence and have individualized desires and intentions. The effects of genes and brain structure on human thought and conduct is vastly outweighed by social norms, rules, education and upbringing as well as socio-economic status.

Genetic and neurological determinism imply that human acts are caused by genes, neurons and developmental factors. Psychological determinism attaches importance to social conditioning in early life. Freudian determinism focuses on subconscious influences. These perspectives imply that compulsion of some sort made a person do what she did. She is not responsible for it, at least not fully. In the extreme, these viewpoints deny the existence of free will and undermine basic moral precepts.

Yet, even under restrictive conditions, people of similar backgrounds do not behave identically. A human deviates from expected behavior in ways a bee cannot. A prison inmate may be obedient or flaunt prison rules. In the face of a burning house, one person jumps in to rescue a child trapped by the inferno while another does not. Codes of morality vary over time and from nation to nation.

Free will is not a binary, either-or issue. It is an issue of the degree, range and quality of freedom. Free will and its actualization vary. No one is totally free and hardly anyone has zero freedom.

2.8 CAPITALISM

Modern societies are capitalist systems. Contrary to popular thinking, it is not a system of autonomous agents interacting freely in the marketplace. Nor is it a system in which the people have the ultimate power to decide how they will be governed. It is a system in which the class with the major economic clout largely determines how the resources of society are allocated, the character of the state, the types of education and health systems, and the dominant voices in the modes of spreading information and ideas. Though transitions from one economic class to another do occur, the basic structure of the system and membership in the dominant class are largely stable.

At the surface, the system appears to exhibit a great degree of freedom in all its domains. Underneath though, effective freedom prevails largely in terms of choices between options that serve the interests of the dominant class, the 1%. The system has powerful institutions based on force—police, courts, prisons and the armed forces—to maintain social stability and protect existent structures against internal and external destabilizing forces. The relative strengths of the coercive and civil structures of governance vary over time and place.

The system has a dominant ideology, neoliberalism, that functions like a religion (RPS 2022). Its key tenet is that capitalism is the best and ideal social and economic system for humanity. Despite transient ups and downs, it promises the greatest good for the greatest many and upholds individual freedoms and rights. While attempts to modify its specific aspects are common, the system as such is inviolable. From family to school, community and the nation, almost all institutions and people subscribe to the supremacy of capitalism.

Yet, room for real free choice, thought and activity exists. Individuals, social organizations and business entities compete for a greater share of the social pie. A minority promotes egalitarianism while another favors authoritarian governance. Political discourse and social life appear like processes driven by non-coerced choices made by millions. Appearance and reality, however, do not overlap.

Consider education. Capitalism does not decide who will do well in school and who will fail. Diligence and tenacity play a major role. Children from poor families excel. But by affecting the allocation of education resources—well-run schools, books, competent teachers, childhood nutrition, family income—it has a major impact on the educational attainment of communities. While the system is not decisive at the micro (student) level, it has a huge effect at the macro (school and community) level.

Capitalism does not determine which driver will cause a road accident. But through the resources allocated for different modes of transportation, manner of land usage that puts many drivers on the road and the relevant rules and laws, it significantly affects the incidence of road accidents. We cannot say which driver or vehicle will be involved in an accident but we can estimate the probability of accidents. It is a probabilistic, not a deterministic phenomenon, and has multiple causative factors.

Capitalism does not determine who will get lung cancer and who will not. Yet, state policies, regulation of tobacco firms, smoking prevention drives, support for people who want to quit smoking, and air pollution control have a major effect on the incidence of lung cancer. On this score, the powerful tobacco companies have for a long time been able to override public health.

Mental health is also strongly affected by systemic factors. Prescriptions for psychoactive drugs have risen sharply risen in the nations of the West. Some 10% of UK residents in 2021 had a prescription for an antidepressant. Mental health care focuses on the afflicted person and perhaps on her family and work issues. Drugs are the mainstay of therapy even though controlled long term evaluation of drug efficacy is not commonly done.

While modern drugs control acute symptoms, levels of full recovery are poor and dependence on drugs with potent side effects ensues. Withdrawal symptoms are more potent than the symptoms of the ailment. Yet, the medical profession is wedded to the drug-based approach. Other approaches exist but are not utilized as frequently as they should be. Pharmaceutical companies make record breaking profits. Drug firms have been convicted of deceptive marketing and research data manipulation. Misuse of psychoactive drugs causes enormous harm including many fatal outcomes and the companies have paid hundreds of millions of dollars in fines and compensation. Yet, it hardly dents their profits.

Many mental ailments are reactions to socially induced distress. The medical profession fixes them as organic, individual problems, and seeks a quick fix via drugs. The reductionist approach sidelines societal causes of mental problems. Neoliberal austerity measures and dependence on the market makes life insecure for most families. Out of pocket costs, and costs of education and other services previously covered by the state have risen sharply. Mental ailments are more common in marginalized communities.

++++

People generally have confidence in their ability to make right choices. A mother feels she can make good decisions about the education of her son and tries her best to remove the obstacles he faces. Though aware of the paucity of textbooks and supplies in his school, when he does not perform well, she reprimands him for insufficient effort and associating with the rowdy crowd. People recognize external constraints they face yet place the primary emphasis on personal effort and initiative. If you try hard, you can overcome any obstacle.

Contrarily, awareness of external constraints makes people complacent. What is the point of spending your energy if you will still be as you are now or worse off at the end of the day? Inculcating a fatalistic mentality, it may make people lose faith in their own power to change their lives. People of your kind will remain where they are, no matter what they do. You cannot change your nature or the system. The mother loses hope; going to college is not for her son. At best, she prays that he stays out of serious trouble while in school.

A reductionist outlook either induces unwarranted optimism or undue feeling of resignation. It makes people self-centered and disregard collective drives for change. You struggle for yourself or resign to the 'reality'. Only a few dedicated activists view external constraints as reasons to redouble their efforts for systemic change. With educational drives and agitation, they seek to organize the people and change conformist, individualistic attitudes. As the spirit of voluntary cooperation flourishes, authentic free will flourishes as well.

Though he held that the subconscious strongly determines personal behavior, Sigmund Freud also argued that psychotherapy can assist people to overcome the unconscious tendencies. They are not all powerful. Following his footsteps, Erich Fromm held that people are unwilling to exercise their full potential to change, individually or collectively, due to fear of ostracism and other adverse effects. Risk outweighs the benefit. Overcoming fear is the main basis for free decisions and actions. Courage to change and free will go hand in hand.

2.9 AN ETERNAL KNOT

This book is based on the premise that religion and science are interlinked social, conceptual and philosophical entities. They are complex, dynamic entities with internal mechanisms of maintaining equilibrium and generating change. Religion derives from, and in turn, affects social and economic structures. It is intertwined with history, culture, science and people's outlooks.

Labeling religion as the ultimate curiosity and science as the penultimate curiosity, Wagner and Briggs (2016) amassed voluminous evidence from across the planet to challenge the thesis that religion and science have always been in conflict. Debunking the perceptions of European archaeologists that ancient non-European cultures did not have a religion or the idea of God, they argue that both religion and science emanated from the basic curiosity: What is the meaning life? Where did we come from? Is there something beyond this world? In a precarious environment filled with dangers and unknowns, religion relied on imagination and mystical experience, and science relied on observation, trial and error, generalization and application. No firm line separated religion from science. In one, what worked was retained and what did not was discarded, while the other, what assuaged the psyche and helped tackle grief persisted and was ritualized.

The philosophies, ideas and institutions of science and religion evolved side by side throughout history. At times at odds with each other and at times in a complimentary mode, they never separated totally. Both are complex emergent phenomena linked by convoluted strands of belief, knowledge, ritual and social, cultural, economic and political structures and by prominent individuals. The Eternal Knot, representing universal interconnectedness, aptly symbolizes the intricate, convoluted and dynamic nature of the relationship between religion and science.

Our study of religion, science and mathematics employs an interdisciplinary (holistic) approach that is complemented with historical and dialectical analyses. We draw liberally from reductionist studies. But, in the final analysis, we ascribe to the dictum that the truth is the whole. Ideas like consciousness, soul, free will, morality, determinism, responsibility, sin, omniscience, omnipotence, divinity, and proofs of the existence of God feature in our explorations. A central aims is to compare the societal functions of religion, science and mathematics. Issues like eugenics and climate change feature as major illustrative cases. As in RPS (2022), our focus will be on Hinduism, Buddhism, Christianity and Islam.

We begin with a survey of the origin, development and manifestations of the eugenics, a doctrine that swept across the Western world in the early part of the 20th century. Eugenics starkly demonstrates that religion, secularism, science, mathematics and politics relate to each other in astoundingly unexpected ways.

CHAPTER 03: EUGENICS

[If] you can breed cattle for milk yield,
horses for running speed,
and dogs for herding skill,
why on Earth should it be impossible to breed humans
for mathematical, musical or athletic ability?
Richard Dawkins

The son shall not suffer for the iniquity of the father,
nor the father suffer for the iniquity of the son.
The Bible, Ezekiel 18:20

Our scientific power has outrun our spiritual power.
We have guided missiles and misguided men.
Martin Luther King

EUGENICS IS a socio-biological doctrine based on two principal propositions: One, the physical and mental abilities of human beings are largely hereditary, and two, the 'quality' of the human species can be enhanced by selective breeding. Human populations can be made healthier, more intelligent and less prone to 'vice' by altering the prevalence of genes that bestow or inhibit health and intelligence.

> *Eugenics might be defined as controlling human reproduction to modify or benefit the species.* (Encyclopedia 2020).

Eugenics proclaims that the genotype (genetic code) directly determines the phenotype (physical, mental, behavioral, biologic and health characteristics). The effects of environmental and random factors are deemed minimal. In practice, eugenics assumes two forms:

Positive eugenics consists of measures to encourage persons with 'good' (desirable) genes to procreate at higher rates. Financial and other forms of inducements are used for this purpose. Positive internal eugenics, promoting marriage among gifted persons, had a poor record wherever it was instituted.

Negative eugenics consists of measures to lower the rates of childbearing among persons with 'bad' (undesirable) genes. Laws are enacted to restrict persons or groups with 'harmful' genes from bearing children by sterilization, confinement and other forms of coercion. 'Good' genes are thereby passed on and 'bad' genes are not.

Degeneracy is a key eugenic concept that covers features like mental illness, neurological deficit, poor health, epilepsy, and physical disability together with behaviors like prostitution, promiscuity, alcoholism, laziness, poor hygiene, criminality, joblessness, selfishness and cowardice. These traits are said to be passed on from parents to offspring. Since degenerates often have many children, their genes, if unchecked, will accumulate, proliferate and threaten the moral, physical and intellectual constitution of humanity. Containment of degeneracy is a prime requirement for the protection of the human race.

Eugenics is conceptualized in terms of the target population as well.

Internal eugenics addresses the genetic worth of individuals within a defined population. For example, people with a mental disability or a physical handicap at birth were said to be 'degenerates' in Nazi Germany.

External eugenics compares entire ethnic, racial or national populations. Thus, the Nazis considered the Roma as a genetically 'inferior' race. External eugenics is an extreme form of racism, xenophobia, nationalism, supremacist religiosity and/or ultra-nationalism.

The treatment of Native American groups in North America is a classic case of external eugenic practice. Unfolding over four centuries, its initial phase was marked extermination followed by isolation, deprivation and partial sterilization. Negative external eugenics at times adorns a deluding mask. Confinement of Africans to Bantustans in Apartheid South Africa was rationalized under the guise of separate but equal. But it was an elaborate hoax; the abject life of Africans in the Bantustans bore no comparison with the prosperity of elegant white neighborhoods of Johannesburg. Separate has invariably meant unequal.

From the middle of the 19th century to the middle of the 20th century, eugenics was a central facet of Western social and political thought. Vigorously implemented, it affected millions. Yet, apart from its horrific manifestation in Nazi Germany, the reality of eugenics has largely been erased from memory.

Modern eugenics signified an unusual confluence of science, politics, ethics, atheism and religion. It amalgamated a diverse array of eminent biologists, social scientists, statisticians, liberals, socialists, right wingers, feminists, writers, theologians and faith leaders to jointly stand behind a program that had egregious social ramifications. In this chapter, we ponder on four questions:

Question 3.1: Is eugenics a science-based doctrine?
Question 3.2: How did eugenics become a popular creed?
Question 3.3: What were the societal effects of eugenics?
Question 3.4: How did eugenics interact with religion?

3.1 ORIGINS

Development of stratified societies in the wake of the agricultural revolution spawned the belief that the king and his elite circle were naturally superior beings blessed by gods and possessed the right and ability to rule. The rest had to obey and trust divine fate. The son of a king or lord had, by birth, the ability and right to be the next king or lord, while the son of a serf or slave was destined to be a serf or a slave. Station at birth determined destiny. People conquered or captured in wars were a lower category of humans for whom normal rules did not apply. Slavery, feudal bondage, patriarchy, draconian laws and mass killing drew their rationale from these beliefs.

It was a logical step then to postulate that the features that make some people more 'elevated' than others are hereditary and are passed on from parents to their offspring. If a superior person is born superior, and if an inferior person is born inferior, then it stands to reason that human society will improve if those with 'good germ plasm' have more children and those with 'bad germ plasm' have fewer children. After all, that is how the domesticated animal stock—cows, horses, poultry, birds, dogs and cats—has been bred and agricultural yield and quality enhanced for centuries.

Eugenics was practiced in ancient Greece, the birthplace of Western civilization. Sparta sought to breed a strong warrior class by killing newborns who were unhealthy or deformed.

Both Plato and Aristotle favored marriage controls that would reduce the birth rate among the lower classes and promote marriage between persons of superior abilities.

Lucius A Seneca was a distinguished Roman philosopher, playwright, and political personality who exercised a major influence on philosophical, ethical and Christian thought. He was one of the three principal proponents of stoicism, a school of philosophy that holds that wisdom is attained by accepting reality, living in a moderate fashion and avoiding overindulgence. Promoting endurance and control of negative emotions, it fosters respect for justice and order. Seneca's writings were celebrated among the upper-class Romans. Yet, despite his stoical acceptance of what nature had brought forth and his exalted ethics, he entertained drastic interventions to snuff out anomalies that were but intrinsic products of nature:

> *We put down mad dogs; we kill the wild, untamed ox; we use the knife on sick sheep to stop their infecting the flock; we destroy abnormal offspring at birth; children, too, if they are born weak or deformed, we drown. Yet this is not the work of anger, but of reason – to separate the sound from the worthless.* Lucius A Seneca (Wikipedia 2020 – Eugenics).

Eugenic goals are actualized by corralling people with 'suspect' genes in their own areas, not allowing them to breed with holders of 'good' genes, performing male and female sterilization and abortion. In its ultimate version, negative eugenics entails terminating the lives of the 'defectives' by withholding food or treatment, inducing lethal disease or gas chamber execution. Colonial domination and brutality often effectively end in eugenic type of outcomes.

Mature Eugenics

The maturation of eugenics had to await the maturation of capitalism in Europe in the 19th century. While increasingly basking under the visions of democracy, individual freedom, scientific rationalism, and technology driven rapid progress, these nations also had to come to terms with the human misery induced by the exploitation of workers and children, slavery and colonial domination—the pillars on which their prosperity was built. How could expanding freedoms and better life quality for some be reconciled with abject poverty and servitude for millions more? The divine rights of the kings and feudal bondage was no more. People had the freedom to improve their station in life. Yet, inequality and pauperism persisted. A rational explanation had to be found.

Scholars absorbed by this conundrum came up with two basic explanations. A minority placed the blame on the social system: It had structures that inevitably privileged a few. To ensure progress for all, systemic changes were needed. But the majority accepted, implicitly or explicitly, that capitalism was a rational system, and argued that the cause lay in the biological makeup of individuals. Those who progressed in life did so because they had inherited a 'good germ plasm' and those who did not, lacked it. The 'lower' elements did not have the physical and intellectual ingredients required to be fully human. The working class was lazy and devoid of the spirit of entrepreneurship and innovation. Prone to immorality and vice, the lower-classes bore too many children. The enslaved and colonized peoples were 'uncivilized, slothful barbarians' who had to learn Christianity, reading, writing, and respect for law and authority. What was being done to them was for their own good. If they did not prosper, it was because of rebelliousness, cultural backwardness and inherent biological defects.

Tenets of Christianity and selective aspects of the expanding scientific knowledge were harnessed to serve the elitist, racist perspectives. *Essay on the Principle of Population* by Reverend Thomas R Malthus was a pioneering tract. Published in 1798, it postulated an intrinsic imbalance between the supply and demand for food. It claimed that as human population grows at a geometrical rate (1,2,4,8,16,32), the supply of food grows at an arithmetical rate (1,2,3,4,5,6). If population growth is not checked, especially for the lower

37

classes who procreate at a higher rate, calamities lie ahead. The theory relied on flimsy evidence. At a later stage, Malthus himself doubted it. Yet, the British Protestant parsons who dominated the discourse on population in that era were generally in agreement with Malthus. Many secular thinkers had a Malthusian disposition as well. At a time when entrepreneurs were venerated and the 'slothful' workers and colonized people were despised by the European elite, Malthusian ideas soon became into a basic pillar of the evolving eugenicist doctrine.

Charles Darwin's works on the Theory of Evolution provided the second pillar for eugenics. His postulation of competition for resources leading to the survival of the fittest as a basic law governing the evolution of plant and animal kingdoms was strongly influenced by the works of Thomas Malthus. Darwin did not pay due attention to the role of cooperation in natural evolution. Yet, the validity of his theory in the biological domain is not at issue. He was a scientist *par excellence*. Drawing on a wide range of evidence, he undertook painstaking critical analysis before drawing a conclusion. The problem arose when laws valid in the biological domain were uncritically extrapolated to human society. Called Social Darwinism, it placed competition as a major force of social development and survival of the fittest as a natural law of human history. Like Malthusianism, Social Darwinism rested on a flawed conceptual and evidence base. But both became ingrained in the capitalist ideology and are now directly or indirectly embraced by mainstream social scientists, economists, politicians and media pundits. As a prominent sociologist in the US proclaimed:

> *A drunkard in the gutter is just where he ought to be, according to the fitness and tendency of things. Nature has set upon him the process of decline and dissolution by which she removes things which have survived their usefulness.*

William G Sumner

Malthusianism and Social Darwinism provided the intellectual foundation for modern eugenics. When envisioning human society, even Charles Darwin set aside his usual intellectual rigor and opined:

> [The] *weak members of civilized societies propagate their kind. No one who has attended to the breeding of domestic animals will doubt that this must be highly injurious to the race of man.* (Darwin 1871).

Another key contention was whether humanity was a single race (monogenesis) or composed of different races derived from different ancestors (polygenesis). Supporters of slavery and colonization favored the latter. Humans races were by birth unequal, they claimed. Darwin was adamantly against slavery and his work significantly punctured the polygenesis doctrine. New fossil evidence and studies of mitochondrial DNA have irrefutably established that all humans are descended from the 'Mitochondrial Eve' or a group of closely related women who lived in Africa 200,000 years ago. Darwin thus played a pioneering role in debunking external eugenics. Nonetheless, the idea has as yet to be banished from the political arena.

Francis Galton

Francis Galton was a 19th century English polymath with an astonishing array of accomplishments to his credit. A first-rate innovative statistician, he formulated major concepts like correlation, regression, standard deviation and median that statisticians commonly employ today, and initiated the use of questionnaires and surveys for social science research. His work in genetics, sociology, psychology, anthropology, meteorology, criminology and cartography had a lasting impact in those fields. Launching a variety of research projects, he set up a highly regarded research laboratory, invented tools for clinical diagnosis, and traveled widely, both for pleasure and geographical inquiry. A prolific author of scientific and popular works, he held senior positions in scientific societies and received high honors for his multiple accomplishments.

Yet, Galton was also the undisputed progenitor of modern eugenics. The role of inheritance in the determination of human traits formed the major theme in his research. The traits he examined included physical features like height, weight, arm span, eye and hair color, vision and hearing, the sense of touch, reaction to stimuli, respiratory capacity and characteristics like mental acuity, intelligence and beauty. Using statistical tools in an innovative fashion, he inquired: What factors affect such human traits? Are they correlated with age, birthplace, residence, occupation, marital status and other personal characteristics? Are they passed on in families? What are the relative effects of nature and nurture?

His subjects were famed families, scientists, visitors to his laboratory, twins and others. He observed, interviewed and measured human subjects, compiled existing data, performed animal experiments and interacted with luminaries of his day.

The complexity and scope of his work defy a simple summary. Profoundly influenced by Darwin's ideas about survival of the fittest, he was the first to use the term 'eugenics,' positing it as *a brief word to express the science of improving stock*. The key conclusion drawn by his studies was that intelligence, beauty, life accomplishments and other characteristics were primarily determined by nature, not nurture. Heredity, not upbringing or environment, primarily affects your status in life. His celebrated work, *Hereditary Genius*—based on studies of famed personalities—posited that intellectual capacity was a measurable, inborn trait that ran in families. He advocated that the prevalence of such positive traits could be increased through scientifically based eugenic measures:

What nature does blindly, slowly and ruthlessly,
man may do providently, quickly and kindly.
Sir Francis Galton

Galton was a scientist of noble intent. He envisioned a society where most people would get high education, engage in professional work and have stable marriage. Incomes would not be too disparate and the 'weak' would live an isolated, celibate but decent existence. Further, people from other lands who were of the 'better sort' would be welcomed. But for the human race to progress towards that advanced state, steps of the sort that nature automatically took to propagate genes with a high survival capacity would have to be consciously implemented. In particular, families of 'merit' should be identified, and their offspring given financial incentives to marry early and bear many children.

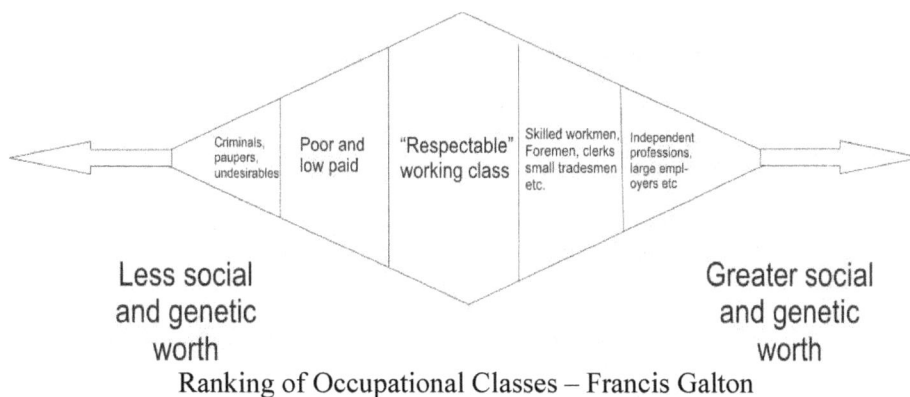

| Criminals, paupers, undesirables | Poor and low paid | "Respectable" working class | Skilled workmen, Foremen, clerks small tradesmen etc. | Independent professions, large employers etc |

Less social and genetic worth   Greater social and genetic worth

Ranking of Occupational Classes – Francis Galton

Galton wrote a total of nine books and about 200 papers. His eugenics papers suffered from key shortfalls. The basic flaw was to assume that complex traits like intelligence, mental acuity, family merit, criminality and beauty could, like age and weight, be quantified. His samples were not representative samples and his inter-generational data were biased and limited in scope. He did not accord due weight to environmentally induced long-term genetic changes. With limited anecdotal and observational data, he drew far-reaching conclusions. But since it had the veneer of scientific rigor, and reflected the prejudices of the elite, his work gathered a wide audience. For example, Galton's ranking of occupational classes, shown above, has no scientific merit.

Galton's outlook was suffused with racial bias. He opined that it would be of benefit to humanity if the Chinese migrated to Africa and took over the continent.

> *[Average] negroes possess too little intellect, self-reliance, and self-control to make it possible for them to sustain the burden of any respectable form of civilization without a large measure of external guidance and support. The Chinaman is a being of another kind, who is endowed with a remarkable aptitude for a high material civilization. ... The Hindoo cannot fulfill the required conditions nearly as well as the Chinaman, for he is inferior to him in strength, industry, aptitude for saving, business habits, and prolific power. The Arab is little more than an eater up of other men's produce; he is a destroyer rather than a creator, and he is unprolific.* (Galton 1873).

Despite his ahistorical, prejudiced ramblings and banal ranking of races and ethnicities, Galton became a celebrated, highly influential person in British and American scientific and political circles. Elected to the fellowship of the Royal Geographic Society and the Royal Society at a young age, he was awarded high civil honors. And it was his scientific credentials and voluminous scholarly output that provided the momentum for the eugenic movement to takeoff and flourish.

The British Eugenics Society (earlier the Eugenics Education Society and later, the Galton Institute) was a major conduit through which eugenics prospered. Established in 1907 with Francis Galton as a co-founder and first honorary president, its basic aim was to promote education and research in eugenics. *The Eugenics Review*, its flagship publication, became the premier British journal in the field. The Society organized public lectures and outreach meetings, printed educational pamphlets and engaged in lobbying for the passage of eugenics-oriented laws. The membership comprised the cream of the British society—scientists, academicians, physicians, political figures, and theologians.

Karl Pearson

Karl Pearson, a follower of Galton, was an intellectual giant who mastered varied disciplines like law, literature, Germanic languages, political theory, history, physics and meteorology. A popular teacher and a fine orator, he published plays, travel diaries and biographies. But he is best known, with Francis Galton and Ronald Fisher, as a founder of the modern science of statistics. Initiating the placement of the discipline of statistics onto a rigorous mathematical foundation, he developed commonly used statistical tools like hypothesis testing, chi-squared test, p-value, correlation and regression coefficients, principal component analysis and the method of moments and extended the use of statistical analysis to investigations in biology, psychology and cosmology. *Biometrika*, the journal he co-founded, was the first high quality statistical journal. And it remains a ranking journal. Running a statistical laboratory, he significantly influenced the expansion of the statistical profession, improved statistical curricula and won scores of prestigious awards. His students became distinguished statisticians. He posited the scientific method—rigorously applied—as the prime tool for acquisition of reliable knowledge in all disciplines.

> *There is no short cut to truth,*
> *no way to gain a knowledge of the Universe*
> *except through the gateway of scientific method.*
> Karl Pearson

Over time, he came to differ from Galton on several scientific, philosophical and political issues, and adopted a more rigorous, circumspect approach to inferring from data. But in relation to eugenics and Social Darwinism, he stood on the same side as his mentor. After Galton retired, he took over the directorship of the Eugenics Records Office at the University College of London in 1907, and in 1911, became the first Galton Professor of Eugenics at the same institution.

Sadly, eugenics dented his reputation as a scientist. In eugenic studies, his customary rigor was set aside. Poor quality data and sloppy methods were used to generate monumental conclusions. He opposed Jewish immigration into Britain on the grounds that on average Jewish males and females were physically and mentally inferior and the immigrants would become 'a parasitic race'. These views rested on flawed inquiry. A recent assessment of the scientific quality and ethics of the studies of Jewish pupils and of the relationship between health and intelligence by Pearson and his colleagues concludes:

> *The methodology used and inferences made by Pearson and his coauthor are sometimes questionable and offer insight into how Pearson's support of eugenics and his own British nationalism could have potentially influenced his often careless and far-fetched inferences.* (Delzell and Poliak 2013).

His scientific perceptions were clouded by racist bias. He aimed to identify social interventions that would make the fertility of the 'white race' outpace that of the 'black race,' and insure the supremacy of the former over the latter. As he stated in a major address to the medical profession:

> *History shows me one way, and one way only, in which a high state of civilization has been produced, namely, the struggle of race with race, and the survival of the physically and mentally fitter race.* Karl Pearson (Pearson 1901).

While he expressed admiration for the Nazi program to purify society and defend the Aryan race, he did not condone mass killing or genocide. But he approved of measures through which the numbers of 'tramps', 'criminals', 'mentally defectives', perpetual 'paupers' and 'blacks' could be controlled.

> *The right to live does not connote the right of each man to reproduce his kind ... As we lessen the stringency of natural selection, and more and more of the weaklings and the unfit survive, we must increase the standard, mental and physical, of parentage.* (Pearson 1912).

Ronald Fisher, the third of the major founders of the modern science of statistics, was also a fervent eugenicist. While critiquing the flaws in the statistical analyses used to backup eugenic claims, he also ascribed to eugenic views and held on to them even after the horrors of the Nazi era holocaust were revealed.

A number of eugenic studies associated bodily measures with behavior, especially 'degenerate' tendencies like criminality. Eugenicists extrapolated these findings to posit a hereditary basis to human behavior. Though they rested on biased samples, diffuse measurements and dubious analysis, these phrenological conclusions entered the established eugenic folklore.

With distinguished scientists and statisticians like Thomas R Malthus, Charles Darwin, Francis Galton, Karl Pearson, Ronald Fisher, and more championing its ideas, the eugenics creed soon rose to scientific, public and political eminence in the Western world. And in no time, it migrated from the drawing board to practice.

3.2 BASIC THEORY

The modern Theory of Evolution developed from the sturdy foundation laid by Charles Darwin and his contemporaries. The theories of heredity formulated by Gregor Mendel and Francis Galton placed it on firmer pillars. The molecular mechanism of heredity was elucidated upon the 1950s discovery of the structure of DNA, the key molecule of life, by James Watson, Francis Crick and Rosalind Franklin. The many path breaking discoveries made since then have given us a good picture of how the astounding variety of living organisms evolved on planet Earth. It is now well established that traits possessed by living beings including humans have emerged from a long, complex process of random genetic mutation and natural selection in a dynamic environment.

At the time when eugenics took hold in the West, knowledge of heredity and genetics was at a rudimentary stage. The DNA had been discovered but its function and structure were unknown. But enough was known to postulate that human characteristics were inherited from parents via germ cells in testes and ovaries. Experiments by Gregor Mendel and other biologists, animal and plant breeding, and observational studies had revealed some of the laws governing inheritance.

This was the basis upon which Francis Galton, Karl Pearson and others constructed the intellectual edifice of eugenics. Their work was premised on the direct applicability of the laws of biology to human society. It assumed that the laws of natural evolution also governed social evolution—Social Darwinism. It accepted the Malthusian doctrine that if human reproduction was not controlled, then the growing numbers would outstrip supply and produce catastrophes like famines, epidemics, social chaos and warfare.

Another key eugenicist premise was that human characteristics, simple and complex, were primarily passed on from parents to offspring and were, if at all, only marginally affected by the natural and social environment. Nature largely overrode nurture. Though eugenic scientists posited it as a tenet based on scientific inquiry, the inbuilt design, data collection, analysis and interpretation related biases in their studies essentially served to produce a foregone conclusion. Today, that viewpoint is called genetic determinism.

What makes oranges from this tree sweeter than oranges from that tree? Why is this cat brown and the other, white? How come this boy is tall but his brother, short? Genetics explains the variation and heritability of such features in terms of probability. For some features, genetic effects are low probability events and for others, high probability events. But genetic determinism goes a step further. It ascribes complex behavioral, health, psychological, physical, educational and life performance characteristics primarily to genes, and at times, to a single gene. Scientists, physicians, psychiatrists with a eugenic predisposition in the UK, USA and Europe connected 'unwanted' traits like deafness, blindness, mental deficiency, physical deformity, epilepsy, tuberculosis, juvenile delinquency, criminality, promiscuity and alcoholism as well as 'positive' traits like good health, physical strength, moral uprightness and high scholastic scores to genes and inheritance. Why did one man become a successful businessperson and another, a criminal? Why did this child achieve high scores in school but that one failed miserably? Why is this woman a diabetic and the other a healthy, robust individual? According to eugenics, it is mainly, if not solely, due to genes. A further, decisive leap was made by ranking not just individuals but also large social groups on a scale of genetic worth (scientific racism).

Eugenics Logo

The primary tenets of the eugenics creed that prevailed in Europe, UK and USA in the first half of the 20th century were: Societies develop through competition and survival of the fittest. The rate of growth of the human population tends to outstrip the scarce resources available to sustain civilized existence. There are too many people of the 'wrong' kind and too few of the 'right' kind. Traits that place people into these groups are heritable. People with mental and physical deficiency, propensity to illness, low moral turpitude, criminal and alcoholic inclination and low intelligence constitute a parasitic stratum. Their higher rate of reproduction threatens the future well-being of the human race. People of African ancestry, indigenous nations and recent immigrants are the 'wrong' kind of people. White people of Anglo-Saxon stock, with Nordic (Aryan) features (blond, fair hair, blue eyes) have innately superior intelligence, health status, ability to excel in life and moral character compared to others. The reproductive rate among the 'right' kind was too low.

The Second International Congress of Eugenics held in 1921 adopted the logo *'Eugenics is the self-direction of human evolution'*. Eugenics, as shown in the eugenics logo above, was a scientific endeavor that unified a broad range of scientific, historical and general disciplines. The fundamental aim of the eugenics creed was to combat the accumulation of harmful physical and mental traits in a population (dysgenics).

With support from a large, influential body of scientists, scholars, social reformers and religious dignitaries, the eugenics tenets gained popularity among policy makers and led to enactment of state sponsored and private programs with adequate funds to make them a reality.

3.3 EUGENICS IN THE ACADEMY

By the turn of the century, eugenics was a mainstream discipline in the West. The US, Canada, UK, Germany, Sweden and other nations had established national eugenic associations. Scientific societies and scholarly journals focusing on eugenics were founded. National and local events and campaigns to publicize and implement the eugenic agenda attracted many. The first International Congress on Eugenics was held in London in 1912.

Eugenics also made major inroads into the education system. Curriculum development experts placed it in biology courses in US and Canadian schools. Biology teachers secured training in the subject and biology textbooks were revised. Emphasis was placed on the notion that feeble-minded individuals were a threat to society. Children were administered psychometric tests of dubious validity and those deemed feeble-minded were sent to specials schools.

Eugenics was integrated into the teaching of biology at the university level as well. Elite institutions like Cornell, Colombia, Harvard, Princeton, Stanford and Yale offered special eugenic programs. At one point, there were 300 eugenics related courses in US colleges and universities. Irving Fisher, an influential, pioneering economist who developed the idea of the price index, was a professor at Yale University for nearly forty years. As a dedicated, leading eugenicist, he supported the establishment of the Race Betterment Foundation, sat on the board of the semi-official Eugenics Record Office and was elected the first president of the American Eugenics Society.

David S Jordan, an authority on fish biology, authored scores of research papers on the subject. A onetime president of Cornell University, he was also the first president of Stanford University. And he was a militant eugenicist. Alarmed about the possible 'degeneration' of the white race, he proposed two ideas to arrest it. One, he posited that humans and cattle are *'governed by the same laws of selection'*. Measures of the type taken by cattle breeders to improve their herd are applicable to human populations. Two, in an influential 1902 document, he said that qualities like intelligence, mental disability and success in life were inherited through blood. None of these assertions emanated from reliable research or plausible data.

Yet, he rose into prominence in eugenic circles and foundations. As one of the guiding lights in the field, he influenced the passage of state laws that legalized forced sterilization and other eugenic practices. After his death, his influence on educational policy and practices was recognized as scores of schools, university campus buildings, and national monuments were named after him. And many animal and plant species bore his name too.

Eugenics studies and programs in the US and abroad were underwritten by major US foundations. The Carnegie Institution funded the Eugenics Records Office (ERO), a premier research facility, at Cold Spring Harbor in New York. Compiling data on millions of Americans, the ERO devised detailed plans, lobbied state legislatures and steered the work of public and private service agencies towards the eugenic agenda. The Rockefeller Foundation funded eugenic studies of American and European scientists. As we detail later, it gave substantial funds for eugenic research in Germany in the 1930s. Wealthy Americans, like the widow of the railway baron EH Harriman, also contributed.

Harvard University stood at the apex of the burgeoning eugenics drive. Two of its presidents, CW Elliott and AL Lowell, advocated that the American stock should be kept 'pure' by restricting immigration of people who did not have a Northern European lineage— Jews, Italians and Asians. They opposed the blending of the 'good' and the 'bad' stock and proposed reducing the 'bad' stock by forced sterilization. Marriage of Irish Catholics with Anglo-Saxon Protestants, Jews with non-Jews, and blacks with whites should be prohibited. Jewish and African American students at Harvard were admitted in low numbers and their campus movements were restricted. Harvard University student records were made available to the ERO as well.

> *Each nation should keep its stock pure.*
> *There should be no blending of races.*
> CW Elliott, 1912 (Cohen 2016).

In churning out proclamations and papers supporting eugenics and making national and international eugenic crusades credible, Elliott and Lowell were in the company of DS Jordan of Stanford, Irving Fisher of Yale and RB Bean of University of Virginia. Other Harvard faculty who championed eugenics were FW Taussig, economics professor and author of a major textbook on economics; EM East, botanist and author of two influential eugenic texts, *Inbreeding and Outbreeding: Their Genetic and Sociological Significance* and *Mankind at the Crossroads*; Zoology professor WE Castle whose *Genetics and Eugenics* was a standard reference for eugenic courses at many universities; RM Yerkes, eminent psychology professor, instructor for psychology and heredity, pioneer of a biased but widely used method of intelligence testing, and author of a major psychology text with chapters on eugenics; W McDougall, a psychology professor and author of openly racist, xenophobic tracts like *The Group Mind* and *Is America Safe for Democracy*; CB Davenport, a zoology professor, whose text *Heredity in Relation to Eugenics* was used in eugenics courses nationwide and utilized for writing high school biology material on eugenics. After leaving Harvard, Davenport founded and became the driving brain behind the ERO. Claiming that 'laziness' was an inherited trait, he opined that when both parents are 'shiftless,' only one out their six children would be 'industrious'; and long-time athletics supervisor DA Sargent claimed that physical exercise was a key to race betterment and promoted pelvic exercise for white female students so that they could bear children with large brains.

Harvard alumni played important, successful roles in organizations whose main goal was to ensure passage of laws to restrict immigration of 'unworthy' people and implement forced sterilization of the 'imbeciles' and 'unfit'. Harvard alumnus TL Stoddard popularized eugenic ideas in the US and beyond. An acclaimed journalist and political scientist, he was also a member of the Ku Klux Klan and the American Eugenics Society and sat on the board of the American Birth Control League. Of the many general audience books he wrote, *The Rising Tide of Color against White World Supremacy* became an international best seller. Its main thesis was that population explosion, unrestrained immigration and growing anti-colonial struggles placed the white race and civilization at risk from being overrun by 'colored' peoples. The book elicited profuse accolades from the segregationist US President Warren Harding and influenced the Nazi regime in Germany. Readily embraced in the prevailing mood of that era, it retains an iconic status in white supremacist circles to this day.

The Harvard-based father and son pair, Oliver Wendell Holmes Sr. and Oliver Wendell Holmes Jr. were also major figures in US eugenics. The former was a physician and distinguished poet who rose to the deanship of the Harvard Medical School. While he held atypically moderate views about the admission of African American students to Harvard, he firmly championed Galton's doctrine and claimed that 'aptitude for learning' as well as moral character were inherited characteristics. In his view, these traits were more prevalent in the elite class to which he belonged. The latter read law at Harvard, served on the Massachusetts Supreme Court, edited the American Law Review and became a professor at the Harvard Law School. His tome, *The Common Law*, influences the practice of law in the US to this day. He is also regarded as the founder of a distinctive school of legal philosophy. His multitude of opinions on civil liberties, constitutional law and democratic norms during his thirty-year tenure on the US Supreme Court are widely cited to this day. Regarded as the most eminent American jurist, Justice Holmes Jr issued pioneering liberal rulings on key matters like anti-trust law, US imperial ventures, law of contempt, citizenship oaths, due process, freedom of speech and admissibility of evidence.

But when it came to eugenics, his liberalism vanished into the thin air. In a 1927 case concerning the forced sterilization of Carrie Buck on grounds of mental deficiency, he authored the majority opinion that upheld the Virginia Sterilization Act of 1924. Not only was she deemed an 'imbecile' without good evidence, but Justice Holmes went on to argue that the interest of public welfare superseded the interest of the individual:

> *It is better for all the world, if instead of waiting to execute degenerate offspring for crime, or to let them starve for their imbecility, society can prevent those who are manifestly unfit from continuing their kind. .... Three generations of imbeciles are enough.* Justice OW Holmes (Wikipedia 2020 – Oliver Wendell Holmes, Jr.)

Eight of the nine justices on the bench concurred with Justice Holmes. Among them were former US president and Chief Justice William H Taft and Justice Louis D Brandeis. The former was a Yale alumnus and the latter a Harvard alumnus. Both had a solidly progressive reputation; Brandeis was popularly dubbed as a 'people's lawyer'. Issued by a distinguished bench, this opinion has not been overturned to this day, and continues to be cited in some cases.

In addition to major academic and scientific luminaries, wealthy Boston area patrons and influential alumni connected to Harvard lent their funding and political clout to the Harvard based eugenics effort. With their backing, Harvard outranked all the US universities in the theorizing, promotion and actualization of eugenics in the US.

> [In] part because of its overall prominence and influence on society, and in part because of its sheer enthusiasm, Harvard was more central to American eugenics than any other university. Harvard has, with some justification, been called the 'brain trust' of twentieth-century eugenics ... (Cohen 2016).

Without going into details, we note that eugenics diffused into the education systems of other Western nations like Britain, Sweden and Germany in a similar fashion and was also championed by leading scholars and scientists in those nations.

3.4 FROM THEORY TO PRACTICE

Despite its dubious scientific and ethical foundations, a generation of scientists, statisticians, scholars, physicians and writers embraced eugenics. Universities in the West offered courses on eugenics. Scientific societies, scholarly journals, and official and semi-official organizations with an exclusive focus on eugenics were established. The data behind their claims were selective and distorted at best and contrived at worst. Key social and economic factors affecting life outcomes were ignored. Even as discerning analysts excoriated the ill-advised conclusions drawn from flawed eugenics research, funds continued to pour in.

It is not an exaggeration to say that by the start of the 20th century, eugenics had become a central preoccupation in the West. With the intellectual foundation laid in the UK, it spread fast into the USA, Canada, Europe, Australia, Japan and beyond. Popular press and magazine articles, emotive pamphlets and enticing books on the topic entranced the public mind. Millions were attracted to the cause of enhancing the moral worth and intellectual caliber of their race and nation.

Eugenics was actualized on six fronts: sterilization, marriage, internment, immigration, procreation and extermination. Though the United States had led the way, its eugenic drive did not reach the ultimate stage of mass murder. Large scale killing of designated groups was mostly done in Nazi Germany. (The German case is dealt with separately).

Forced Sterilization: The US state of Indiana set a global precedent in 1907 by passing a law decreeing sterilization of 'unfit' individuals. While the Indiana Supreme Court nullified it in 1921, by 1925, twenty US states had legalized involuntary sterilization for inmates of mental hospitals, prisons and other places. These laws received the imprimatur of the US Supreme Court in 1927. By 1942, compulsory sterilization was legal in 30 states. Between 1907 and 1963, over 64,000 US residents were sterilized without their consent or knowledge. About two-thirds of the victims were 'oversexed' women. Some 45,000 were persons with a mental disability. Epileptics and deaf and blind people were sterilized in some states. A few states forcibly sterilized incarcerated 'habitual criminals'. Not surprisingly, poor African Americans and Native American were represented disproportionately in these programs. A few of the Japanese Americans placed in camps during WWII were also sterilized. Support from bankers, business moguls and the University of California ensured that more than a third of the coerced sterilizations were done in California. These anti-humane laws remained on the books in many US states even after the Nazi Nuremberg trials. But they were rarely invoked. The last forced sterilization in Oregon occurred in 1981.

Starting from the late 1920s, compulsory surgical sterilizations were done in two Canadian provinces. The last law permitting it was repealed in 1972. At least 3,000 cases were officially acknowledged and about 850 were compensated. Medically coerced sterilization of women of the First Nations endures. Over 60 First Nation women filed a lawsuit in 2017 stating that they were prevented from seeing their newborn child unless they gave consent for sterilization.

Forced sterilization was legalized in nations of Northern Europe—Denmark, Estonia, Finland, Iceland, Sweden and Switzerland—but did not occur in the UK, the birthplace of eugenics. Between 1935 and 1976, more than 60,000 women were sterilized without their consent in Sweden. Some 5,000 men were sterilized in that program as well. Some victims later received compensation.

Marriage Restrictions: Laws and rules prohibiting marriage and/or sexual relations between people from different racial or ethnic groups have a long history. Fearing that it would endanger Christian 'purity,' the Catholic Church had disallowed marriages between Jews and Christians. Violators faced harsh penalties. Miscegenation between whites and blacks in the US, by law or custom, dated to the early colonial days and slavery. The first such law was passed in Maryland in 1691. After the abolition of slavery in 1865, some states repealed these laws.

The trend was reversed after eugenics groups lobbied for the enactment of miscegenation legislation. It was the area where eugenics recorded its first major success. By the late 1800s, 38 states had laws prohibiting inter-racial unions. In some states, the laws were repealed after a few years. Still, miscegenation was illegal in 29 states in 1924.

These laws differed in detail. Some included stronger anti-incest laws. A 1920 Louisiana law banned unions between whites and African Americans or Native Americans. A 1924 Virginia law made marriage between a white and a non-white person a felony. These laws prevented tens of thousands of loving multi-racial couples from marrying.

Internment: Another eugenic stratagem was to confine people with mental illness, mental disability and physical handicaps, and other 'unfit' persons in institutions or locations so as to ensure that they would not bear children or contaminate the 'good' gene pool. Segregating camps were common in the Scandinavian nations. But cost considerations made it a minor part of the eugenic drive in the US. Here the emphasis was placed on identification and sterilization. African Americans and Native Americans were, however, already effectively isolated in ghettos and reservations.

Immigration: Controlling the number and type of immigrants was a critical part of the eugenic drive, especially in the US. Eugenicists further whipped up the prevailing xenophobic and racist sentiments, propagating the view that the early immigrants from Europe, Anglo-Saxon Protestants, had a high prevalence of 'good' genes since they had made *'notable contributions to civilization'*. The influx of foreigners—Jews, Asians, Catholics, Italians and Eastern Europeans—threatened the 'racial purity' of the nation. That peoples of African origin and Native Americans had to be kept in check was an integral, but not always an explicitly articulated tenet of all eugenic thought.

A fiery proponent of racial hierarchies, Harvard botanist Edward M East was a fervent backer of strict immigration control. The formation of the Immigration Restriction League in 1894 by an elite group of Harvard alumni was a milestone in the realization of immigration admission based on eugenic criteria. It was led by wealthy descendants of the first groups of migrants from the UK to the US. Living in the exclusive neighborhoods of Boston and Cambridge, these 'Boston Brahmins' regarded themselves as the highest caste of the American society. Hundreds of like-minded Harvard graduates joined them. Branches in other states were set up. Deeming immigration an issue of race, their goal was to exclude people *'who did not share their northern European lineage'*. (Cohen 2016).

They first promoted literacy testing as a criterion for entry to the US. It would lower the number of entrants, especially from southern and eastern European nations. Their persistent lobbying effort saw the insertion of a clause mandating proven literacy a condition for immigration to the US in 1917. Providing expert testimony and spreading posters and material on the dangers of race neutral immigration, the League played an instrumental role in the passage of two landmark laws: the Emigration Quota Act of 1921 and the Immigration Act of 1924. Consequently, strict racial and national immigration quotas were enacted, and the entry tax was raised, with dramatic effects. The overall numbers declined sharply, and the composition of the immigrants changed markedly:

As the percentage of immigrants from northern Europe increased significantly, Jewish immigration fell from 190,000 in 1920 to 7,000 in 1926; Italian immigration fell nearly as sharply; and immigration from Asia was almost completely cut off until 1952. (Cohen 2016).

Most of these restrictions remained in place until after WWII. Historians feel they had formed a major barrier for the Jews attempting to escape from Nazi Germany. Tens of thousands of 'undesirables' already in the US, including Jewish children, were also deported to face an uncertain future.

Publicity and Procreation: Proponents of eugenics devised campaigns to spread the eugenic message across rural and urban America. Books and pamphlets written for the mass audience complemented public talks, radio programs and placement of posters bearing eugenic messages in public venues. The Society for Racial Hygiene in Sweden printed pamphlets on eugenics and its social goals that were freely distributed in large numbers to health personnel, teachers, clergy and others. Donations from media firms and wealthy individuals funded a traveling eugenic exhibition whose aim was to depict the 'ideal' humans, the 'good' Nordic persons with specific physical and mental features. They were contrasted with persons of 'low quality,' namely, criminals, gypsies and vagrants. The exhibition—photos, sculptures, paintings and reading material—drew large crowds as it went from town to town. Its basic message was that the Nordic race was the superior race, and among them, the Swedes stood at the apex. A 1926 US poster that repudiated persons with 'defective' genes and encouraged procreation among those with 'high-grade' genes is illustrative.

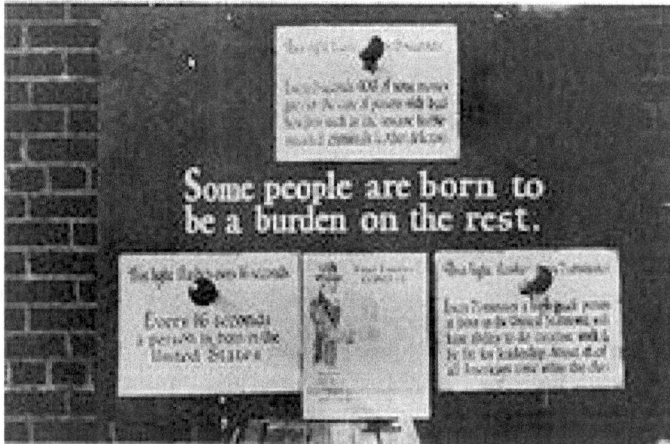

Eugenics Poster, USA, 1926.

Another forum for the spreading the eugenics message was devised by Mary de Garmo, a teacher, official in the St. Louis Chamber of Commerce and an activist for mother and baby welfare, children's education, veterans and the war dead. A loyalist of the eugenics cause, she authored the influential *World's Baby Eugenic Almanac for Parents.* She initiated the Better Babies Contest, the first of which was held at the Louisiana state fair in 1908. State fairs, an American tradition, were the prime venue for exhibiting and marketing farm products and livestock. Cows, sheep, dogs, pigs and horses brought to the fair were judged in terms of quality, ranked by ribbons and awarded prizes. Entertainment and educational activities for children and adults were also at hand. There were sheep shearing and other competitions. Diverse goods were available at a low price. Hundreds of thousands, if not more, usually thronged to the annual event.

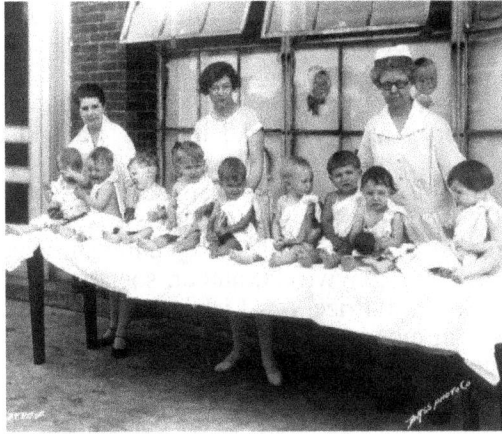

Better Babies Contest, Indiana, 1927

De Garmo saw it an ideal venue for a baby contest. Backed by well-connected individuals, she managed to convince mothers to bring their 6 to 48 months old babies to the fair. Ranked by a panel of health experts in terms of physique, and mental and physical health, high scoring babies got awards and prizes. Advocated by the *Woman's Home Companion* magazine and major women's groups, the contest featured in many state fairs. It combined education to improve nutrition, reduce infant mortality and promote child health with eugenic ideas. And the contests also served as a marketing venue for the manufacturers of baby foods.

Using a score card for varied attributes, babies were ranked on a 1000 point gender-specific scale devised by expert statisticians. Babies with high scores got medals and prizes. From 1916 on, 50,000 babies were judged in this manner. In places, the contests expanded to Better Families Contests. Later they declined in popularity and were replaced by specific public health programs. But information from these contests provided another 'scientific' pillar for the eugenic doctrine and helped set eugenic standards for 'fit' and 'unfit' babies.

That Better Baby Contests were held in a venue where farm animals and plants were also judged reflected in part the link between prominent eugenicists and the agricultural industry. And it spoke to the key tent of Social Darwinism: Laws and practices applying to flora and fauna apply to human populations as well. Mendelian genetic laws that predicted inheritance of attributes like color and size of peas and beans applied also to features such as disability, intelligence, alcoholism and criminality. A dairy farm selectively breeds the best milk yielding cows. A garden is kept beautiful and viable by careful pruning and replanting. Just as genes of the 'disabled' animal or plant are removed from circulation, the 'bad' human genes must be deleted in a likewise fashion. It was an essential step to ensure that the human population remained healthy and prosperous.

Eugenics Certificate

Among other innovations aimed to make a fast buck, some charlatan physicians provided eugenic certification to persons seeking a marriage partner.

Extermination: The problem of 'bad' germ plasm harming the human race had an ultimate solution—extermination of individuals who had it. The 'fit' had the right to survive and propagate; the 'unfit' did not.

Though eugenics reached this stage to a decisive extent only in Germany under the Nazis, genocidal eugenics was not just the ravings of a few extremists. It permeated mainstream Western thought and was articulated by eminent social reformers and literary figures as well. Bernard Shaw, the playwright, author, socialist of the Fabian mold, anti-war activist and winner of the Nobel Prize for Literature was an avowed eugenicist. State overseen selective breeding, in his view, would be a basic feature of a socialist society. He was also not averse to placing '*defectives*' in a '*lethal chamber*' (Freedland 2012).

DH Lawrence, a British poet and author whose novels and work broached issues of sexuality, mental health and freedom and critiqued the demeaning aspects of modernization in a controversial yet pioneering way, provides another case. He was ostracized by officialdom, and his books were censored. But on the question of eugenics, he went in the opposite direction. In a letter to a friend, he wrote:

> *If I had it my way, I would build a lethal chamber big as the Crystal Palace, with a military band playing softly, and a Cinematograph working brightly; then I'd go out in the back streets and main streets and bring them in, all the sick, the halt and the maimed; I would lead them gently, and they would smile me a weary thanks; and the band would softly bubble out the 'Hallelujah Chorus'.* DH Lawrence (Grue 2010).

Cases of involuntary euthanasia, withholding lifesaving treatment to children and adults who were deemed 'unfit,' did transpire in the US and Europe. But apart from Germany, the practice did not catch on. A few doctors withheld treatments from seriously ill children but were not held to account. In one mental institution in Illinois, doctors fed milk from tubercular cows to the inmates. The eugenically fit patients were expected to survive. As a result, the annual death rate at this institution rose up to 40%.

3.5 NAZI EUGENICS

Capitalism needs strong and complex state institutions—legislature, civil service, police, legal system and military—to function and expand. Primarily a protector of the interests of the capitalist class and the capitalist economy, the state is also the arbiter between competing capitalist enterprises, and between the capitalist class and the other social classes (workers, small business owners and the middle class). It protects national firms against foreign firms and projects itself abroad to secure markets, resources, investment opportunities and profits for national firms.

Capitalist economies are unstable economies characterized by cycles of boom and bust. Economic inequality usually expands during the crisis phase, and the lower and middle classes bear the brunt of the losses. When faced with a crisis, one response of capitalist states has been to expand arms manufacturing, the military, police and other security arms and launch military ventures abroad. Such actions provide a secure venue for investment and higher profits.

Serious crises and extreme inequality propel the capitalist state towards authoritarianism. The repressive state institutions come to the fore in social and civic affairs. Stronger controls on large and small businesses are instituted. Force contains the efforts of the working classes and disaffected segments in society. Conservative forces and politicians espouse populist and xenophobic ideas that hitherto had festered in the background and redirect popular anger onto the perceived internal and external 'enemies' of the nation.

Germany after WW I was a society in deep crisis. Prewar exuberance was replaced by a clear defeat, national humiliation, a damaged economy and the loss of colonies. Punitive reparations were imposed. A militant labor uprising, though brutally suppressed, chilled the capitalists and the political establishment. The conditions were ripe for a demagogic, ultra-nationalistic, extreme right political movement to take hold.

The Nazi Party, initially espousing anti-big business rhetoric yet mainly aiming to drive workers away from socialist movements, rose with rapidity in this political arena. Led by Adolf Hitler, a fiery upstart, it glorified and called for the protection of the Germanic (Aryan) master race.

Nazi Swastika

Under the official name the National Socialist German Workers' Party, and with the swastika as its symbol, the Nazi Party program called for welfare programs and restrictions of civil rights. Neither a socialist nor a worker-friendly party, it was a far right, imperialist party allied with the major German corporations. That reality was masked by the racist, anti-Semitic and eugenic vision stridently espoused by Hitler.

Hitler's outlook was shaped by a childhood diet of American Indian tales by American and German authors, including Karl May, whose bestselling books idealized the Native American way of life. Later he came to admire the manner by which the 'Indians' had been conquered and decimated by the 'Nordic' white settlers, corralled in desolate reservations, and forced to submit. He approvingly noted in a major speech:

> [America] *gunned down the millions of Redskins to a few hundred thousand, and now keeps the modest remnant under observation in a cage* ... Adolf Hitler (Miller 2020).

Hitler considered the segregationist, genocidal features of US history and laws a model for Germany. Just as white America exercised its dominance over Africans and 'Redskins' so should Germany deal with Jews and others who endangered the purity of the 'Aryan race'. Embracing the Darwinian idea of the survival of the fittest, Hitler and fellow Nazis believed that the German nation had been weakened by *'the infusion of degenerate elements into its bloodstream'*.

A considerable interest in eugenics in Germany had existed even before the Nazis. Influenced by Galton and other British eugenicists, German scholars and scientists operated under the banner of the German Society for Racial Hygiene formed in 1905. Its journal was their venue for scholarly discourse. Hitler was enthralled by the sensational, pseudo-scientific tracts of two leading American eugenicists: *The Passing of the Great Race* by Madison Grant —famed zoologist, lawyer and conservationist—and *The Case for Sterilization* by Leon Whitney—author of books on horse and dog breeding and president of the American Eugenics Society. Sending letters to both, he profusely praised their ideas, and called *The Passing of the Great Race* his 'Bible'. In turn, when the Nazis began their mass sterilization program in 1935, Whitney was one of his cheerleaders on American soil. Emulating the Nazis, he advocated sterilization of ten million Americans in order to prevent *'rapid societal disintegration'*.

Elected the Chancellor of Germany in 1933, Hitler anointed himself as the *Fuhrer* (Supreme Leader) a year later. The Nazi regime spared no time in putting his words into deeds. The ensuing eugenic program had five components: new laws, sterilization, medical experiments and research, seclusion and ultimately, extermination.

Eugenic, Racist and Anti-Semitic Laws: A keen student of the state and federal American laws that endorsed racist, segregationist and eugenic—miscegenation, sterilization, immigration—measures, Hitler sought to enact laws that would similarly protect the racial purity and health of the German people. Drawing from American scholars, he claimed that science was on his side. The idea of *de facto* second-class citizenship for some racial and ethnic groups buttressed by legal mandates fitted the Nazi mentality.

German legal scholars had a longstanding interest in US law, especially the portions dealing with racial, ethnic and indigenous people. Attention to such laws grew in the Nazi era. The doctrine of Manifest Destiny had a particular attraction: That the US was an exceptionally virtuous nation with a God given mandate and power to remake the world in its own image. They were also fascinated by the claim that the land in the continental US was 'free land' at the disposal of the settlers from Europe. Within that framework, they saw the 1830 Indian Removal Act passed by the US Congress and other race laws as a precedent for their expansionist polices towards Eastern Europe and Russia. Nazi lawyers studied the Jim Crow laws of different US states that separated and subordinated African Americans. Dubbed *Lebensraum* (open land), Hitler and other Nazi leaders justified their colonial policies with explicit branding of Poles, Ukrainians, Russians and other Slav peoples as well as Jews as 'Indians'. These 'savages' had to be treated like 'American Redskins' and displaced to make room for millions of German settlers.

Nazi lawyers, including the leading legal minds in Germany, produced voluminous tomes based on in-depth studies of relevant US laws. Their aim was to devise analogous laws for their own nation. For example, the Nazi law on sterilization extensively drew upon the writings of California based eugenicists, the state's sterilization programs and the relevant legislation. The lead personality in this effort, a German legal scholar named Heinrich Krieger, studied at the University of Arkansas Law School and wrote a major thesis on American laws on race and Native Americans. His writings were the primary source for a June 1934 meeting of seventeen '*German jurists, lawyers, scholars, and [Nazi] party officials*' concerned with drafting the new laws.

> *Three major themes of American federal and state laws were presented in the scholarly materials prepared for the meeting and were analyzed and discussed at great length: the anti-miscegenation laws of thirty American states; the federal and state laws creating second-class citizens in the United States; and American immigration and naturalization laws.* (Miller 2020).

> *America was so much the leading racist jurisdiction that even Nazis looked here for inspiration. They wanted nothing to do with American liberal and humanitarian currents, but when it came to race law, they saw the United States as a country that had started down a path that they wanted to walk to the final destination.* James Whitman (Richardson 2019).

A 1935 educational visit of 45 Nazi lawyers to New York organized by Krieger received an enthusiastic welcome from representatives of US law organizations and prominent lawyers who were undoubtedly aware of the anti-Semitic and racist designs of the Nazi regime. Some Nazi scholars felt the US laws to be '*too harsh*' for German society. But they were regarded as a model from which German laws could be formulated. In turn, US eugenicists and celebrities envied, praised and publicized German race laws and worked to devise similar statutes for the US.

The German parliament passed three laws in 1934, the Nuremberg Laws, codifying the Nazi philosophy. These included the Reich Citizenship Law and the Law on the Protection of German Blood and Honor. The former law and its extension turned Jews, Gypsies, Africans and their children into second class-citizens and deprived them of their civil rights. Sexual relations and marriage between Aryans and these groups were proscribed.

> *The Blood Law took the surprising legal step of criminalizing marriages and sexual relations between Jews and Germans. Except for the United States, and South Africa in regards to premarital sex, no country in the world imposed criminal sanctions on miscegenation (marriages between people of different races). This American development was too harsh for the Nazis at first, but they debated the idea for a long time and finally adopted the American position to criminalize miscegenation.* (Miller 2020).

Sterilization of people with 'bad' genes was legalized too and deportation and extermination of 'backward' races and peoples received the blessings of the law as well.

Sterilization: As of January 1934, it was permissible to forcibly sterilize any person deemed to have one of the 'genetic' disorders listed in the Law for the Prevention of Hereditary Diseased Offspring. But consent of a Genetic Health Court was needed for the procedure. Framed by prominent German eugenicists and lawyers, this law was directly based on the Model Eugenic Sterilization Law drafted by the American eugenicist HH Laughlin. Earlier, his model had been adopted by the US state of Virginia. Laughlin took pride in the influence he had exercised on the German law and printed an English translation in a US eugenics magazine. In turn, the University of Heidelberg bestowed an honorary degree on Laughlin for his contributions to the '*science of racial cleansing*'.

The German drive started expeditiously: Some 30,000 to 40,000 people were sterilized within a year. Eventually, about 400,000 people were sterilized in German hospitals. People with physical and mental impairments, deafness, blindness, muscular dystrophy, cerebral palsy and epilepsy were selected for the procedure. Also ensnared were criminals, 'lazy' persons, alcoholics, political opponents, welfare recipients, homosexuals, sex workers, and Jewish, Roma, Afro-German and mixed-race persons. While the law permitted sterilization for persons over the age of 14, younger children were also affected. Women were sterilized by injecting the uterus with a toxic chemical while men had high doses of X-rays directed at their sexual organs. Many suffered injuries and quite a number died.

Eugenic Research and Experimentation: As in Britain and America, research played a crucial role in the development of the eugenic outlook in Germany. German doctors and biologists conducted and published eugenic investigations whose aim was to devise scientific solutions for removing 'inferior' genes from human society. Taking their cue from the US eugenic studies, they collaborated with American eugenicists and also conducted their own health and social studies.

The US Eugenics Record Office headed by the Harvard trained biologist Charles Davenport established ties with German scientists. The Rockefeller Foundation was a key source of funds for German eugenic research. By 1926, it had awarded $410,000 in research grants to over two hundred German scientists for eugenic studies. And, in the following years, it gave over $400,000 for the expansion of the Kaiser Wilhelm Institute for Psychiatry. The Foundation gave stipends to students from other nations to study in Germany during the Nazi era. Other non-profit US agencies funded racial science research in Nazi Germany as well. The Kaiser Wilhelm Institute for Anthropology, Human Heredity and Eugenics was also a beneficiary of the Rockefeller Foundation largess.

Germany-based psychiatrist and geneticist Ernest Rudin secured major grants from the Rockefeller funds. A co-founder of the German Society for Racial Hygiene, he came to head the Kaiser Wilhelm Institute of Psychiatry and the International Federation of Eugenic Organizations. His elaborate, large sample studies analyzed with novel statistical methods were widely cited and earned him high honors from international medical and psychiatric organizations. Despite the major methodological flaws in these studies, he is still viewed as the founding father of the field of psychiatric genetics.

An avid eugenicist from his early days, Rudin advocated sterilization of alcoholics and withholding medical treatment from the mentally infirm, epileptics and alcoholics. He played a key role in the formulation of the 1933 Law for the Prevention of Hereditary Diseased Offspring. He was also an expert advisor for the racial policies of the Nazi regime and supported its euthanasia program. A portion of his specimens and data came from the atrocious experiments done on Jews and Gypsies by Nazi doctors. Awarded a medal by Adolf Hitler for his contributions he no doubt was a key scientific architect of the Nazi crimes against humanity.

Twin Studies

Human twins were a major area of Nazi eugenic research. Eugenics asserts that nature (genes, germ plasm) overrides nurture (environment, social class, upbringing, education) in the determination of physical, mental and behavioral human traits. Studying the birth features, growth and lives of twins, especially (genetically) identical twins, thus constitutes an important aspect of the nature versus nurture debate.

The History of Twins, a 1875 paper by Francis Galton, laid the basis for such studies. His sample started with 600 twin families, and he got responses to about 64 twin pairs for the questionnaires he had mailed. He was particularly interested in long term comparisons of physical and mental features of the twins. What was their eventual concordance in appearance, physical abilities, illness, lifespan and character? Similar twins, he found, continued similar until old age and dissimilar twins remained dissimilar. What you are at birth decides your fate. Nurture cannot override nature. That was his basic conclusion.

His study had numerous flaws. A part of his data was anecdotal and vague; alternative explanations were ignored; cases that challenged his hypothesis were explained away with a rhetorical flourish; and observer bias was not considered. Yet, while imbued with racist assumptions, in terms of scientific quality, his work stood above the patently wild pronouncements of the proponents of eugenics of his day and laid a foundation for later studies. It also brought home the notion that such issues had to be investigated through methods of science and statistics, not by theological inquiry.

Twin studies were a major component of the research at the Eugenics Records Office in New York, especially under the directorship of Charles Davenport. Several biological and epidemiological studies of twins were conducted in that time. The ERO also established links with scientists in Germany. With a built-in bias, their unsurprising conclusions were: That how we think, feel, behave and perform in life is determined at birth, and is unaffected by upbringing; that such traits run in families; and that certain races are intrinsically superior to other races in terms of such traits.

Eugenically inclined German scientists took up twin studies in earnest. They included Ernst Rudin, his students, Franz J Kallmann, a psychiatrist, and Professor Ottmar F von Verschuer, a geneticist at the Kaiser Wilhelm Institute for Anthropology, Human Heredity and Eugenics. As respected experts in their fields, they directly or indirectly benefited from the American funds for eugenic studies pouring into Germany.

Twins research in Nazi Germany reached its apex through the work of Josef Mengele. He studied at the University of Munich and then at the University of Frankfurt to qualify as an anthropologist and physician. Attending lectures by Ernst Rudin on heredity and eugenics reinforced his eugenicist and racist beliefs. Like his teacher, he joined the Nazi Party. He also served in the Nazi paramilitary force. His career as a researcher started at the KW Institute for Heredity, Biology and Racial Hygiene. Performing twin and other studies under von Verschuer, he was praised for his meticulous effort and clarity of exposition. As WW II started, he enlisted in the army and rose to the position of a battalion medical officer. But after being wounded on the front, he was appointed as the senior medical officer at the rank of a captain at Auschwitz II, one of the 40 Nazi concentration camps in Poland.

It was an ideal setting to get many twin-research subjects in a confined setting and continue the twin studies begun earlier. His contact with von Verschuer enabled him to get funding for the research from the Reich Research Council. The medical studies he conducted on twins and other subjects represent a horror-filled litany of gross abuse, a nadir in bio-medical ethics that remains unrivaled to this day.

Auschwitz II received train loads of victims for extermination every week. Mengele and his assistants searched the arrivals for twins. Most they found were children. Siezed from their parents, they were placed in a separate facility with slightly better living conditions than in the rest of the camp and were not dispatched to the gas chamber. In all, he used about 1,500 twin pairs (3,000 individuals) in his experiments.

Mengele was an astute scientist. To control systematic error, he usually used a paired design: one twin was treated while the other was made a control subject. If the treated twin was on the verge of death, both were killed, and their bodies compared by dissection. His experimented included infecting the subject with pathogens, injecting dyes into their eyes, and removal of organs without anesthesia.

One project attempted to identify 'specific proteins' involved in a genetically determined response to infection. For this experiment, Jewish and Gypsy twins were injected with identical amounts of typhoid bacteria, then blood samples were taken at regular intervals, and sent to Berlin for analysis. (Stahnisch 2013).

Detailed records were kept, and the reports and specimens—human body parts—were sent to von Verschuer at the Institute for Anthropology, Human Heredity and Eugenics. Only about 200 of the 3,000 twins survived the grisly ordeals.

Mengele's twin studies were a part of large-scale medical study program at different concentration camps. They included surgical studies of regeneration of bones, muscles and nerves to uncover the biology of regeneration and transplantation; keeping subjects naked in sub-zero temperatures on the ground or in cold water tanks for hours to determine the effects of hypothermia and rewarming by different methods; infecting healthy subjects with material containing the causative organisms of malaria, typhus, tuberculosis, typhoid fever, yellow fever and viral hepatitis to determine the course of the diseases and then treating them with experimental drugs; inflicting battle-field like wounds, including severe burns, on subjects, contaminating them with dangerous bacteria and testing various treatments for such wounds; determining cheap but effective procedures for sterilization by using female subjects; and assessing the effects of electro-shock therapy for mental illness. All the experiments were done without anesthesia.

Research Subjects

This is but a partial list of the egregious experiments done by Nazi doctors and nurses. Some had over a thousand subjects. Consent was not an issue: A recruit had to participate; the alternative was the gas chamber. In terms of numbers, one assessment states:

There were some 70 such 'medical-research' programs at Nazi concentration camps involving some 7,000 prisoners and some 200 physicians, who worked directly in the concentration camps, but they were not alone. They maintained close professional and research contacts with leading medical institutions and universities and an ongoing relationship with research laboratories. Indeed, the German medical establishment was involved in this work. JVL (2020).

A more comprehensive evaluation has that about 28,000 subjects were used in Nazi medical experiments. Some 4,000 died and many survivors suffered grave life-long disabilities. The subjects were 6,900 Poles, 4,000 Yugoslavs, 2,400 Germans, 2,000 Hungarians, 1,300 Czechoslovakians, 1,000 Soviet nationals and other nationalities in smaller numbers. Jews formed the main ethnic group. (Weindling et al. 2016). Those who were reluctant to participate were killed; their number is unknown.

A major portion of the German medical establishment (doctors, nurses, pharmacists, laboratory personnel, administrators) was, by direct participation, open approval or discreet silence, complicit in these crimes. Major German companies collaborated in or benefited from the Nazi studies. After the War, a few perpetrators were tried and sentenced but most, like Josef Mengele, evaded justice. During the Nuremberg trials, lawyers for the Nazi doctors cited American eugenic laws and wartime experiments about malaria and other diseases on its prison population in their defense. In the years leading up to WW II, American foundations, research organizations and eugenicists had provided critical support for and lauded the scientists and doctors in the German sterilization, euthanasia and other eugenic efforts. Many of these doctors were later involved in the war time experiments in the death camps. Further, during WW II and earlier, Japan conducted equally horrific medical experiments in China and other places. But here, US influence ensured that no trials, accounting, apology or compensation for these crimes took place.

Extermination: The Nazi era exterminations occurred in three phases: Beginning with the total silencing of the political opposition, it proceeded to the sterilization program, then to euthanasia and climaxed in mass incarceration and murder in the concentration camps during WW II.

Silencing the Opposition: A socialist uprising led by Rosa Luxemburg and her compatriots, and backed by the strong labor movement, posed a major threat to rule of capital in Germany in the chaotic aftermath of WW I. It was suppressed with ferocity and the leaders were killed. The Nazi Party led by Adolf Hitler arose in the ensuing political vacuum. The main political parties on the scene were the Social Democratic Party of Germany (SPD) and the Communist Party of Germany (KPD). Determined to secure total control, the Nazis systematically employed their paramilitary wing, the Brown Shirts, to attack members of other parties and disrupt their gatherings.

Notwithstanding its initial dallying with socialism, the Nazi Party was not a socialist party in any sense of the term. An anti-worker party allied with major German corporations, it was anti-business only to the extent of closing Jewish businesses. Aiming to replace Jewish capitalists with Aryan German capitalists, the Nazis paradoxically portrayed Jews both as greedy capitalists as well as carriers of the 'Bolshevik virus'.

The German economy was severely affected by the post-1929 economic depression. Thousands of businesses closed, joblessness grew stupendously, wages shrank drastically, and hunger took hold among the poor. Using populist rhetoric and speaking in the name of the downtrodden and women, the Nazis deflected the blame onto Jews and communists. The astutely packaged words of hate, fear and hope, that were projected by the media, films and posters, brought ordinary Germans in large numbers under Nazi influence. Hitler was depicted as the only person who could restore stability and make Germany great again.

Hitler attained power not by a coup but via the ballot box. He was voted into office by elections held in 1932 and 1933. But once in power, the Nazi Party systematically neutralized democracy and established absolute rule. The political opposition was the first to go on the chopping block. Deemed a Jewish doctrine—Judeo-Bolshevism—communism became the main threat to security of the German nation. Leaders and members of the KPD and anyone faintly linked to them—dissident writers, artists and intellectuals, labor union leaders, anarchists, socialists, social democrats—were rounded up and confined. The intensity of the anti-communist drive brought the traditional political elite and the business world closer to the Nazis. A purge of the civil service to remove 'non-Aryan' enemies of the nation, mainly Jews, was instituted. Judges not supportive of Nazi rule were removed. After securing a large electoral mandate, the *Fuhrer* ruled as an absolute dictator. His decrees became the law of the land. Civil rights were suspended, trade unions were banned, books were burned, and the press placed under strict censorship. Thousands of newspapers were closed because their editors were not 'Aryan'. And the Nazi Party became the sole legal party.

In 1933, some 200,000 people considered hostile to Nazism and guilty of 'socialist activity' were placed under arrest by the Nazi militias. That year the first concentration camp was built in Dachau. Its express purpose was to confine leading opponents of Nazi rule. Over 4,000 communists and their sympathizers were locked up by July 1933. Living conditions were filthy; torture was routine. Fearing the autonomy and pro-poor sympathies of some leaders of the Brown Shirts militia, Hitler ordered the murder of over a hundred militia leaders. The SS, the elite paramilitary squad, and the Gestapo, the German secret police, carried out the killings.

In a short time, the SS established four camps for political opponents. By the end of 1933, they housed about 27,000 prisoners—priests, lawyers, aristocrats, economists, engineers, scientists, doctors, historians, teachers, professors, politicians, resistance fighters, military officers, artists, writers, journalists, comedians, actors, musicians, and athletes. Germans, Poles, Austrians, Slovenians, Frenchmen, Russians and others, both Jewish and non-Jewish, were confined. Some 80% were communists, 10% were social democrats, and 10%, unaffiliated trade unionists. A slight expression of anti-Nazi sentiments risked confinement in a camp.

Of those captured in this phase, many were released within a year. But those in senior positions of their organizations remained in the camps for a long time. Some of them were inducted in the gruesome medical studies, a few escaped, but many later perished in the death camps. Ernst Thalmann, the communist leader, spent eleven years in the camps before being executed by the SS in 1944.

The same pattern was repeated as German armies conquered territories in WW II. The priority was to neutralize communists and leftists. Tens of thousands of communists, socialists, anarchists and Soviet political commissars were rounded up, and were immediately executed or sent to concentration camps.

Euthanasia: By the start of WW II, the Nazi regime had forcibly sterilized about 375,000 persons. About 6,000 died from the complications. It was then raised to a new level. Designated German doctors were authorized to identify individuals who were terminally ill, had major physical and mental ailments or were 'too old,' for 'mercy killing'. They were the 'unworthy' who constituted a parasitic group posing an undue economic burden on society. The victims of this T4 Akton program were patients in psychiatric and geriatric facilities and general hospitals.

Collection Bus for Killing Patients

Health facilities across Germany sent information on potential candidates to T4 headquarters in Berlin. Those selected by the doctors were transported to one of six extermination centers —four in Germany and two in Austria—by the SS and swiftly put to death. At first it was done by withholding food and treatment, or injection with a lethal chemical. Later, a more efficient method, the gas chamber, became the method of choice. Over 90,000 hospital beds were cleared and until the end of the war in 1945, over 200,000 people were exterminated at these centers. A medical facility in occupied Ukraine was another site for the Nazi T4 program type of killings. In all, more than 100,000 infirm non-Germans were euthanized by the Nazis.

It was not a project of a few deranged doctors. Caregivers and doctors across the land selected the victims. A majority of the German mental health specialists cooperated with the program. German chemists and companies provided lethal gas and injection material. Some participating professionals became specialists in the field and joined even bigger Nazi killing programs.

Yet, these doctors and health professionals saw themselves as upright, ethical beings. What they were doing was for the good of the patients: To free the seriously ill and disabled patients from lives of excessive suffering, misery and mental confusion.

Mass Murder: The detention camps for political opponents and euthanasia facilities for the mentally retarded and others served as prototypes for a major expansion of the camp system after the late 1930s. The population of the camps expanded rapidly once WW II began. Operated mostly by the SS, the camps were of two kinds: Concentration Camps and Extermination Camps.

Concentration Camps: Housing Jews, political opponents, criminals and prisoners of war, the main purpose of these camps was to form a supply of ultra-cheap labor for economic and military purposes. Camps were located near stone quarries, brick-works and factories run by major German firms. By 1945, 27 major camps and over 1,000 smaller satellite concentration camps had operated in Germany and parts of Europe occupied by the German army. The total number of detainees at any point was over 1.5 million.

Barely minimal food was provided. The policy was to work the prisoners to the point of utter exhaustion. Then they were left to die or killed, in gas chambers or by other means. Killing was done by traveling euthanasia doctors. Soviet prisoners were butchered in large numbers. Hundreds of thousands died *en route* to the camp site. Yet, a large number of inmates survived the war.

Extermination Camps: While death was a probable but incidental outcome in a concentration camp, in an extermination camp, it was the planned outcome. The Nazis established six death camps in Poland to implement what they called the *Final Solution to the Jewish Question*. Jews, Gypsies, Poles and others were rounded up in Germany and elsewhere, transported usually by train to a camp and systematically put to death. At the Auschwitz II camp, Josef Mengele, the lead medical investigator, was also responsible for selecting people to be killed. It earned him the title the *Angel of Death*.

Killing was done in specially designed gas chambers or gas vans using Zyklon B gas or carbon monoxide. In a couple of camps, work until demise by starvation was also deployed. As noted earlier, some inmates were recruited for medical experiments. Over 2.7 million died in these camps; some 90% of them were Jews.

Overall, the fatalities in the concentration camps, extermination camps, euthanasia programs, mass shootings, enforced starvation in villages and towns, and other non-battlefield settings exacted by the Nazis and their allies were:

> **Soviet Citizens**: About 8 million (prisoners of war, civil officials, villagers and townsfolk).
> **Jews**: About 5 million (extermination camps and other mass killings).
> **Non-Jewish Polish Citizens**: About 1.9 million.
> **Roma People**: About half a million.
> **T-4 Euthanasia Victims**: About 400,000 (in Germany and beyond).
> **Varied Groups**: About 200,000 (communists, political dissidents, Jehovah's Witnesses, homosexuals).

The racist, eugenicist, master race ideology of the Nazis not only caused millions of deaths at the battle front and tremendous destruction on a global scale but also put an end to the lives of

16 million people who did not wield a weapon. And this is not counting the civilians killed in massive air raids, and indiscriminate bombings conducted by both sides.

3.6 THE EUGENIC CIRCLE

Eugenics, as Francis Galton proclaimed, was *'the science of improving the inherited stock'* aimed at reducing the burden of mental and physical disabilities, illnesses and anti-social behavior in the present and future generations. A message that reverberates to this day, it resonated widely in Western nations in the late 19th and early 20th centuries. A broad section of people from varied political, economic, educational, racial and religious backgrounds— commoners and elite—embraced eugenic ideas, often even in their extreme form. In Britain, the United States, Sweden, Norway, France and Nazi Germany, eugenics was accepted by a wide segment of eminent members and institutions of 'respectable' society.

We named some of these eugenicists earlier. Now we extend the list of prominent supporters of eugenics.

Major Political Leaders: Adolf Hitler, Warren Harding, Theodore Roosevelt and William H Taft (US presidents), Winston Churchill (UK prime minister), Lord Balfour (senior UK politician), Elihu Root (US Secretary of State, Nobel laureate).

> *The multiplication of the feeble-minded is a very terrible danger to the race.* Winston Churchill (Brignell 2010a).

> *I wish very much that the wrong people could be prevented entirely from breeding; and when the evil nature of these people is sufficiently flagrant, this should be done. Criminals should be sterilized, and feeble-minded persons forbidden to leave offspring behind them.* US President Theodore Roosevelt (Brignell 2010b).

Eminent Jurists: US Supreme Court Justices William H Taft, Louis Brandeis and Oliver W Holmes. Arthur MacDonald, reputed criminologist.

Eminent Economists: Irvig Fisher, John M Keynes, Harold Laski (LSE professor, British Labor Party chairman), Gunnar Myrdal, William Beveridge (social reformer, champion of the welfare state).

Notable Writers and Poets: DH Lawrence, HG Wells, Virginia Wolf, WB Yeats, TS Elliot, GB Shaw.

> [The] *only fundamental and possible socialism is the socialization of the selective breeding of man.* GB Shaw (Freedland 2012).

> *It is in the sterilization of failures, and not in the selection of successes for breeding, that the possibility of an improvement of the human stock lies.* HG Wells (Baker 2015).

University Presidents: CW Eliot and AL Lowell (Harvard University), David S Jordan (Cornell University, Stanford University, famed marine biologist).

Philosophers, Sociologists and Historians: Bertrand Russel, Martin Heidegger, Friedrich Nietzsche, Herbert Spencer, Oswald Spengler, William G Sumner (American Sociological Association founder), Martin P Nilsson (historian).

Pioneering Statisticians: Francis Galton, Karl Pearson, RA Fisher, Frank Yates, Fredrick S Crum.

> *The nations whose institutions, laws, traditions and ideals, tend most to the production of better and fitter men and women, will quite naturally and inevitably supplant, first those whose organization tends to breed decadence, and later those who, though naturally healthy, still fail to see the importance of specifically eugenic ideas.* Ronald Fisher (Clayton 2020).

After WW II, RA Fisher praised the professionalism of von Verschuer, Josef Mengele's mentor, and said he had no doubt *'that the [Nazi] Party sincerely wished to benefit the German racial stock, especially by the elimination of manifest defectives, such as those deficient mentally'*. (Clayton 2020).

Prominent Scientists and Inventors: Charles Darwin, EM East (botanist), CB Davenport (zoology professor), Alexander G Bell (inventor), Madison Grant (conservationist), Herman Lundborg (physician), Herman Nilsson-Ehle (geneticist), Robert Larsson (founder, Swedish Mendelian Society), Nils von Hofsten (zoologist), Jacques Cousteau (ocean explorer), Julian Huxley (geneticist, first director of UNESCO), Hermann J Muller (radiation medicine specialist, Nobel laureate, educator).

> *I believe in such a selection of immigrants as shall not tend to adulterate our national germ plasm with socially unfit traits.* Charles Davenport (Wikipedia 2020 – Charles Davenport).

Prominent Psychologists and Physicians: Alexis Carrel (cell biology, organ transplant pioneer, Nobel laureate), Henry H Goddard, Robert Yerkes (psychometrics), Carrie Derrick, W McDougall (psychology professor), Paul Popenoe (venereal disease specialist), Herman Lundborg (physician), John H Kellogg (breakfast cereal inventor, public health reformer).

Education Specialists: Edward L Thorndike and G Stanley Hall (curriculum development experts), Horace D Taft (lawyer, educator).

Business Magnates: Andrew Carnegie (industrialist, philanthropist), Henry Ford (auto-manufacturing), Paul Gosney (agri-business tycoon), Charles Goethe (banker), Mary Harriman (widow of railway magnate).

Feminist Leaders: Marie Stopes and Margaret Sanger (women's health and contraception champions), Hellen Keller (socialist, feminist, pacifist).

> *It is the possibility of happiness, intelligence and power that give life its sanctity, and they are absent in the case of a poor, misshapen, paralyzed, unthinking creature.* Hellen Keller (Baker 2015).

Labor and Socialist Leaders: William Green (president, American Federation of Labor), Will Crooks and Archibald Church (labor MPs), Sidney and Beatrice Webb (Fabian Society founders).

Foundations and Research Institutions: The Carnegie Institution, the Rockefeller Foundation, the Cold Springs Harbor Laboratory, Kaiser Wilhelm Institute, French Foundation for the Study of Human Problems.

This partial list of prominent eugenicists suffices to demonstrate that far from being a fringe movement of right-wing fanatics, eugenics was an eminently respected, pervasive, popular component of the Western society. Conservative as well as liberal media (*The Manchester Guardian, New Statesman, New York Times, Los Angeles Times*) heralded eugenic perspectives; Hollywood movies celebrated the doctrine; it entered school and university curricula. Feminists, labor leaders, and as we shall see later, even leaders of religious and marginalized groups were mesmerized by it.

And not just extreme fascistic governments took up eugenic programs but even the social democratic governments in the Scandinavian countries funded eugenic institutes and adopted the programs they devised. On the one hand, individual freedom, democracy and the rule of law were glorified as the defining banners of Western civilization. On the other hand, some people were deemed genetically deficient 'sub-humans' whose civil rights could be curtailed with blessings of the law and forced to endure treatment that for others needed free consent. No discordance was seen between the two visions.

The noted eugenicist luminaries did not hold identical ideas. The more liberal eugenicists opposed negative eugenic measures like forced sterilization and euthanasia. Hard line eugenicists saw state assistance to the marginalized and genetically 'unfit' as a waste of resources and a danger to the future generations. Others favored assisting them but keeping their numbers in check through some form of social intervention.

Eugenics, Racism and Colonialism

Eugenics is a repulsive doctrine. Yet, it was not that long ago that it had a popular following in the West. Learning its history usually elicits a major surprise in a modern Westerner. How could it have happened?

There are two basic reasons why. One, despite recent corrections, the teaching of history in schools and universities remains biased. Most media accounts of major historical events leave out critical matters. Historians are tainted with the ideas of exceptionalism, nationalism and a pro-capitalist disposition. Major but inconvenient events are often pushed under the rug. Thus, it is common for a person to earn a doctorate in statistics and devote a lifetime in the field without getting an inkling of the eugenic foundations of the discipline.

Two, media writers and scholars in the West generally depict eugenics without considering imperialism, xenophobia and racism associated with the rise of capitalism. Yet, the depiction of Africans and peoples of the Global South as inferior beings has been a central part of the ideology of conquest, slavery, and colonial rule since ships from Portugal, Spain, Holland, England and other Western nations ventured towards Africa, Asia and the Americas. The difference was that instead of being cloaked in Biblical garb, the eugenics ideas were presented by men of science, and in 'scientific' terms.

Absence of remorse towards mass killings of humans was ingrained in the American psyche since the days of genocidal pogroms against Native Americans. Employing duplicitous diplomacy, military brutality, starvation, and deliberately spreading dangerous ailments, indigenous nations were virtually decimated. As a US General famed for his racist views and brutality declared:

The only good Indian is a dead Indian.
General Philip H Sheridan

Impressed by that history, Adolf Hitler not only sought to emulate the laws that dealt with Native Americans but also repeat it during his incursions into Eastern Europe and Russia. Oppression of the Australian Aborigines by the settlers from Europe also reflected the American experience.

When the eugenic movement rose into prominence, the capitalist order in US, Germany and other European nations was facing a strong challenge from leftist worker parties. Strikes, boycotts and uprisings were common. The Bolsheviks had triumphed in Russia. The capitalist states responded with distinct fury against threats from working class militancy. From 1919 to 1921, US Attorney General AM Palmer launched raids to arrest immigrants who were prominent in socialist and anarchist politics. About 3,000 were arrested and over 550 were deported. Most were Italians or Jews. As in Nazi Germany, they were depicted as carriers of 'Bolshevik germs'. These moves formed the backdrop of the eugenically framed, racist immigration laws passed in the 1920s. From that time red baiting, fomenting red-scares and launching anti-communist crusades on flimsy grounds became a standard practice for American politicians and US agencies both for internal and foreign policies.

The 1920s in the US were also years of renewal of vile forms of racism. The Ku Klux Klan marched with torches in their white hoods. Lynching went on. Each year over 30 persons were lynched on average. Over 70 percent of the victims were African Americans. The Klan advocated euthanasia by force for black people. When African American tenant farmers strove to improve their conditions, white landowners instigated a ferocious backlash. One massacre took the lives of nearly 240 farmers. Elsewhere white racist rioters vandalized or burned down thriving African American businesses and killed over 250 persons with a dark skin. Prejudice against Asians and Mexicans fed the Californian eugenics drive. In the southern states, hospital sterilizations disproportionately victimized women of color. And medical students performed unnecessary hysterectomies to develop their surgical skills.

When an American visitor inquired into the harshness of the actions against the Jews, Adolf Hitler referred to the lynching tradition in the US. To him, the Americans were the '*great rope and lamppost artists of the world*'. He said that his program was inspired by the actions against '*Indians*', Africans, Chinese and Mexicans in the US (Nock 1941). Yet, while '*enemies of the state*' were openly lynched in the streets, the Nazi Party did not officially approve such extra-legal acts. It focused on devising efficient techniques of mass elimination.

The eugenic era was an era of ongoing, enhanced brutalities and massacres under colonial and imperial intrusions in Africa, Asia and Latin America. The barbarism of the Belgians in the Congo, the French in West and North Africa, the British in Kenya, the Middle East and Tasmania, the Dutch in Indonesia, the Germans in Botswana and Tanzania, and the Portuguese in Mozambique were forerunners of the German genocidal program during WW II. Between 1904 and 1908, the German colonial army perpetrated the first genocide of the 20th century. Some 100,000 Herero and 10,000 Nam peoples in Namibia were killed by armed assault, enforced starvation and dehydration, and confinement in concentration camps where hygiene was atrocious, and abuse was commonplace.

The barbaric manner by which European and American armies fought against each other during WW I hardly contributed to respect for the dignity and lives of 'the others'.

Seen in the broader context of slaughter of native peoples, racism, suppression of the working class, colonial perfidy, imperial barbarity and war, the eugenic movement was not an aberration. It was not an anomaly of the 'glorious' Western civilization. Its ideology and programs reflected the elitist denigration and abuse of human life common in the capitalist world. Only now the victims were closer to home. That people with physical and mental deficits and chronic illnesses, Jews, Poles, the Roma, criminals, and alcoholics were 'sub-humans' unfit to live and procreate reflected the other side of Darwin's prognosis that 'civilized' races would eventually replace 'savage' races across the world.

The broad appeal of eugenics among ordinary citizens and the well-to-do stemmed from the fact they shared the white supremacist vision that the West had a Manifest Destiny to rule the planet. On domestic issues, conservatives and liberals tooth and nail. But on the foreign arena, morality was cast aside and the imperial projects were embraced by both sides. At times, they disagreed about the tactics, but never about the necessity.

That tendency is epitomized by Charles Dickens, the British literary giant whose elegantly crafted prose picturesquely highlighted the miserable lives of the poor, excoriated the greedy, cruel gentry and merchants, and illuminated the complexity of human psychology. He supported a variety of liberal programs to improve the conditions of the downtrodden in the British society. He stood against child labor and the death penalty and called for improved sanitation in the poor areas. Nonetheless, he was also a hard-core racist, xenophobe, anti-Semite imperialist. With an ambivalent stand on anti-slavery efforts, he opposed giving African Americans the right to vote and saw Native Americans as violence-prone, dirty 'savages'. Proclaiming the superiority of Western civilization and values, Dickens supported British colonial ventures in India and elsewhere to the hilt.

His reaction to the Indian freedom uprising of 1857 tells all. The uprising was caused by the extreme hardships imposed on the farmers and craftsmen by the rapacious practices of the British India Company. It resulted in the death of about two hundred British. The Company forces retaliated by slaughtering thousands of Indians. While Dickens criticized exploitative practices at home, in the case of India, they were of no concern to him. He crudely opined that if he was the commander of the British forces in India, he would issue a proclamation stating:

> *I, The Inimitable, holding this office of mine, and firmly believing that I hold it by the permission of Heaven and not by the appointment of Satan, have the honor to inform you Hindoo gentry that it is my intention, with all possible avoidance of unnecessary cruelty and with all merciful swiftness of execution, to exterminate the Race from the face of the Earth ...* Charles Dickens (Poly 2012).

Winston Churchill, one of the most celebrated politicians of the 20th century, was a major presence in the British politics for fifty years. Twice the Prime Minister, for decades, a member of parliament, several times a minister and high official, and leader of the Conservative Party, he is most famous for his leadership against Nazi Germany during World War II. His stirring speeches inspired the British public and encouraged the Allied forces to fight on when their position appeared precarious.

Yet, Churchill was an unrepentant, callous imperialist. He called for the use of poison gas in British military intervention in the Middle East and called the Kikuyu of Kenya, a *'brutish people,'* Kurdish people, an *'uncivilized tribe,'* and Native Americans and Australian Aborigines, *'inferior races'*. Dickens and Churchill were liberal reformers on domestic issues. But for the colonies, they sang an identical racist melody:

> *I hate Indians. They are a beastly people with a beastly religion.* Winston Churchill (Hari 2010).

Eugenics was a key part of the overall ideology that rationalized capitalism, imperialism, elitism and white supremacy.

3.7 RACE, GENDER AND EUGENICS

The allure of eugenics was so strong that it captivated its victims as well. In the US, eugenics compounded the racist indignities African Americans had endured for a long time. Yet, eugenics was embraced by African American movements, intellectuals and activists fighting for racial justice.

Established in 1909, the National Association for the Advancement of Colored People (NAACP) has been the premier organization fighting for racial equality in the US. The eugenic era enhanced discrimination, violence and lynching against 'colored' people. The NAACP rejected any doctrine that branded a person inferior on the basis of racial or ethnic origin. It repudiated race-based eugenics, but did not reject eugenic ideas outright.

The notion that mental and physical prowess and a healthy disposition were hereditary was taken up by African American leaders as a useful idea for promoting racial equality. The black race should adopt measures to improve its genetic stock because that would enable them to more effectively counter the domination by the white race. Hence, since 'colored' babies were excluded from the main Better Baby Contests, the NAACP initiated its own contests. External eugenics was condemned while internal eugenics was adopted and followed the lead of white America.

Baltimore Branch NAACP Baby Contest Winners, 1946

In 1924, fifteen branches of the NAACP held Better Baby Contests. Serving as a fundraising venue for civil rights campaigns, some of the contests were supported by Gerber Foods, the baby foods manufacturer.

Among African American writers, intellectuals, physicians, biologists, and social scientists, supporters of eugenics outnumbered the opponents. African American newspapers and magazines carried generally favorable articles. Vocal critics were a few. But unlike the white eugenicists, African American eugenicists stressed the role of upbringing and social environment in educational success and the formation of behavioral traits. Prominent scholars at the major historic Black universities like Howard, Hampton and Tuskegee declared that gifted Black persons were as good as gifted Whites and that the eugenics drive should jointly promote the prevalence of 'good' genes among all races. Eugenic and birth control measures should be framed to counter racial discrimination and promote racial equality. The key message to the white majority was: Like you, we have 'bad' and 'good' genes. Hence, as equals, we should combine our forces to combat the curse of 'bad' genes. Scientific racism was to be tamed by multi-racial eugenics. Elitism prevailed in both circles.

WEB Du Bois, a Pan-Africanist stalwart of social justice and civil rights, holder of a PhD from Harvard, author of the landmark work, *The Souls of Black Folk*, and a founder of the NAACP, was ensnared by eugenics in his early activist days. He opined that the '*Talented Tenth*' of African Americans should procreate in greater numbers so as to uplift the moral and intellectual qualities of their race. Comparing humans with vegetables, he sagely declared that quality, not quantity, counted (Singleton 2014).

> *The Negro race, like all other races, is going to be saved by exceptional men.* WEB Du Bois (Nuriddin 2017).

WEB Du Bois, 1919

Espousal of eugenics by stalwart African American thinkers marks a low point in the struggle for racial justice and equality in the US. Fairness for one group was sought at the expense of denial of the basic rights of another group. Yet it was African Americans who were targeted in greater numbers in the US eugenic programs. In the 1930s to 1940s sterilization drives launched in North Carolina, about 60% of the victims were African Americans.

> *Eugenics created dual realities of racial improvement and racial violence for African Americans in the twentieth century. African Americans sought to mobilize eugenics for racial improvement but were ultimately disproportionately targeted by eugenics legislation.* (Nuriddin 2017).

In March 1929, the Chicago Forum Council, a multi-racial organization, staged a major public debate on race and eugenics. It was the only debate of this kind ever held. The contention was: '*Shall the Negro Be Encouraged to Seek Cultural Equality?*' And the sub-theme was '*Has the Negro the Same Intellectual Possibilities as Other Races?*' On the opposing side was TL Stoddard, a white supremacist and internationally bestselling author of *The Rising Tide of Color against White World-Supremacy*. In the 1930s, when his book had approval of the Nazi leaders, Stoddard visited Germany and wrote articles praising the Nazi eugenic measures.

Appearing for the affirmative side was none other than WEB Du Bois. Before an audience of some 5,000 people, Stoddard diluted his extremist views by promoting the notion of separate but equal. Racial integration would do more harm than good, and each race should be left, as it was being done, to promote its own development. Du Bois astutely dissected these claims and exposed their hypocritical foundations. He prevailed on the stage. Stoddard's ideas were reduced to the point of ridicule. But the vision of Stoddard held sway among the white majority. Racism, buttressed by pseudo-scientific claims, remains a central social malady in the modern era as well.

Du Bois had a complex career. His views evolved as he confronted racism in theory and practice. He later came to believe that it was more than a matter of skin color. Capitalism was the basic cause of the racist, exploitative conditions faced by African Americans. He also said that the US policies towards Africa undermined the progress of the continent. As he leaned towards socialism, he and his close associate, Paul Robeson, were hounded by the US government.

Eugenics and Feminism

The oppression of women has taken multiple forms. In the home, women were denied autonomy, endured the drudgery of domestic work, were denied property rights and were sexually abused. As men ate and drank, they stood in servitude. With restricted educational opportunities, they were not allowed to join certain professions or run their own businesses and got lower pay than men for the same type of work. In the civil arena, they did not have the right to vote or become members of legislative bodies. Marriage, divorce and property laws favored men. And so on.

The eugenics movement and the struggle for women's rights in the West were linked in paradoxical fashion. Francis Galton, the founder of scientific eugenics, claimed that women were generally inferior to men. Yet, because of their role as 'breeders,' they were key to the progress of the human race. Among other things, he advocated that women from talented families should be provided financial incentives to marry early.

In his time, Western women could not study at major universities, were barred from practicing medicine and law, and did not have the right to vote. Rules for divorce and property laws made it very difficult for them to escape abusive relationships or chart their own course in life. Laws protecting them from sexual abuse were weak and poorly enforced. Working class women suffered further through exploitation in factories at lower wages than men and domestic abuse. Minority women bore the added burden of racial and ethnic discrimination. As a former slave and champion of women's rights noted:

> *And ain't I a woman? Look at me! Look at my arm! I have ploughed and planted, and gathered into barns, and no man could head me! And ain't I a woman? I could work as much and eat as much as a man—when I could get it—and bear the lash as well! And ain't I a woman? I have borne 13 children, and seen most all sold off to slavery, and when I cried out with my mother's grief, none but Jesus heard me! And ain't I a woman?* Sojourner Truth, 1851 (History Editors 2020).

In the first two decades of the 20th century, women's movements in the West stridently fought for the right to vote and stand in municipal, state and federal elections. In the US, after a bitter struggle marked by police violence and public ostracism, they finally won it in 1920. The suffragists had also campaigned for women to have greater control over their own bodies, make decisions about children and childbearing, and for public health measures of specific benefit to women.

Champions of women's rights were also invested in promotion of health and welfare of mothers and children. Voting rights would give women the power to affect laws favoring maternal and child health, female education, better access to social amenities, nurse training, and funding health programs for rural women. Curiously, eugenics, appropriately applied, was taken by the feminists as a basis for achieving these goals. Accepting that 'bad' genes were, to a major extent, the cause of women's problems, they held that a program of preventing transmission of 'bad' genes and facilitating proliferation of 'good' genes would benefit women.

> *Early female public health reformers believed that eugenics reform was critical to national growth, and that eugenics itself should be focused on more than just the bearing of children. Rather, eugenics required mothers who could bear 'healthy' children, raise intelligent citizens, and be engaged in scientific motherhood methods.* (Gibbons 2020).

Eminent feminists cooperated with prominent eugenicists and public health experts and leaders to support Better Baby Contests, pass laws restricting marriage, immigration control, sterilization and even euthanasia. Among the ardent women eugenicists were:

United States: Gertrude Davenport (accomplished scientist), Hellen Keller (advocate of the disabled, writer), Florence Kelly (journalist) and Margaret Sanger (nurse, educator, founder of Planned Parenthood).

Britain: Marie Stopes (botanist, contraception pioneer), Margaret Pyke (leading feminist), Beatrice Webb (socialist, co-founder of the Fabian Society).

Canada: Five acclaimed women's rights activists: Henrietta M Edwards, Nellie McClung, Louise McKinney, Emily Murphy and Irene Parlby.

Educated as a nurse, **Margaret Sanger** began her activism as a socialist in New York City, supporting worker strikes and interacting with the leading socialists in America. Her interactions with working class women made her see how their lives blighted by frequent unwanted pregnancies, miscarriage, bearing many children and associated maladies. They suffered serious effects of self-induced and back-alley abortions. The lack of knowledge about sexuality and pregnancy prevention, and laws making contraceptives and information about their use illegal were, in her view, the main causes of this state of affairs. Contraception was the key for empowering women.

In response, she launched a life-long strident drive to popularize family planning, work for safe motherhood, prevent unneeded pregnancy, and legalize contraception. Her many achievements included pioneering sex education via articles, pamphlets and speeches that candidly laid out reproductive issues and provided information on family planning; combating the censorship of sex education; legalization of birth control measures and education; the first birth control clinic in the US that moreover had all female health personnel; wider distribution of contraceptive devices; and the Planned Parenthood Federation of America. She suffered social ostracism, official harassment, months spent behind bars, and self-imposed exile. Yet, she did not change course. The tribulations she faced catalyzed birth control drives across the US.

Sanger was not against abortion when needed to save the mother's life. But she strongly opposed abortion as a means of birth control and argued that prevention of unplanned pregnancies would largely eliminate the need for abortion. Unlike most white feminists, she worked with African American leaders and managed to establish a birth control clinic in Harlem that later employed black physicians and nurses. WEB Du Bois, who supported birth control as a measure for African American advancement, wrote for Sanger's magazine, *Birth Control Review*.

Sanger promoted her message internationally, along with Marie Stopes, the British birth control and contraception campaigner. She visited China where, with the help of Pearl S Buck, she set up a birth control clinic. In 1935, she spent nearly three months in India lecturing on birth control to packed audiences in several cities, setting up fifty birth control information centers and interacting with women's and medical organizations. She also held talks with Indian luminaries, namely, Rabindranath Tagore, Jawaharlal Nehru and MK Gandhi. Spending two days with Gandhi at his *ashram*, she engaged in a contentious but cordial discussion on birth control.

Holding that population control was essential for guarding national freedom and eradicating poverty, Sanger said that women needed the means to control conception. Gandhi agreed that domination of women by men was the basic cause of many social problems and that women should have the right to make decisions regarding childbirth. They had to stand up for their rights. Yet, he said abstinence, not artificial birth control methods, was the solution. He also felt that Western feminists were not in touch with the social and cultural realities of India. Sanger responded that her interactions with poor Indian women in Bombay had revealed that they did not want to have too many children. Gandhi countered that education was the key. Sanger responded that education and birth control had to be jointly pursued, and that would enable women to improve their standard of living as well. Gandhi, with his aversion towards sex for its own sake, did not agree. A meeting of minds did not prevail, and Sanger could not persuade Gandhi to take up the cause of birth control.

M Sanger and MK Gandhi, 1935

Despite her internationalism and non-racist stand, Sanger did not escape the spell of eugenics. Her belief in the Malthusian dogma—that poverty, famine and conflict were due to overpopulation—was reinforced while exiled in the UK. As she collaborated with ardent eugenicists like L Stoddard, her concern for working class women was eclipsed by issues of importance to middle class women.

In her 1922 book, *The Pivot of Civilization*, Margret Sanger writes that feeble-minded people pose a serious financial burden on modern society. Not accepting the use of lethal chambers, she argued for separation.

> *Every feeble-minded girl or woman of the hereditary type, especially of the moron class, should be segregated during the reproductive period. ... [We] prefer the policy of immediate sterilization, of making sure that parenthood is absolutely prohibited to the feeble-minded.* Margaret Sanger (Horvath 2020a).

She debated three long-term options. Compassion for the mentally deficient and dependent people addresses the symptoms, not the root causes. It is an inadequate emotional response. Marxian socialism, which seeks to revolutionize society and promote equality, has a shallow analysis of human nature and is unsuited for tackling the biologic deterioration of the human race. Having dismissed these options, Sanger declared that negative eugenics holds some promise, but positive eugenics is not a viable method. Her 1932 work, *A Plan for Peace*, lays out a three-step negative eugenic program for developing a healthy human society:

> *The first step would thus be to control the intake and output of morons, mental defectives, epileptics.*

> *The second step would be to take an inventory of the secondary group such as illiterates, paupers, unemployables, criminals, prostitutes, dope-fiends; classify them in special departments under government medical protection and segregate them on farms and open spaces as long as necessary for the strengthening and development of moral conduct.*

> *Having corralled this enormous part of our population and placed it on a basis of health instead of punishment, it is safe to say that fifteen or twenty million of our population would then be organized into soldiers of defense— defending the unborn against their own disabilities.* Margaret Sanger (Hovrath 2020a).

Sanger was not alone. Many champions of women's civil and reproductive rights held such draconian views. Marie Stopes, Margaret Pyke and Mary Stocks, three towering personalities in the British women's struggles, favored reducing the hordes of '*defectives*' in society. In 1921, Stopes set up the first birth control clinic in England. The focus was on abstinence, not contraception. She favored eugenic measures like compulsory sterilization to control birth and admired the Nazi doctrine of racial purity. Generally, '*British feminists were more enthusiastic than German feminists for advocacy of eugenic measures*'. (Allen 2000).

The immoral entanglement of eugenics with promoting the rights and welfare of women had serious consequences. It marked the deviation of the main feminist movements from working class to middle and upper class interests. The eugenic measures they proposed would much more affect the former. It expressed the elite anxiety for the rising numbers of 'drunks,' unemployed, and the 'riffraff' who were deemed a major drain of governmental and private resources. And it also reflected the prevailing mainstream racist and anti-immigrant views. African American women who were prominent in the struggle for civil rights like Thelma B Boozer and Rebecca S Taylor as well as leading Canadian feminists adopted similarly prejudiced views against the most vulnerable people in society. Fighting to liberate one's group, yet sanctioning another gross form of oppression, has been too frequent an occurrence in human history.

Gertrude Davenport, wife of Charles Davenport, the eminent American eugenicist, was a qualified zoologist in her own right. In collaboration with her husband, she conducted in-depth studies of the heredity of skin pigmentation, eye color and hair form in humans and animal embryology. They jointly authored popular Zoology textbooks for high school and advanced students. She also independently wrote scientific works and managed a biological research facility. As a partner of equal scientific repute with her husband and given the paucity of women in science at that time, she was heralded as an inspirational model for girls, one whose life exemplified that a woman can be both, an accomplished professional and a devoted mother and wife.

Nonetheless, concerned about the danger posed by the 'feeble-minded' to the humanity, she ardently propounded strong eugenic measures just as her husband did. Using cases from Europe to support the thesis that habitual criminals breed habitual criminals, she wrote papers on the high cost of 'deficient' individuals to society. She typified a conservative brand of feminism which encourages women to subscribe to mainstream views as the means of advancing their status. These feminists advocated eugenic measures like sterilization, incarceration and quotas for immigrants that disproportionately harmed working class and disabled women.

The endorsement of eugenics and bland racism by pioneering feminists cast an ugly stain on gender politics which reverberates to this day. Their eugenic links are now invoked by conservative voices to castigate their stellar roles in the drive to attain equality, autonomy and justice for women, and are deployed to undermine the ongoing struggles for women's rights.

3.8 BUDDHISM, HINDUISM, ISLAM AND EUGENICS

What has been said thus far provides the context for the main theme of this chapter, namely, the relationship between religions and eugenics. This issue can be tackled at two levels: scriptural and practical. The key questions are: Does religion endorse the proposition that the physical, mental, and social traits and abilities of individuals are hereditary? Does religion endorse the practice of barring or encouraging procreation by certain groups of people? Does religion endorse forced sterilization, segregation and genocide? This section deals with Hinduism, Buddhism and Islam. The case of Christianity is pursued in the next section.

Eugenics and Hinduism

Hinduism, the most ancient of the four major religions, has four primary tenets—*samsara*, *karma*, *dharma* and caste—that relate to eugenics. *Samsara* is the cycle of rebirth. *Karma* implies that one's status in the present life is affected by past life. To attain spiritual liberation, one must abide by one's *dharma*, rules that are appropriate to one's caste. Violating *dharma* and performing bad deeds in the past life relegates a person into a low caste in the next one.

Eugenics holds that the past—inherited parental germ plasm—controls the present. But Hinduism says that you, not your parents, determine your fate. Unlike eugenics, Hinduism says that by abiding by your *dharma* you can change your future. It holds that the past determines the present but not necessarily the future.

No Hindu holy text condones forced sterilization, euthanasia and genocide. Hinduism does not call for elimination of low caste persons but condones caste-based social segregation. Caste purity must be protected. Low caste and high caste persons must abide by their own *dharma* and not intermingle.

Hinduism and eugenics share the notion of prohibiting marriage between different social groups. While the US eugenic laws banned interracial marriage, Hindu custom and beliefs, as enshrined in *The Laws of Manu,* look down on inter-caste unions:

> *Hinduism's outlook is very similar to that of nineteenth- and early twentieth-century Western eugenics, which proposed controlling reproduction to prevent what were considered undesirable unions. Although the specific rules for regulating marriage and reproduction were different from those proposed by Western eugenics, the spirit is the same: to protect the human species from degeneration due to unsuitable matches. Hinduism does not define suitability for marriage according to scientific understanding of genetics, but by caste membership, which is hereditary, and by physical traits, which are correlated with astrology.* (Encyclopedia 2020).

Eugenics and Hinduism share a deep anti-female bias. Eugenic sterilization disproportionately targeted women. Hindu women in India face discrimination at all levels of life including participation in religious rites and ordination as priests. A horrific consequence of that bias, which has eugenic connotations, is infanticide and abortion of female fetuses. But it is not a purely religious issue. Social and economic factors play a large role in its persistence.

Hindutva, the supremacist form of Hinduism formulated by VD Savarkar in the 1920s, saw India as a unified nation based on Hindu culture, a culture that included Sikhism, Jainism and Buddhism, but not Islam and Christianity. The latter were foreign doctrines that threatened the integrity of the nation. He opposed the multi-cultural policies of the Indian National Congress and the call by MK Gandhi's for Hindu-Muslim unity.

As the president of the ultra-nationalist Hindu Mahasabha, he expressed admiration for how Adolf Hitler dealt with Jews and claimed that the Muslims in India were akin to the Jews in Germany and had to be excluded from the political and social life of the nation. The RSS, another far right Hindu group formed in the colonial era, adopted Savarkar's *Hindutva* doctrine. Its founder also admired Hitler and praised the idea of a master race.

Today, the RSS is the dominant far-right, Hindu nationalist front in India. The parent of the BJP, the ruling party, it operates schools and charities, and a large, militant paramilitary force. Under BJP, Indian Muslims have faced increased marginalization, discrimination and violence. Mahasabha leaders have advocated compulsory sterilization for Muslims and Christians and justified attacks on churches and mosques. The *Hindutva* doctrine has influenced several Hindu states to consider laws to restrict marriage between Hindus and Muslims. Muslim men are accused of 'love-jihad,' that is, preying upon Hindu girls and converting them to Islam. External eugenics is rearing its ugly head once again, but now shrouded in a religious form.

The swastika, a four-twisted-arms figure with several variations, is a cultural symbol in Mayan, Aztec, Hopi and Navajo cultures of America, Buddhist, Jain and Hindu cultures of South Asia, the Ethiopian empire, ancient European communities and Islamic nations. A symbol for divinity and the magnificence of nature, particularly the sun, it projects hope, well-being and luck. The Nazis held that European peoples originated from an ancient, racially pure, light-skinned Aryan civilization in India. After German archaeologists found it in ancient sites, and noted its appearance in the Vedic texts, the Nazis adopted the swastika as the prime symbol of the (mythical) Aryan race. It became the symbol of the Nazi Party and was later incorporated into the German national flag.

The Hindu far-right declares that the pure Hindus are from this master race. But Muslims are not Aryans. *Hindutva* extremists have a paradoxical relationship towards Christianity. They abhor its presence in India but their followers in the US, UK and Canada diaspora make common cause with the extremist White nationalist organizations that are allied with evangelical Christian groups. Singing identical Islamophobic melodies, both advocate the reduction of the birth rate of Muslims in their nations. Hindu nationalists fervently support NM Modi of India and the former US president, Donald Trump.

Admiration for the Nazis has transcended political circles. Nazi thought and symbolism have seeped into popular Indian culture. Consumer items—tea mugs, laptops, T-shirts, sweaters, computers, helmets, jewelry, decorations—with the likeness of Hitler and Nazi swastika are available in the marketplace. Cafes, stores and production facilities in some cities carry the 'great leader's' name. A Hitler ice-cream brand exists and his major work, the *Mein Kampf,* sells well and is used in some management courses.

Fascistic imagery that is illegal in European nations has an open field under the *Hindutva* reign (Sharma 2018). Given the murderous history of the partition of India, and the series of anti-Muslim violent pogroms since then, this hardly augurs for a harmonious future between religious groups in India.

Eugenics and Buddhism

Emerging within a Hindu society, Buddhism accepted the ideas of *samsara* and *karma* but rejected the caste system. The Buddhist concept of *dharma* is embodied in *The Eight Fold Noble Path* whose tenets show the road to spiritual salvation. Violating them continues dukkha (suffering) in the present life and rebirth into a life of suffering or relegation into a hellish realm. Buddhism also decries harming sentient life:

> *One is not called noble who harms living beings.*
> Gautama Buddha

Buddhism accords with eugenics by teaching that the past determines the present. But while eugenics declares genes inherited from parents as the causative factors, it places the responsibility on conduct in the past life. Strong eugenics holds that for preventing the harmful causative genes from being transmitted, a child born with major physical disabilities should be euthanized. Supporting the child is a waste of public resources. Liberal eugenics says the disabled child may be assisted but be prevented from procreating through sterilization or isolation. One Buddhist perspective is that interfering with *karma*, a fundamental law of nature, will only generate future suffering. Let the child suffer. That will neutralize its misdeeds of the past. Another perspective holds that the principle of compassion is primary and overriding, and that one should do the best to ensure a decent life for the child. Neither Hindu nor Buddhist scriptures provide a firm guide on this matter. But in line with modern ethical tenets, most Buddhist and Hindu families would find the former option abhorrent and adhere to the latter.

Buddhist scriptures do not have other eugenic precepts. There are no caste-based marriage prohibitions. Yet, the flowering of Buddhism occurred in feudal-type societies where the monasteries and monks were wealthy landowners who engaged in usurious and exploitative activities and served the princes and the kings. In Tibet, monks became an exclusive caste with special privileges and treated the tenant farmers and peasants with utmost brutality. The ordinary people were dispensable. Infanticide was common in some places where Buddhism, Confucianism and Taoism were practiced, often in a syncretic manner. The same held true for discrimination against women. But these practices did not have a eugenic rationale.

Buddhist monks have often blessed wars against 'enemy' peoples and thus have effectively embraced external eugenics. That thousands were killed in these ventures was not seen as a departure from the *Eight Fold Noble Path.*

The fascist era in Japan provides an egregious case of the association between power, violence and Buddhism. The Samurai warriors of Japan had a monk-like status. They strove to master meditation in order to perfect their combat skills.

The linkage between Buddhism and war became more solid at the start of the 20th century. Prominent Zen Buddhist and Shinto monks praised the nationalist spirit that underpinned Japanese militarism. When Japanese armies invaded China, Korea and other nations, unleashing mayhem and massacres, they did so with the blessings of the leading monks. When Japanese doctors experimented on the prisoners in China in ways as horrid as the Nazi doctors, the monks stood by in silence.

A host of Zen masters, Sawaki Kodo, Sakai Tokugen and DT Suzuki among them, played a multi-faceted role. In the early phase, they provided moral support to right wing nationalist elements in the military and political circles. Later they worked with the state authorities to promote martial arts and physical education in schools and for the public. The aim was to prepare them for military duty. Zen masters also held meditation classes for the police. Others lectured in colleges, denouncing Marxism and Western ideas as *'foreign ideologies'*. An eminent Zen master likened the Buddhist notion of detachment from the self to having full loyalty to the emperor.

> *Discarding one's body beneath the military flag is true selflessness. ... It is in doing this that you immediately become faithful retainers of the emperor and perfect soldiers.* Sawaki Kodo (Victoria 2015).

Some Zen masters joined the military and fought in the front lines. Others played a public relations role in conquered territories. Respected masters like Suzuki and Sawaki visited China to connect with the Buddhist monks in China. By noting their shared religious heritage, they sought to appease their Chinese counterparts and win support for Japanese policies. Some real or fake Zen masters worked as spies for the Japanese forces.

In the 1930s, Japan and Germany forged an anti-communist, anti-Soviet alliance. As in Germany, alleged anarchists, socialists, pacifists, labor leaders and members of the Japanese Communist Party, some 60,000 in all, were rounded up on spurious grounds. The leaders were tortured, and some died in custody. At the same time, some Zen masters began to cultivate ties with the Nazis. Count Graf Durkheim, a German official based in Japan, and who was later indicted for war crimes, became a student of DT Suzuki.

Suzuki was aware of the discriminatory laws and harsh treatment of Jews and political dissidents in Germany in the 1930s. In his articles for a Japanese outlet about his 1938 visit to Germany, he praised Hitler for uniting a fractious nation. While calling the policy on Jews 'cruel,' he rationalized it as an expedient for a nation facing an emergency. Quoting Hitler in positive terms, the anti-Jewish pogroms were depicted as actions *'taken in self-defense'* against outsiders and depicted the *'wandering nature'* of Jews in *karmic* terms. Suzuki favored emulating the Nazi focus on disciplined youth brigades, especially volunteers working on farms in the countryside. The methods of character training used by the Nazis was in line with the traditional training of monks and both shared an antipathy towards communism.

Hitler and his henchmen believed that the European stock came from a master race in the East. Appropriating the swastika symbol of Asian religions, they sent delegations and scientific expeditions to Tibet to study Buddhism and conduct anthropocentric studies. Nazi biologists concluded that the Tibetan nobility had features like those of the North Europeans. Theosophists and Freemasons were persecuted by the Nazis. But, apart from a few brief skirmishes, Buddhism was tolerated. In 1933, the main Buddhist center in Berlin hosted the First European Buddhist Congress. Some Buddhists at the center served as Chinese and Japanese language translators for German officials. Tolerance for Buddhism also helped the formation of a stronger alliance with Japan.

Many Zen masters and Buddhist monks supported the imperial effort. Only a handful opposed it. Abbot Takenaka Shogen was a major opponent. At a 1937 event to send off army reservists to the battle zones in China, he courageously proclaimed:

> *War is both criminal and, at the same time, the enemy of humanity; it should be stopped. In both northern and central China,* [Japan] *should stop with what it has already occupied. War is never a benefit to a nation, rather it is a terrible loss.... From this point of view, I think it would be wise for the state to stop this war.* Takenaka Shogen (Victoria and Muneo 2014).

His words angered many, but he continued expressing his sentiments. After being reported to the police by two monks, he was arrested, convicted for spreading *'wild rumors'* and given a suspended sentence. Ostracized and deprived of his right to teach Buddhism, he never recanted. Opposition to war in the modern imperial era Japan was a distinctly rare occurrence among the Buddhist clergy.

Criticizing the ties with Nazi Germany was also taboo in the Japanese Buddhist circles. At a 1933 conference of over 450 Japanese Buddhist youth groups, only one group, the Youth League for Revitalizing Buddhism, declared that Nazism was an anti-Buddhist foreign ideology. Calling Hitler's suppression of Jews and others '*inhumane,*' it proposed a motion condemning his policies. In response, the hosts changed the venue and expelled the Youth League from the national umbrella body. There is no evidence that DT Suzuki, a professor of Buddhism at the Otani Buddhist College, the conference host, voiced concerns about the treatment meted out to the Youth League.

After WW II, DT Suzuki authored widely translated educational books on Buddhism, taught at several Western universities and became one of the best-known promoters of Buddhism in the West. Seen as an innovative scholar, he was nominated for the Nobel Peace Prize in 1963. It was a seamless transition from an exponent of fascistic nationalism to a champion of cultural harmony.

In sum: There is little in the Buddhist holy books and principles that can be taken as a justification of eugenics. But given the long history of the association of Buddhist monasteries and orders with power and wealth, their support for wars waged by kings, and the 20th century links with Japanese and German fascism their brutal pogroms, there is no cause to believe that Buddhism has special immunity to eugenics. Had Japan engaged in the Nazi type of internal eugenic practices, the Zen masters would probably have gone along. The participation of Buddhist monks in the ongoing ethnic cleansing of the Rohingya people in Myanmar supports that proposition (RPS 2022).

Eugenics and Islam

In the pre-Islamic societies of the Middle East, sickly or deformed babies were often killed. Marriages that would yield brave, hardy offspring were promoted. The aim was to create a strong tribe. Under the Quranic tenet that all children are equal in the eyes of *Allah*, such practices declined considerably once Islam took hold in the region.

> *It is God who brought you forth from your mothers' wombs* ... Quran 16:78
> (Sachedina 1995).

Islamic laws allow abortion when the mother's life is endangered but not for other reasons. Muslims are urged to marry Muslims and advised to select pious, healthy spouses of similar social standing. The Hadiths reflect varying views on marriage between close relatives, marriage within the social clan and marriage with outsiders. Clear guidance and realization of the long-term consequences of these choices is absent.

Some Sharia texts apparently assign a hereditary basis to alcoholism, insanity, prostitution and 'fatuous' behavior, particularly among women. Women who display such characteristics should not bear children or provide breast milk for the infants of other mothers.

While eugenic practices like sterilization, euthanasia, forced abortion and isolation of the disabled do not have sanction in Islamic traditions and texts, marriage prohibitions for the purpose of having fit and healthy children would elicit a degree of support in Islam.

The assignation of non-Muslims as *dhimmi* who have different rights and privileges than Muslims could in theory lead to external eugenics. Historically, non-Muslims were denied some freedoms such as practicing their faith in public and were taxed differently in Muslim majority nations. In family and community affairs, they could follow their own laws and customs but in business, property and criminal matters, they had to abide by identical state laws. This general picture varied over time and place. Religious minorities, especially Jews, enjoyed a better status in many early Islamic empires than they had done before and after Muslim rule. Of recent, progressive Muslim theologians have clearly rejected the *dhimmi* system.

The Armenian genocide, and the repression and massacres of Kurds in Turkey are two instances of external eugenics in a Muslim majority nation. Between 1915 and 1923, the

73

Turkish government embarked on a thoroughgoing program of rounding up, expelling and exterminating ethnic Armenians in Turkey and adjoining areas. Able-bodied male Armenians were killed at the outset. Later women, children, the aged and infirm were sent on army-supervised, brutal death marches to remote areas. The vast majority perished. In all, some one million Armenians were killed in this pogrom. It is only recently that the Armenian genocide has received publicity and official disapproval in the Western nations.

The Kurdish minority in Turkey (and adjacent nations) has for decades endured systematic denial of civil rights, military attacks, suppression of their language, demolition of hundreds of villages, food embargoes, and detention and torture of its leaders. Tens of thousands of Kurds have died, and hundreds of thousands became internal or external refugees. Turkey has benefited from massive supplies of military gear and diplomatic silence from the US for its the onslaught against the Kurds. The policy has popular support and enjoys the backing of both the secular and Islamic parties in the nation.

Muslims and the Nazis: In the run up to and during WW II, the Nazis and Western nations competed, using similar tactics, for support from Muslim leaders and organizations in the Middle East. Nazi propaganda portrayed Muslims and Germans as allies in a fight against common foes—the British Empire, the USA, France, the USSR, Zionism, and communism. Deploying Muslim deserters from the Soviet Army, they started a public relations drive to win over Muslim soldiers from Soviet and allied forces to their side.

> *In the war zones, Germany engaged with a wide range of religious policies and propaganda to promote the Nazi regime as the patron of Islam. ... German military authorities also made extensive efforts to co-opt Islamic dignitaries.* (Shtrauchler 2017).

In the Soviet areas with Muslim populations, the Nazi army rebuilt mosques and Islamic schools, encouraged Islamic practices, cleverly used Islamic ideas and trained its soldiers to temper their racism and show respect for Islam. The program paid dividends. Many Muslims joined their fighting units. One of these units was active in the harsh suppression of the 1944 Warsaw uprising. Yet, most Muslims and Muslim soldiers remained loyal to the Soviet army and fought against the Nazis.

As they rose to power, the Nazis sought allies across the globe, including Turkey, Iran and the Arab nations. The discriminatory laws passed in the 1930s thus exempted individuals from these nations who were not Jews. That policy gained momentum during WW II.

In most colonized nations, whatever their religious coloration, the post-WW I era was a time of burgeoning anti-colonial struggles. Under the anti-imperial ideology of self-determination, the resistance was organized within secular, nationalist, socialist and religious parties. That was also the case in the Muslim nations of North Africa, the Middle East and Asia, including Indonesia. The growing encroachment of Zionist settlers in Palestine reverberated across the Islamic world. The Balfour Declaration issued by Britain spurred Zionist settlers to take over, by means fair and foul, the fertile lands and buildings owned by Muslims and Christians. The drive to establish an exclusive homeland for the Jews was an affront to Muslims as Jerusalem was a holy city of Islam as well. This conflict was another basis for an alliance between some anti-colonial groups in the Muslim areas and the Nazis. The Nazis supported the nationalist cause as a way to weaken Britain. And since they advocated a total annihilation of Jews, they were also opposed to the establishment of a national homeland for the Jews, be it in Palestine or elsewhere.

While expressing a modicum of admiration for Islam, the Nazis held on to their racist views about non-Europeans. Their real stand on the people of the Middle East and Asia was hardly laudatory:

> *Hitler had told his military commanders in 1939, shortly before the beginning of World War II: 'We shall continue to make disturbances in the Far East and in Arabia. Let us think as men and let us see in these peoples*

at best lacquered half-apes who are anxious to experience the lash'.
Wikipedia (2021 -- Relations between Nazi Germany and the Arab World).

Opportunistically projecting themselves as friends of Islam, the Nazis were able to gain strong support from two prominent Muslim leaders, Amin al-Husseini, the Grand Mufti of Jerusalem, and Rashid A al-Gaylani, the Prime Minister of Iraq. The Mufiti, a respected figure in the Muslim world, was accorded an *'honorary Aryan'* status when he visited Berlin. On his part, he encouraged Arabs in Germany to join the Nazi military machine. In Iraq, besides giving military support to the government of al-Gaylani, the Nazis embarked on a broad cultural drive:

> *The German embassy purchased the newspaper al-Alam al-Arabi ('The Arab World') which published anti-Semitic, anti-British, and pro-Nazi propaganda, including a serialized translation of Hitler's Mein Kampf in Arabic.* (Shtrauchler 2017).

The Nazi army and the SS recruited tens of thousands of Muslims in North Africa and beyond who became front-line cannon fodders for the Germans. Three minor political parties in the Arab nations adopted a Nazi-style supremacist ideology.

Nonetheless, in North Africa and the Arab nations, practical support for the Nazis was an exception. The SS operated a network of labor camps whose aim was to coral and kill the Jewish population. Thousands of Jews perished in these camps. But hardly any Muslim leader cooperated with the program and King Mohamed of Morocco refused to issue discriminatory directives against local Jews. Apart from a few incidents of strife, Muslims and Jews had coexisted harmoniously for centuries. When the Nazi army marched in, many Muslims sheltered their Jewish neighbors. The same thing occurred in Albania.

In the early 1930s, the Nazis permitted German Jews to migrate to Palestine. Over 60,000 Jews thereby landed in Palestine. But later, as the gas chambers were built, they altered this policy. It was not a concession to the Muslim leaders who opposed the influx of the Jews in Palestine. The Nazis marched to their own tune, giving at best tepid support, in ways consistent with their own interests, to the anti-colonial drives in North Africa and the Middle East.

Exaggerating the scope of the Muslim-Nazi collaboration, Zionist scholars have depicted it in a purely anti-Semitic framework, and sidelined the anti-colonial ideas and struggles under which it occurred. The freedom struggle in Muslim lands had a variety of banners— liberal and secular nationalism, socialist nationalism and religious nationalism. As in India, when some parties reached out to the Nazis, it was more of a case of 'the enemy of my enemy is my friend' and not of ideological affinity. In the then prevailing political climate, fascism and Nazism were seen to be incompatible with Islam and the freedom struggle. Nationalism and economic development, not religion, was key to the programs of the main political parties. MN Sidiqi, a Palestinian socialist, produced a strong critique of the Nazi doctrine, branding it as a violation of Islamic principles. It reflected the view of many leading Muslim scholars.

Finally, we take note of a branch of Islam, the Nation of Islam, that began in the USA in the 1930s. WF Muhammad, the founder, said that his religion stood on three main pillars: *'Allah is God, the white man is the devil, and the so-called Negroes are the Asiatic Black People, the cream of the planet Earth'* (Wikipedia 2021 – Religion in America). These eugenicist tenets were an extreme reaction to the horrors African Americans endured in those days. Black supremacist eugenics emerged as a counter point to White supremacist eugenics.

Active in civil rights causes, the Nation espoused economic and cultural autonomy for African people under the banner of Islam. But its version of Islam in which the founder was deemed a prophet of Allah was rejected by mainstream Muslims. In the 1960s two eminent US civil rights leaders, Malcolm X and Muhammad Ali, joined the group. But realizing the multi-racial composition of the global Muslim community, both later left it and joined mainstream Sunni Islam. Today, leaders of the Nation of Islam are still in the forefront of protests against police brutality and discrimination of African Americans, and the Nation enjoys a reasonable degree of support from Africans Americans and Africans in the US, Muslim and non-Muslim. But it is political support, not religious support. And the Nation of Islam remains loyal to the eugenic-like tenets of its founder.

3.9 EUGENICS AND CHRISTIANITY

Modern eugenics was born, developed and reached its zenith in Christian majority nations—USA, Britain, Sweden and Germany.

The Bible

Does the Bible support eugenics? As with slavery, depending on the section of the holy book you consult, the answer can be yes or no. Taking their cue from the linkage of eugenics to animal and plant husbandry, the eugenicists invoked selective Biblical verses to support their doctrine:

> *A good tree cannot bear bad fruit, and a bad tree cannot bear good fruit. Every tree that does not bear good fruit is cut down and thrown into the fire. …. For every tree is known by its own fruit.* Wikipedia (2021 – Matthew 3:10, 7:17--18, Luke 6:43--45).

On the other side, invoking the verses Psalm 41:1, Psalm 82:4, Acts 20:35, Matthew 25:35 and Mathew 25:36, a pro-Christian outlet holds that eugenic practices are not consonant with Christian values of mercy and charity for the weak and the needy. Using John 9:1--4, it relates the following parable:

> *One day as Jesus and His disciples were walking in Jerusalem, His disciples asked about a man born blind. They wanted to know 'who sinned, this man or his parents, that he was born blind?' … Jesus replied, 'Neither this man nor his parents sinned, … but this happened so that the works of God might be displayed in him'. Who are we to decide who does or does not display the works of God?* (GQ 2021).

This is taken to mean that human disabilities are neither hereditary nor the fault of the individual. The Bible does not support the *karmic* notion that a person's current condition is due to past life sins. Further, the multitude of the versions of the Bible present somewhat different renditions of such passages. In general, Christian scriptures do not present an explicit endorsement of eugenics.

The birth of Christianity fostered a decline in practices like female infanticide, forced abortions, polygamy, promiscuity, incest, divorce, adultery and forced marriage that had been common in pre-Christian times. The status of women in society rose to an extent. Yet, once Christianity became the official religion of the Roman Empire, patriarchy prevailed and some of the progressive tendencies were, to a degree, nullified.

We fast forward to the nineteenth century. Europe and America were in the throes of rapid development of capitalism. New opinions flowered. The ideological supremacy of Christianity was challenged by science-based, rationalist outlooks. The evolving doctrine of eugenics was among them. Disenchanted with Biblical teachings from an early age, Francis Galton saw the dominance of the Church in the education system as a barrier to intellectual development and regarded Darwin's *The Origin of the Species* as a decisive refutation of Christian theology. Later he published a statistical study that cast doubt on the power of prayer. He also faulted Christianity, especially Catholicism, as a major force behind the decline in the quality of the human stock in Europe. The Church had corralled men and women of compassionate, meditative nature, and those endowed with artistic and literary skills into celibate orders. It had persecuted, imprisoned and exiled thousands of innovative scholars and thinkers. Both measures had reduced the transmission of the germ plasm of the gifted and ethical members of society. By active default, the Church had raised the prevalence of the '*servile*' and '*stupid*'.

> *[As] the Church brutalized human nature by her system of celibacy applied to the gentle, she demoralized it by her system of persecution of the intelligent, the sincere, and the free.* Francis Galton. (Pearson 1930).

Yet, Galton did not disavow religion. The moral stability of society needed a system of beliefs and values that people adopted as a matter of blind faith. Noting the '*religious significance*' of the Theory of Evolution, he envisioned a reformed faith that would integrate science and biology, particularly eugenic values and ideas, within its teachings. An advanced human society would have a scientific religion with eugenics at its core and with eugenicists manning the priesthood.

> *Religious precepts .... require to be reinterpreted in order to make them conform to the needs of progressive nations.* Francis Galton (Pearson 1930).

Julian Huxley, the eminent biologist, public intellectual and avid eugenicist who became the first director of UNESCO, excoriated prevailing religions, their rituals, clergy and institutions and called for an ethical, rational religion that dispensed with the belief in a personal god. Ronald A Fisher, eminent statistician and evolutionary theorist but ardent eugenicist and anti-Semite, was, on the other hand, a devout Christian. In his eyes, eugenics not only complimented Christian faith but was also a basis for spiritual salvation.

Their counterparts in America married the eugenic creed with Christian faith or declared it as the religion of the modern era. In his fervent 1916 propaganda pamphlet, *Eugenics as a Religion,* Charles Davenport, the lead American eugenicist, posited 12 edicts, of which one reads:

> *I believe that I am the trustee of the germ plasm that I carry, that this has been passed on to me through thousands of generations before me; and that I betray the trust if...I so act as to jeopardize it, with its excellent possibilities, or, from motives of personal convenience, to unduly limit offspring.* Charles Davenport (EA Editors 2021).

In 1926, the American Eugenics Society published *A Eugenics Catechism*, a fervently proselytizing pamphlet featuring over a hundred issues tackled in an organized, clear question and answer format. Some of the questions and answers were:

> *Q. Does eugenics contradict the Bible?*
> *A. The Bible has much to say for eugenics. It tells us that men do not gather grapes from thorns and figs from thistles.*
> *Q. Is eugenics antagonistic to the Bible?*
> *A. No. The aim of eugenics is to insure the totality of human welfare in the long run. .....*
> *Q. What is the most precious thing in the world?*
> *A. The human germ plasm.*
> *Q. How may one's germ plasm become immortal?*
> *A. Only by perpetuation through children.*
> *Q. What is a person's eugenical duty to civilization?*
> *A. To see that his own good qualities are passed on to future generations provided they exceed his bad qualities. If he has, on the whole, an excess of*

> *dysgenic qualities, they should be eliminated by letting the germ plasm die out with the individual.* (AES 1926).

The pinnacle came with the eugenic formulation of the Ten Commandments by AE Wiggam. A prominent publicist for the eugenic cause, and author of *The New Decalogue of Science*, he drew upon biology, psychology and anthropology to present a strong argument for the case that the human race faced a grave risk of mental and physical decay. Too many weak germ plasms were being passed on. His remedy: A Christianization of humanity based on *Wiggam's Ten Commandments* that invoked eugenics to advocate a reformed ethic, socialization of science, humanization of industry and elevation of artistic and intellectual pursuits.

These Commandments replaced the Biblical injunction to have no other gods but God with the fundamental duty to abide by eugenic precepts, the injunction to not make or worship idols with the duty to perform scientific research and the injunction '*Thou shall not kill*' with promoting preferential reproduction.

Eugenics challenged the hegemony of religion over human thought by science to a new level. Religion was being upstaged on its own turf by a doctrine that was rapidly capturing not just the higher circles in the land but also the common folk.

Britain

From inception, the British Eugenics Society aimed to win over the medical profession and the clergy to its cause. The terminology and ideas expressed in its main journal, *The Eugenics Review*, and pamphlets were tailored so as to not alienate doctors and priests. Society members also gave lectures to church congregations and clerical groups. Reverend WR Inge, a prominent Christian theologian, Professor of Divinity at the University of Cambridge and later, the dean of St. Paul's Cathedral in London, was an early convert to eugenics. His large scholarly output was complimented by his papers for *The Eugenics Review* and a weekly column for a major British newspaper. He urged the church to set aside the issue of miracles and concentrate on experiencing God through prayer. He held that Christian moral teachings were consistent with eugenics. While not supporting the burning of heretics at the stake, he faulted the church for having burnt '*the wrong people*'. Inge's erudite efforts to reconcile eugenics and the Christian outlook earned him international fame.

Reverend JHF Peile, lecturer in Divinity at Corpus Christy College and later Archdeacon of Warwick and Archdeacon of Worcester, was a major Christian ally of the eugenic cause. He wrote that while there were good doctrinal grounds for clerical support for eugenics, it was limited by the fact that priests led busy lives and eugenicists had neglected distributing their material to the clergy. His words spawned a drive to distribute eugenic material to churches across Britain. Ernest W Barnes, a mathematician, Fellow of the Royal Society and pacifist, later embraced theology and became the Bishop of Birmingham. He held that while physical and mental capacities were mostly hereditary, they would improve by appropriate upbringing. Unlike many eugenicists, he was not a racist and deemed racial inter-breeding an unavoidable global phenomenon. Under right conditions, it could benefit humanity. He remained an advocate of voluntary sterilization of the 'feeble minded,' even after the end of WW II.

Though eugenics garnered a measure of expressed or tacit support from Protestant churches, the British Catholic Church disagreed. GK Chesterton, a popular novelist, art expert, social commentator and amateur theologian gave a loud, resplendent voice to the Catholic angst towards eugenics in the UK. His 1922 book, *Eugenics and Other Evils*, is a biting satire that adroitly bares its inconsistencies. To him eugenics, in whatever form or scope, was an abhorrent doctrine.

> *Eugenics itself, in large quantities or small, coming quickly or coming slowly, urged from good motives or bad, applied to a thousand people or applied to three, Eugenics itself is a thing no more to be bargained about than poisoning.* (Chesterton 1922).

He excoriated the idea that individuals should face restrictions in the choice of marriage partners:

> *The Eugenists' books and articles are full of suggestions that non-eugenic unions should and may come to be regarded as we regard sins; that we should really feel that marrying an invalid is a kind of cruelty to children.* (Chesterton 1922).

Contrasting the religious inquisitor with the ardent eugenicist, he noted:

> *The devotee boasted that he would never abandon the faith; and therefore he persecuted for the faith. But the doctor of science actually boasts that he will always abandon a hypothesis; and yet he persecutes for the hypothesis. The Inquisitor violently enforced his creed, because it was unchangeable. The savant enforces it violently because he may change it the next day.* (Chesterton 1922).

Eugenics advocated an ideal society of gifted people. Chesterton opined that it would entrench a state enforced, science-based tyranny of the upper classes over the weak and the multitude.

Chesterton's opinions were expressed in an engaging style. But he was short on facts and social theory. While he critiqued eugenicists for positing a hereditary basis to all human traits without having adequate evidence, he at times appeared to take issue with science itself. Nonetheless, his warnings were prophetic.

As many British Catholic Church leaders opposed eugenics, the alliance of Catholics and progressive elements in the Labor Party was a major factor that prevented the spread of American style forced sterilization in the UK.

The divergent views of Protestants and Catholics on eugenics reflected a general phenomenon. It was in the Protestant nations—USA, Canada, Sweden, and Germany—that negative eugenics was extensively actualized. Mussolini, the Italian fascist dictator, accepted the idea of cultivating a superior Italian race. Yet, forced sterilization, euthanasia or marriage prohibition did not receive state sanction and were not implemented in this predominantly Catholic nation during his rule.

United States

In the United States, eugenic advocacy gained a wider, less-controversial ground among the Protestant clergy. The American Eugenics Association had a specific program of cooperation with religious groups. Among other things, it held an annual competition for the best sermon on eugenics. Influential Christian leaders sat on the board of the Association. Despite some discord, they largely stood by its programs. Clergymen sermonized on eugenics. Some favored policies to encourage healthy families to have more children. Others endorsed sterilization, confinement and euthanasia as well. Methodist, Presbyterian and Episcopalian churches hosted meetings to discuss eugenics, promoted Better Baby Contests and helped organize Race Betterment Conferences. Eugenics literature, including a book on the subject endorsed by two prominent Methodist bishops, circulated in many churches.

Reverend Harry F Ward, a liberal Methodist and an academic specialist in Christian ethics, opined that the goal of eradicating *'causes that produce the weak'* made eugenics compatible with Christianity. Reverend Clarence T Wilson, an eminent Methodist administrator, believed that *'only the white Aryan race was the descendant of the lost tribes of Israel'*. (UMC 2016).

In 1929, the Methodist Review published the sermon 'Eugenics: A Lay Sermon' by George Huntington Donaldson. In the sermon, Donaldson argues, 'the strongest and the best are selected for the task of propagating the likeness of God and carrying on his work of improving the race'. (UMC 2016).

Augustus H Strong was one of the most eminent American theologians of the 19th and 20th centuries. *Systematic Theology*, his main treatise, was often used for theological training. Strong also accepted eugenics and phrenology. Adapting the Biblical idea of stigmata—the crucifixion wounds on the body of Jesus Christ—he raised alarm about the encroaching stigmata of degeneracy. He viewed human development as a competition between a divinely driven morally and physically enriching process and a human driven debasing process that leads to moral depravity and hastens its Bible-predicted downfall. Prayers and eugenic interventions would help curb that decline.

Conservative evangelicals and Catholics were circumspect on the issue. The espousal of birth control, abortion and sterilization by eugenicists produced but a shaky support from these churches. Several theologians with a footing in biology, law or social sciences found its ideas wanting. James M Boddy, a physician and Presbyterian thinker of African American descent exposed the race and anti-minority bias in phrenologic studies and their eugenic verdicts. Father John A Ryan and Father John M Cooper, both active members of the American Eugenics Association, sought to reconcile eugenics with Catholic teachings. But after spending over a decade on its board, they parted company with the Association.

The anti-Semitic concomitants of eugenics generated much opposition within Judaism, yet, even among the Rabbis, there was a degree of support for eugenics.

Casti Connubii, the encyclical on marriage and birth control issued by Pope Pius XI in 1930 firmly rejected eugenics and galvanized the Catholic resistance.

Pope Pius XI declared that any eugenic method is an attempt to assert power and judgment over individuals that civil authorities can never legitimately possess. (Brian 2016).

Previously, the Catholic clergy had supported parts of the eugenic program but it had been a vacillating stand. With Papal authority behind them, they condemned it in stronger terms. Catholic magazines stressed that each human—physically and mentally able or disabled—was an image of God. Accusing eugenicists for 'playing God,' they castigated the inhumane treatment of the weak. Catholic experts queried the validity of the studies done by eugenicists and lamented the racist implementation of eugenic measures. The latter critique resonated in the US where the Southern European, Mexican and South American immigrants ensnared by eugenic measures were mostly Catholic. Catholic leaders not only criticized eugenics but also mobilized to defeat the sterilization bills in several states. In Ohio, their efforts paid off.

At a time when eugenics was being hailed by all the centers of influence in the nation, it took courage to oppose it. To its credit, the Catholic Church was one of the few that did. And unlike others, its opposition was organized and effective. But it had limitations. The US Catholic Church has a history of slave ownership, gross abuse of Native Americans, maltreatment of Latinos, and segregation in services, schools and hospitals. While 4% of the about 70 million Catholics in the US are African American, only 1% of the Catholic priests are African American. Despite its placement of equal value on all human life, the Papal encyclical hardly altered the status of African Americans in the Catholic institutions. Racism is a key pillar of external eugenics. Yet, the resources devoted by the Catholic Church to its anti-abortion crusade pale in comparison to what it has devoted to combating racism. Spurred by the Black Lives Matter campaign, Pope Francis has recently called for a full accounting of this ugly part of the history of the Church. Measures to redress the remnants of this ugly history in the religious and secular activities of the Church are at a nascent stage.

Without going into details, we note that the role of the church in relation to eugenics in Canada was similar to that in the United States.

Germany

At the start of the 20th century, Germans were 95% Christian, and 3%, Jewish. Nearly two-thirds of the former were Protestant, and one-third, Catholic. During the late 1800s Catholics had suffered official discrimination and were barred from key governmental jobs. Many priests were exiled or jailed, and their media outlets were shut down. The post-WW I constitution reversed these policies by declaring that Germany had no official religion and enshrining the freedom of religion.

The advent of the Nazi Party signified a major change. It was rooted in the views of Adolf Hitler who grew up as a skeptical Roman Catholic, did not attend church after the age of 18 but considered himself a Catholic till his death. In his public speeches, he mentioned God and the Spirit, and invoked religious imagery. But what he and the Nazis advocated was a reformed nondenominational version of Christianity. Positive Christianity, as it was called, sought a Bible expunged of its Jewish elements, with Jesus as an Aryan, and which was consonant with restoring the greatness of the German nation.

> *We hold the spiritual forces of Christianity to be indispensable elements in the moral uplift of most of the German people.* Adolf Hitler, 1933 (Wikipedia 2021 -- Religious Views of Adolf Hitler).

Was Hitler a genuine Christian or an opportunist? Scholars continue to debate the issue. His private correspondence and what his close confidants have written indicate that he was strongly critical of the Vatican and the Church and thought that the Christianity of the Jews was infused with 'false' ideas. Yet, he was not anti-religious and regarded religion as a tool for control in the current state of social and political development.

> *The political leader should not estimate the worth of a religion by taking some of its shortcomings into account, but he should ask himself whether there be any practical substitute in a view which is demonstrably better. Until such a substitute be available, only fools and criminals would think of abolishing existing religion.* Adolf Hitler, *Mein Kampf* (Wikipedia 2021 -- Religious Views of Adolf Hitler).

The *'practical substitute'* envisioned by Hitler linked Positive Christianity with science, namely the Nazi eugenic version of science.

Hitler had nothing but contempt for atheism, which he branded as animal like thinking of the uneducated. It was not just a philosophical position but emanated from his firm hostility to Marxism, communism, *'Judeo-Bolshevism'* and *'international atheism'*. He was a fierce foe of trade unions and militant working-class parties in Germany that represented the most viable opposition to Nazism. One of his first acts as the Chancellor was to round up political opponents and dispatch them to the Dachau concentration camp. The German Free Thinkers League, with about half a million mostly atheist members, was banned in 1933. A number of its leaders landed in Dachau. Subsequently, Hitler declared that:

> *For eight months we have been waging a heroic battle against the Communist threat to our Volk, the decomposition of our culture, the subversion of our art, and the poisoning of our public morality. We have put an end to denial of God and abuse of religion.* Adolf Hitler (Wikipedia 2021 -- Religious Views of Adolf Hitler).

High Nazi officials, notably in the SS, had anti-Christian views, dabbled in occult practices like sun worship and desired to suppress Christianity. But Hitler needed votes and loyalty from Christians. Accordingly, he exploited the post-WW I disillusionment among the working class and middle class by artful appeals to Christianity and supremacist nationalism. He also forbade Nazi officials from openly espousing anti-Christian views or leaving the church.

> *We tolerate no one in our ranks who attacks the ideas of Christianity ... in fact our movement is Christian.* Adolf Hitler, 1928 (Wikipedia 2021 -- Religious Views of Adolf Hitler).

The words did not reflect principles but astute tactics. Religious rhetoric served to facilitate a working relationship with the churches and pacify the Vatican and the German churches that opposed the Nazi doctrine. At the outset he promised to protect religious freedom. Yet, he went against his words soon after his political opponents had been subdued.

> [As] *Hitler consolidated his power, schools became a major battleground in the Nazi campaign against the churches. In 1937, the Nazis banned any member of the Hitler Youth from simultaneously belonging to a religious youth movement. Religious education was not permitted in the Hitler Youth and by 1939, clergymen teachers had been removed from virtually all state schools. Hitler sometimes allowed pressure to be placed on German parents to remove children from religious classes to be given ideological instruction in its place, while in elite Nazi schools, Christian prayers were replaced with Teutonic rituals and sun-worship. By 1939 all Catholic denominational schools had been disbanded or converted to public facilities.* (Wikipedia 2021 -- Religious Views of Adolf Hitler).

Bringing the multitude of the Protestant churches in Germany under the umbrella of a centralized German Evangelical Church was also on the cards. Though most churches agreed with the proposal, they wanted autonomy in the selection of the leaders. But they were outmaneuvered by a devious tactic of the Nazis. Bishops sympathetic to the Nazis held the leadership of the central church. Subsequently, priests with Jewish ancestry or who were married to non-Aryans were defrocked. Priests who criticized the official policies risked imprisonment or worse.

Many Protestant bishops and theologians had no qualms about Nazism. Paul Althaus, a distinguished Lutheran theologian, preeminent expert on the life and doctrine of Martin Luther and professor of theology, promoted the eugenic idea of a hereditary pure nation. An anti-Semite and advocate of the death penalty, he painted Adolf Hitler in glorious terms:

> *Our Protestant churches have greeted the turning point of 1933 as a gift and miracle of God.* Paul Althaus (Wikipedia 2021 – Paul Althaus).

Emanuel Hirsch, a Protestant theologian and professor at Gottingen University joined the Nazi Party and the SS. A pronounced anti-Semite, he declared that Jesus was not a Jew but an Aryan. When Jewish academics were expelled from German universities, he remained silent. His racial theology echoed the Nazi hatred of Jews. His 1933 verdict on Hitler was:

> *No other people of the world has a statesman who is so serious about Christendom; when Adolf Hitler concluded his great speech on May 1st with a prayer everybody could feel the wonderful candor therein.* Emanuel Hirsch (Wikipedia 2021 -- Emanuel Hirsch).

Gerhard Kittel, the senior editor of the ten-volume *Dictionary of the New Testament* was closely associated with the Nazi Party. Though his writings before 1933 reflected a pro-

Jewish stand, once the Nazis were in control, he reversed his views and became a Jew-hating member of the Nazi party.

Endorsements by distinguished theologians and senior clergy gave spiritual legitimacy to the Nazi cause and enhanced the loyalty of ordinary Christians to Hitler and his programs. The Christian vote was critical in the 1932 and 1933 elections that brought the Nazis to power. After 1933, many clergy and church institutions participated in the selection of the 'unfit' persons for internment and euthanasia.

Protestant pastors were divided over the secular and spiritual aspects of the Nazi policy towards Christianity. At one end was the German Christian faction. Espousing Positive Christianity, it fervently backed Nazi policies, claimed that Jesus was not a Jew but an Aryan, and called Hitler the new Messiah. The Confessing Church faction stood at the other end. Though it was firmly opposed to state interference in doctrinal matters, so long as freedom of religion was respected, it had no qualms about the general Nazi program. Of the about 18,000 Protestant pastors in Germany at the time, 3,000 were allied to the German Christian faction and 3,000 to the Confessing Church faction. The rest, some 12,000 pastors, stood on neutral ground. This silent majority did little for the victims of Nazi atrocities. At best, they privately appealed to the state bureaucracy and modified the selection criteria for euthanizing the 'unfit'. But the Nazis tolerated nothing short of absolute loyalty. Anything less was an invitation for the SS to pay you a visit. In 1935, 700 pastors linked to the Confessing Church were arrested and sent to concentration camps (Wikipedia 2021 -- Confessing Church).

Numerous autonomous Freemason lodges operated in Germany at that time. Branding them a part of the Jewish-Masonic conspiracy, lodges that did not echo the Nazi program were forced to close and their members were harassed by the police. In 1935, all Masonic lodges were closed down and their property was seized.

Jehovah's Witnesses in Germany, about 30,000 in 1930, had faced harsh repression. Abiding by what they say is the original Christianity of the first century, the Witnesses are a tightly organized and highly disciplined sect. Though basically respectful of state authority, they do not pledge allegiance to any nation, salute national flags or sing national anthems. As conscientious objectors, they refuse to join the military or work in factories linked to the military. In the jingoistic environment of Nazi Germany, these practices made them *'enemies of the state,'* and set them on a collision course with the SS. The secret police vandalized their offices, disrupted their meetings, confiscated their literature and harassed them as they went about their daily lives. About 10,000 Witnesses were either forced into exile or ceased to practice their faith. A grim fate awaited the 20,000 Witnesses who remained active.

> [About] *half were convicted and sentenced at one time or another during the Nazi era for anywhere from one month to four years, with the average being about 18 months. At least 3,000 Jehovah's Witnesses were sent to concentration camps ...* (USHMM 2021).

Numerically, the Seventh Day Adventist Church (SDA) is a minor Christian sect. Yet, it has a substantial global presence in missionary work, health, education, social services and business. The German branch of the SDA was set up in 1888. By 1930, it had 40,000 adherents and 500 pastors. With about 800 staff employed in the publishing division, it was heavily invested in the advocacy of the Christian doctrine and education.

The climate of fear under the Nazis made the Church apprehensive about its future in Germany. The observance of the Saturday Sabbath which it shared with Jews and other unconventional tendencies made it a potential target. To protect their church, the SDA leadership thereby went overboard in praising Hitler and his ideas.

> *A fresh enlivening, and renewing reformation spirit is blowing through our German lands . . . The word of God and Christianity shall be restored to a*

place of honor. Adolf Minck, President of the Adventist German Church, 1933 (Alomia 2010).

Projecting Hitler as a diligent leader who moreover adhered to an Adventist type dietary regime of vegetarianism and avoidance of stimulants and harboring the illusion that Hitler would protect religious freedom, Adventist leaders advised their members to adopt a friendly attitude towards and vote for the Nazi Party.

In the Adventist town of Friedensau, Germany, 99.9% voted for the Nazi parliamentary state. (Schroder 2003).

The German SDA was caught up in the blanket ban on religious and secular groups issued by the Nazis in November 1933. But as a result of intense lobbying by its leaders and professing loyalty to Hitler, the ban on the SDA was lifted within two weeks. (As an aside we note that John Harvey Kellog, the American SDA luminary who was a pioneering expert on health and fitness, and the inventor of the breakfast cereal, was also a fervent eugenicist).

Subsequently, the German SDA expanded its support for the Nazi Party. From 1933 to 1945, SDA leaders in Austria, Germany and the US as well as the US-based SDA General Conference not only accepted the idea of Aryan supremacy but also actively cooperated with it. Aiming for tolerance from the Nazis, SDA publications endorsed Nazi repression of Jews and others. Jews were labeled as '*bloodhounds*', '*vermin*', and 'aliens'. SDA leaders backed forced sterilization and euthanasia programs for the weak and infirm, and lauded them in Adventist publications. Adventists with a Jewish background and dissident Adventists were reported to the Nazis. Some had to flee, and some wound up in concentration camps. Adventists of Jewish ancestry were expelled. SDA church practice was modified so as not to offend the Nazis. SDA leaders in the US and elsewhere in Europe helped to provide positive publicity for the Nazis.

A reformist faction within the SDA refused to abide by the blanket support for the Nazis. During WW I, the German SDA split over the issue of military service. A minority opposed the decision to permit Adventists to bear arms. In the Nazi era, the schism was more acute. The reformists opposed any change to the Sabbath schedule and armed military service. It sought to strictly follow traditional Adventist teachings. Many reformists were reported to the Nazi authorities and suffered direly. As the majority of the Adventists openly supported or silently accepted Nazi repression, a few Adventists assisted their Jewish neighbors in a variety of ways to escape capture.

Besides the Protestant Confessing Church faction, the Catholic Church was the other significant religion-based opponent of the Nazis. Catholic leaders openly attacked sterilization and euthanasia projects, both when they were legal and when they were being done in secret. Archbishop Van Galen and Archbishop Worm condemned the policies in their sermons. Copies of these sermons were sent to parishes in Germany. The Catholic stand derived from the encyclical on marriage and birth control issued by Pope Pius XI in 1930. The voices of the Catholic leaders carried sufficient weight to bring these projects to a temporary halt. The worst the Nazis could do to these dissidents was to put them under house arrest for a short time. But lower-level Catholic clergy did not fare well. According to records found later, a total of 2,771 junior and senior religious figures were sent to the Dachau camp. The vast majority, 2,579, were Catholics. The rest were Protestant (109), Orthodox Christian (30) and Muslim (2). The religion of the remaining 59 could not be determined. At least 1,034 of the clergy died in the camp (Garver and Garver 1992). As the Nazi forces marched across Europe, many clergy, including nuns, joined the resistance movements. Many were captured, tortured and killed.

In the first 16 months of the war, 700 Polish priests died at the hands of the Nazis, and 3000 more were sent to concentration camps; more than half did not return. (Garver and Garver 1992).

Reverend Dietrich Bonhoeffer, a Lutheran pastor and theologian was one of the few Christian leaders in Germany who denounced all aspects of Nazi policies and valiantly defended the Jews. Interned in a concentration camp, he was charged with treason and hanged in 1945.

Synopsis

Overtly or by tacit silence, the vast majority of religious denominations in Germany accepted the Nazi policies. Some eagerly joined its vicious projects. About half of the euthanized persons came from Protestant or Catholic managed institutions. Only a few religious groups defied the Nazi policies. The opposition ranged from rejection of state interference in religious matters to denouncing the Nazi program in its totality. Many clergy and lay members of the religious opposition lost their freedom or were killed. The Nazis did not have intrinsic affinity to an existing religion. Espousing Christianity was a tool for securing total control. However, they considered themselves as a unifying force standing for national salvation and fighting for God against evil. Nazism was evolving into a fanatical, religious doctrine (Positive Christianity) whose essential tenet was the God-blessed supremacy and racial purity of the (Aryan) German nation (*volk*). Adolf Hitler was deemed the Messiah of the emergent religion whose symbol was the swastika and whose rituals included the mandatory Nazi salute 'Heil Hitler'.

With the exception of the Catholic Church and a few Protestant clergy, the relationship between religion and the Nazis generally was a symbiotic one. The Nazis used Christianity to consolidate their power. In turn, most of the Christians in Germany were mesmerized by Hitler's ultra-nationalist tirades:

> *It was Christians who voted Hitler into power and it was the same Christians who praised his arrival in 1933 to the chancellery as a new beginning and renewal of hope for Christianity in spite of his racial tirades. It was Christians who stood idly by as the rights of the Jewish minority that had contributed so much to German culture were stripped away with very little or no protest by the Christian majority.* (Alomia 2010).

Through its role in the Atlantic slave trade, decimation of Native peoples and colonial conquest and rule, Christianity assisted the propagation of the white supremacist doctrine that European culture epitomized civilization. It is the will of God; all cultures must Christianize and emulate European mores and thought. Eugenics gave a modern, rational sounding twist to the same claim. Not God, but the immanent laws of nature made the world as it was. Social problems must be remedied by scientific, eugenic, management of procreation and society.

The eugenics creed founded by Francis Galton reached its apex in Nazi Germany. It was accepted and promoted by eminent scientists, writers, scholars, liberal and conservative politicians, feminists and social reformers but also by a good portion of the Christian clergy. Depicting the (white) Nordic race as the superior race was not seen as an anti-Christian notion. While rejecting the Theory of Evolution, Christians of varied denominations adopted a doctrine that was based on it. Original sin complemented heredity in the degeneration of the human race. Sin was passed on from generation to generation via the germ plasm. Just as God would save the spiritually virtuous (white) humans from descent into hell, so would eugenics protect the people with fit germ plasm from being polluted by degenerates, inferior races and immigrants. The goal of race protection and betterment, spiritual and physical, attracted Protestant and Catholic ministries and laity alike. Jewish ministries in the US also adopted basic ideas of eugenics. Invocation of Biblical verses and usage of terms like the stigmata of degeneracy made it more appealing to Christians.

Eugenics challenged Christian morality and ethics by devising its own moral code and commandments. But it also appealed to the Bible to back up its values and morality. Eugenics and Christianity were presented as the twin machines for saving humanity from physical, moral and spiritual degeneracy. Church sermons often became intertwined with eugenic ideas. Eugenics was scientific racism. It integrated science, liberal thought and religion with elite prejudice into a single package. As such it captivated the minds of tens of millions of religious people in the Western world. It was an elitist creed that appealed to elites and commoners, including its victims. At one point in time during the first half of the 20th century, public life was inundated with eugenics.

Yet, a spectrum of opinion towards eugenics prevailed, ranging from firm adherents, partial believers and neutral observers to dedicated opponents. Despite the atrocious ideas and deeds that engendered eugenics, the last group generally was in a minority. Scientists opposed to eugenics pointed out the multiplicity of basic flaws in the studies conducted by eugenicists and the shallow nature of their conclusions. Theologians who decried it said that eugenic ideas, especially the assignments of a decisive role to heredity, were fundamentally antithetical to the laws of God. At the outset, the Catholic Church was partial to eugenics. But later it became a major force standing against forced sterilization and euthanasia. The anti-eugenic stand of the Catholic Church reflected its agenda of equating abortion with murder and opposing abortion and family planning when even done with the consent of the mother and the family. Nations with a Protestant majority were generally more likely to adopt eugenic programs than nations with a Catholic majority.

3.10 MARXISM AND EUGENICS

Is Marxism compatible with Eugenics? We compare the views of Thomas Malthus and Francis Galton, the two primary eugenic thinkers, with those of Marx and Engels.

Malthus declared that while human population grows at a geometric rate, food supply increases at an arithmetical rate. At some point, survival needs outstrip the supply. This disjuncture produces famine, disease, social strife and war. These effects together with natural disasters lower the population size and restore the balance between supply and need. The propensity of the poor to have many children was the source of the malady. More children mean more stomachs to fill. Raising wages is counterproductive; it only encourages promiscuity, illegitimacy, sloth and more offspring. Charity likewise fosters evil. Malthus held that wages have a natural tendency to settle at a level that aligns with the basic subsistence needs of the workers. He promoted celibacy, delayed marriages and family planning to lower population growth but did not favor contraceptives. Influenced by Malthus, Galton extended Malthusian thinking by positing the centrality of germ plasm and heredity and controlling procreation among those with the 'bad' germ plasm.

Marx and Engels held that food supply and population size are related to the mode of production and level of development of productive forces. While the Malthusian proposition may apply to early human formations, it did not hold in the capitalist age, an age of technological innovation. Noting the multiplicity of factual flaws of Malthusian thought, they unreservedly derided its *'fundamental meanness'* and the aspersion of immorality it cast on the working people. The very low wages paid by capitalists force the entire family to work. Children are seen as an asset who contribute to the family income. Poverty and high infant mortality also raise the birth rate. As wages rise and living conditions improve, the birth rate tends to decline, and population growth is lowered. The elites have fewer children; the poor have more.

Marx and Engels saw the long-term solution in the replacement of the social system that artificially keeps wages low and generates poverty and inequality (capitalism) with a system in which the working-class exercises power, controls the means of production and exchange, and where a fair distribution of the social output prevails (socialism). Humanity is not at the mercy of iron laws of nature but has the capacity to understand nature and exercise control over it.

> *The real problem wasn't too many people or too little food, said Marx, but that private capitalists owned the means of meeting men's needs.* (Weisman 1970).

Malthusianism was an element of the prevailing conservative, pro-capitalist ideology justifying the dominance of capital over labor, The Marxian vision was a grounded, coherent doctrine for the emancipation of labor.

Marx and Engels had high regard for Charles Darwin's masterpiece, *On the Origin of Species,* and felt that it complemented their own work on social development. As Engels noted at Marx's burial:

> *Just as Darwin discovered the law of development of organic nature, so Marx discovered the law of development of human history.* Friedrich Engels, 1883 (Angus 2009).

In 1872, Darwin received a copy of the second edition of *Capital*, Volume 1. The first edition had appeared in 1867. It was from the author and had the inscription '*on the part of his sincere admirer, Karl Marx*'. (Angus 2009). Thanking him for the book, Darwin wrote to Marx that they were united in a common drive to disseminate knowledge. Yet, Darwin did not read *Capital*. Had he done so, he would have realized that Marx shared his disdain for linking natural evolution with societal evolution and socialism.

While he revolutionized biology and demolished teleological and Biblical explanations of living nature, Darwin was not immune from the dominant sociopolitical prejudices of his time. While repelled by slavery and averse to drastic policies to control human population, he believed the white race was by evolution more advanced than other races and that men were superior to women. His formulation of the theory of evolution was affected by Malthusian ideas on population and survival. The centrality of competition between species he posited reflected the prevailing ideas on the dominant role of competition in economic development. And the importance of cooperation within and between species as an engine for natural evolution was sidelined.

Marx and Engels recognized that Darwin had provided a firm scientific depiction of natural evolution. But they decried a simplistic extrapolation of the laws of natural evolution to the development of human society. Laws governing the operation of a complex system cannot be reduced to the laws governing its constituent parts. Though humans are subject to the laws of biology, the structure, history and functioning of human societies cannot be explained by just resorting to the laws of biology. Humans differ from other animals in one critical manner. While animals acquire their subsistence from nature by foraging and hunting, humans develop tools that allow them to produce things not found in nature.

Social Darwinism took its cue from extrapolation of the laws of nature to society. Herbert Spencer, the sociologist who pioneered this view, visualized human history through the lens of the '*survival of the fittest*'. Like animal species, different forms of human societies and different social strata within societies competed for dominance. The fit rose and others declined. This was the Social Darwinist explanation for the chasm between the rich and the poor and the levels of development of nations. Fitter nations rule over weaker nations. That emanates from immutable laws of nature. It was vision dear to business magnates and imperial powers.

> *[The] growth of a large business is merely a survival of the fittest...merely the working out of a law of nature and a law of God.* John D Rockefeller (Angus 2009).

The major proponent of Social Darwinism in the American academy, who was also an Episcopalian minister and avowed detractor of trade unions and state intervention in the economy, concurred:

> *The millionaires are a product of natural selection.* William G Sumner (Angus 2009).

87

In contrast, Marx and Engels studied the structures of social formations to derive a more grounded vision of society in which economic structures, state institutions like the laws, army and the police together with the education system, the media and the church interacted in a dynamic manner to buttress the dominance of capital over labor. Capitalism was not the final stage of history. It was a product of class struggle, and it would be undone by class struggle. Social Darwinism rationalized a system of inequality and racism. Marxism provided a general, analytically derived direction for the struggle towards a system of equality and social justice.

> *The conception of history as a series of class struggles is already much richer in content and deeper than merely reducing it to weakly distinguished phases of the struggle for existence.* Friedrich Engels (Angus 2009).

Eugenics took Malthusian and Social Darwinist perspectives onto a higher plane by positing a hereditary mechanism for explaining social variation and providing a program for race betterment. It was a vision at odds with the ideas of class analysis and class struggle. Marx and Engels did not directly address eugenics. But it is apparent that had they done so, they would have excoriated it in as pungent terms as they had used for Malthusianism and Social Darwinism.

VI Lenin and Leon Trotsky

VI Lenin, the leader of the 1917 Russian revolution, is deemed the next most influential Marxist after Marx and Engels. He did not write about eugenics. But, noting his close theoretical attachment to the ideas of Marx and Engels, we can say that had he done so, he would have denigrated it. Leon Trotsky, the co-leader of the Russian revolution, briefly wrote about it. His proclivity to mathematics would have enabled him to venture into the technical writings of Francis Galton and Karl Pearson, though we do not know if that is what he did. Yet, he understood that eugenics was not based on '*genuine scientific methods*'. In a 1935 paper, he speculated about the possibility of communist transformation in the US.

> *While the romantic numskulls of Nazi Germany are dreaming of restoring the old race of Europe's Dark Forest to its original purity, or rather its original filth, you Americans, after taking a firm grip on your economic machinery and your culture, will apply genuine scientific methods to the problem of eugenics. Within a century, out of your melting pot of races there will come a new breed of men – the first worthy of the name of Man.* Leon Trotsky (Trotsky 1935).

Trotsky abhorred racist eugenics practiced under fascism yet saw eugenics as a potential tool for improving the human stock, physically, mentally and culturally, in a society based on economic justice and equality. Eugenics was acceptable if it was a multi-racial program within a democratic, socialist future based on '*genuine scientific methods*'.

JBS Haldane

JBS Haldane, a significant contributor to genetics, physiology, evolutionary theory and biostatistics, was a passionate public intellectual. Interestingly, his degree studies were in classics and mathematics. Drawn to socialism during WW I, he later joined the Communist Party of Britain and became a frequent feature in *The Daily Worker*. His output of magazine articles, papers and books—on diverse issues in science, society and politics—was prolific. But his critical disposition irked academic and state authorities, earning him official surveillance and loss of academic opportunities. A firm internationalist, he detached himself from his nation of birth in 1956, took up Indian citizenship and spent the rest of his life working in India. It was a response to his revulsion at the British plan to forcibly take over the Suez Canal from Egypt.

His main interests covered quantitative analysis of natural selection, the mechanisms of heredity in plants and applications of plant breeding. He also tackled eugenics. He countered the specter of imminent mental decline of the human race raised by the eugenicists by observing that while there has been stupendous progress in knowledge and standard of living in the past 50,000 years, there was little evidence to show that babies born then had a different level of intelligence than babies born in the 1900s.

In a 1934 essay entitled *Human Biology and Politics*, he tackled the main tenets of eugenics. One, the human race would benefit by a total elimination of rare, deleterious traits. Two, encouraging some social groups to have more children and making other groups have fewer children would improve the human stock. Three, tight racist immigration policies were of general benefit (Haldane 1934). Using numerous examples and genetic principles, he demonstrated that each tenet has major flaws. Setting aside the issues of fairness and justice, he painstakingly demonstrated that even in its own terms, eugenics is an erroneous doctrine.

> *Curiously enough eugenic organizations rarely include a demand for peace in their programs, in spite of the fact that modern war leads to the destruction of the fittest members of both sides engaged in it.* (Haldane 1934).

Referring to Frederick T Trouton, a famed physicist who discovered a law linking boiling points of liquids with vaporization energy, Haldane observed:

> [We] *have no adequate criterion of mental defect. The late Professor Trouton did not learn to read until the age of 12. If he had been an elementary school child, he would have been sent to a special school for defectives. He was so far from being defective that at the age of 17 he discovered the law which bears his name.* (Haldane 1934).

Haldane emphasized the impact of environment, nutrition and upbringing on mental and physical abilities and noted that the rate of population growth was inversely related to the level of economic development. He argued that under the capitalist system, any eugenic program would be a discriminatory program. But he did not rule out eugenic measures for congenital diseases that were fatal. He at first supported what he felt was a socialist program in the USSR. But disagreeing with its blind allegiance to the Stalinist doctrine, he eventually resigned from the British Communist Party. He also critiqued the perverted theories of biology favored by the Soviet authorities.

Lancelot Hogben

Lancelot Hogben, a pioneering zoologist and biostatistician of conservative Methodist parentage, is best known for his highly regarded popularization of mathematics, science and language. He also co-developed an improved test that remained the most widely used test for pregnancy for one and a half decades. His interests took him to universities in the UK, South Africa, Canada and Guyana. Like Haldane, he was a social activist. His pacifism earned him a three-month jail term during WW I. And his abhorrence of racism propelled him to resign from his academic position in South Africa. A dedicated socialist attracted by Marxism at first, he eventually adhered to what he called 'scientific humanism'.

As the first Chair of Social Biology at the London School of Economics, Hogben continued his innovative theoretical and experimental studies in genetics and applied statistics, and became one of most outspoken, scientifically grounded critics of the popular doctrine of eugenics. Like JBS Haldane, he crossed swords with RA Fisher, the statistical luminary who had inherited the eugenic mantle of Francis Galton.

Eugenics holds that human physical and mental characteristics are hereditary. The environment at best plays a subsidiary role. The critics of eugenics held that the environment in which the person grew up was as, if not more, important than genes. Hogben's enduring contribution was to point out that nature and nurture functioned in an interdependent, not an exclusive, manner. The same gene can produce differing effects in different environments. The statistical models for the gene-environment interaction he constructed still form the basis of research on the issue. Pointing out the pitfalls of simplistic use of average values in medicine, Hogben noted the centrality of individual variation and pioneered what is now known as the 'N-of-1 clinical trial'. In an interview for a newly published book, he summed up his worldview:

> *I like Scandinavians, skiing, swimming and socialists who realize it is our business to promote social progress by peaceful methods. I dislike football, economists, eugenicists, Fascists, Stalinists, and Scottish conservatives. I think that sex is necessary, and bankers are not.* (Wikipedia 2021 -- Lancelot Hogben).

In the first half of the 20th century, the vocal proponents of scientific racism and eugenic ideas outnumbered their vocal opponents. Swimming against the dominant current invited ostracism and censorship. Yet, besides Haldane and Hogben, several eminent natural and social scientists, scholars, philosophers, mathematicians and popular writers, mostly of the leftist persuasion, took up the anti-eugenic cause. They included: JD Bernal (crystallographer, historian of science, expert on x-rays, peace activist), Franz Boas (anthropologist, ethnographer, quantitative analyst, pioneer of cultural relativism, anti-racist, anti-fascist activist, internationalist, author of many popular and scientific books, critic of scientists who were spies and informers), Enid Charles (demographer, international expert on population statistics, epidemiologist, WHO consultant, socialist, feminist, and research collaborator of Lancelot Hogben), Benjamin Farrington (classicist, historian of science and Greek civilization, author of pamphlets on socialism and philosophy), Julian S Huxley (socialist, biologist, first director of UNESCO and first president of the British Humanist Association), Herman Levy (mathematician, author of statistics texts and popular works on science), Margaret Mead (anthropologist, anti-racist, critic of IQ tests, president of the American Association for the Advancement of Science), Hermann Muller (Nobel laureate, anti-nuclear war and socialist campaigner, pioneer of mutagenesis, evolutionary and mathematical biology, and president of the American Humanist Association), Joseph Needham (biochemist, Christian socialist, preeminent authority on science, culture and civilization in China and Korea, author of books on science, philosophy, religion and history), George D Thompson (classicist, philosopher, expert on Irish language, author of books on the Chinese revolution, Marxism and poetry, science and art and history of capitalism), and Jack Lindsay, a prolific multi-disciplinary author. The words of Hermann Muller reflected their general stand:

> [Eugenics was] *a hopelessly perverted movement...lending a false appearance of scientific basis to advocates of race and class prejudice, defenders of church and state, fascists and Hitlerites generally.* Hermann Muller (Slorach 2020).

As WW II began, a group of fourteen opponents of scientific racism issued the *Eugenics Manifesto* which addressed the question of improvement of the physical and mental characteristics of humanity. Not totally ruling out the need for it, the Manifesto categorically said that such an improvement would first require a major transformation of social and cultural conditions, equalization of economic and social opportunities, and firm control of racial and national discrimination and antagonism.

[The] *effective genetic improvement of mankind is dependent upon major changes in social conditions, and correlative changes in human attitudes. In the first place, there can be no valid basis for estimating and comparing the intrinsic worth of different individuals, without economic and social conditions which provide approximately equal opportunities for all members of society instead of stratifying them from birth into classes with widely different privileges. The second major hindrance to genetic improvement lies in the economic and political conditions which foster antagonism between different peoples, nations and 'races'.* Eugenics Manifesto (Crew et al. 1939).

The Manifesto emphasized economic security for parents, special support for pregnant women, legalized voluntary birth control, wide dissemination of scientific knowledge, cooperation between scientists and physicians, and democratic decision making. These would be a prerequisite for the adoption of any measures to improve population health. In essence, it was an anti-eugenic manifesto, and negated the eugenic practices that were popular across the Western world including Nazi Germany.

Eugenics in the USSR

Around the year 1900, Russia was a semi-feudal society under tyrannical Tsarist rule. Together with the monarchy, a class of large landowners and the Orthodox Church owned or controlled most of the land. Some four fifths of the population—tenant farmers, independent peasants, artisans and domestic servants—lived in the rural areas. Deprived of the fruits of their labors, their lives were marred by hunger, disease and early death. Urban industrial and clerical jobs occupied some 15% of the working age group. Factory work was hazardous, hours were long, wages, abysmally low, and benefits, nonexistent. The people were also denied basic civil and legal rights. The majority had little access to health services and education.

From the 1860s, a series of reformist and militant movements began to fight for change. A few minimal gains were attained. Each movement was ruthlessly suppressed by the Tsarist police. The astute tactics of the Bolshevik party under VI Lenin changed the balance of forces. After tumultuous two decades of ups and downs, the Bolsheviks hammered the final nail in the Tsarist coffin in 1917.

Inheriting a society in ruins and a bankrupt state apparatus they also faced an armed opposition funded by Western capitalist nations. Tuberculosis, venereal diseases, typhoid fever and cholera were rampant. Thousands of doctors and nurses inducted in the Tsarist army had died. A major challenge was to develop institutions able to provide affordable, effective health services to the broad mass of the people as well as to initiate a disease prevention program.

Initially, the medical workers in Russia were polarized along class lines. The junior health staff including nurses generally supported the Bolshevik demand of free, comprehensive health care for all while the doctors did not. After the 1905 failed revolt against Tsarist rule, many doctors also called for free health services and advocated political reforms and civil liberties. Linking health and social issues, they joined the Bolshevik Party. A small group of Marxist physicians had a lead role in the national medical organizations. Penetration of progressive ideas in medical practice enabled the Soviet government to embark on far reaching changes in the health system.

Among the initial steps taken were to establish commissariats of labor and health and begin construction of clean water supply systems, drainage canals and electrical grids. Industrial and domestic effluence was monitored and controlled. Food and medicinal standards were enacted. National health and social insurance plans were formulated. Employees gained family and maternity benefits. Health resorts for workers were built. Most pharmacies and factories making medical drugs were taken over by the state. Medical and pharmaceutical training schools were expanded. The enrollment in medical schools rose from about 20,000 in 1913 to 76,000 in 1932. Some 90% of the doctors worked for the state. Two thousand hospitals were built between 1928 and 1932. A significant expansion of the preventive and curative health services occurred in the first one-and-a-half decades of Soviet rule. Medical care was free and accessible. The incidence of infectious ailments like tuberculosis, typhoid fever, cholera, venereal diseases and typhus declined dramatically in those years.

Care for the physically and mentally disabled was a key concern in the early Soviet years. Years of war, social strife, hazardous work conditions, extreme economic hardship and loss of loved ones had left major scars in the nation. In the Tsarist era, mental illness among the lower classes was linked to criminality. The mentally infirm, alcoholics, tramps, petty criminals and 'undesirable' persons (political dissidents) were confined to asylums where isolation, neglect and punishment, not rehabilitation, were the goals. The psychiatric profession had to abide by the dictates of the police. Again, it was the Marxist psychiatrists, few that they were, who led the charge against this setup. They held that the social system was diseased. Social stability, improved living standards, and appropriate health care were important for reducing psycho-social problems. The Tsarist system, which could not provide these essentials, had to go. It was not the insane but the dictatorial state that was the real threat to society. Psychiatry had to serve the masses, not just the elite. An increasing number of psychiatrists viewed mental problems in a broader context, joined the revolutionary cause and declared that the suffering and injuries caused in the wars launched by the Tsarist regime were a major contributor to the rise of mental diseases.

Outspoken psychiatrists and neurologists were hounded by the police. Many were locked up. Professional meetings were shut down. But by 1917, the regime was on its knees. In many mental asylums, leftist doctors and the junior staff ejected the administrators and took over the running of the place. More liberal, non-punitive methods of treating the mentally ill were explored. Due to such changes as well as the material hardships and food shortages of those days, the number of inmates in mental asylums in 1923 was about a fourth of what it had been in 1917.

++++

What roles did eugenics play in Tsarist Russia and the USSR? In Russia of the early 1900s, psychiatrists, health experts, physicians and scientists in large numbers fell under the spell of British and American style eugenics. Harboring elitist perspectives, they advocated measures to reduce the 'unfit' and increase the 'fit'. The Tsarist regime had its own draconian ways of controlling the people, and hardly heeded their voices. But it monitored the professional bodies, giving them little room to operate and debate as they wished.

The 1917 revolution lifted these restrictions. Physicians and scientists set up new professional associations and reinvigorated the existing ones. The Russian Eugenics Society, established in 1920, focused on research on heredity. Exchanging ideas and documents with Swedish and American eugenicists, it fostered a vibrant internal debate in scholarly outlets. Unlike in Europe, the UK and the US, Soviet physicians, biologists, anthropologists and sociologists had by then largely accepted socialist ideas and saw the negative eugenics measures like incarceration, sterilization and euthanasia as the products of racist and elitist prejudices.

The dominant view looked at humans as social animals whereby social factors (nurture), not heredity, primarily determined behavior. Culture, environment, education and economic opportunity were the key modifiable factors affecting human health. They touted better education, living and work conditions, and preventive health measures, not sterilization, segregation and denigration, as the ways of improving the quality of life in the nation. Russian jurists favored crime prevention through effective measures to tackle poverty, social inequality and joblessness. Convicted criminals were to be rehabilitated. Scientists, biologists, educationists and health experts with a eugenic stand favored eugenics measures that were quite distinct from those prevailing in the US. Eugenicists in the US targeted the destitute and the downtrodden. But their Soviet counterparts attacked class privilege and sought to uplift the masses and raise the hereditary potential via state funded improvement of health, nutritional and social conditions. That there was state funding for eugenic research and institutions did not mean it was officially endorsed. It reflected the wide latitude for intellectual discourse in the early years of Soviet rule. In the process, racist and anti-Semitic views were also expressed.

Mikhail Volotskoi, a founding member of the Russian Eugenics Society, at first supported negative eugenic measures like forced sterilization. But later he developed an erudite Marxist critique of eugenics and emphasized public health measures and improvement of health services. It was not a question of protecting race hygiene through control of heredity but one of raising social hygiene via egalitarian economic development. The focus should be on the masses, not the elite. This view was encapsulated in the leftist slogan: '*Not from the upper ten thousand, but from the lower millions*'.

Support for positive eugenics existed in the scientific community. Alexander S Serebrovsky, an internationally famed geneticist who also had a senior position in the Bolshevik Party promoted artificial insemination with carefully chosen sperm as a way of reducing hereditary illnesses. But his proposal was never actualized. And no program with forced sterilization, confinement and euthanasia was ever implemented in the USSR. The disabled were accorded state funded assistance. Abortion by consent was legalized and made freely available. Reformed laws set 18 years as the minimum age for marriage and, among other things, required the couple to inform each other in advance if they had a disease that was deemed to be hereditary.

In contrast to the US, the Soviet government abolished the restrictive immigration rules of the Tsarist era. National origin, health status and mental or physical disability were no longer criteria for admission. With an internationalist foreign policy, the Soviet state gave wide support to nationalist and socialist movements in Asia and Africa that were fighting for freedom from Western colonial and racist domination.

But, as Joseph Stalin consolidated his power, the atmosphere of relative freedom in scientific and general discourse came to an abrupt halt. In biology and genetics, the ideas of Trofim Lysenko, based on a dogmatic perversion of science and Marxism, superseded Mendelian genetics and the Darwinian Theory of Evolution. An agronomist and biologist, Lysenko claimed that natural evolution was not governed by random mutation and natural selection but by direct transmission (inheritance) of acquired characteristics from a parent to the offspring. If long necks are beneficial to giraffes, then the tendency to develop long necks is passed on to giraffe babies. It was an echo of the pre-Darwinian views of the eminent French naturalist JB Lamarck. Further, Lysenko proclaimed that probability theory, statistics and controlled trials were of no use in agronomy.

Espousal of strong loyalty to the Communist Party led to his rapid rise in the political establishment. Placed in charge of agricultural projects and biological centers, he was projected by the Party as a genius innovator whose techniques would dramatically improve agricultural productivity. His loud denunciations of the scientists who disagreed with him as anti-Marxist reactionaries won applause from Joseph Stalin. About 3,000 biologists, geneticists and other scientists who opposed his policies were caught up in the Stalinist dragnet. Many were sacked from their jobs; some were imprisoned, and some were executed.

Lysenko claimed that wheat yield would rise if the plants were exposed to cold. But such claims were exaggerations based on manipulated or contrived data. They prevailed for decades only because no scientist with a credible stature was there to critique and expose them. And the effects on agriculture and food production in Russia were nothing short of a majestic failure.

Stalinism emanated from a rigid, ossified interpretation of Marxism. Even a slight deviation from the official line was not tolerated. Scientists had to toe that line as well. Though there was major progress in various scientific fields and technology under Stalin, it was marked by critical gaps filled by pseudo-scientific theories. Genetics suffered heavily as research and education in genetics and related fields were censored. Independent geneticists could only function by disguising their work in other terms. The Russian Eugenic Society, one of the first to get the ax, was banned in 1930.

++++

While a murderous form of eugenics based on Social Darwinist thinking was on the ascendance in fascist Nazi Germany, in Stalinist Russia, it was replaced by Lysenko's anti-Darwinian theory that had a brutal impact on the well-being of the Russian people. Both were pseudo-scientific doctrines serving and supported by authoritarian state power.

3.11 REFLECTIONS

Eugenics aims to improve the genetic quality of the human species through control of reproduction. Though it has a long history, it took the pioneering efforts of Francis Galton to become a formal, formidable movement. Invoking Malthusian ideas on population and Darwinian laws of natural selection, Galton and his disciples conducted observational and clinical studies and applied innovative statistical methods. The work was designed to place their ideas on a credible scientific foundation.

The principal eugenic claim was that the physical, behavioral and mental characteristics are determined by heredity. A person is intelligent, physically strong or has good habits because he/she has the genes that enhance these traits. Likewise, if a person is lazy, mentally deficient or a thief, it is due to possession of 'bad' genes. As genes are passed on from parents to children, such traits run in families.

The eugenicists proclaimed that if human reproduction was allowed to proceed unchecked, 'bad' genes would dominate 'good' genes, and the quality of the human stock would decline precipitously. Human civilization and culture were in grave danger. Pointing to the successes of plant and livestock breeding, they said that the solution was at hand. Only resources and political will were lacking.

Eugenic policies came in two strands. Positive eugenics enjoined people with 'good' genes to procreate at higher rates; negative eugenics focused on ways of controlling breeding among those with 'bad' genes. The latter aimed at corralling people with suspect genes in their own areas or institutions, not allowing them to breed with the holders of 'good' genes, male and female sterilization, and abortion. In its ultimate form, negative eugenics espoused terminating the lives of the 'defectives' by withholding treatment, inducing lethal disease, starvation or placement in gas chambers. It was all done for the good of humanity.

> *What nature does blindly, slowly and ruthlessly,*
> *man may do providently, quickly and kindly.*
> Sir Francis Galton

Internal eugenics applied the control measures to individuals of a given community, race or nation. External eugenics targeted other races, ethnicities, and nations. It included immigration control and violent assaults. Operating under the broad umbrella of scientific racism, it encompassed anti-Semitism, xenophobia and brutal racism. External eugenics rested on the foundation of enslavement of African peoples, genocidal pogroms against native peoples, colonialism and murderous imperial interventions. It divided humanity into two groups, the 'superior' stock emanating from the white Nordic (Aryan) races and the 'inferior' stock from the rest of humanity. Historically many nations have claimed a superior status over all other nations. The American proclamation of Manifest Destiny and American Exceptionalism, the Zionist idea of a Chosen People, or depiction of British colonial ventures as civilizing missions fit into the mold. The Apartheid doctrine in South Africa posited the superiority of the white race and a total separation of the races. It was the epitome of external eugenics.

Despite the presence of prominent scientists, statisticians, physicians, psychiatrists and social scientists in its ranks, eugenics was a pseudo-scientific doctrine resting on studies with major flaws in design, conduct, analysis and interpretation. Evidence demonstrating the harmful effects of eugenics was ignored and no evidence to show the genetic benefits of an eugenic program was presented.

Positing the dominance of nature over nurture, eugenics ignored the key social, cultural and economic factors that affect life outcomes. In humans, social evolution augments and modifies natural evolution. If you are born into a wealthy family, your chance of living a long, healthy life is higher than if you grow up in a family that struggles to put food on the table. In essence, it was a racist, elitist doctrine, a venue for entrenching an aristocracy, It was draped in a scientific garb by automatic extension of biological laws into society, and a distorted perception of the mechanism of heredity. Providing handy scapegoats for the multiplicity of the ills of capitalism, it effectively deflected attention from the misdeeds of the corporate and financial aristocracy.

The Eugenicists

The most astounding feature of 20th century eugenics was the range of eminent personages—scientists, physicians, psychiatrists, mathematicians, statisticians, anthropologists, sociologists, philosophers, writers, journalists, social reformers, jurists, educationists, feminists, advocates for minority rights and politicians—in the Western world who firmly backed it. Uniting conservatives, liberals and socialists under one roof, it linked peace activists with imperialists, and theologians and priests with atheists and humanists. Elite circles in Brazil and Mexico fell under its spell. The otherwise critical thinker somehow overlooked the fact that it was pseudo-scientific, dubious and anti-humane doctrine.

Eugenicists from Francis Galton to Charles Davenport branded charity, welfare programs and public health measures for the poor and infirm a waste of public and private funds. Public health agencies embraced eugenic programs, sometimes to an extreme degree. Thus, the 1934 annual meeting of the American Public Health Association included a laudatory display of the eugenic ideas and programs of Nazi Germany.

The presence of prominent persons associated with progressive causes in the eugenic bandwagon had a profoundly negative, lasting effect on the visions they promoted. Modern day conservative opponents of abortion, birth control and contraception point to the dastardly eugenic views of feminists like Margaret Sanger. They overlook the distinction between voluntary, informed birth control and state enforced birth control. Referring to Nazi eugenics, they link abortion, birth control and contraception to racism and murder. Despite their hypocritical stand on human rights in general, their anti-eugenic words carry weight.

Similarly, the fact that eugenics was embraced by many early stalwarts of the welfare state, public health organizations and by Scandinavian social democratic nations is used by conservatives to support their own agenda. They claim that state funded social programs like nutritional support for poor expectant mothers and babies, public education, health care for the poor and uninsured and legal aid are expedient measures to serve a hidden eugenic, racist agenda. The liberals are called double-talkers out to undermine family values.

Apologies

In this era of apologies, egregious misdeeds of the past have entered the mainstream discourse in the media and the political domain. Statutes and symbols of racism and imperialism are coming under attack from students and other activists. Eugenics has also captured the spotlight. First in Germany and later in the Netherlands, Sweden, Italy, Hungary, Canada and the Vatican, governments issued apologies for their dastardly roles in the Nazi Holocaust. Other institutions followed in their wake. Japan has also issued apologies for its eugenic type of war crimes in China and Korea.

Pioneering statisticians and eugenicists Karl Pearson and RA Fisher were also the first two professors of eugenics at University College London (UCL). The professorship was initially funded by Francis Galton, a pioneering statistician and the father of modern eugenics. Yet, after a sustained campaign by progressive students, staff and anti-racist activists, UCL, like other Western academic institutions, had no choice but to acknowledge its historical linkages with racism and colonialism. An official committee to look into the matter was appointed. In January 2021, UCL formally apologized for the '*fundamental role* [it had played] *in the development, propagation and legitimization of eugenics,*' and declared that:

> [Eugenics was a] *dangerous ideology* [that] *cemented the spurious idea that varieties of human life could be assigned different value. It provided justification for some of the most appalling crimes in human history: genocide, forced euthanasia, colonialism and other forms of mass murder and oppression based on racial and ableist hierarchy.* (Adams 2021).

Buildings, a lecture theater and a research center named after Galton and Fisher were renamed. Yet, there was a lacuna here. It did not acknowledge that a eugenics conference in which white supremacists had participated had been secretly held at UCL as late as in 2017.

Anti-racist activism has prompted similar soul searching in the American academies. Cornell University and Stanford University administrators are coming to terms with the fact that DS Jordan, a past president, had not only been a famed marine biologist and educational reformer but also a rabid, influential eugenicist. After his death, schools, campus buildings and scholarship programs came to bear his name. But in 2010, two California schools named after him changed their names. In 2020, similar changes transpired at Indiana University and Stanford University. The Sierra Club, a premier conservation society in the US, had honored Jordan for leading the organization and being a firm conservationist. It too came to recognize his advocacy of white supremacy and eugenics, and publicly distanced itself from his work.

The anti-eugenic awakening also affected religious denominations. In 2016, the Methodist Church in the US issued a comprehensive report on how its leaders and ministers had participated in propagating eugenics ideas and race betterment programs, sterilization drives and interracial marriage restrictions. It had also provided senior officers in the American Eugenics Association. The 2016 document stated:

> *The United Methodist General Conference formally apologizes for Methodist leaders and Methodist bodies who in the past supported eugenics as sound science and sound theology. We lament the ways eugenics was used to justify the sterilization of persons deemed less worthy. We lament that Methodist support of eugenics policies was used to keep persons of different races from marrying and forming legally recognized families. We are especially grieved that the politics of eugenics led to the extermination of millions of people by the Nazi government and continues today as 'ethnic cleansing' around the world. We urge United Methodist annual conferences to educate their members about eugenics and advocate for ethical uses of science.* (UMC 2016).

Likewise, the Seventh Day Adventist church, which had actively taken part in the Nazi eugenic programs in Germany and Austria, issued an apology for the role it had played in the Holocaust. It expressed regret for the fact that '*our peoples became associated with racial fanaticism destroying the lives and freedom of 6 million Jews and representatives of minorities in all of Europe …*' (Kellner 2005). The Christian denominations that had supported eugenics had ignored the distinctly unchristian views of the intellectual father of the doctrine.

> *I do not believe in the Bible as a divine revelation, & therefore not in Jesus Christ as the son of God.* Charles Darwin (Rejon 2018).

Many institutional and state actors who actively featured in the dastardly eugenic drive have not publicly acknowledged the part their luminaries played in it. This is the case, for example, with the statistical associations. Thus, the web-based biographical accounts of RA Fisher, the venerated statistician, rarely take note of his anti-Semitism, and espousal of eugenics, including that of Nazi Germany.

Eugenicists were reborn after WW II as respected, apolitical geneticists, scientists, social scientists and physicians. Their roles in the eugenic movement were, until very recently, blotted out from history and science textbooks.

Limitations

Apologies for past misdeeds of powerful institutions and leaders in society are always important. But they must be followed up by informational redress and compensation for survivors and descendants who continue to shoulder the burden of those misdeeds. Apart from the billions of dollars paid by Germany to Jewish survivors and the state of Israel, this type of follow up is a rarity.

Another concern is that acknowledging and apologizing for the crimes of yesterday ought not rationalize or justify the crimes of today. It has to be an accounting of all egregious misdeeds and a firm commitment to respect and protect the fundamental human rights of people everywhere.

Take the case, for example, of the sanctions imposed on Iraq after the 1991 Gulf War. It was an unprecedented, comprehensive, military enforced sanctions regime that lasted for more than ten years. Led by the American and British governments, it was a barbaric regime based on absolute lies about weapons of mass destruction that consumed the lives of more than half a million Iraqi children. Whilst US President Bill Clinton issued apologies for the grievous human rights violations of the US military and spy agencies in Guatemala and Korea in the 1950s, he rigidly enforced the sanctions against Iraq and authorized years of periodic bombing campaigns on civilian targets in Iraq, destroying grain silos and rural agricultural and veterinary stations. He destroyed a pharmaceutical plant in Sudan. These were criminal deeds. Most churches, civil rights groups, scientific organizations, academics and human rights bodies in the West feigned ignorance and kept quiet. Acknowledging past misdeeds thus becomes a license to perpetrate new misdeeds.

Another issue is anti-Semitism, a basic feature of eugenic movements. The resurgence of white supremacist nationalism in the US and Europe has been accompanied by a rise in the popularity of racist, xenophobic and anti-Semitic rhetoric. But while eugenic apologies display awareness of that trend, they ignore the crimes of the Israeli state. The United Methodist apology, UMC (2016), lists the cruelties in Bosnia, Cambodia Rwanda and Sudan as instances of modern-day ethnic cleansing, the systematic ethnic cleansing in Palestine is not mentioned. The mainstream view conflates anti-Zionism with anti-Semitism. Valid criticisms of Israel get the anti-Semitic label and are dismissed. An effective code of silence prevails. Apologies for eugenics issued thus far remain biased and incomplete.

Opposition to Eugenics

Just as the supporters for eugenics came from diverse backgrounds and philosophies, so did its detractors. But the opponents were few. Between 1900 and when the horrors of the Nazi death camps became widely known, the supporters of eugenics far outnumbered the opponents.

Opposition to eugenics took varied forms. Some rejected its theoretical foundations (Social Darwinism, Malthusianism and genetic transmission of behavioral and mental traits) and lambasted practical eugenic programs. Some critics just decried specific eugenic programs like euthanasia. Fabian socialists, political liberals, the Catholic Church and some feminists favored voluntary positive eugenics but did not support coerced negative eugenics. In general, the Protestant Churches were less opposed to eugenics than the Catholic Churches.

Advocates of people with physical and mental disabilities and infirmities staunchly castigated the notion that some human lives had a higher moral value than other human lives. Everyone has an equal right to live and the mentally or physically challenged deserve support to enable them to live meaningful, socially beneficial lives.

NAACP, the main African American civil rights organization in the US, embraced positive eugenics and held its own Better Baby contests. But it castigated the racist eugenics.

All races had similar distributions of 'good' and 'bad' genes. Humanity needed an inter-racial program of increasing the prevalence of 'good' genes.

In Britain, the land where modern eugenics was born, several proposals to enact eugenics programs were tabled in the parliament. But they were all defeated due to joint opposition from the trade unions and socialist MPs. A similar anti-eugenic drive by socialists was launched in the Scandinavian countries. Eugenics had a shaky, late takeoff in France due to opposition from the Catholic Church. It was against sterilization and the state policy of encouraging population growth. Once the Nazis occupied the country during WW II, eugenics gathered wider support. Catholic majority Italy, Spain and Portugal did not enact eugenic laws. But it was a mixed story. Many Catholic Churches in Nazi Germany acquiesced to the vile anti-Semitic doctrines and pogroms.

The eugenics creed generated strong opposition from several biologists, scientists, social reformers and social justice activists. Discerning scientists critiqued its evidentiary and methodological foundation. While accepting reproductive intervention to improve the health of successive generations, they rejected coercive, state-driven methods. Franz Boas, the pioneering American anthropologist, was a leading detractor of eugenics and scientific racism. Human cultures cannot be ranked. All races have a rich, unique cultural heritage. He favored interaction between cultures on the basis of mutual respect as a means to enhances the general human culture.

The most comprehensive, grounded and lucid attack on the theory and practice of eugenics was mounted by political activists, natural scientists and social scientists who had adopted the Marxist perspective on social change. Their worldview had no room for the notion that human behavioral traits are fixed by inheritance. They blamed poor health, poverty, criminal conduct, and alcoholism among the working class on the exploitation of workers. Capitalism reduced their chances of getting education, good health care, appropriate nutrition, better living and sanitary conditions. It was not that there were too many people, but there was too much exploitation. The solution to humanity's problems did not lie in eugenic measures but in a just and progressive social and economic order. Being ardent anti-imperialists, the Marxists were anti-racists as well.

Putting the responsibility for poverty on the poor themselves, eugenicists of the conservative mold decried state welfare programs, taxation for the rich and state intervention in the economy. But the Marxists struggled not just for such measures but also for abolishing the capitalist economic order. The rich were rich not because they were inherently better but because they used capitalist institutions to alienate the workers from the fruits of their labor. Marxist scientists were the most effective and persistent critics of eugenics within the scientific community.

Eugenics existed as a theoretical and research field in the early years of Soviet Russia. But unlike in the West, eugenicists were in a minority and their voices were drowned out by the dominant Marxist views and Soviet era public health, workplace and other social reforms. Eugenics was banned in the early 1930s. But what followed was not the socialist society Marx and Engels had envisioned but a dictatorial, state-capitalist order in which patently erroneous biological theories prevailed.

A Complex Knot

The singular lesson of the eugenics era is that an automatic contrasting of religion and science is simplistic and erroneous. Science was used both to shore up and critique eugenics. And it was the same with religion. Eminent scientists, priests and theologians supported eugenics, and other eminent scientists, priests and theologians condemned it.

Neither religion nor science originated eugenics. The roots of this doctrine lay in the configuration of the capitalist system of the late 19th and early 20th centuries. Capitalism and its colonial, imperial facets and the ideological drive to justify and find scapegoats for the problems facing the system formed the basis of popularization of eugenics. Its roots were class exploitation and colonial racism. Religion and science were harnessed to serve that order. Only a few bold scientists and religious figures swam against the current and attacked what they considered was an abuse of science and religion.

EUGENICS

Eugenics fell into disrepute after WW II, but not due to religious or scientific reasons. It was banished from public discourse because Germany and Japan were defeated. Humanity was horror struck by the revelation of the Nazi death camps. Though the socialist system expanded, capitalism and imperialism gained a new lease of life under US hegemony. Brutal external eugenics flowered as US bombs rained down on Korea, Vietnam and Cambodia, military dictatorship prevailed in South and Central America, European colonizers fought vicious battles, as in Algeria and Kenya, to retain the colonies. Portuguese colonial rule and Apartheid South Africa had firm Western support. African governments were overthrown and replaced by friendly dictators as in Congo, Nigeria and Ghana. Capitalism strove to suppress the emergence of independent, vibrant economies—socialist or capitalist—in the Global South and keep them under their thumb. In the process, Malthusianism and Social Darwinism were reborn in newer garbs.

As long as capitalism exists and economic inequalities between and within nations expand, the system will require explanations for why the poor are poor, the rich are rich, and how the national economic pie is sliced, why corporate rule is in the long run for the benefit of all and to justify the global neoliberal order. It will need ideological support for its remedial measures. Scientists, economists, historians, scholars, medical experts and development specialists will get funds to produce scientific justifications for neoliberal measures. The influence of neo-Malthusian development theories exemplify this assertion.

Espousal of eugenics has been banned from public and scientific discourse. Yet, as a philosophy and modified practice, eugenics continues to survive, but now in the context of the science of genetics and the human genome project. The lessons of eugenics of the yesteryear do not receive the attention they deserve. We return to eugenics later. The relationship between eugenics and mathematics is discussed in Chapter 5.

99

CHAPTER 04: SCIENCE

Science tells us how the heavens go.
Religion tells us how to go to heaven.
Galileo Galilei

Religion is a culture of faith;
Science is a culture of doubt.
Richard P Feynman

There may be a conflict between
soft-minded religionists and tough-minded scientists,
but not between science and religion.
Martin Luther King

❖

IN THE EGALITARIAN but insecure hunting-gathering era, knowledge of the reality was intertwined with veneration of divine entities. This began to change as settled societies took root and production expanded. As a small upper class gained social control, religion acquired an elaborate pantheon of gods, rituals and priests, and a new function, to bless the unequal social order. A gradual process that extricated knowledge of reality from religion was set in motion. Today, religion and science are two separate pillars of society. Religion, with its holy texts, is overseen by trained priests and has elegant places of worship where rituals old and new continue to attract large congregations. Science is taught in schools and colleges, has a diversity of disciplines, trained scientists, teachers and professors, professional groups, journals and texts together with laboratories and structures. Daily life is awash in items produced using scientific knowledge.

We began exploring the evolution of the social relationship between science and religion in RPS (2022). Our focus was on Buddhism, Hinduism, Christianity and Islam and some minor faith systems. The dialog between Albert Einstein and Rabindranath Tagore, for example, contrasted faith with science. The previous chapter on eugenics revealed that the roles of religion and science in society are complex, with unexpected, disturbing moral and factual connotations.

Now we expand our purview to the conceptual dimension of the relationship between religion and science. We inquire: As ways of perceiving reality, are religion and science compatible? Do they share a common ground? Is one religion more scientific than another? Does God exist? Do discoveries in science negate the existence of divinities? Does free will exist? And not forgetting the social dimension, we explore the relationship between the science, religion and socioeconomic structures, and particularly, the global neoliberal system.

4.1 WHAT IS SCIENCE?

Our definition of 'religion' appears in the Introduction. For a formal definition of science, we first note that science is both a body of knowledge and a method of acquiring knowledge. It has arisen from a dynamic interplay between theory and practice using inductive and deductive approaches. It develops by questioning, posing alternatives, gathering evidence and rational analysis. Intuition plays a part but ultimately each claim has to be verified repeatedly. For each claim by a scientist, we inquire: Is it reproducible?

Religion is based on faith, belief, scriptures and rituals. It undergoes change but that largely arises from social factors, contesting interests and divergent interpretations of scriptures. Facts are secondary to religion. Yet, real life is more complex. At times, doubt and selective facts feature within religion and faith makes inroads into science.

The present scientific corpus fills hundreds of thousands of encyclopedic volumes and is distributed among the numerous subdivisions of science. Broadly, the subdivisions are classified as basic science or applied science, and natural science or social science. The basic natural sciences include physics, chemistry, biology, Earth science and space science while the applied natural sciences include engineering, agronomy, medicine and materials science. The basic social sciences cover economics, political science, anthropology, sociology and psychology while the applied social sciences include business administration, jurisprudence, public policy, marketing and pedagogy. Mathematics with 'pure' fields like pure mathematics, mathematical statistics and logic, and applied fields such as applied mathematics, applied statistics and computer science is regarded as a formal science.

Many definitions of science exist. Consider two. According to Wikipedia:

Science .... is a systematic enterprise that builds and organizes knowledge in the form of testable explanations and predictions about the Universe. Wikipedia (2021 – Science).

Elsewhere it is stated that:

Science is the pursuit and application of knowledge and understanding of the natural and social world following a systematic methodology based on evidence. SC (2021).

These definitions need elaboration in terms of the diversity of investigative methods and types of information gathered in different scientific disciplines. Astronomy, for example, is not amenable to repeatable experimentation in the way chemistry is, and the information gathered by historians is usually not as amenable to quantitative analysis as that gathered by economists. Some fields emphasize case studies and qualitative analysis while others focus on systematic sampling and quantitative analysis. Some fields do observational studies, others, experimental studies, and some combine both. These definitions are but starting points for understanding the nature of the scientific enterprise.

Study designs and analytic methods vary between natural and social sciences, and between physics, cosmology, archaeology, biology and economics. But the essence— systematic evidence gathering, open critical evaluation, and possible enhancement or rejection on a continuous basis—remains.

In the case of biology and medicine, we conceptualize the method of science as an activity pursued along the following systematic scheme: Formulation of an hypothesis; designing an experimental, observational or records-based study to test it; collecting relevant data using the study protocol; summarizing and analyzing the data with statistical and mathematical tools; interpretation of the data analysis; concluding whether the evidence supports the hypothesis or not; writing the report and submitting it for publication; peer review and possible modification of the report; and, if published, discussion and critical reactions by other scientists. Often, several studies test the same or a similar hypothesis. Some may support it; others may not. As such studies accumulate, a systematic review analyzes the totality of the evidence: Does it support the hypothesis, or are more studies needed before it becomes a part of the scientific pantheon?

Science is generally seen as a value free body of knowledge. But that is an oversimplified view. Science depends on four key ethical tenets: cooperation, respect for the truth, personal integrity and free flow of information. These tenets are often compromised in practice by commercial, personal or political factors.

Scientific praxis deviates from ethical norms and methodological tenets for varied reasons. Errors emanate from a lack of knowledge of or an improper use of the methodological tools. Scientists violate study protocols, take short cuts, sideline quality control measures, alter or smoothen the collected data, perform misleading or selective data analysis, and do not present a balanced picture of their findings. Notably in medicine, public health, psychology, sociology, history, education and genetics, scientific praxis is rife with implicit and explicit biases. Even error correcting measures like peer review of study proposals and submitted papers, post-publication review, and independent replication studies are biased or sub-standard. Commercial, personal and political biases, and fraud plague all areas of science. We shall have more to say on these issues later.

Despite its limitations, science is the sole venue we have for acquiring reliable knowledge of natural and social realities. In this book, as in its prequel, RPS (2022), we rely on the scientific method as applied to social phenomena. Its three pillars are interconnectedness, historical analysis and regard for contradictory tendencies, that is emergence, change and dialectics. Using material from diverse secondary sources, we invoke these tools to discern the conceptual essence of religion and science and extend the analysis in RPS (2022) to further elaborate on the comparative roles of religion and science in society. Special attention is paid to the ramifications of the neoliberal socioeconomic structure prevailing across the globe. We begin with a common query.

4.2 DOES GOD EXIST?

Monotheistic religions generally assert that there is a being (God, Allah) that is the supreme, most perfect being with supernatural powers. He created the Universe and the laws governing it. Polytheistic religions have a constellation of gods, goddesses and divine beings with different powers and roles. Richard Dawkins formulated this assertion as an hypothesis:

The God Hypothesis

There exists a superhuman, supernatural intelligence who deliberately designed and created the Universe and everything in it, including us. (Dawkins 2006).

This hypothesis forms the principal bone of contention between the scientific and religious outlooks. The contention has become more acute in the current era. But it existed, in some fashion, in virtually all cultures where an organized, literate form of religion emerged. Agnostic and atheistic views appeared in Hindu, Buddhist, Islamic and Christian societies from the early days. Apostasy in that era had dire consequences. A total rejection of belief in God or divinities was therefore rare. Instead, criticism of religious tenets was presented in veiled terms. Such restrictions still exist in many parts of the world. At the same time, open debates on this issue have also become more common.

Dawkins regards the God Hypothesis as an hypothesis that can be tested by the methods of science. This section deals with some of the arguments that have been advanced to assert, query or deny the existence of God and divine, supernatural beings. The affirmations and denials have transpired in Hindu, Buddhist, Christian, Muslim, Jewish and other cultures.

The Cosmological Argument

The Cosmological Argument for the existence of God proceeds from our empirical and sensual perceptions of the existence of objects and a state of order in the Universe. Objects move and undergo change. Each change has a cause. Starting from the present state, this argument goes back in time through a chain of causality to arrive at an ultimate cause.

Aristotle was an early proponent of this argument. As propounded by Muslim theologians Al-Kindi and Al-Ghazali, it was argued:

103

> *Every being which begins has a cause for its beginning; now the world is a being which begins; therefore, it possesses a cause for its beginning.* Wikipedia (2021 -- *Kalam* Cosmological Argument).

Thomas Aquinas, an influential 13th century Catholic theologian, formulated an elaborate version of the cosmological argument in *Summa Theologica*. Called Five Ways, it has five components:

The Unmoved Mover: Objects move. An object does not move itself. A moving object was set in motion by some mover. What moves the mover? Another mover. And so on. Yet, an infinite chain of movers is not possible. Hence an unmoved first mover started the process.

The First Cause: Every emergent entity has a cause. An entity cannot cause itself, and an infinite sequence of past causes is impossible. Hence, an uncaused first cause ultimately caused all to be.

Contingency: No thing *must* exist. Everything is contingent. Hence there was a time when nothing but the essential being existed.

Goodness: Goodness is judged in relation to perfect goodness. Thus, a maximal goodness that engenders all goodness exists.

Order and Design: Orderly design arises from intelligent beings. Nature abounds with order and design. Hence a designer that created order in nature and propels it towards a higher order exists.

God is the unmoved mover, the first cause, the necessary being, the perfectly good being and the prime designer of order in the Universe.

The Five Ways arguments spawned varied, extended arguments to demonstrate the existence of God. Some are cosmological arguments, others tackle the issue from the vantage of morality, and yet others invoke the findings of science.

Moral arguments for the existence of God derive from the perception that religion is the sole foundation of morality and ethics. It is claimed that religious codes of morality emanated from a supremely moral being, God. The fourth way of Aquinas is an instance of that argument. *The Stanford Encyclopedia of Philosophy* depicts the basic modern moral argument for the existence of God as:

> There are objective moral facts.
> God provides the best explanation of the existence of objective moral facts.
> Therefore, (probably) God exists. (Evans 2018).

Another argument proceeds from the notion of human dignity:

> *Human persons have a special intrinsic value that we call dignity.*
> *The only (or the best) explanation of the fact that humans possess dignity is*
> *that they are created by a supremely good God in His own image.*
> *Probably there is a supremely good God.* (Evans 2018).

Couched in terms of probability, these arguments do not assert or claim to prove the existence of God. Instead, they claim that the existence of God provides the most reasonable explanation for the particular state of affairs.

We now present a series of points and counterpoints for other arguments that have been proposed to assert the existence of a supreme divine being.

Universality of Belief

Point: Religion and belief in a supreme being have existed in virtually all societies in history. That level of pervasiveness could not have been due to chance. It is a sign that God exists.

Counterpoint: Belief in supernatural powers arose from evolutionary and sociological processes. People in the past felt that mountains represented gods. Such beliefs no longer have much currency. Beliefs in divine powers, however, persist because they serve psychic needs and social purposes.

Belief, however popular or apparently self-evident, does not imply fact. The majority in the past and millions to this day believe that the Earth is flat. Of recent, their ranks have been swelling in Britain, USA and Brazil. The Flat Earth Conference is an annual event. Seven million people including Ernesto Araujo, once the Foreign Minister, in Brazil accept that faulty idea. Flat Earth believers dismiss the photos of Earth taken from space and other evidence showing the spherical shaped of the Earth as doctored evidence, a product of governmental conspiracy. To some, the Earth is a flat, circular disk, Yet they do not mention its diameter.

Diversity of Belief

Counterpoint: Over 4,000 religions have been recorded in the world. Each religion has its own deity or deities, creation tale and moral code. They have varied rites and rituals, ideas about the fate of people who violate religious strictures, and possibly belief in rebirth. Often, religions have conflicted with one another, and some religions have declared their superiority over the rest. That is hardly evidence for the existence of a single supreme being. Or does it point to the presence of competing gods establishing their own religions?

Point: Humans cannot fathom the mind of God. There is no reason why He should not manifest Himself in different ways in different places. However He does it, it behooves upon people to worship him and follow the guidance they have been blessed with and not cast aspersions on the faith of others.

Counterpoint: Since the 1870s, over three hundred new religions or faith-based movements have sprouted into being. One of them is the Unification Church founded in 1954 in Korea by Sun Myung Moon. It is a Christian sect that proclaims Reverend Moon as the Messiah tasked by Jesus Christ to establish a pure Kingdom of God on Earth. Apart from the Bible, it follows the teachings in the book *Divine Principle* co-authored by Moon. It has rites and rituals that differ from mainline Christian denominations. It also has a significant presence in the business, political, media and cultural arenas, and espouses a right wing anti-socialist agenda. Accused of being a cult, its teachings have been called blasphemous by other Christians.

Is the Unification Church an indicator of the existence of God? Or is it a cult with a questionable agenda? Where is the line dividing a cult and a religion? If one religion is based on false beliefs, is it not possible that all are?

Messages from God

Point: At crucial junctures in history, humanity was blessed by the presence of morally unblemished, wise personages, divine revelations and holy texts. They proclaimed virtually identical basic moral tenets like the Golden Rule that remain indispensable guides for human conduct. That can only be if they originated from a compassionate supreme being seeking to drive humanity onto the right path.

Counterpoint: Codes of conduct have an evolutionary and sociological basis. Cooperation and compassion boosted the ability of early societies to survive in a harsh environment. That societies formed similar moral codes is understandable. As societies became stratified, the codes of morality were modified under the imprimatur of organized religions. Explanation for moral codes requires detailed sociological and historical investigation. Invoking a divine origin is an escape from this intellectually challenging endeavor. For example, suppose the universal edict, Thou Shall Not Kill, is discarded. Which social system can survive if random killing is permitted?

Miracles

Point: The myriad of extraordinary events (miracle, premonition, improbable coincidence, paranormal ability, spontaneous healing, extrasensory power) experienced by many across time and place are not illusions or fabrications. In some cases, several people have simultaneously experienced a miracle. Doctors, psychologists and scientists have attested to some miracles. Yet, no plausible scientific explanation has come forth. Such extraordinary happenings indicate a reality beyond the normal, beyond science, a sign of the presence of a supreme divine being.

Counterpoint: A variety of psychiatric and medical conditions can give rise to unusual or abnormal visions and experiences. Even though they may be illusory, for the patient, they are concrete and real. Mental health specialists agree that people without a mental disorder have such experiences as well.

A disjuncture between perception and reality is caused by several factors. Sleep deprivation or low blood glucose can induce an uncoupled trance like state. Mental perception is also distorted by varied cognitive biases, suggestions from respected sources and abnormal fears. Confirmation bias is common among well-educated persons. People see what they expect to see, whatever the fact. An occurrence contradicting our belief is rationalized as a rare chance aberration. Mental short cuts and superficial first impressions sway our perception. People fail to see a major change right in front of their eyes and at times, a minor change is elevated into a disturbing event. Inaccurate or imaginary memories shape how we view the present. In ways more than one, our minds can create a vast chasm between what we feel or see and what exists and explain it away. An inexplicable healing episode may arise from the pervasive placebo effect, the natural ability of the human body to combat disease, a short-term effect, or a deceptive claim.

The operation of psychological and physiological effects has been found in thousands of observational and experimental studies. Before ascribing an unusual event to a divine cause, we need to rule out the play of such effects. No credible scientific investigation has verified a paranormal event or miracle. Mother Teresa was beatified because she was certified by the Catholic Church to have performed miracles. Investigations by independent persons, however, found no credible evidence for her miracles (RPS 2022).

Mystical Experience

Point: People have a mystical or spiritual experience. It is said to engender a humble, composed, kind personality and detachment from material vices. To the affected, it gives meaning and purpose to life. This is a sign of being touched by God.

> *In mystic states we both become one with the Absolute and we become aware of our oneness.* William James (Wikipedia 2021 -- Scholarly Approaches to Mysticism).

Counterpoint: Mystical experiences occur in Hindu, Buddhist, Christian, Muslim and other traditions, among sages and lay people. This is not in doubt. It is not a pathological or dream state. It is experienced by people who are awake and functional. A mystical event may be preceded by prolonged meditation or contemplation, living a spiritual, ascetic life, or it may occur in the absence of such conditions.

The experience may inculcate a transient disconnect with space and time, euphoria or ecstasy, and feelings of sublime tranquility. For a religious person, it is a union with the divine. Yet, a secular meditator or a Buddhist who does not believe in a supernatural creator can also have a mystical experience.

Mystical feelings span a spectrum. For some it is the awe induced by a breathtaking natural phenomenon like a starry sky, a majestic canyon, a flock of ten thousand flamingos at a grand lake; for some it is the joy of a new discovery; for some, it is musical elation, and for some, a sublime state of consciousness. Hallucinogens and epilepsy may cause an out-of-the-world state of mind. A mystic may feel a divine presence, a deep sense of unity with nature, or a state of pure emptiness and nothingness.

There is a broad consensus that mystics tend to be humble, self-confident, compassionate and altruistic. They also are less attached to material possessions.

A mystical event is a subjective event. Psychologists and neurologists note that empirically verifiable evidence is difficult to gather. Studies of brain function and structure for mystical experience have not produced solid conclusions. How does one distinguish between a false claim from a true one? How can a thing that is beyond description be tested? Local culture affects the nature and conveyance mystical events. Despite the challenges, scientific studies have generally ruled out supernatural causes. To say that a mystical experience is an indicator of the existence of God has no sound factual basis. It is more like an in-built feature of the human brain that developed via evolution and is manifested occasionally.

Complexity and Order

Point: Complexity and order come from purposeful design. A watch implies a watchmaker. Human organs like the heart and the lung are much more complex than a watch. The complexity, order and regularity of natural phenomena imply the existence of an intelligent designer. This proposition is known as the doctrine of **Intelligent Design**.

Counterpoint: Living entities—microbes, plants and animals—with their complex organs emerged through a gradual, long-term process of natural selection. It took millions of years of random mutations and accumulation of beneficial features for a complex organ like the heart to assume its current form. The amazing diversity of life forms—from mango trees and ants to elephants and whales—is a product of a non-goal driven process of evolution. The fossil record supplemented by genetic studies support what is now one of most established propositions in science. Scientists differ on some aspects of the evolutionary process but there is no scientific basis to the Biblical claim that God created everything in six days, and humans were created on the sixth day.

Religions vary widely in terms of acceptance of the Theory of Evolution. A year 2009 survey in the US found that about half the population affirm that the Darwinian theory provides the best explanation for the origin of life on Earth. But variability by religion was vast. The affirmation rate was around 80% for Buddhists and Hindus, 60% for Catholics, 50% for mainline Protestants, 45% for Muslims but only about 25% for evangelical Christians and 20% for Mormons. Jehovah's Witnesses stood last with less than 10% affirmation rate.

Buddhists are most deferential towards the Theory. Opinions about it are not fixed among Christians. Most fundamentalist evangelicals strictly subscribe to the doctrine of Intelligent Design and the depiction of creation as exactly given in the Bible. The Catholic Church had a similar stand until recently. In 2014, Pope Francis faulted the doctrine of Intelligent Design and stated that evolution and creation are complimentary processes driven by laws created by God. Religions that doubt the Theory of Evolution have their own creation stories. And they are quite varied. Different parts of the scriptures of the same religion may have a different creation story. It begs the question: Which is correct, the creation tale of a Bantu religion, a Native American religion or Christianity?

Life on Earth

Point: Life on Earth is possible due to a timely combination of many factors and events with specific characteristics that are just within the correct limits. These include distance from the sun, atmospheric pressure, presence and mix of atoms and molecules on Earth and in the atmosphere, amount of water and the land mass needed to sustain life. The fundamental laws governing the Universe also support the existence of life on Earth. Such a manifold amalgam of conditions, fine-tuning, cannot be fortuitous.

> *The incredible fine-tuning of the Universe presents the most powerful argument for the existence of an immanent creative entity we may well call God. Lacking convincing scientific evidence to the contrary, such a power may be necessary to force all the parameters we need for our existence—cosmological, physical, chemical, biological and cognitive—to be what they are.* (Aczel 2014).

Counterpoint: Astronomers estimate that billions of planets circling other stars have the basic conditions—water, elements and molecules, and a non-toxic atmosphere—required for some form of life to exist. Many have been identified. Life can be carbon based, as on Earth, but it could also be non-carbon based. It can be extremely diverse and survive under very harsh conditions. Life is found in very hot deep-sea vents as well as on very cold mountains.

According to one estimate, there are 40 to 80 billion rocky planets with Earth like temperatures and surface water in our galaxy. A conservative reckoning posits that 40,000 may be life sustaining. But the number with living organisms may be smaller and the number with intelligent life, much smaller, perhaps fewer than 10. Yet, taking the billions of galaxies into account, the number of planets in the Universe that possibly have life becomes appreciably large (Siegel 2017).

Consider two scenarios. A thorough astronomical investigation with advanced instruments finds that among the millions of planets with conditions favorable to life either none has life, or one or more have life. Does the first outcome signify the existence of God? Does the second negate the existence of God? Can we rule out the possibility of life emerging in the future? If God created life on Earth, can He not create it elsewhere? Is the fine-tuning for life the only possible fine-tuning? Can God not create another set of laws of nature and structure of matter for life to exist? Clearly, the argument from fine-tuning and special combination of life sustaining features on Earth raises more questions than it answers.

The Big Bang

The Big Bang theory states that the Universe came into being 13.8 billion years ago when an extremely hot and dense point mass began to expand. It lays out the stages of this process in detail and shows how galaxies, stars and planets were formed. Supported by extensive evidence and basic cosmological considerations, it is now regarded as the standard model for the origin and state of the Universe and has displaced other theories like the steady-state model of the Universe.

That the Universe has a beginning has for long formed the basis for arguing that God exists. Greek and Christian philosophers, including Thomas Aquinas, utilized it to frame their views. Refined versions of the argument were presented by Muslim theologians, notably Al-Ghazli. GW Leibniz, the co-inventor of calculus, also argued along similar lines. With the acceptance of the Big Bang theory as the model for the origin and state of the Universe, these cosmological arguments gained a new lease of life. The currently most cogent argument in this tradition has come from the Christian theologian William L Craig. Inspired by Muslims scholars, he calls it the Kalam Cosmological Argument (KCA).

Point: According to the KCA, the Big Bang theory supports the view that God created the Universe at some point in time. A basic version of the KCA is:

Premise: Whatever begins to exist has a cause.
Observation: The Universe began to exist.
Conclusion: Therefore, the Universe has a cause.
(Wikipedia 2021 – Kalam Cosmological Argument).

Accepting that the Universe has a cause, Craig proceeds to argue for the existence of *'an uncaused, personal Creator of the Universe exists, who .... is beginning-less, changeless, immaterial, timeless, space-less and infinitely powerful'*. (see Craig (1999) and Craig (2000) for the details).

For Craig, the Big Bang signifies the beginning of the material Universe and space-time. A space-time-based version of the KCA is:

Premise 1: Anything that exists in space-time has a cause of its existence.
Premise 2: Nothing in space-time can be the cause of its own existence.
Observation: The Universe exists in space-time.
Conclusion: Therefore, the Universe has a cause of its existence outside space-time. (Adapted from Romero and Perez 2012).

Counterpoint: How the Universe began and reached its current state are complex issues. Questions about the Big Bang theory persist. One concerns the origin of the mass-energy of the Universe. The total amount of mass-energy in the Universe is estimated at 4.4×10^{55} grams. This is equivalent to the mass of about

$$73 \times 10^{26} \ (7,300,000,000,000,000,000,000,000,000)$$

Earth-size planets. What was the origin of this prodigious mass-energy? Was it created from nothing when the Big Bang occurred, or was it already there in the form of an incredibly dense, maximal temperature point mass?

The law of conservation of mass-energy, one of the most established laws of physics, says that the total mass-energy of a closed system is constant. In other words, the total mass-energy in the Universe cannot be increased or reduced. This law posits the existence of a dense, high energy point mass at the start of the Big Bang. Yet, other physicists contend that this law may not have operated at the Big Bang. It is an unresolved issue involving quantum fluctuations, dark matter, dark energy, the primal particle known as the Higgs Boson, and the possibility of multiple universes. The KCA posits the Big Bang as creation of something out of nothing. But that supposition oversimplifies current science.

Hinduism, Buddhism and Jainism hold that the Universe follows a cyclical pattern of creation and destruction. Hindu scriptural texts denote one cycle as a day in the life of Brahman. Roger Penrose, a Nobel Prize winning physicist, is a proponent of cyclical cosmology. An endless series of big bangs followed by big crunches does not accord with the unitary process of creation and ending of the Universe found in monotheistic religions, Judaism, Christianity and Islam. It runs contrary to the KCA.

Many Christian denominations, notably the Catholic Church, accept the Big Bang theory, holding that it confirms the Bible. Speaking to a scientific audience, Pope Francis declared that the prevailing scientific theory about the expansion of the Universe is not with inconsistent the existence of God. But fundamentalist Christian denominations who adhere strictly to the creation story of the Bible reject the theory of the Big Bang.

Muslim scholars—Shia, Sunni, Ahmadiyya and Sufi—go the furthest in supporting the Big Bang theory. They even claim that the singularity of the Big Bang and the notion of an expanding Universe appear in the Quran. The Bahai faith followers claim that the Universe is timeless, without a beginning or an end. It has internal changes only.

The diversity of religious stories on the beginning of the Universe poses a conundrum: Which is right? Can all of them be consistent with the KCA?

The logicians and scientists who have evaluated the KCA generally do not fault its logic. But they query its premises and the interpretation of its conclusion. Can cause and effect be conceptualized outside of space-time? What does it mean to say that A caused B in a condition devoid of space and time? Another concern is that the KCA exempts God from its logic. It does not ask who or what caused God. Is that a tenable proposition?

> *Every premise of the Kalam Cosmological Argument requires a suspension of belief and accepting the physical world as it appears as representative of universal processes when in reality, every premise can absolutely be doubted on reasonable means. The Kalam Cosmological Argument, which attempts to prove that there certainly is a God that spurred the Universe into existence, is logically indeterminate at every step; the conclusion that a beginning-less Creator exists cannot be accepted as true with these conditions.* (Vaughn 2021).

4.3 DETERMINISM

Free will and determinism, two important ideas of philosophic, scientific, and theological nature, are central to the relation between science and religion. They were introduced in Chapter 2. The key issue is: Is the future determined by the past? Several basic answers to this question exist.

Scientific (Causal or Hard) Determinism: The future is wholly determined (caused) by the past. Simon Laplace, the 19th century mathematician, physicist and philosopher was a major proponent of this position. Inspired by Newton's laws of motion, he postulated that the position and state of all elementary particles in the Universe at any point in time together with the complete set of laws governing their behavior suffice to determine their configuration with full precision at any future time point. He also assumed that all events have prior cause(s). Nothing happens just like that. Given the state and position of each particle in the Universe, the basic laws of nature fully determine the future. Starting from the Big Bang, the entire history of the Universe, including the emergence of life on Earth, human beings and society are fully determined. There is only one course of development that the Universe can follow. No room for uncertainty exists.

Applied to humans, hard determinism implies that human behavior and features are fully determined by basic causal forces and factors operant in the physical and biological domains (nature). Another version includes factors in the familial, environmental and social domains (nurture). Human acts in any setting are determined by the constellation of these forces. If the setting is precisely specified and the causal forces fully known, we can predict people's reactions (behavior) exactly. Humans choices are not free or autonomous. They are fully determined by circumstances beyond their control. There is no true free will, no freedom. Even one's beliefs and acts are predetermined.

> *Determinism ... affirms the inevitability of the actual.* (Narain 2014).

One objection to scientific determinism is that even in a simple system with few basic parts, the potential number of interactions between them is so large that it would need a computer with more power than exists in the Universe to model the behavior of the system. Deterministic prediction is not practical even in simple cases, let alone for a complex system like the human brain. Determinists counter that the inability to actualize it does not negate determinism. That science has failed to replicate species formation in the laboratory does not mean that it does not occur in nature.

Fatalism is a form of hard determinism applied to humans: What happened was destined to happen. Our decisions cannot influence the outcome. Why not sit back and just let the river of life flow? Fatalism is not uncommon among people enduring multiple, chronic problems. Losing hope, they say: It is the will of God. Or, you cannot change the system. In practice, hardly anyone is a complete fatalist. Even the hardcore determinists do not just sit back.

Compatibilism holds that prior conditions affect the future but not fully. Events, physical or mental, occur due to a combination of material causes and chance. In the human domain, it implies that free will and determination are not mutually exclusive. Free will exists amidst strong biologic, genetic, psychological and social forces. But the degree of freedom depends on the specific circumstances. Broad generalizations—the laws of physics or my genes made me do it—cannot, for example, cast a meaningful light on why one man married while his twin brother remained single. Compatibilism occurs in two forms, weak and strong.

Weak Compatibilism: Humans can make real choices, but choices are largely constrained by external conditions arising from prior causes and chance events. Free will exists but in a muted form.

Strong Compatibilism: Humans make real choices. External constraints can often be overcome. Free will makes a real difference if conscious effort to control external factors is expended. Prior causality plays a minor role.

Libertarianism (Complete Indeterminism) holds that human acts arise from inner contemplation and are not subject to prior causes. Free will reigns supreme. Humans can and do make real choices based on their feelings and desires even in most restrictive conditions. And as autonomous authors of their acts, they are fully responsible for them. Hardly any scientist takes the strict libertarian stand. Strong determinism and libertarianism are theoretical constructs. In relation to human behavior; they are not seen in practice. The basic choice in real life is between weak and strong compatibilism.

++++

Probabilistic Causation: Physical and mental phenomena are generally tinged with uncertainty and variability. Nothing repeats exactly, though the deviation from the norm may be small or large. Pascal's strict deterministic formulation does not adequately represent reality.

An alternative formulation represents natural and social phenomena in stochastic (probabilistic) terms. Using probability theory to model the individual units, this method often gives a reasonably accurate model for the behavior of large collections of units. By operating at the systems level, it also effectively counters the computational conundrum afflicting strict determinism.

One case is the statistical models of gases. The exact motion of billions of gas particles in a closed box cannot be computed fully. But it is possible to apply the theory of probability to determine the behavior of the system as a whole. The formula relating the temperature, pressure and volume of the gas in a closed container derives from holistic probabilistic analysis.

Take the case of smoking. Cigarette smoking is the most important risk factor for lung cancer. Epidemiologists estimate that elimination of cigarette smoking can reduce the incidence of lung cancer by over 80%. Yet, some non-smokers also get lung cancer and some life-time smokers do not get it. That non-smokers develop lung cancer is attributed to factors like age, air pollution, radon, asbestos, tuberculosis, family history, secondhand tobacco smoke, industrial and organic chemicals, radiation therapy and beta carotene supplements. Taking the prevalence of these risk factors into account, we can predict with reasonable accuracy the proportion who will get lung cancer in a large population. A probability model computes the chance that a smoker with a given risk profile will develop lung cancer. Yet, exact prediction is not possible.

Suppose health researchers have identified all the risk factors for lung cancer, quantified their main and interactive effects and uncovered the exact mechanism of carcinogenesis. Will it then be possible to predict who will get lung cancer and who will not? Proponents of causal determinism reply in the affirmative.

Simon Laplace, the father of scientific determinism, was also a founder of probability theory. Modern adherents of his philosophy value probability models but hold that their deployment arises from incomplete knowledge. It does not reflect the intrinsic behavior of a system but is an approximate substitute whose utility will diminish as science progresses.

The emergence of **quantum mechanics** at the start of the 20th century posed a fundamental challenge to this viewpoint. Quantum mechanics examines the properties of light and basic particles of matter at the microscopic level. It has three basic postulates: Nature at its core is a discrete entity. Energy travels in packets. Two, light can assume both particulate and wave forms. Three, matter also behaves as a wave. While the interpretation of the laws and findings of quantum mechanics are subject to disputation, no dispute about the utility of its theories and results exists. Quantum mechanics forms the foundation for modern electronic technology.

The Heisenberg Uncertainty Principle, a basic quantum physics tenet, states that the position and momentum of a particle cannot simultaneously be found with certainty. Exact determination of the position of a particle will alter its momentum while exact determination of the momentum will alter its position. Hence, apart from positing an insurmountable limit on knowledge, it stipulates that nature at its basic level behaves in a probabilistic manner. The path or position of an electron orbiting the nucleus of an atom cannot be mapped with exactitude. Only a probability distribution for its cloud-like spatial trajectory can be given. Uncertainty and randomness are built into the nature of phenomena, a proposition contrary to Newtonian mechanics. Disputes over whether nature is intrinsically probabilistic have held the center stage in the philosophy of science since the 1920s. A famous series of exchanges between Albert Einstein and Niels Bohr over the issue is encapsulated as follows:

God does not play with dice. Albert Einstein
Stop telling God what to do with his dice. Niels Bohr

The counter intuitive quantum findings impelled eminent scientists to dive into Hindu and Buddhist religious texts for a conceptual resolution. Theorists like Erwin Schrodinger and Niels Bohr felt that these texts provided a philosophy integrating determinism with uncertainty (free will). Among other things, they delved into the *Upanishads*, a set of Hindu holy texts written over 2,500 years ago. These texts expound on the nature of the relationship between the supreme reality (Brahman) and the human soul (atman) and describe how spiritual reflection, meditation and following one's *dharma* lead to *karmic* transformation and salvation. It is proclaimed that there '*is only one universal self, and we are all one with it*'. (Kulkarni 2020).

To Schrodinger and Bohr, this principle provided the clue to matters like the commonality and synchronicity of reality for multiple observers even when their observations altered reality in dissimilar ways. For underneath all phenomena lay a unified consciousness, a universal mind. Discrete observers are an illusion (*Maya*). Physical reality arises from consciousness. Perception determines existence. It was an idealist interpretation of quantum mechanics which squarely challenged the materialist view that had dominated the world of science since the advent of Newtonian physics. Instead of holding that existence is independent of perception, it was claimed that perception changes reality. It does not just reflect reality.

Myriads of suns, surrounded by possibly inhabited planets, multiplicity of galaxies, each one with its myriads of suns... According to me, all these things are Maya. Erwin Schrodinger (Kulkarni 2020).

Irked by the metaphysical denials of the existence of an objective reality, Einstein affirmed it in plain terms. In a letter to a colleague, he wrote:

Do you believe the Moon exists only when I look at it? Albert Einstein (Kulkarni 2020).

The scientists who favored the idealistic interpretations of quantum phenomena did not abandon the scientific method, logical precision and experimental verification to embrace a religious philosophy. Texts like the *Upanishads* were a spiritual guide in a world abounding with moral ambiguities. They filled a void created by alienation from Christianity. These scientists were not primed to worship Brahman or other gods. Spirituality was a psychic anchor in the confounding microscopic terrain in which not just the basic tenets of science but even common sense were falling apart. Everything appeared subjective.

> *Change the way you look at things and the things you look at change.* (Stevenson 2019).

Eminent physicist Carlo Rovelli faults this (philosophically) idealist vision by noting that the 'observer' is conflated with a conscious human observer present at the site. And objective reality is seen as a malleable, subjective entity. But the 'observer' can as well be an inanimate object.

> *In quantum physics parlance an 'observer' can be a detector, a screen, or even a stone. Anything that is affected by a process. It does not need to be conscious, or human, or living, or anything of the sort...*(Rovelli 2022).

Interjections of scholars and theologians from various faith traditions into the interpretation of quantum mechanics has added a further twist to our view of the nature of nature.

4.4 FREE WILL

The issue of determinism versus free will relates to the role of nature and nurture in human behavior. Does nature dominate nurture in the causation of physical human traits and human behavior or can nurture partly or mostly override nature? Can humans choose and behave in ways not constrained by nature, or nurture, or both? These questions have profound implications for all major areas of life such as law and justice administration, education, economy, politics and religion. And they are fundamental to assignation of personal responsibility.

With a banana and an orange at hand, you can eat any, both, or neither. You decide to only have the orange. To you, it is your own free preference; no outside influence or coercion was involved.

You walk on a busy street. A man drops his wallet on the pavement but does not realize it. You pick it up, discreetly. It has a wad of money. You can run after him or keep the cash. No one has seen you. The choice is your own. In this case, it is also a moral issue, a choice between right and wrong.

We value autonomy. Disliking impositions, we seek to be masters of our own destiny and trust our capacity to act independently in diverse situations. Even if an external force leaves us no option but to make a specific choice, we still hold that had the force not been present, we could have made a free and appropriate choice.

Free will denotes the human capacity to deliberate and make choices about life actions independent of natural, social or supernatural restraints. It also implies bearing the responsibility for the consequences, good or bad, of one's choice. Mentally incapacitated persons and children are usually not held responsible for their actions the way normal adults are. You are elated when your choice leads to your goals but blame yourself when it does not. Free will is associated with the sense of personal freedom, dignity and responsibility.

Free will entails initiative and self-control. If we succumb to repetitive or addictive behaviors that are socially disparaged—alcoholism, craving, drug abuse, gambling, overeating, procrastination—we feel guilty. The extent to which we can exercise self-control and overcome such habits makes us proud. Free will raises self-esteem and sense of self-worth.

Free will presupposes an entity that wills, a conscious 'me'. What is this 'me'? Is it the mind, brain, soul or the entirety of being? What is consciousness? These queries have perplexed theologians, scientists, legal scholars, philosophers, sociologists and psychologists. And they bear strongly upon the relationship between religion and science.

Hard causal determinism postulates that every event has cause(s), and that the past fully determines the future. In the human domain, it is the obverse of free will, implying that human actions are determined by external forces and past events. You think you have the freedom to choose, but actually you are propelled towards a specific choice. Whether you are aware of it or not, and whether you like it or not, a myriad of forces—biological and genetic, social norms, peer pressure, and/or a divine power—propel you in a given direction. Recent controversies about determinism and free will pertain to two areas of science, genetics and neurology. These issues were introduced in Section 2.7. We now take them further.

4.5 GENETICS AND FREE WILL

Genetic Determinism is a specific form of scientific determinism. It posits that for humans, not just elemental physical and biological attributes like height, hair color and weight, health, but also mental characteristics and complex behaviors have strong genetic bases. We saw in Chapter 3 that studies done by prominent scientists and statisticians in the eugenics era concluded that complex behaviors and traits like intelligence, criminality, 'imbecility' and diligence are heritable, and pass on from generation to generation by 'germ plasm'. Nature dominates nurture. Humans behave in a determined fashion. Later it was seen that virtually all these studies had fundamental flaws.

For example, a lauded study of a single family that supposedly found that criminal tendencies run in families was endorsed by the influential Eugenics Record Office in the US. Subsequently it was revealed that not only were family members misidentified but also that the data on criminality were fraudulent. Yet, the findings of this and other studies were used to justify the eugenic sterilization programs. Even though it was backed by prominent scientists, eugenics was a pseudo-science based on despicable ethical standards.

The term eugenics fell into disrepute after WW II. Yet, its essence lived on. Eugenicists were reborn as geneticists. For the initial two decades, when social democratic parties held in office in the West, nurture secured greater weight over nature in scientific literature and public policy. With the discovery of the DNA and advances in theoretical and applied genetics, theories espousing the primacy of hereditary (genetic) factors again rose among biologists, physicians and policy makers. With the sequencing of the entire human genome in 2000, the rhetoric reached a feverish pitch.

It is postulated that a single gene or many genes operating in unison not only affect physical traits like height, weight, eye, hair color and sensitivity to toxins but also our life span, personality, sexuality, impulsive tendencies, risk of varied mental and physical illnesses, intelligence, educational attainment, mathematical ability, religious choice, economic status, political preference, cultural inclinations, circle of friends and higher cognitive functions. Our ability to plan, act and adapt under changing conditions has a genetic basis. Genes affect the brain circuitry of babies in the womb that later raise the life-time risks of autism spectrum disorder, bipolar disorder, attention deficit hyperactivity disorder, major depressive disorder and schizophrenia. Previous observations that family history raises the risk of cancer, heart disease and diabetes are being validated by gene studies. Political and religious preference and formation of friendship circles are said to be affected by genes. The structure and circuitry of the brain, genetically laid down in the womb, are hypothesized to affect adult behavior. Three forms of genetic determinism exist.

Strong (Hard) Genetic Determinism: Genes almost solely determine the physical, mental and behavioral characteristics of humans.

Moderate Genetic Determinism (Compatibilism): Environmental factors, upbringing and genes jointly determine the physical, mental and behavioral characteristics of humans. For some characteristics, genes play the dominant role while for other characteristics, they play a subsidiary role.

Libertarianism: Human choice and chance primarily govern human life. Free will is dominant. Genes exercise a minor, if any, effect on their behavior.

Hard genetic determinism asserts that complex human behaviors are inherited and hard-wired into our brains. Parenting style, social status and education have little if any influence. Possible accurate prediction at birth of a person's mental ability and future psychological profile is around the corner. Genes, not nurture, make us who and what we are (Plomin 2019). Free will is an illusion.

Moderate compatibilism allows environmental and random determinants to play a role, but genetic effects outshine other effects. Genes decisively dominate nurture. An educational psychologist thus notes:

> *DNA isn't all that matters but it matters more than everything else put together. .... Nice parents have nice children because they are all nice genetically.* Robert Plomin (Comfort 2018).

On the other hand, strong compatibilism accords greater weight to environmental and random factors. Genes play a subsidiary role. Nurture dominates nature.

These beliefs have major social and political consequences. In 2013, the UK education secretary was informed by his senior advisor that 70% of educational performance is due to genes and the student's IQ. The quality of teaching has a minor effect. Special government programs to improve general attainment and the vast educational bureaucracy are a wastage of public resources. Instead, the government should target talented students, allocate more funds for private schools, give private schools greater leeway in their curricula and operation and give parents greater choice in determining the fate of their children. Genetic determinism in social matters is usually associated with a conservative agenda.

Genome-Wide Association Study

Rapid sequencing of the whole genome has opened a new methodological venue for investigating the relationship between genes and personal traits. Called a genome-wide association study (GWAS), it compares the presence of genetic markers at specific locations in the genomes of the groups with and without the trait under study. The simplest, most frequently done study, the single-nucleotide polymorphism (SNP) study, utilizes one genetic marker. The design, conduct and interpretation of GWAS studies are affected by complex genetic, practical and statistical issues. Setting them aside, we present a typical GWAS scenario:

Example 4.1: Consider a sample of 21,000 individuals, among whom 1,000 have been convicted of violent crime and 20,000 are otherwise law-abiding persons. Genomic analysis for a specific SNP abnormality shows that in the law-abiding group, 500 persons have the abnormality but 19,500 do not. Further, in the group of violent criminals, 100 persons have the abnormality but 900 do not.

Thus, 10% (100/1,000) of the criminal group has the SNP abnormality, but only 2.5% (500/20,000) of the non-criminal group has it. The SNP prevalence difference is 10% - 2.5% = 7.5% and the prevalence ratio is 10%/2.5% = 4. By further data analysis, we can show that such a difference between the two groups is a statistically significant difference and implies at the 95% level of certainty that in the population at large, the SNP prevalence difference lies between 6% and 9%. Such evidence supports the claim that commission of violent crime has a genetic foundation. Often multiple SNPs are associated with a particular trait or behavior.

Example 4.2: Now consider an actual study conducted with a sample 883 male and 888 female Han Chinese university students in Singapore. The aim was to check the association between the dopamine D4 receptor encoding gene, DRD4, and political attitude. Citing previous studies done in the US that had found such an association, the investigators queried if that finding would hold up elsewhere. Their data analysis did not reveal evidence for a genetic effect among male students. Nonetheless, they concluded that:

> [our] *results provided evidence for a role of the DRD4 gene variants in contributing to individual differences in political attitude particularly in females* .... (Ebstein et al. 2015).

By early 2021, nearly 4,000 GWAS studies covering a variety of physical, health, and behavioral traits had been published. While the sample sizes in most studies were small, some had more than 200,000 subjects. Pooled analyses of studies of similar or identical traits were also published. In most, strict criteria for statistical significance were used. They generally reported highly significant correlations between the varied traits and genetic markers.

Example 4.3: The largest to-date pooled analysis of studies of intelligence, psychiatric disorders and genes included nearly 280,000 subjects. Testing 9,400,000 markers, it found 206 genomic loci (positions) and 1,041 genes to be significantly associated with intelligence. Strong genetic correlations with a range of neuro-psychiatric disorders were also indicated. Using an array of statistical methods, the authors affirm:

> *Intelligence is highly heritable and a major determinant of human health and well-being.* (Savage et al. 2018).

Antisocial behaviors (ASB) are behaviors that harm others. Examples include hostile or violent acts, theft, cheating and felonies. Varied ASBs have also been examined in GWAS studies. A recent review paper in this field observes:

> *In addition to the monetary effects, violent criminal behavior also has significant social and emotional costs. Communities with high rates of crime often face high rates of unemployment, drug and alcohol abuse, poverty, and other social pathologic conditions. Survivors of crime often experience emotional trauma and can develop serious mental health problems, such as post-traumatic stress disorder. In addition, ASB has high comorbidity with other psychiatric traits and maladaptive behaviors.* (Tielbeek et al. 2017).

Violent crime has serious and wide-ranging sequela. Thus the authors stress the importance of identifying its causes in order to prevent its occurrence and treat the perpetrators. Like in the studies of the hereditary nature of criminality done in the eugenics era, it is claimed that:

> *Accumulated evidence from quantitative and molecular genetic studies reveals the substantial influence of genetic factors in the etiology of ASB.* (Tielbeek et al. 2017).

Female genomes have two X chromosomes while male genomes have an X and a Y chromosome. In 1965, a British geneticist noted the presence of an extra Y chromosome in 7 out of the 12 inmates of a high security facility. Replication of this finding in other prisons and psychiatric facilities generated a media furor and created the perception that a scientific way of early detection of criminals had been discovered.

In the early 1970s, to intervene in the genetically inevitable, hospital nurseries in England, Canada, Denmark, and Boston began checking for the telltale extra Y among boy babies and following up with social worker visits to offer 'anticipatory guidance' to the parents on dealing with their toddling future felons. Reason prevailed and by 1974 the screening programs stopped. (Lewis 2020).

Genetic coupling of criminal or potentially criminal tendencies with a specific gene or several genes gained momentum in the 1990s. The mood regulating MAOA gene is high on the list. Mutations of this so-called warrior gene are said to predispose a person to violence. Numerous other genes were later added to the predictive pool through GWAS analyses. With fast growth of genetic data bases, and as gun violence and terrorism grab the headlines, research on identification of potential mass murderers and terrorists secures ample funds.

++++

Assertions of genetic bases of complex human behavior raise ethical and social concerns. Yet, concerns raised by critical scientists and scholars are lost in the ensuing media hype. Not only is the scientific validity of the relevant studies in doubt but the practicality of the claims is also questionable.

For illustration, we return to the extra Y chromosome screening program. It is estimated that 1 out of 1,000 individuals on average has an extra Y chromosome. Assume that the claim of criminal proclivity is valid and say the felony rate among those with the extra chromosome is 1% while among the rest it is half of that, that is, 0.5%. Suppose a random sample of 19,820 babies is screened, and the families of the babies with this abnormality are counseled and the child is monitored by social workers until adulthood. The number of persons convicted of a felony in each group is noted.

Assume the findings are as follows. Of the 19,800 persons without an extra Y chromosome, 999 (5%) committed a felony and 18,981 (95%) did not. And in the 20 persons with an extra Y chromosome 2 (10%) committed a felony but 18 (90%) did not. Thus, out of the 20 persons monitored from a young age, 18 will have been needlessly subjected to long-term violations of their privacy and other rights. As people with an extra Y chromosome tend to have physical and communication problems, these intrusions may impel them towards anti-social behavior. And out of the 1001 felons, this program will have nabbed only two. Though the apprehension rate of 0.2% is hardly anything to boast about, the industry spawned by the program will make major financial gains. As the history of medicine indicates, common sense rarely prevents the adoption of screening programs that are greeted with euphoria at the outset but later quietly recede into the oblivion.

GWAS studies generally have small samples, perform multiple analyses, and are plagued by high rates of false positive findings. The presence of substantial false positivity was clearly indicated in the case of genetic studies of general intelligence. Chabris et al. (2012) checked the predictive ability of 12 SNPs for general intelligence data from three well-designed, independent large sample studies. Instead of the expected 12 to 15 associations at the 5% level of statistical significance, they found only one. Reflecting the spirit of the seminal paper of Ioannidis (2005), their paper was titled: *Most Reported Genetic Associations with General Intelligence Are Probably False Positives*, and cautioned that unlike for some physical and medical traits,

[different] *approaches than candidate genes are needed in the molecular genetics of psychology and social science.* (Chabris et al. 2012).

Nonetheless genetic euphoria lives on. Current studies using new techniques to decode the genome and continue to link genes with not only physical traits and health outcomes but also complex tendencies. Despite their methodological problems, they claim a link between genes and traits like childhood conduct, personality, intelligence, laziness, alcoholism, addiction and anti-social behavior. Even for studies without major problems, implementation of their findings is fraught with the potential to cause serious harm to blameless persons.

Example 4.1 (continued): We return to the hypothetical data on the relation between an SNP and criminal behavior. A person with the SNP has $100/600 = 16.7\%$ risk of criminal behavior while a person without it has $900/20,400 = 4.4\%$ risk of such behavior. The SNP raises the risk of criminality by $16.7\% - 4.4\% = 12.3\%$. In other words, in every 100 persons with the SNP, there will be some 12 additional criminals. Or in every 8 persons with the SNP, on average, one person will be a criminal. But seven will be law abiding citizens. Yet, they too will come under a cloud of suspicion for years, if not decades, with possibly ruinous consequences for their lives and families.

Even ardent proponents of genetic determinism see that using variations at a single gene site to predict complex behaviors is a foolhardy exercise. For example, the effect of the MAOA gene on antisocial conduct is found only in persons who have suffered childhood abuse and problems. It is more of a proxy than a cause for antisocial behavior. If there is a genetic effect, then it arises from possibly hundreds of genes. A polygenic score based on the GWAS-identified genes is now advocated as the predictor of the behavior in question. Behaviors like aggressiveness and traits like learning disability are correlated with tailored scores based on multiple genes. Yet, these are statistical constructs devoid of direct interpretability. Using genetic scores for monitoring needs a cut-point to classify the at-risk persons. The ensuing binary measure would have the same concerns as noted in relation to Example 4.1. The multiplicity of genes also undermines feasibility.

When the identification of candidate SNPs has a high false positive rate, use of scores based on multiple SNPs does not obviate the problem. Further, GWAS studies focus on statistical significance but rarely on effect size, its variability and practical significance. Genetic determinism faces a strong challenge from the emerging field of epigenetics that show that cellular and external environmental play a critical role in the switching on or off of genes. Epigenetic effects are also transmitted across generations.

In the light of such crucial effects and major societal influences, to hold that genes markedly determine educational attainment, social conduct, personality and other complex traits is not warranted. To claim that the decision of a person about what to wear at work or the politician she will vote for is influenced by genes is to stretch the imagination. Will she follow the crowd or choose upon independent reflection? Choice and free will transcend genes. A psychological and sociological analysis is needed. A narrow focus on genes and the individual detracts attention from the potent socioeconomic factors that critically affect behavior and life attainments.

4.6 NEUROSCIENCE AND FREE WILL

Scientific determinism first emerged in relation to the Newtonian formulation of the laws of mechanics. It then surfaced in biology with the developments in heredity and genetics. And most recently, it has taken center stage in psychology and neurology.

BF Skinner's theory of radical behaviorism expounded that human acts do not arise from unfettered decisions. The ramifications of past actions significantly influence present decisions. Personal experience and the social and natural environments condition behavior. Humans feel they are making free choices. But it is not so. Free will is largely missing.

Some acts are voluntary, arising from a conscious choice. Shall I wear a white or blue shirt? Some acts are reflexive reactions wired into our brains. If I accidentally place my hand on a hot pot, I automatically and immediately withdraw it. Some reflexive acts are learned acts that in the past were conscious acts. Suppose a white shirt is obligatory at the workplace. When you started work, you bore it in mind every day. But now it is done without apparent thought. The potential consequences of a non-white shirt have conditioned you. It is not fully a free choice.

Skinner proposed creating an ideal society by operant conditioning. Just as pigeons can be trained to do certain tasks, humans can be conditioned to reject antisocial and criminal tendencies and adopt altruistic modes of action with appropriate positive and negative rewards and penalties. Conditioning, he said, is a fact of life. But it mostly operates in undesirable, unsystematic ways. The point is to bring it under conscious control and reorient it. But who will lead the effort and what kind of goals will it entail? Left unsaid, it was assumed that the process would unfold in conformity with the dominant capitalist system.

++++

Neuroscientists use the electroencephalogram (EEG) and magnetic resonance imaging (MIR) to monitor the pattern, intensity, location and timing of brain electrical activity in humans. From 1960s onward, such techniques have been used to monitor brain activity of people placed in different settings and exhibiting varied behaviors. The findings of some of these well-publicized studies have lent credence to Skinner's claim that free will does not exist. It is also argued that the brain is a physical organ with specific functional laws and our thoughts, hopes, memories, and dreams are but biochemical reactions arising from electric currents in the brain. Acts like raising one's hand are preceded by electric signals in the brain. Yet, a series of studies from the 1960s showed that the signals arise before the person has consciously made the decision to raise the hand.

Consider an experiment. You hold two switches, one in each hand. You need to press the left-hand or the right-hand switch by your own choice and use the second hand of a clock to note the time when you made it. The neuronal signals in your brain are being monitored. We expect that conscious choice determines action. The brain signal inducing the pressing should come after you decided which switch to press. Yet, many experiments based on variations of this scenario and different techniques found that the peaks of the brain signals precede the decision times.

A striking study in the 1960s recorded the following average scenario: At 0 milliseconds, the brain gives the signal. At 350 milliseconds, the subject is aware of the signal. At 500 milliseconds, the action is initiated. The result seems to indicate that the brain first decides what to do and the feeling of making a choice comes later. Of the many experiments that followed, some showed that it was possible to predict personal choice using brain signals even prior to that person's conscious awareness of her choice.

> With contemporary brain scanning technology, scientists in 2008 were able to predict with 60% accuracy whether 12 subjects would press a button with their left or right hand up to 10 seconds before the subject became aware of having made that choice. (Wikipedia (2021 -- Neuroscience of Free Will)).

In another set of experiments, subjects were placed in a rapidly changing situation and told to make a choice. Later they were asked if they had made the correct choice, the wrong choice or were unable to decide. The pace of change was such that a factual choice was not possible. Yet, more than the expected number said they had made the correct choice. Belief that a free choice was made came after the choice had been made. Rewriting memory, the brain reversed the order of events.

A recent study found that the timings of the peak of the brain signal and of making a conscious choice were predictably related to the breathing cycle. It was taken as evidence to support the idea that physical bodily processes determine mental choice, and that free will is a perception without an objective basis.

The general interpretation of such studies is that the brain determines the act and does so before the person has consciously chosen it. Free will is an after-the-fact feeling. According to Benjamin Libet, a pioneer in this field:

> [Cerebral] *initiation of a spontaneous, freely voluntary act can begin unconsciously, that is, before there is any (at least recallable) subjective awareness that a 'decision' to act has already been initiated cerebrally. This*

> *introduces certain constraints on the potentiality for conscious initiation and control of voluntary acts.* (Libet 2007).

As such findings were replicated in hundreds of studies of varied designs, many neuroscientists accepted the idea that brain decides before the mind does. Media headlines injected that notion into the public mind.

Another argument invokes causal determinism. It notes that human acts are precipitated by the firing of neurons in a sequential manner. Hence, a chain of firings links present thoughts and acts with past thoughts and acts. In principle, the present is exactly predictable by the past. There is no free will. Freudian determinism stresses subconscious influences deriving from early life as major deterministic factors.

Another position stresses societal influences. Sociological investigations have identified social and economic factors like quality of life, childhood poverty and nutrition, education level, job status and income as predictors of drug use, criminality and violent conduct. Social conditions, not free choice, make people behave the way they do. Psychological determinism implies that mental compulsion of some sort made a person do what he did. He is not responsible for it, at least not fully. Alcohol, narcotics and anti-psychotic drugs alter brain chemistry and behavior. Brain diseases and surgery induce major changes in personality and conduct. Material changes cause psychological changes. We do not choose our desires; they are chosen for us by external conditions.

One neuroscientist argues that what you are is a matter of luck. Even murderous psychopaths:

> *didn't pick their genes. They didn't pick their parents. They didn't make their brains, yet their brains are the source of their intentions and actions.* Sam Harris (Cave 2016).

Yet, in this cacophony of determinism, some critical scientists remain doubtful. Postulating that the brain generates wave like patterns in all conscious acts, they hold that the decision to act is not instantaneous. It is a drawn-out process in which the brain continuously collects and processes sensory input data. Using control groups and refined techniques this hypothesis was tested in 2012 by Aaron Schurger and others. The apparent occurrence of the peak signal before the time of having a feeling of making the decision was both a chance event and a result of how the data were analyzed. As further research backed such findings, the support among neuroscientists for the idea that the brain decides before the mind feels declined markedly. Evaluating the statistical methods used to process neurological data, another paper found that the predictive ability of postulated timings in large sample real data was poor. A better model to explain the linkage between thought and conduct was needed.

4.7 CONSCIOUSNESS

Building upon the foundation laid in Section 2.6, we take that the mind is a mental superstructure built upon the physical infrastructure of the nervous system, notably the brain. Our mind is associated with our ability to think, reason, imagine, make choices and experience a myriad of emotions—sadness, joy, love, hate, anger, disgust, hopefulness—and our conscience (morality), the sense of right and wrong.

The notion of mind is linked to a parallel notion, consciousness, namely, our sense of awareness, identity and self-worth. Being conscious is to be aware of yourself as a living, breathing, autonomous entity functioning in an external environment.

We cannot see, hear, touch, smell or taste the mind or consciousness. We cannot measure either with clarity. While many areas that relate to the different aspects of the mind have been mapped within the brain, the mind and consciousness remain elusive, controversial ideas.

The current approaches to explicate consciousness fall into three major categories: theistic, pantheistic and scientific.

The theistic view of human consciousness posits that it is not reducible to the physical processes of the body or brain. It is beyond science as it comes from the spiritual realm, and signifies the linkage of humans to divinity. Humans are conscious because they have souls. Self-awareness, discernment of right from wrong, emotions and free will are soul induced characteristics. Nonetheless, devoid of divine blessing, human consciousness is defective. True awareness (salvation) is attained by devotional worship and abiding by scriptural guidance. The pinnacle of this process is submergence within an all-encompassing supreme being.

> *Consciousness, fully awake to itself, is the Light of God.*
> Maharishi Mahesh Yogi

Theists adduce human consciousness as evidence for the existence of God.

The pantheistic view of human consciousness dispenses with divinities. Positing the interconnectedness of all phenomena in the Universe and the linkage of the observer and the observed, it claims that there are no discrete entities, no discrete souls, no God or gods. Appearances to the contrary are illusions. Only an all-pervasive energy field (universal consciousness), of which humans are a key part, exists. Erwin Schrodinger, an eminent quantum physicist, wrote:

> *The total number of minds in the Universe is one. In fact, consciousness is a singularity phasing within all beings.* Erwin Schrodinger (Wintjen 2022).

Like the theistic view, the pantheistic view asserts that consciousness is not reducible to the mechanisms of the body and brain and cannot be explicated by science. It can be comprehended and attained in its highest form through mystical contemplation. True self-awareness arises from exploring our inner sanctum and discerning our integrality with the Universe.

> *The inner world is very potent for me –*
> *I don't ascribe to any God or Jesus or Buddha*
> *- I just have a sense of it and revere it along*
> *with the natural world and human consciousness.*
> Annie Lennox

The scientific view holds that consciousness can be discerned with the methods of science. But there is no consensus on how that is to be done. Sigmund Freud, a pioneer investigator of the nature of consciousness, held that the mind operates at three levels: conscious, subconscious and unconscious. The conscious mind encompasses the thoughts, emotions and behaviors we are directly aware of. The subconscious mind represents the automatic, and often conditioned, thoughts, emotions and behaviors that we can become aware of with due effort. The unconscious mind consists of the past thoughts, emotions and behaviors that are virtually beyond our conscious reach even as they affect our thoughts and acts (Farnsworth 2020). Modern neuroscientists have ventured beyond Freudian explanations. Their general point of departure is that:

> [consciousness] *is the function of the human mind that receives and processes information, crystallizes it and then stores it or rejects it with the help of the five senses, the reasoning ability of the mind, imagination, emotion and memory.* (Vithoulkas and Muresanu 2014).

But the scientists in this field are divided into two camps: the reductionist camp and the system theory camp.

The neuro-deterministic (reductionist) camp views consciousness in terms of a distinctive multiplicity of emotions, modes of thoughts, and awareness associated feelings, each with a unique pattern of signals from a specific center in the brain. Consciousness is explained through identification of the centers and the relevant patterns of signals. Voluminous as the task is, positing consciousness in terms of the detailed biology of the brain is the only valid way to deal with the issue. All other approaches are deemed unscientific or experimentally untenable.

Sophisticated instruments and innovative study designs show that varied emotions, sensations, memory and experiences map to specific parts of the brain. There is confidence that the primary location of consciousness in the brain will be located someday. Current evidence shows that some brain regions are unlikely to harbor the seat of consciousness. Injury to the cerebellum, which has four times the number of neurons than the rest of the brain, affects motor control and gait but hardly affects one's sense of self, wellness, memory, hearing, sight, and feelings. One expert posits the cerebral cortex as the probable seat of consciousness:

> [It] *appears that the sights, sounds and other sensations of life as we experience it are generated by regions within the posterior cortex. As far as we can tell, almost all conscious experiences have their origin there.* (Koch 2018).

Locating the seat of consciousness, if there is one, is but the first step in a long journey. Phenomena like self-awareness, anger and joy are not either-or entities, but occur along multidimensional continua. The electrical signals they generate tell us next to nothing about the diverse manifestations of such entities in real life. Another foundational dilemma is: What makes a physical system made up of cells and molecules possess awareness of itself and the surroundings? Some scientists consider love, hate and similar feelings to be too subjective to yield to scientific study. But both eminent physicist Stephen Hawking and Francis Crick, the co-discoverer of the structure of DNA, asserted that the laws governing the functioning of basic entities would one day unravel the mystery of consciousness.

The emergence (systems, holistic) theory camp posits consciousness a function of the nervous system as a whole. Over twenty conceptual and mathematical models to explain consciousness in systemic terms have been proposed and many experimental and theoretical investigations along these lines have occurred. We describe two major theories.

The Global Neuronal Workspace Theory posits that consciousness arises from the joint operation of a complex, across-the-brain, non-linear web of a myriad of neurons. Consciousness states reflect the brain areas from which information has been accessed and the way in which it has been processed. Interwoven within the web are collections of neurons that represent recall, perception, motion, assessment and attention. The global web is connected as well with the non-consciousness functions of the brain. Dispensing with locating consciousness states in specific brain centers, this theory asserts that any system with sufficient interlinkages and data processing power can attain consciousness. It projects that futuristic quantum computers with massive parallel processing ability will be intelligent and conscious.

The Integrated Information Theory regards the mind and consciousness as emergent phenomena. Thoughts or emotions do not occur through the activation of a single neuronal strand unfolding in a linear way. Each arises from interactive activation of multiple neuronal strands to produce a multi-faceted picture of reality. A person is aware of the taste, aroma and temperature of the food she is eating, the music playing in the background, other people and objects at the table as well as recalling her childhood in a conjoined manner. Consciousness is a dynamic construct based on both the hardware and the software of the brain. The former consists of structure of the brain and the neurons of its functional regions. The latter consists of the signals deriving from bodily and environmental sensations, memory and complex, dynamic interactions among the billions of neurons in the brain.

> [Any] *complex and interconnected mechanism whose structure encodes a set of cause-and-effect relationships will have ... some level of consciousness. ... [Consciousness] is intrinsic causal power associated with complex mechanisms such as the human brain. ... Consciousness cannot be computed: it must be built into the structure of the system.* (Koch 2018).

Using information theoretic mathematical analysis, it demonstrates that:

> *coarse-grained macroscopic states of a physical system (such as the psychological state of a brain) can have more causal power over the system's future than a more detailed, fine-grained description of the system possibly could.* (Wolchover 2017).

This theory links subjective awareness to the complex, integrated state of information attained by interconnected ensembles of neurons in the brain. It posits that higher levels of a system have more predictive power than lower levels. Random variation at a micro level emerges as a pattern at the macro level.

Consciousness is an emergent entity with its own laws of causality that operate at the macroscopic level. Humans have agency that arises from the macroscopic laws of the mind. Language, literacy, learning and life history impinge upon the character of consciousness. Conscious action is based on intuitive and/or reflective thoughts. The reflections may be rational, irrational or random.

To reiterate, the Integrated Information Theory explicates the panorama of human thought, feelings, desires and choices by positing multiple levels of causation. Founded on special mathematical analysis of neural networks, its key tenet is that a group of neurons possesses more effective information than the individual neurons combined and has its own mode of causation. When neurons are grouped further to form higher level groups, predictive ability increases at the higher levels.

Group level casual laws emerge from the neuronal interconnectivity between semi-groups combined with the presence of neural redundancies and elemental randomness in neuronal firings within groups. As laid out by one of the pioneers:

> *Causal emergence is possible ... because of the randomness and redundancy that plagues the base scale of neurons. ... [Imagine] a network consisting of two groups of 10 neurons each. Each neuron in group A is linked to several neurons in group B, and when a neuron in group A fires, it usually causes one of the B neurons to fire as well. Exactly which linked neuron fires is unpredictable. If, say, the state of group A is {1,0,0,1,1,1,0,1,1,0}, where 1s and 0s represent neurons that do and don't fire, respectively, the resulting state of group B can have myriad possible combinations of 1s and 0s. On average, six neurons in group B will fire, but which six is nearly random; the microstate is hopelessly in-deterministic. Now, imagine that we coarse grain over the system, so that this time, we group all the A neurons together and simply count the total number that fire. The state of group A is {6}. This state is highly likely to lead to the state of group B also being {6}. The macro state is more reliable and effective; calculations show it has more effective information.* Erik Hoel (Wolchover 2017).

Presently, the Integrated Information Theory is the principal challenger to hard neurological reductionist determinism. But it is a young theoretical edifice in need of further development and experimental support. One drawback that needs rectification is that the societal factors it currently incorporates do not adequately reflect socioeconomic realities.

++++

Despite the impressive, experimentally backed models of consciousness that have been developed of recent, the more-than-a-decade-old verdict of a prominent neurologist stands.

> *At present, however, no single model of consciousness appears sufficient to account fully for the multidimensional properties of conscious experience. Moreover, although some of these models have gained prominence, none has yet been accepted as definitive, or even as a foundation upon which to build a definitive model.* (Seth 2007).

4.8 FREE WILL AND RESPONSIBILITY

The major or minor choices people make in life are generally constrained by personal and external factors. They include biological (genetic, neurological) factors, health, age, family and family history, economic status, ethnicity and education. External factors include the general culture, laws, racism, history and socioeconomic structure.

There are three basic positions regarding people's ability of people to make autonomous choices: hard determinism, compatibilism and libertarianism. The first position asserts that all is predetermined, and people cannot make a free choice in a meaningful way. The third position asserts that people can make free, unfettered choice in all circumstances. We earlier argued that these two positions are purely theoretical positions of little relevance to social reality.

The middle position, compatibilism, asserts that a degree of freedom in making choices in real life always exists.

> *Compatibilism is the belief that free will and determinism are mutually compatible and that it is possible to believe in both without being logically inconsistent.* ((Wikipedia 2022 – Compatibilism).

The practical contention is: How large is the degree of freedom? Is it sufficiently large to render the person responsible and accountable for her choice and action? **Weak compatibilism** says no. She is unable to make a meaningfully unfettered choice due to major external or past impositions. But **strong compatibilism** says yes. She has the capacity to overcome any restrictions and freely decide and act. She is responsible and accountable for the consequences of her choices.

In this section, we compare the views of different scientists on the nature of free will and consider the practical implications of their ideas.

Richard Dawkins, an evolutionary geneticist, makes a compatibilist case for free will by first asserting that living entities emerged from a competitive process of survival among genes. Their biological constitutions were affected by the interactions between genes and the environment. Genes program their hosts to propagate themselves, but that programming is both deterministic and statistical. The complexity of living beings and, especially the human brains, provide ample room for variability in form and behavior beyond that governed by genes. Genes are selfish entities. But they can give rise to beings that demonstrate altruistic actions towards members of their own and other species.

Thoughts and behaviors do not emanate from specific parts of the body, but from the person as a whole. Humans are not enslaved by biology. They are conscious beings who can

124

overcome biological constraints by exercising free will. The size and structure of our brains has enabled us to develop intelligence, conscience and values. We have the capacity to *'depart from the dictates of the selfish genes and to build for ourselves a new kind of un-Darwinian life'*. Richard Dawkins (PBS 2021).

Darwinism—the law of the jungle—is a retrograde theme for society. Humans need to organize social, educational and legal institutions that encourage moral behavior. By exercising free will, they can establish a world based on moral, not biologically driven, principles.

For example, sexual desire, a key survival trait, emanates from genes. But dressing nicely to impress the opposite sex is not programmed by genes. Use of contraceptives is an anti-Darwinian practice. And why one person prefers to read a book and another, to watch a movie, is not based on genes. There are human behaviors—gazing at the stars or writing an email to friend—whose connection to genetic survival is remote at best. All human behavior cannot be attributed to genes.

Dawkins declares that generally everyone, including a strict determinist, feels he or she has free will. Free will is not just a sensation. It exists. But how it came about and operates is yet a question in need of extensive study.

Matt Ridley, geneticist and philosopher, has similar views:

> *Once we see the genome as something that interacts with the environment rather than pulling the strings on a puppet, then we can begin to grope our way to a much more free-willed personality. We are a product of our background and our genes, but we are also capable of taking spontaneous decisions.* Matt Ridley (Ahlstrom 2001).

Philosopher **Bruce Waller** provides a distinctly intriguing perspective on free will. Adopting a naturalistic stance, he argues that everything is caused and fully determined. No room for deviation from laws of nature exists. Though free will is commonly linked to moral responsibility, he claims that such a linkage is not scientifically valid and leads to harmful social policies. In his opinion, a better conceptualization of free will is to regard it as freedom of choice: It is

> *the capacity to effectively explore alternative paths in response to a combination of environmental contingencies and internal motives.* Bruce Waller (Caruso 2016).

Free will is the ability to formulate a range of options and choose among them without external influences, Accordingly, it has to be disassociated from the moral, religious and mythological baggage of the past.

Waller notes that extensive evidence from several disciplines shows that free will, as he defines it, is an evolutionary, adaptive entity that is also present in other animals. The flight paths of a foraging bee, for example, vary according to the availability of food. In times of scarcity, the bee is likely to explore widely while in times of plenty, it follows fewer paths. Free will is a flexible trait in that the choices can be altered as circumstances change. Free will is not a uniquely human feature. Decades of scientific work is needed to unravel its nature.

Free will and determinism are not incompatible. But they describe behavior at different levels. Humans use their brains to formulate a large range of options and are freer than animals in significant ways. Ordinary people view free will as the ability and opportunity to pursue one's desires in the absence of coercion.

In a similar vein, a physicist who ascribes to humanism opines:

We are the cosmos made conscious
and life is the means by which
the Universe understands itself.
Brian Cox

Dennett versus Caruso

Daniel C Dennett is a neuroscientist, philosopher and a prominent New Atheist who has written popular and scholarly works on science and religion. **Gregg D Caruso,** a legal scholar and philosopher, has authored a series of papers on free will in relation to crime and punishment and public health. Recently, they locked horns on the issue of free will.

Holding a compatibilist viewpoint, Dennett holds that free will exists. As such, humans have moral responsibility. Evolution has endowed humans with exquisite mental and reasoning capacity. Their ability to make autonomous choices is an observable entity. At birth, humans are not autonomous agents. Infants largely act under forces beyond their control. But as they grow up, they learn to think and act independently. Unlike wasps, they can avoid futile repetitive behavior, learn, plan and switch to other tasks when they so feel like. That is indicative of the existence of free will. Biology is not destiny.

Yet, free will has limits, though they do not imply compulsion. The key issue is to decide when a person can exercise free will and to what degree. That one is not able to attain what one wills does not negate free will. Humans can change the degree of freedom—in thought, feeling and action—they possess. Choice is freedom, and freedom is choice.

Consciousness, memory and free will are whole-brain entities. When we make a choice, the 'we' includes the brain. A factually founded model for free will based on neuroscience and which integrates deterministic forces has yet to be developed. Compatibilism is a coherent, scientific, empirically tested and practically useful idea.

Free will is also a reliable basis for moral, legal and social precepts. Society imposes rules to control the conduct of individuals in sports, school, work, family life and public places. The rules are designed to maintain order, fairness and the general state of freedom. Violations may necessitate mild to grave penalties. Individuals understand this and are thus responsible for their actions. When someone breaks a rule, he or she should be held to account. Non-adherence to this tenet will lead to a breakdown of the social order. There will be no morality. Rules of conduct do not negate free will. Learned self-control is not equivalent to determinism.

He dismisses the use of ideas borrowed from quantum mechanics to support the existence of free will as a ridiculous endeavor. Quantum level fluctuations cannot affect human choice and behavior. He as well disparages the notion of unfettered libertarian free will.

As humans, we have the best chance to produce good behavior. We should be satisfied with what we have and not fret over our lack of libertarian free will. DC Dennett (Dennett 2015).

In sum, Dennett advocates enhancement of individual freedoms within the context of social responsibility and accountability.

Gregg D Caruso accepts the existence of free will but holds that it is largely curtailed by social conditions. Focusing on personal responsibility detracts from dealing with social conditions that promote immoral choices and conduct. At birth, humans have different states of physical health and land into life conditions that vary widely. Unequal birth status is compounded by racial, economic and educational inequalities in society. Inequality engenders more inequality.

> *Since our genes, parents, peers and other environmental influences all contribute to making us who we are, and since we have no control over these, it seems that who we are is at least largely a matter of luck. …. [Low] socioeconomic status in childhood can affect everything from brain development to life expectancy, education, incarceration rates and income. The same is true for educational inequity, exposure to violence, and nutritional disparities.* G Caruso (Dennett and Caruso 2018).

Life conditions circumscribe human thought, choices and conduct. What humans are and what they do is ultimately the result of factors beyond their control. Humans are thereby not morally responsible for their actions.

Punishment is a major pillar of the legal system. If people willingly do a bad thing, then a bad thing needs to happen to them. They deserve the punishment. Such a retributive approach to crime is neither morally defensible nor a panacea to deter crime. It only breeds more crime.

The justice system in the US is strongly afflicted with racial bias. Effective crime rate reduction requires a public health approach and dealing with systemic social, racial, environmental and economic inequalities. For example, the racially biased 'War on Drugs' promoted by the US needs to be ended and replaced with a public health approach to the problem of drug use. Rules and penalties in sports are acceptable because they protect players and encourage fair play.

++++

Neither Dennet nor Caruso accept divine causation in human affairs. Both espouse secular compatibilism. Occasional rhetorical flourishes aside, both regard hard determinism and libertarian free will as unrealistic descriptions of the human condition. Dennett places personal freedom over social factors. Individual responsibility overrides life circumstances. Caruso holds that life circumstances severely constrain if not almost extinguish free will.

Both address key problems of human society: crime, responsibility and punishment. Stripped of terminological niceties, their discord reflects the discord between a right-of-center liberal (Dennett) and a left-of-center liberal (Caruso) in the US. Should the tough-on-crime policy or the War on Drugs continue? It is a scientifically flavored ideological debate within the confines of the capitalist system. Dennett finds the system fundamentally fair but Caruso holds that it is basically unfair. Yet, Caruso does not entertain displacement of the system by one based on equality and justice. Though their focus was on the US, the implications of their debate are global.

Impulsivity

Suppose a man picks up a bag of beans from a supermarket shelf. After looking at it for a minute, he suddenly casts it aside. Unfortunately, it hits a nearby child who starts wailing. The distraught mother speaks angrily. The man apologizes. A security guard appears and detains him. Later, he is charged and tried for willful assault. His lawyer argues that he had no intent of injuring anyone and calls a psychologist who testifies that the man is genetically prone to uncontrolled impulsive behavior. Is he guilty?

An impulsive act is a rapid response to a stimulus that occurs without forethought and concern for the consequences. People react instantly in ways they later regret. It is a personality trait linked with mental disorders, aggression, drug addiction, alcohol abuse, and tendency to self-harm. Some studies have concluded that it is heritable. A recent study examined the association between scaled measures of impulsivity and genetic variation in a realized sample of 426 young adults. A tailored questionnaire and a series of computer-based tasks were used to assess impulsivity. The authors concluded:

> We identified various genes and gene regulatory pathways associated with empirically derived impulsivity components. Our study suggests that gene networks implicated previously in brain development, neurotransmission and immune response are related to impulsive tendencies and behaviors. (Khadka et al. 2014).

Though about 2,100 individuals were initially recruited, genome sequencing was done only for about a fifth. Impulsivity was measured in a laboratory setting. The reported link between genes and behavior was not necessarily a causal link. The genetic and impulsivity scores used were numerical entities lacking concrete interpretation. Many such scores exist. How the implicated genes interact with each other and generate impulsivity remains a mystery. Yet, today such studies readily capture headlines: Heritable impulsivity implies diminished responsibility for unwilled conduct. Impulsivity inhibits free will. Conclusions based on flawed studies which ignore other factors that affect impulsivity can hardly generate meaningful approaches to ascertain it causes and deal with the consequences.

Smoking

Tobacco smoking is dangerous for human health. According to the WHO:

> Tobacco kills more than 8 million people each year. More than 7 million of those deaths are the result of direct tobacco use while around 1.2 million are the result of non-smokers being exposed to second-hand smoke. ... Over 80% of the world's 1.3 billion tobacco users live in low- and middle-income countries. (WHO 2020).

A life-long smoker who has contracted lung cancer sues the US tobacco company, whose brand she smoked, for compensation and medical costs. Her lawyers argue that glamorous media ads and free samples induced her to start smoking at a young age. And once addicted, the serious withdrawal effects made her unable to stop even when she tried. The tobacco company had known about the health hazards of tobacco smoke and its addictiveness for a long time but had hidden the facts from the regulators and the public. It still continues mounting misleading, deceptive campaigns to market its products. The company also uses its financial power to sway legislators not to enact stringent anti-tobacco laws.

The company lawyers argue that the plaintiff was not compelled to smoke cigarettes. To smoke or not was her own choice. If anything, she was influenced by her friends and bears the responsibility for her choice. It is not the company's fault that she has genes implicated with addictiveness. The company is selling a legal, not an illicit, product. It markets its products using advertising firms that are employed by many manufacturers. No deception is involved.

Further, there are many causes of lung cancer. A good proportion of smokers do not get lung cancer and a good proportion of people with lung cancer have never smoked. It is a play of genes, other causative factors and chance.

The basic facts about the social impact of tobacco smoke are: In the US, nearly 9 out of 10 smokers start smoking before the age of 18. Some 3 out of 4 high school smokers become adult smokers. It is a highly addictive behavior that creates social bonds that reinforce the behavior. A circular from the American Cancer Society (ACS) reads:

> The tobacco industry's ads, price breaks, and other promotions for its products are a big influence in [the American] society. The tobacco industry spends billions of dollars each year to create and market ads that show

smoking as exciting, glamorous, and safe. Tobacco use is also shown in video games, online, and on TV. And movies showing people smoking are another big influence. Studies show that young people who see smoking in movies are more likely to start smoking. (ACS 2020).

Besides lung cancer, smoking increases the risk of cancer of the mouth, throat, digestive organs and other organs as well as of emphysema, chronic obstructive pulmonary disease and heart disease. Smoking poses a great burden on the health care system and disproportionately affects the poor, marginalized, less educated communities. These risks are difficult to reduce because the nicotine in tobacco smoke is as addictive as cocaine and heroin, and alters personal mood and emotions. Quitting smoking is more difficult than quitting alcohol drinking and cocaine or opiate use. Nicotine in the tobacco smoke is quickly absorbed in the blood stream, alters mood and induces a mental cycle for more nicotine. Its withdrawal symptoms are so unpleasant that most people who stop smoking resume the habit. Studies indicate nearly that 70% of smokers would like to break the habit. But on average it takes 30 attempts before a person can stop smoking for good.

Despite external assistance, smoking is a decidedly hard habit to break, even by otherwise strong-willed individuals. But the common perception is that people cannot quit because they lack will power and discipline.

> *Addiction is marked by the repeated, compulsive seeking or use of a substance despite its harmful effects and unwanted consequences. Addiction is mental or emotional dependence on a substance.* (ACS 2020).

Nicotine addiction is a difficult to treat illness, an illness that compels the addict to act against his own financial, health and familial interests. It is a major public health issue. Measures like restrictions on advertising, ban on sales to minors and warning labels on the packages are in place, but they do not suffice. They were enacted after a long drawn-out process in which the tobacco companies used their financial and political power to block every move to reduce smoking rates.

Smoking is more than an exercise of free will. It is a determined process which denies people their ability to exercise free will. People are captivated by habit from which escape is very difficult. It also causes illnesses that drastically reduce people's quality of life and capacity to make meaningful choices. As one commentator puts it:

> *Let's not be blinded or confused by political ideology and believe those who market, brand, and reframe the issue to muddle what freedom of choice is. Those addicted to nicotine are sick and lack the ability to make that choice.* (Ann 2019).

For the smoker suing the company, the relevant questions are: Was she a free agent when she began to smoke? Was she exercising free will when she went on ingesting tobacco smoke daily? Who bears the responsibility for her cancer—herself, the company, both or no one? What actions are needed to dramatically reduce smoking related health problems?

Such questions also apply to the actors on the other side: Were company executives exercising free will in their decision to market a harmful product? Should they be tried for criminal conduct and, if convicted, be fined or imprisoned? Or were they inexorably driven by the profit maximizing logic of capitalism? Is the corporate capitalist system at fault? Do not the companies factor in the potential legal damages in their cost benefit calculations? Can their power and influence be curtailed without a basic transformation of the capitalist system?

Resolving the dangerous problem of smoking cannot be done with remedial measures. It requires complete transformation of neoliberal capitalism. And that effort requires the people to exercise collective free will.

4.9 FREE WILL AND MORALITY

Does belief in free will make you more of a moral and responsible person? Does it affect your stand on issues like crime and punishment and welfare policies?

Scientific studies generally show that the higher their level of belief in free will, the more responsible the people feel for their own actions. And the more they feel that their action was affected by forces beyond their control the less they feel that they deserve blame for the negative outcomes. Lack of belief in free will makes people less inclined to control their base drives and more disposed towards socially inappropriate behavior.

Specifically, strong belief in free will makes students more likely to assist their fellow students. Students exposed to scientific denigration of free will are less likely to assist homeless persons. Doubts about the existence of free will correlates with higher stress level, low index of happiness, lower sense of loyalty towards friends and loved ones, reduced gratefulness and a negative attitude towards the meaning of life. Free will skepticism tends to enhance conformism, reduce creativity, lower the ability to learn from errors, diminish academic performance, and lower the quality of work performance.

Furthermore, studies generally indicate that free will believers have more stringent attitudes towards actions of other people and tend to hold them more responsible for their actions. They support harsher penalties for people convicted of crimes since they feel that the crimes were committed on a voluntary basis. They favor reduced unemployment assistance because the unemployed have opted to be lazy and live off the public coffers. If you work hard, you progress in life; if you do not, you remain at the bottom. Your future rests in your own hands. Good behavior deserves reward and bad behavior deserves punishment. There is little middle ground.

On the other hand, free will skeptics are generally more tolerant of the actions of others. They are more aware of the barriers imposed by nature, nurture and social circumstances that constrain people. They diminish the responsibility a person has for his choices and actions. Hence, they advocate reformation of the criminal justice system by reducing focus on punishment and incarceration and paying more attention to prevention and rehabilitation. Punishment or the threat of it does not deter crime or recidivism. Deviant conduct is alleviated by education, job training, and reduction of poverty and unemployment rate. Other necessary acts are mental health services to marginal communities and prisoners, fewer prisons, long term counseling for prison inmates and people showing anti-social behaviors, and public health-based control of drug abuse. Punishment should not be retributive. That is said to be the optimal way to protect individuals and society.

The studies indicating a positive link between belief in free will and a higher sense of morality and personal responsibility had numerous methodological limitations. Most were laboratory-based, not real-life studies. Many recruited students. Outcomes were observed on a short-term basis. Issues of samples size, control group, over analysis of data were not adequately addressed. Correlation was at times conflated with causation. Almost all the studies were done in the affluent nations of the West. Concerns about the validity and generalizability of their findings thus remain.

Yet, the broad consistency of their conclusions is striking. Belief in free will is linked to better moral values, conduct and life outcomes and free will skepticism is linked to undesirable values, conduct and outcomes. Determinism undermines the tradition of reward and punishment and denigrates key social institutions. What does it matter whether a person murdered someone or rescued a baby from an inferno, if in both cases, forces beyond his control made him do what he did.

Such implications have prompted scientists and scholars to opine that advocacy of belief in free will benefit individuals and society. Deterministic ideas may induce morally irresponsible conduct and worsen social problems. A greater popularity of belief in free will encourages people to behave in a morally prudent manner and promote social stability. It will also strengthen the major social institutions and norms founded on the belief in freedom and free will.

We must believe in free will, we have no choice.
Isaac B Singer

Yet, hard determinists are faced with a fundamental quandary. If free will is indeed a delusion, is it ethical to promote belief in the existence free will just because it may benefit society? Should goodness be delinked from truth?

Determinists generally accept that even though it may be illusory, belief in free will is critical for social stability and harmony and must be protected. But they disagree on how that is to be done? Saul Smilansky argues that interests of society trump truth. People naturally tend to believe in free will. Let them hold on to their illusions. No steps to educate them otherwise should be taken. The scientific evidence for determinism, and even the idea of determinism should remain within academic and research circles, like privileged information.

Promoting determinism is complacent and dangerous. S Smilansky (Cave 2016).

Sam Harris, a staunch determinist, disagrees. The truth should not be withheld from public view. Individuals and society people can cope without believing in free will. Determinism is not fatalism. Only the truth can set you free. Society must rethink the traditional approach to reward and punishment and focus on addressing the modifiable causes of anti-social behavior. Crime cannot be tackled on the basis of blame, hatred or retribution. People need to be educated about science, how to weigh scientific evidence and reflect on preventive policies for addressing social issues. Hard determinists like BN Waller also hold that greater freedom of choice is facilitates improved moral conduct. Practical resolution of this important conundrum is difficult to conceive in the confines of the unjust, unequal, freedom-depriving neoliberal capitalist system.

4.10 MARXISM, FREE WILL AND CONSCIOUSNESS

Consciousness, or the individual's awareness of the world around her and the choices she makes in life are affected by a multiplicity of factors. These include biological factors (genes, brain structure, physical and mental health); childhood nutrition, family and education; psychological factors (personality and character); media, economic status and social norms. People associate with occupational, professional and religious groups that affect their perceptions and decisions. The extent to which a person can transcend the effects of these restrictive forces in her choices reflects the strength of her power to exercise free will.

Marxism accords a central emphasis to individual freedom. Identifying the causes that limit people's ability to make free choices and develop ways to overcome the barriers is central to the philosophy developed by Karl Marx and Friedrich Engels. Like conventional social theories, it generally accepts the play of the above noted factors. But where it distinguishes itself from the other theories is its emphasis on economic relations as a major determinant of how people look at themselves and the world, make decisions and act. Consciousness is strongly affected by membership in economic class and by the interests of the dominant class in society. What the common people feel are views and ideas of benefit to the society as a whole are often views and ideas that protect the long-term interests of the dominant class.

The Marxian vision of free will derives from considering development as a process that counters the forces of necessity (deterministic forces). Based on rudimentary technology, early human societies were at the mercy of nature. Individual and collective choices were restricted largely by necessity. History abounds with endeavors to enlarge choice. As knowledge grew and better tools developed, conquest of nature-imposed restrictions expanded as well. Creative modalities enabled by the larger human brain—language, writing, art, song, music, dance, sculpture, religion, philosophy, science, mathematics—developed, human imagination and creativity bloomed, and the room for individual and collective choice and free expression expanded. It was a nonlinear dialectical process reflecting the interplay of subjective (conscious) factors and objective (natural, economic and sociological) factors.

Yet, the development of productive forces also brought along a growing degree of social stratification. In a diversity of ways, new social formations emerged in which a small group of rulers exercised control over the majority, the societal resources like land and the wealth generated by the productive activities. The dominant class controlled the state institutions and was able to largely dictate what people could do. The exercise of effective free will for the upper classes enlarged as it shrunk among the lower classes.

Exclusive utilization of force is not a viable means of maintaining rule. It is more cost-effective and sustainable if the dominated accept that the rulers protect the interests of the society as a whole. Force is a last resort. Thus, development of hierarchical class societies spawned an elaborate apparatus of ideas and visions and the institutions to propagate them that injected a worldview ultimately protecting the interests of the ruling class into popular consciousness.

> *The ideas of the ruling class are in every epoch the ruling ideas, i.e. the class which is the ruling material force of society, is at the same time its ruling intellectual force. The class which has the means of material production at its disposal, has control at the same time over the means of mental production, so that thereby, generally speaking, the ideas of those who lack the means of mental production are subject to it. The ruling ideas are nothing more than the ideal expression of the dominant material relationships, the dominant material relationships grasped as ideas.*
>
> Karl Marx

In the Marxian framework, holding outlooks at variance with the actuality of class oppression is called **false consciousness**. It is akin to visualizing a tree with a camera obscura whereby the resultant image is blurred and inverted, a form of **cognitive dissonance.** The mechanism generating false consciousness, the ideological apparatus of the system, has evolved over centuries and operates in the education sphere, mass media, political and legal institutions, religious entities and the prevailing ensemble of conventional wisdom. It does not arise from a central conspiracy and it does not operate smoothly or in a routine fashion either. Contradictions and controversies are the norm. An ever-present small group of activists exposing aspects of distorted consciousness gives the image of free, unbiased discourse. At times, they even manage, but temporarily, to gain the upper hand.

False consciousness arises on a foundation of mental and material **alienation**. Under capitalism, the producers no longer control production. They just perform minor, repetitive tasks that make them lose perspective, are delinked from fellow workers by time, place and distance and denied a creative role in the design and marketing of the fruits of their labor. They work for the wage, that is it. Caught up in the consumerist rat race, they lose touch with humanity's general concerns. In Marxist parlance, they are alienated. Neoliberal globalization with its global supply chains, exaltation of individualism, and single-minded pursuit of profits and generation of astounding levels of inequality has carried alienation to unforeseen high vistas. But the capitalists are alienated as well, though in quite different forms.

Marxism asserts that free will and alienation are incompatible. Genuine free will is associated with conscious, productive, creative and unfettered activity.

> *We presuppose labor in a form in which it is an exclusively human characteristic. A spider conducts operations which resemble those of the weaver, and a bee would put many a human architect to shame by the construction of its honeycomb cells. But what distinguishes the worst architect from the best of bees is that the architect builds the cell in his mind before he constructs it in wax.*
>
> Karl Marx

Free will is catalyzed by the existence higher levels of freedom in different walks of life. And freedom is a socially determined concept that changes over time. It is seen among other animals but in a rudimentary form.

> *The first men who separated themselves from the animal kingdom were in all essentials as unfree as the animals themselves, but each step forward in the field of culture was a step towards freedom.* Friedrich Engels (Spirkin 1983).

Yet, in class societies, freedom does not always entail rational or laudatory moral outcomes. Patently harmful and irrational outcomes have occurred in the name of freedom. Freedom for some has often accompanied added bondage for others. Despite its ideology of promoting individual freedom and democracy, capitalism restricts the exercise of free will in a multiplicity of ways. Though individuals view major choices through the lens of morality, conscience, and duty to oneself, family and the state, the manner in which they do it reflects the mental grip of false consciousness.

A key aim of socialist transformation is to combat alienation and promote unity among the common people. Socialists aim to educate the masses about the social realty. foster values that serve humanity and enlarge the scope for genuine free will. In other words, the aim is to promote **class consciousness**.

The Marxian approach to social change, in theory and practice, is neither deterministic nor libertarian. While positing a key role to economic factors (productive forces and relations of production), it also allows for operation of non-economic forces and mental factors as drivers of change. It views social change as a dialectical process (dynamic compatibilism) cognizant of the restriction imposed by material conditions yet allowing an instrumental role to the collective exercise of popular will.

> *Men make their own history, but they do not make it as they please; they do not make it under self-selected circumstances, but under circumstances existing already, given and transmitted from the past.* (Marx 1852).

History is a complex process in which interactions between a multitude of objective and subjective forces as well as the play of chance (accidental events) leads to an astonishing diversity in politics, culture, belief systems, and social relations between societies with similar economic structures. For example, humans engage in artistic endeavors for reasons ranging from pure expressions of special abilities to commercial considerations. Art may be conformist, neutral or revolutionary, but goal driven art is hardly artistic. Economic conditions affect the material used by the artist and the types of artistic activities. But strict determinism hardly prevails. As Marx opined:

> *As regards art, it is well known that some of its peaks by no means correspond to the general development of society; nor do they therefore to the material substructure.* (Marx 1859).

Alexander Spirkin

The Soviet psychologist and philosopher Alexander Spirkin presented a distinctly cogent formulation of the Marxian concept of consciousness, self-awareness and free will. Asserting that pure freedom does not exist, he observes that the exercise of individual free will is affected by natural, biologic, psychological factors as by legal and moral codes, personality and character and artistic proclivities. The inter-linkages between individuals, the demands society imposes on them, and their dependence on the natural and social environment world constrain free will. Free will is a social, and not a personal construct. Social restraints are acceptable provide they occur within a just and democratic social structure and not a structure in which a privileged minority imposes its will over the rest.

Choice and responsibility are interconnected. Without free choice, there is no personal responsibility. Intent and existent conditions also circumscribe personal responsibility. Social standards governing conduct make it a moral, psychic, legal and sociopolitical entity. Instilling responsibility and a sense of shame (conscience) as well as awareness of the effects of one's choice are parts of the mechanism for social control too. The key issue is: In whose interests are those mechanisms operating? Spirkin connects free will to goal driven action.

A human being is not a piece of driftwood on the waves of cause-effect connections. He is active. Free will manifests itself precisely in purposeful activity. (Spirkin 1983).

No person can or does exist apart from society. One cannot live in society and be free of society. (Spirkin 1983).

Free will evolves in a dialectical fashion. Circumstances condition human life but humans change the circumstances of life; people are a product of social relations, yet social relations emerge from human activity. Free will relates to subjective and objective factors. A one-sided emphasis on subjective factors (libertarianism or idealism) is as flawed as a one-sided emphasis on objective (material) factors (determinism, materialism). Subjective and objective factors occur in two forms, a stabilizing form and a transforming form. The former preserves the structure of the social system; the latter undermines it. They interact in dialectical, flexible ways. Generally, the preservative forms dominate but in times of major crisis, the latter prevail. Ultimately, people make history. Revolutionary education, and organizing and mobilizing the masses attests to the importance Marxism attaches to subjective forces. A revolution is not a coup d'état. It is an exercise of collective will by the people. Fundamental change in economic relations and state power is not a product of the automatic development of material conditions.

False consciousness inhibits fundamental transformation. People regard what exists as destiny or fate, an outcome beyond their control. People rationalize: It was my bad luck; I had to do it; It was meant to be; Whatever you do, it will remain the same; We cannot change human nature. Oblivious to human resilience and creativity, it fosters pessimism that absolves people from responsibility and belittles human dignity.

Yearning to overcome bondage, injustice and discrimination has been central to progress everywhere. People persevered despite the odds and sacrificed their lives in the name of freedom. What starts off as a minority movement often mushrooms into a majority one.

[No] coercion, even of the most violent nature, rules out the possibility of freedom, although it may severely restrict that possibility. ... The degree to which personal freedom is restricted by compulsion on the part of the ruling classes in a state based on exploitation has varied historically. (Spirkin 1983).

Criticism and Response

According to the majority of scholars today and the mass media, Marxism is the polar opposite of freedom. It has no room for human creativity because Marx and Engels reduced all social phenomena and history to economic forces and gave a dogmatic scheme for the development of society. History has demonstrated that Marxism is a flawed and failed recipe for humanity. The nations that embraced Marxian socialism—USSR, China, North Korea, Vietnam—banned and sternly dealt with free expression. Opponents were imprisoned or killed in the hundreds of thousands. It reflected the Marxist notion of 'dictatorship of the proletariat'. Marxism aims to violently overthrow capitalism, a system based on freedom and democracy. It is a materialist philosophy in which the idea of free will has no space.

Marxists generally counter these criticisms by observing that the notion of a uniform scheme for development of human society emanates not from the works of Marx and Engels but from latter day Soviet literature and a biased reading of Marxist literature. The socialist revolutions in Russia, China, North Korea, Vietnam and Cuba displaced not democratic systems but brutal, murderous regimes supported by major capitalist powers.

Western capitalism was founded on genocide of indigenous populations, international slave trade and colonialism that extinguished millions of lives. Capitalist democracy emerged over centuries of bitter struggle against retrograde forces. Yet it is a pseudo-democracy as the influence of money in politics, racism, media monopolies, and the two party duopolies make voting a symbolic choice between parties with almost identical policies on basic economic, military and political issues. Differing on a few hot-button social issues, they engage in dirty mudslinging that gives an illusion of a vibrant, free polity. Capitalist democracy effectively guards a plutocracy.

History does not proceed in a smooth, progressive fashion. Freedom comes and goes. Faced with intense external opposition and internal class formations, the revolutionary socialist states degenerated into authoritarian state capitalist regimes. The initial attempts to develop socialism have withered. But socialism remains the best option to tackle the major problems facing humanity and ensure genuine freedoms. Struggles by the exploited classes continues to put socialism on the horizon. But, victory is not guaranteed.

Knowledge and Free Will

Knowledge is a perquisite for meaningful choice and responsibility. If you are aware of the possible consequences of your deeds, your responsibility for them is greater than if you are not. Well intended but ill-informed actions can have disastrous consequences. Ignorance leads to bondage, not freedom.

> *One cannot desire what one does not know. The core of freedom is conscious necessity and action, governed by the extent to which we are aware of that necessity, of the possibility of its realization.* (Siprkin 1983).

Information is power. Promoting awareness, it catalyzes change. An education system, media and state institutions that instill critical, scientific thought, and inculcate values of universal compassion and social justice are antithetical to a system based on exploitation, vast inequality and a biased system of justice.

Education has made major strides everywhere today. Literacy is high. People learn a wide range of subjects. Electronic, broadcast and print media outlets disseminate news and views that seem to span a broad spectrum of opinion. Robust laws protecting media freedom and organizations devoted to protecting free speech exist. Critical gaps need to be filled and overt censorship continues in quite a number of nations. But, the scope for exercise of meaningful choice based on adequate information seems to be higher than any time in history.

Nations of Europe and North America have no censorship boards and ideas that span from the extreme left to centrism and onto the extreme right float freely. The Internet is a game changer in this respect. But overall, a systemic mechanism of filtration of ideas that places especially leftist ideas into a small corner exists. It is an automatic, subtle process that generally ensures that ideas favorable to capitalism attain broad circulation and prominence. The prevailing ideas mask the fact that a small class has disproportionate access to resources and centers of decision making and power.

This mechanism instills relative ignorance, biased ideas and a skewed code of ethics in the popular psyche. Powerful economic and social incentives generate widespread conformism. Journalists, editors and media commentators understand the limits. A few hardy ones broach them on occasion. But in the long run their careers and financial security suffer.

One example sends the message to the rest. Ultimately, a nationalist, pro-capitalist vision rules the world of ideas.

Edward L Bernays, the 'Father of Public Relations' was the principal innovator in this field. Refined by psychologists and specialists, his methods are employed by major corporations, non-profit agencies, politicians and political parties, science organizations, churches, and governmental bodies, including the military. And, as meticulously documented by the studies of Edward Hermann, Noam Chomsky and other scholars, these methods are spectacularly effective.

Herbert Marcuse, a prominent radical scholar of the 1960s proposed that capitalism does not require force or a central authority to ensure broad conformity to the system. Critical thought is dampened by exaggerating the level of individual freedom. The market is flooded with consumer goods. A common national identity that covers the rich and the poor, the powerful and powerless, effectively curtails political discourse to that consonant with the existence of the system. The result is a one-dimensional human. Consider a crucial example.

War on Iraq

From 1991 onward, the US and UK governments peddled the line that Iraqi President Saddam Hussein was building up an arsenal of weapons of mass destruction that posed a grave danger to the Middle East region and the West. This perception was enhanced after the airline attacks on the Twin Towers in New York in 2001. Weapons of mass destruction and the 'war on terror' formed the basis for a draconian regime of economic sanctions and a sustained low level and high intensity bombing incursions over Iraq that culminated in the year 2003 invasion of Iraq. The propaganda crescendo against Iraq attained its climax in the speech at the January 2003 meeting of the UN Security by US Secretary of State Colin Powell. Armed with elaborate PowerPoint slides, he made an apparently solid case that Iraq, among other things, had 18 mobile biological weapons sites. It set the stage for the invasion of the nation.

Sanctions and sustained war consumed over a million Iraqi civilian lives, mostly children and the elderly. A prosperous nation was plunged into chaos, civil strife, malnutrition and rampant poverty. Yet, after Iraq was defeated, not a single nuclear, biological or chemical weapon was found by the invading forces. It had been a deliberate campaign of gross deception carried out by British and American governments from both sides of the political spectrum. No evidence that Saddam Hussein had anything to do with the Twin Tower attacks has ever been found. Yet, the majority of the American public without question accepted such allegations. How could such falsehoods dominate public perceptions in a highly educated nation that has the highest number and rate of college graduates in the world? Had blind patriotism made them take at face value the words of the politicians whose utterances on domestic issues deeply divide them? Was it an instance of nationwide cognitive dissonance? Support for the war effort increased dramatically once the invasion was mounted. But was it based on informed free will?

Iraq was one of the egregious instances of the falsehoods that have driven US wars and foreign policies for two centuries. Deceptive but effective campaigns—conducted by the Federal and state governments, major corporations, and deep pocketed vested interests—to mold public opinion on major and minor issues are a standard feature of life in America. On key issues, the truth struggles hard to prevail in the public mind. Is this an atmosphere in which informed free will can exist?

What about responsibility and accountability: By any objective assessment, American presidents—George Bush Sr, Bill Clinton, George Bush Jr—and British prime minister Tony Blair need to be tried in an international court for launching a systematic, premeditated murderous assault carried out under false pretenses on the people of Iraq. The citizens of the USA and the UK bear some responsibility for permitting their leaders to carry out such a barbaric deed and ignoring the truth when it came to light. The US doctrine that 'politics stops at the shore' implies a double standard of morality, one for domestic policy and another for foreign policy. The life of an Iraqi child has less worth than as that of an American child. Freedom conjoined with deficient morality is not a freedom worth having.

A Synopsis

Millennia of cultural evolution have embedded the striving for freedom and dignity firmly into the human psyche across the planet. Affirming the right to self-determination, communities struggle against external domination. People seek to be autonomous beings in charge of their own lives. In personal and family life, people want effective means to meet family needs and pursue goals. In civic and national affairs, they desire freedom and democracy.

Free will is linked to morality and responsibility. It is the basis for law and public policy, reward and punishment. Hard determinism and libertarianism, of secular or theistic variety, are unsound and morally wanting perspectives. Free will operates within the confines of biological, psychological, social, cultural, political and economic forces. Free will and action are in a state of perpetual struggle with determinism. Freedom entails awareness of and effort to transcend necessity.

Free will exists in a continuum. At times, the room for choice is wide and at times, it is narrow. It is constrained by genetic, neurological, psychological, social, cultural and economic factors. The Marxist analysis of free will includes these factors but also structural factors and class relations in a historic, dialectical fashion. Individuals and capitalists are afflicted by alienation and their free will is compromised, though in different ways. Qualitative enhancement and equality in actualization of free will requires structural transformation.

The Marxist depiction of morality and free choice as purely human features have not incorporated the findings of elements of choice and morality in animal behavior. It has not grappled with the evolutionary foundation of free will. With a few exceptions, the Marxist work on free will remains chained to repetitive, obsolete jargon.

Determinism and free will are pitted against each other in a yin-yang way. Free will loses its import in the absence of freedom of action. Noble free will entails thoughtful acts based on high moral standards in the face of restraints. It needs self-control to block impulsive acts and untoward habits and is intertwined with one's obligation to one-self, family, community and humanity. It is predicated upon knowledge essential to make meaningful choices. It entails overcoming fear and acting in ways that may displease the powerful. It means controlling biological instincts, narrow interests and mob-type behaviors. It embraces altruism and compassion as the standard operating procedure in life. Above all, it means collectively struggling against capitalism for a just, equal, environmentally responsible social order that will serve humanity, not the elites. Enlarging one's circle of responsibility to humanity and the ecosystem now and in the future and sacrificing personal interests for the common good are the most noble forms of free choice.

4.11 RELIGION AND FREE WILL

Formulations of free will and moral agency in varied religions range from strict determinism to libertarian free will. The religious visions on free will resemble the secular visions. The difference lies in the nature of causative factors.

Strict (Theological) Determinism: A supreme divine entity fully determines everything. Life is completely programmed. There is no free will.. Virtuous deeds and sins derive from destiny; humans have no agency. Only fate exists; there is no freedom.

Weak (Theological) Compatibilism: Human behavior is largely constrained by the will and plan of a divine power. Humans have little latitude of choice. Yet, they are responsible for their choices. Fate dominates freedom.

Strong (Theological) Compatibilism (Concurrence): A divine power set the Universe in motion according to certain rules. Humans have considerable latitude to act in the context of those rules. Humans are largely free agents with the moral responsibility for their own free acts. Freedom dominates fate.

(Theological) Libertarianism: A divine power created the Universe. Other than providing a code of ideal conduct, it does not intervene in human affairs. It does not influence human choices. Humans are free moral agents who are responsible for what they do. Freedom prevails; fate is immaterial.

The religious notion of free will is afflicted with a key paradox:

The Paradox of Free Will

An omniscient supreme being knows the past, the present and the future, including all human desires, choices and actions. Hence, at any moment, the future is predetermined. Yet, humans are bestowed with the ability to choose between right and wrong. They have some degree of free will. But if only the future known to the supreme being is possible, is that free will. Is not free will an illusion?

There are two related puzzles. If humans have unrestrained free will, then the supreme being does not or cannot exercise real control over them. Why should they pledge obeisance to him or pray to him? But if absolute divine determinism prevails, why should they worry about right and wrong, sinful and good deeds? Now we discuss how the four major religions navigate these logical conundrums.

Hinduism and Free Will

The diverse traditions within Hinduism generally subscribe to the idea of reincarnation and that a person's present state is a product of her actions in the previous life, *karma* (RPS 2022). They worship a supreme omniscient being (Brahman) who fully comprehends the past, present and the future. Yet, these deterministic effects on present life do not preclude the existence of free will. By abiding by her *dharma* (sacred duty) and making wise choices in her present life, a person can affect her future. Good deeds lead to a good life hereafter and bad deeds, to a *karmic* downfall. Fate and free will are reconciled within a temporal perspective:

> *Fate is past karma, free-will is present karma. Both are really one, that is, karma, though they may differ in the matter of time. There can be no conflict when they are really one.* CB Swaminah (Wikipedia 2021 – Free Will in Theology)

By positing that uncertainty and randomness were built into the nature of phenomena, and that light had both particulate and wave-like forms, the emergence of quantum mechanics in the twentieth century posed a strong challenge to deterministic thinking in science (Chapter 2). Such counter intuitive findings impelled several eminent scientists to dive into Hindu and Buddhist religious texts for a conceptual resolution. Quantum theorists like Erwin Schrodinger and Niels Bohr felt that these texts provided a philosophy integrating determinism with uncertainty (free will). Among other things, they delved into the *Upanishads*, a set of Hindu holy texts written over 2,500 years ago. These texts expound on the nature of the relationship between the supreme reality (Brahman) and the human soul (atman) and describe how spiritual reflection, meditation and following one's *dharma* lead to *karmic* transformation and salvation. It is proclaimed that there '*is only one universal self, and we are all one with it*'. (Kulkarni 2020).

To Schrodinger and Bohr, this principle provided the clue to matters like the commonality and synchronicity of reality for multiple observers even when their observations altered reality in dissimilar ways. For underneath all phenomena lay a unified consciousness, a universal mind. Discrete observers are an illusion (*Maya*). Physical reality arises from

138

consciousness; perception determines existence. It was an idealist interpretation of quantum mechanics which squarely challenged the materialist view that existence is independent of perception which had dominated the world of science since the advent of Newtonian physics. Perception changes reality, it does not just reflect it.

> *Myriads of suns, surrounded by possibly inhabited planets, multiplicity of galaxies, each one with its myriads of suns... According to me, all these things are Maya.* Erwin Schrodinger (Kulkarni 2020).

The scientists who favored the idealistic interpretations of quantum phenomena saw the *Upanishads* as a spiritual guide in a world abounding with moral and philosophical ambiguities. It filled a void created by alienation from Christianity. But they were not primed to worship Brahman or other gods. Spirituality was a psychological anchor in the confounding microscopic terrain in which not just the basic tenets of science but even common sense were falling apart.

In sum, Hinduism blends predetermined fate (*karma*) with the ability to make independent choices in life (free will) that will affect a person's future. But it does not satisfactorily resolve the incompatibility between free volition and omniscience. Brahman knows the future. Only one future is possible. Does that not make even a modicum of free will an illusion, an element of *Maya*?

Buddhism and Free Will

Like Hinduism, Buddhism accepts the notion of reincarnation. But whereas the former recognizes the existence of discrete inner souls (atman), Buddhism deems it a mirage (RPS 2022). No separate, permanent souls exist. Only an interconnected universal soul, a dynamic essence connected with all beings and the Universe prevails. Buddhism is posited as a middle way between objectification of self and total denial of self and reality.

> [The] *sense that each of us is an autonomous, non-physical subject who exercises ultimate control over the body and mind without being influenced by prior physical or psychological conditions is an illusion.* BA Wallace (O'Brien 2021).

The dynamic individual beings are subject to the doctrine of *karma*. Their present state was caused by their past actions and their present and future lives will derive from their deeds performed of their own volition in the present life.

Buddhism qualifies that genuine free will is limited by a myriad of social influences like greed, avarice, lust, acquisitiveness, hate, jealousy, despondency and selfishness. These influences engender suffering (dukkha), the main plight of humanity. To overcome it, people should follow the Eight Fold Noble Path. As the venue for attaining enlightenment, it requires being aware of one's negative proclivities, abandoning them and dispensing with egoism.

By jointly granting a room to determinism and free volition and action, Buddhism resembles Hinduism. While an element of predetermination to our lives exists, we also bear responsibility for our actions.

> *To study the Buddha Way is to study the self.*
> *To study the self is to forget the self.*
> *To forget the self is to be enlightened*
> *by the 10,000 things.*
> Japanese Zen Master Ehei Dogen (O'Brien 2021).

The Buddhist compromise between determinism and unrestrained free will is denoted as Dependent Origination. Human acts are affected by physical needs, external forces and internalized ideas. Free will is not exactly free. But it combats fatalism by teaching that people can improve their state, not just in a future life but even the present one. *Karma*, a strong conditioning factor, is not immutable destiny. It presents a range of actions and choices within which either continued dukkha or prospects for *moksha* are realized. Where you go depends on you, not fate. Preoccupation with the self leads people astray. The ultimate aim is to overcome self and attain a higher, eminently selfless state of being.

Many scholars have drawn parallels between some of the formulations in Buddhism and scientific analysis. The process of identifying the causes of human distress and prescribe remedies has been likened to performing medical diagnosis and giving treatment. Dispensing with the supernatural trappings of Buddhism, the dynamic interconnectedness of phenomena and dependent origination are sees as tenets resembling modern scientific thought. Some pronouncements of the Buddha are posited as the germs of the scientific method based on theory, experiment and validation.

> *Accept as completely true only that which is praised by the wise and which you test for yourself and know to be good for yourself and others.* Gautama Buddha (DK Publishers 2013; page 144).

Apart from the ideas of soul, a divine overseer, focus on future life and lesser emphasis on determinism, the Buddhist perspective on free will has much in common with Hinduism. In both, free will is limited free will. Buddhism delves on the moral consequences of current deeds, on personal responsibility, in the present life. And as it does not entertain the idea of an omniscient divinity, it circumvents the Paradox of Free Will.

Islam and Free Will

Islam is founded on the belief that Allah is the most powerful and most merciful supreme being who created all that exists. Its key tenet is absolute faith in Allah, his Prophet and the Quran. The doctrine of al-Qadar (Divine Predestination) asserts that Allah planned and has full knowledge of the past, present and future. He is cognizant in advance of all human acts, thoughts and desires.

> *Allah has created and balanced all things and has fixed their destinies and guided them.* The Quran 87:2-3 (Bitesize 2013).

Muslims are urged to study the Quran, abide by the five pillars of Islam and desist from bad deeds. Humans are endowed with the power to choose what to do in their lives. If they abandon virtue and embrace evil deeds, they will be sent to Jahannam (hell) on the Day of Judgment. Such beliefs led Muslim theologians to reflect on divine predestination from the early days of Islam.

> *How can we reconcile the two apparently contradictory facts that Allah has absolute power and sovereignty over all creation, and that at the same time we are responsible for our actions? Are we forced to do what we do, or are our choices meaningful?* (Parrott 2017).

One Islamic school stressed absolute free will while another focused on the paramoutcy of determinism. The former deemed humans as controllers of their destiny. Allah does not drive them toward evil. Influences of the wayward angel Azazil and shaytans (satanic devils) do. As free authors of their acts, they bear the responsibility for them. Upholding the supremacy of Allah, the latter school says Allah decrees everything. He has willed good and evil. All human acts arise from His plan. Humans are not accountable for what is beyond their power and control.

He leads astray whom He wills and guides whom He wills. The Quran 16:93
(Afsaruddin 2005).

Disputations between free will and determinism hav raged throughout the history of Islam.
Later day Islamic thinkers reconciled these disparate stands by stating that holding humans
responsible for their deeds does not negate the supremacy of Allah. Humans can make a
difference, good or bad, in their lives through their deeds. It is a power bestowed upon them
by Allah. We see a logical contradiction because our limited reasoning faculties cannot
comprehend the workings of the divine mind which exists beyond space and time. Instead of
futile speculations, we should have faith in Allah and read the Quran for guidance. Inspired
by its wisdom, we should take actions that accord with the path of righteousness. For
example, if we see a person who has fallen, we should not just say that it was the will of Allah
but help him stand up.

> *Let not one of you refrain from working for his provision, supplicating to*
> *Allah to provide while he knows that the sky does not rain gold and silver.*
> Al-Ghazzali (Parrott 2017).

Birth and death, illness and health, calamity and good fortune—Allah decrees all. Whatever
befalls us we should be thankful to Allah and instead of succumbing to resignation, lead our
lives with diligence and actions that are in accordance with His divine guidance.

Disputations over free will had significant connotations in the political realm. What
should Muslims do when the Caliph became an unjust, corrupt ruler? The deterministic view
saw him as the choice of Allah. His acts emanated from Allah and his fate would be
determined by Allah. Proponents of free will challenged this conformist position by arguing
that Allah has given humans the ability to judge between right and wrong, between justice
and injustice. Islam requires the Caliph to rule with compassion and fairness. Hence, people
can act with Allah's blessings to remove him if he is an unjust and oppressive ruler.

Muslim scholars extended the arguments over free will beyond the personal level. Al-
Ghazali, a prominent Islamic thinker and legal scholar of the Golden Age of Islam, held that
human acts were acts of free will and a part of Allah's divine plan. Humans did not have the
ability to perceive what Allah does, why He does it, or for whose benefit. But they have the
capacity to change their behavior and their surroundings. Ignorance, copying others and
temptation lead to evil acts. Education and instilling self-control are essential. Societal
measures are meaningful only in the presence of free will. Appropriate exercise of free will
enables humans to advance spiritually.

The diversity of opinion on free will was in part due to varied interpretation of different
passages from the Quran and the Hadiths. Finely parsed arguments and cryptic terminology
clouded the vision. Sunni and Shia scholars presented the issue differently. Yet, today a broad
consensus on the Islamic notion of free will has emerged.

> *Modernist Muslim commentators insist that the Quran should be read*
> *holistically. Taking certain verses out of context and interpreting them*
> *atomistically has been conducive to the view that the Quran encourages*
> *belief in predestination. Read as a whole, the Quran endorses, however, the*
> *concept of human freedom in choosing one's belief and of human*
> *responsibility for their actions. God has foreknowledge of human actions,*
> *but this divine knowledge does not compel humans to commit sin.* (Al-
> Jubouri 2004).

The consensus fine tunes but does not resolve the Paradox of Free Will. Yet, to the believer,
faith supersedes human logic and divine logic is beyond human comprehension. The Paradox
is thereby not a paradox.

Christianity and Free Will

Christians believe that there is one God who is manifested as a Trinity—God the Father, Jesus Christ, the Son of God and the Holy Spirit. Jesus Christ was sent to Earth to save humanity from damnation and death. He died at the cross but was resurrected. He will return to Earth on the Day of Judgment to save the faithful who will ascend to heaven. The sinners will be relegated to Hell. The Bible, the authentic word of God, provides an essential guide for life. Worship God. Life exists beyond bodily death. Love your neighbor. Life is sacred. And so on.

God, who created and controls the Universe, is supreme and perfect in power and knowledge. He loves everyone unconditionally. Christians can know Him through faith, prayer, spiritual experience and his grace. But not all Christian denominations and sects have identical stands on these beliefs (RPS 2022).

Humans have a God-given ability to freely choose right from wrong. The story of Adam and Eve—the first humans—in the Garden of Eden, is apropos. God allowed them to do what they pleased but also told them not to eat the forbidden fruit. They did and were cast out. The successive human generations are tainted with original sin, inclining them towards sinful actions. In His divine grace and love, God sent Jesus Christ to Earth to rescue humanity from sin.

Humans possess free will. God has not imposed restraints to sway them one way or another. Yet, He also knows the future and the choices humans will make. But he does not directly influence them. Humans are responsible for what they choose. He has provided a guide—the Bible—to enable them to make virtuous choices and take good actions in life.

> *For you have been called to live in freedom, my brothers and sisters. But don't use your freedom to satisfy your sinful nature. Instead, use your freedom to serve one another in love.* The Bible (Galatian 5:13).

How do Christians resolve the contradiction between foreknowledge and predestination, on one hand, and free, unrestrained choice, on the other? The Catholic Church does not address it directly. The Orthodox Church accepts free will but declares that to attain salvation humans must submit to the grace of God. The human is like a drowning man. He cannot save himself. God, in His divine grace, has thrown him a rope. The person is free to take it or not. If he takes it, he will be pulled to safety. It is his own free choice, to be with God or not. Lutherans say that humans possess free will in worldly affairs but cannot make rightful spiritual decisions without God's help. The inherited original sin inclines them to sin unless God acts to save them. Some Calvinists stress absolute divine determinism while others call for self-determined free will to overcome the burden of sin. Mormons say that strict predestination does not prevail. Humans are free moral agents under the grace of God. Jehovah's Witnesses claim that even as God is supremely powerful, he does not use his power to control all the events in his realm. Citing the Bible, they assert the primacy of free will in emphatic terms:

> *God dignifies us with free will, the power to make decisions of our own rather than having God or fate predetermine what we do. ... Our success or failure is not determined by fate. If we want to succeed at an endeavor, we must work hard. ... To a great extent, we can determine our future. ... God created humans in his image. And like our Creator, we have free will.* (JW 2021).

> *Remove grace, and you have nothing whereby to be saved. Remove free will and you have nothing that could be saved.* Anselm of Canterbury

Other denominational interpretations of free will and predestination also exist. The literature is sprinkled with arcane terminology, questionable logic and convoluted reasoning in attempts to reconcile complete foreknowledge and free will. The voluminous intellectual gyrations therein notwithstanding, the Paradox of Free Will remains unresolved.

4.12 FOES OR ALLIES

When the religious outlook governed life, science played the second fiddle. Scientists kowtowed to ecclesiastical authorities and took pains to stress the compatibility of advances in science with religious doctrines. The development of capitalism and intrusion of science and technology in all walks of life in the West empowered scientist and philosophers to mount strident challenges to the religious world view. Today, the pendulum has swung in the other direction. There is no dispute about the indispensability of science for modern society. Religion remains a key facet of life but more questions about its relevance and import are being raised.

Theologians, scientists, philosophers and lay persons give varied responses as to whether religion and science irreconcilable or compatible? Some says that religion is a relic of the past with no positive role in modern life. Some reject outright any scientific idea that seems to contradict what their holy books say. But most people allow the coexistence of religion and science.

The spectrum includes: Religious belief is outdated in the scientific age; The existence of God cannot be verified objectively; Humanity needs to accept rational thinking, not blind faith in God, miracles, prophesies and outmoded rites and rituals; Science cannot explain everything; It cannot explain what happens to us after death; As a source for moral guidance and spiritual solace, science can never replace religion; Religion gives meaning and purpose to life, science does not; Religion and science serve complementary ends, humanity needs both. Some of these issues were explored earlier. Now we explore them further starting with the views of three prominent geneticists. But before doing that we explain three ideas the often permeate the discussion on science and religion.

Creationism, Intelligent Design and God-of-Gaps

Creationism is the general label for doctrines claiming that nature and life were created in their present form by a divine being. Evolution is a myth. The doctrine of Intelligent Design argues that the complexity of life implies the existence of an intelligent supreme designer. Young Earth Creationism posits that what exists was created less than 10,000 years ago, and that humans were created in a span of six 24-hour days. Creationism derives from a literal reading of the Bible and deploying selective facts. US-based right-wing Christian denominations and think tanks are its principal drivers. Thus, it has more currency in the US, the most scientifically advanced nation, than anywhere else.

Assume a process A is attributed to divine origination in the holy book of some religion. And suppose that science has not been to explain it until now. The faithful had taken that as an indicator of the superiority of religion over science. Now suppose further that recently a scientific explication of A has been attained. How do the faithful react? Often they point to another process X in the holy book for which a scientific explanation is still lacking and use it to restate their claim of the superiority of religion over science. This mode of reasoning is called the God-of-the-gaps argument. Since science is replete with phenomena that are unexplained, it states that science will always be incomplete while religion is complete. Science has many gaps. Only God can fill all the gaps.

To the faithful, it is an emotionally appealing way of thinking. But in the long run, it is a self-defeating stand. Science is not static; what it cannot explain today, it may explain tomorrow. The God-of-the-gaps stand puts the faithful on an ever receding path that progressively damages the credibility of their faith. To invoke God as a filler of gaps is to demean God.

The God of the Bible, however, is much more than a god of the gaps. Christians believe that God is always at work in the natural world, in the gaps as well as in the areas that science can explain. (BL Editors 2019).

But even the faithful who reject the God-of-the-gaps line of thinking make a few exceptions. Among them is the fact that science has not been able to explain why the basic laws and constants of the Universe are so fine-tuned as to make a fragile phenomenon like life possible. They see such issues not as temporary gaps, but ones that science can never fulfill. They lie beyond the purview science, and are a pointer to the existence of a divine being.

Three Geneticists

Francis Collins, an honored and award-winning molecular geneticist and physician, has directed the US National Institutes of Health since 2009. He contributed to the uncovering of the genetic basis of Huntington's disease, cystic fibrosis and other diseases, led the international consortium that in June 2000 decoded the entire human genome, and organized the establishment of a comprehensive catalog of human genetic variations. These developments, he feels, have laid the basis for precise elucidation of the molecular basis of disease, estimating the genetic risk of disease, and tailor-made individualized therapies. He cautions that such potentialities of modern genetics must address associated ethical and legal implications. He has also promoted genetic research in Africa.

Collins was raised in a setting where religious observance was minimal, and music, arts and science took the center stage. By the time he joined college, he was an atheist wedded to a scientific outlook on life and the Universe. In graduate school, his views reversed. His clinical experience showed that faith mattered to many of his patients. He then undertook a study of world religions, in the course of which CS Lewis's influential evangelical text, *Mere Christianity,* exercised a strong influence on his outlook. Exposure to religious ideas and reflecting on the wonders of nature induced a change of heart. He became a devout Christian. Yet, that change did not dent his attachment to science. Religion and science are, for him, complimentary pursuits.

Faith and science are two ways of knowing. They have to answer different questions. Science answers questions about 'how,' faith answers questions about 'why'. Francis Collins (Liu et al. 2009).

Consider some of the key tenets of science: The Universe began 13.6 billion years ago from a point mass with the Big Bang. The finely calibrated basic constants of physics led to the formation of stars and planets. The Universe is governed by the basic laws of physics. Complex life forms and intelligent life on Earth emerged from the process of evolution. To Collins, such findings point to the presence of a divine power. Thus, at the announcement of the decoding of the human genome in 2000, he stated:

It is humbling for me, and awe-inspiring to realize that we have caught the first glimpse of our own instruction book, previously known only to God. (Wikipedia 2021 -- Francis Collins).

Collins regards creationism an unfounded, pseudo-scientific idea, and with a few exceptions, rejects the God-of-the-gaps doctrine. Both undermine the supremacy of God. Generally ascribing to the Darwinian Theory of Evolution, he holds that science and religion are eminently compatible. His bestseller, *The Language of God: A Scientist Presents Evidence for Belief,* criticizes secular atheism and anti-science fundamentalists and urges people of faith to applaud science since it enables us to appreciate the grandeur of God.

The key tenets of his position are: God is beyond the natural world, beyond space and time. It is irrational to apply the logic of science to God. Automatic recourse to God to explain what science cannot currently explain is foolhardy. It is logically untenable to assert that there is no God. The Theory of Evolution is a solid scientific theory. The Bible creation story should not be interpreted literally. God set in motion a process that led to the appearance of intelligent life with free will and a sense of religious morality. The existence of similar moral codes across time and place point to a divine origin of morality. Altruist, self-sacrificing behavior is not an evolutionary trait but one engendered by God. When does embryonic life begin? That is a religious question, not a scientific one. He cares about humans and intervenes in their lives. He can perform miracles. Science and religion are two sides of the same coin.

> *The God of the Bible is also the God of the genome ... He can be worshiped in the cathedral or in the laboratory.* Francis Collins (Paulson 2016).

Collins uses the case of Mother Teresa, the most respected Christian personality of the 20th century, to demonstrate the influence of God in human affairs. He admires her for comforting, giving refuge to and tenderly healing thousands, and performing miracles. But on this score, Collins is ill-informed. The saintly image of Mother Teresa was a media driven false image. The reality was grotesque. She cavorted with dictators and fraudsters. Many of her centers were dens of intense suffering and neglect. And her miracles were disputed by the people at the scene (RPS 2022).

Francis Crick was a scientific luminary whose interests spanned molecular biology, biophysics and neuroscience. His role in the discovery of the structure and function of the DNA molecule earned him and two other scientists the Nobel Prize for Physiology or Medicine in 1962.

He read science books and did chemistry experiments, photography and glass blowing with his uncle from an early age. After the age 12, he refused to attend church services. His main aim was to explore the differences between life and non-life, and elucidate the nature of consciousness—two issues on which he felt religion had provided misleading answers. As a self-declared humanist, he was an outspoken critic of religion, especially Christianity. Unlike Francis Collins, Crick felt that there was no room for coexistence of science and religion.

> [Human] *problems can and must be faced in terms of human moral and intellectual resources without invoking supernatural authority. .... [Due to progress of science the] simple fables of the religions of the world have come to seem like tales told to children. .... The God hypothesis is rather discredited.* (Wikipedia (2021 – Francis Crick), Anonymous 2003).

Crick said the idea that a non-material soul survives after death is '*an imagined idea*' and argued that religious education and religion inspired topics like creation science should have no place in public school systems. Together with 21 other Nobel Prize winners, he cosigned the Humanist Manifest issued in 2003 (RPS 2022).

James Watson, a famed molecular biologist and zoologist, was one of the four people who discovered the structure and function of the DNA molecule. From 1968 to 2007, he occupied leadership positions at the renowned Cold Spring Harbor Laboratory (CHSL) in the US. His three textbooks on molecular biology set a new standard for clarity of exposition, and remain in print.

As a peace activist, he protested the US aggression on Vietnam and advocated ending nuclear proliferation. Yet, he abandoned leftist causes because he felt that the Marxist view on genetics was misguided. Of recent, he has earned notoriety for his insensitive remarks about homosexuals and people who are overweight. Deeming 'stupidity' a disease, he has called for measures to cure '*really stupid*' people. A firm advocate of genetic screening and genetic engineering, he says that beauty, especially among women, can be genetically engineered. He is also confident that full knowledge of the genetic variations will lead to effective personalized preventive and curative medical therapies.

James Watson says that he is a scientist, not a racist. Yet, he believes that IQ differences between blacks and whites are due to genes, Jews are intelligent, Chinese are intelligent but not creative, and people with darker skin have a more active libido. On account of his racist comments, he has faced bans from public appearances. The CHSL has disassociated itself from Watson and stripped him of his honorary titles.

Like Francis Crick, his fellow associate in genetics, Watson is an unapologetic atheist. Though born a Catholic, his father's doubts about the existence of God tempered his religiosity. After becoming aware of the Catholic Church's support for General Franco's fascist regime at the start of WW II, he stopped attending church services. In his view, learning science makes one less likely to accept religious stories, which he brands as *'myths from the past'*. His opinion on the potential benefits of genetics is remarkably optimistic.

> *Only with the discovery of the double helix and the ensuing genetic revolution have we had grounds for thinking that the powers held traditionally to be the exclusive property of the gods might one day be ours.*
> James Watson (Anonymous 2003).

Together with Francis Crick, Watson signed the Humanist Manifesto of 2003.

Here we have a trio of towering figures in molecular biology and genetics whose perspectives on religion are widely disparate, and whose moral tenets also bear unpleasant stains.

Three Scientists

In a year 2018 interview, three faith inclined scientists indicated their views on the concordance of religion and science (Salleh 2018).

Jennifer Wiseman is an astrophysicist whose fascination with the starry sky and space exploration began in childhood. Credited with the discovery of a comet, her research focuses on the formation of stars and planets. Even as science is a marvelous, indispensable tool for exploring and learning about the cosmos, it cannot provide satisfactory answers to profound questions like: What is our significance in the Universe? What is the meaning of life? Only religious belief gives the answers. God, who is responsible for the Universe, loves humans. The ability of humans to conduct complex investigations is a gift from God. Current pictures of a major conflict between science and religion are media generated exaggerations. Scientific work can deepen faith. The Bible is a historic document that needs to be interpreted in the context of the era in which it was written. It is not a science text.

Andrew Harman is an immunologist who directs medical research projects in Australia. His areas of focus are HIV and Crohn's disease. After rejecting the idea of a creator God and Christianity, he spent his teen years in a cynical, anchor less mood feeling that life was futile. But that dissipated after he discovered Buddhism during a trip to Asia. Today he is both a scientist and an ordained Buddhist minister. A major appeal of Buddhism is that it does not entertain a creator God. Buddhism is a faith based on experience that acquaints you with the true nature of reality. Religions based on blind faith are incompatible with science.
Attachment to things, exacerbated by modern lifestyle, is the main source of human suffering. Buddhist practices combat psychological conflict and instill love and spiritual peace. The emphasis on detachment enables you weather the ups and downs of life. Buddhism is not in line with a purely reductionist mode of science. Teaching mindfulness and mediation at his place of work and elsewhere, he opines:

Science is about learning, Buddhism is about living. ... I think Buddhism and science are absolutely in tune with each other fundamentally. They're both driven by the idea that you can't just believe something without any evidence. Andrew Harman (Salleh 2018).

Fahad Ali investigates plant genetics. He feels that genetic modification is a potent tool for developing effective therapies for disease and enhancing global food production. He grew up in a strong Islamic culture. His school textbooks disparaged the Theory of Evolution. But when he read Darwin's *Origin of the Species*, he accepted its validity. As a gay man, he was disturbed by the unfair treatment of women and homosexuals by Muslims. Subsequently, he disavowed his faith. Yet, when his mother contracted cancer, he felt a spiritual vacuum and returned to Islam. Now he holds that the Quran and the Bible should not be interpreted literally. Belief in God provides meaning and a sense of purpose to life. The Quran teaches basic values like kindness and generosity and promotes rational thinking. God is not a God of gaps. It is inadvisable to invoke God when science has not been able to explain certain phenomena. Historically, Muslim scientists and philosophers had made major contributions to mathematics and all areas of science. Now, even though the products of science are commonly used, science and critical thinking are disparaged in many Muslim nations. Militant atheists and religious fundamentalists have created a false chasm between science and religion. Evolutionary theory and the Big Bang theory are in consonance with the idea of an all-powerful creator God.

Here we have the case of three working scientists who, in their own ways, see an essential harmony between religion and faith. But each looks at a single religion. Exploration of harmony between faith systems is not paid due attention.

NOMA

Stephen Jay Gould was a distinguished biologist and paleontologist who together with colleague Niles Eldredge proposed the theory of punctuated equilibrium as an addendum to the conventional Darwinian evolutionary model. The latter posits that species formation occurs through gradual accumulation of minor changes; the former holds that evolution is marked by long periods of relative stability occasionally interrupted by a sudden emergence of species. Beside several highly regarded specialist treatises on evolution, Gould authored numerous widely read general books and magazine articles on evolution and science. A celebrated media personality, he held high positions in scientific societies and received numerous awards for his service to science.

Gould grew up in a secular Jewish family. His father had Marxist political views, but he was at ease with the ideas of C Wright Mills and Noam Chomsky. As a university student, he was active in the US civil rights and social justice causes. Rejecting theories that ascribed an evolutionary basis to complex human behaviors, he held that evolutionary psychology and socio-biology were pseudo-scientific, deterministic doctrines that rationalized racism and patriarchy. A man of multiple talents and interests ranging from music to baseball cards and rare books, he was conversant in English and four other European languages.

Convinced that the existence of God could not be proved or disproved, Gould grew up as an agnostic and remained one throughout his life. He felt it fruitless to try to reconcile religion and science into a single worldview or to posit them as either-or entities. They are disjoint domains of human life addressing separate but essential needs. His seminal book, *Rock of Ages: Science and Religion in the Fullness of Life*, converted Gould into the most celebrated non-religious scientist advocating peaceful coexistence between religion and science. The book presents an erudite, historically grounded case for the principle of Non-Overlapping Magisteria (NOMA), namely, that religion and science are disjoint, non-antagonistic domains of human purpose. Science connotes logic and evidence; religion connotes morality and meaning—disparate topics governed by disparate methods. Like water and oil, they do not mix.

As a scientist dedicated to rational thought, Gould did not accept notions like God, soul and life after death. Yet, spiritual beliefs are deeply ingrained in the human psyche. Scientists should not deny or demean them. And theologians should accept that issues like the emergence of the human species cannot be explained via recourse to holy texts.

The schism between religion and science is due to the influence of minority, fundamentalist tendencies in the religious world. The association of religion with secular power in the past:

> [generated] *the historical paradox that throughout Western history organized religion has fostered both the most unspeakable horrors and the most heart-rending examples of human goodness in the face of personal danger.* (Gould 1997).

Advocacy of NOMA does not entail accommodating all faith-based views. He rejected creationism, had no sympathy for the literal interpretations of the Bible and publicly campaigned against introducing 'Creation Science' in school curricula. Pointing out that in general lay people grant separate spaces to religion and science, he noted that the Catholic Church, most Protestant denominations and major global faiths do not reject the Darwinian evolutionary theory. The opposition to evolution theory is mostly driven by a few hard line, evangelical denominations in the US.

The NOMA principle received a mixed reception in the religious world. Those who interpret the Bible in a literal fashion rejected it outright. But many people of faith appreciated the notice of respectful coexistence with science from an eminent scientist. Some otherwise liberal minded theologians were uneasy. By ruling out a personal God who intervenes in human affairs, answers prayers and performs miracles, NOMA put science above religion. It selectively deploys the ideas of some theologians to construct an abstract religion and does not consider the major scriptural and moral differences between different faith systems. For a person of faith, they said, religion must reign over science:

> *The First Commandment was and is and will remain that we shall have no other gods before the one God, and that means that the conversation between science and religion is not and never can be a debate among equals.* (Rhodes 1999).

NOMA did not generate any widespread opposition in the scientific circles. Compared to people not directly engaged in scientific pursuits, scientists are more likely to be non-religious. But a significant portion of the scientists tend to be religious. Among the non-religious scientists, many adhere to a non-theistic form of spirituality. Most scientists see no cause for open warfare with religion. Live and let live, that is their creed. The US National Academy of Science considers religion and science as two separate realms of human experience.

Alfred North Whitehead

Mathematician, philosopher Alfred North Whitehead, who collaborated with Bertrand Russell to produce *Principa Mathematica,* a premier work on logic and the foundation of mathematics, formulated a singularly distinct perspective on the connection of religion to science. His basic tenet was that reality is formed of processes, not a collection of discrete entities. Processes interact with other processes to generate new realities.

Whitehead declared that religion and science express two fundamental facets of human life. Both are ever developing processes, though science is intrinsically more adaptable to change than theology. He blamed the manner of expression of religious ideas for the conflict between religion and science. Thus, he faulted the presentations of the shape and motion of Earth, the delineation of geological time, and natural evolution in religious texts. One example he sites is from *Christian Topography*, a 535 AD book by Cosmas, a well-traveled monk, that declared that the world is a flat parallelogram with length twice its breadth.

Nonetheless, religion is not always in error. Galileo held that the Earth moves, and the sun is fixed. The Inquisition said that the Earth is fixed and the sun moves. Newtonian cosmologists pointed out that both of them move. Yet, according to the Theory of Relativity, depending on the frame of reference, any of the three assertions can be true. What is critical is not the assertion of fact but the manner of arriving at it. Religion and science are not distinguished by details but as distinct ways of thought.

He called for tolerance within and between science and religion. The faithful should not explain away new scientific discoveries. A self-defeating endeavor, it demeans both science and religion. Religion will always remain on the defensive, making undignified retreats.

The idea of a supremely powerful God who judges and imposes his will on humanity is not an essential feature of religion. It expresses a desire for mental comfort, moral order, disdain for evil deeds, and sanction for wrongful conduct. Religions can be a vehicle for good deeds as well as for immoral deeds that harm many humans.

To have a future, religion must steer away from alleged facts, rites, rituals, blind worship of divinity and sanctifying worldly authority. It should modulate a deep message of spirituality with a system of truths that mold character. An ossified religion is a dead religion.

> *The clash* [between religion and science] *is a sign that there are wider truths and finer perspectives within which a reconciliation of a deeper religion and a more subtle science will be found.* (Whitehead 1925).

Like some other eminent scientists, AN Whitehead singled out Buddhism as a particularly ennobling spiritual entity.

> *The future course of history would center on this generation's resolving the issue of the proper relationship between science and religion, so fundamental are the religious symbols through which people give meaning to their lives and so powerful the scientific knowledge through which we shape and control our lives.... And it is in regard to this troubling issue, I think, that Eastern religions, particularly Buddhism, are seen to hold out the promise of achieving some resolution.* (Wikipedia 2021 – Alfred North Whitehead).

And like his fellow like-minded scientists, the Buddhism he talked about was the idealized, Westernized Buddhism that was not grounded in the realities of Buddhism in Asia (RPS 2022).

While SJ Gould envisioned a state of harmony between science and existent religion, AN Whitehead looked forward to peaceful coexistence between science and a stripped down, spiritual version of religion. Both are forms of NOMA.

Against NOMA

NOMA is rejected by scientists who identify with the creed known as New Atheism (RPS 2022). To them, it is a tactical sophism that abrogates the basic principles of science. We consider the views of two prominent New Atheists.

Richard Dawkins, an eminent evolutionary biologist, author of best-selling popular works on evolution and a firebrand New Atheist sharply excoriates the NOMA principle. He holds that all faith systems derive from superstition. There is no moderate faith system. Concessions to beliefs in divine powers empowers religious extremists whose agenda is nothing short of absolute control of the political process, education and imposing their own version of science in the public mind. Assessing the danger to be acute in the US, he calls upon scientists and scholars to eschew complacency, stop appeasing *'sensible'* religious tendencies and firmly combat the *'American Taliban'*. To him religion is an infectious virus that infects the human brain from early childhood.

Dawkins holds that the religion is not an inviolate entity from a separate magisterium. It can be interrogated with the methods of science. He also stresses the importance and nature of hypothesis testing in science.

Science and the Negative

Suppose A is a proposition of a factual nature. It may be true or not true. Not-A is a negative proposition—that A is not true. In general, science cannot prove a negative proposition with one hundred percent certainty. But it can investigate it, amass the associated evidence and estimate the probability that it is true.

Science does not proclaim eternal truths. All laws and propositions of science are subject to revision. Evidence based uncertainty is a strength, not a weakness that permits you to keep an open mind and refine your knowledge, or at times, adopt a very different theory.

Many people believe in flying saucers. Some people claim that they have been visited by extraterrestrial beings. Science cannot definitively prove that these are false beliefs. But it can investigate them. And from what we know thus far, the probability that they are true is extremely low. Similarly, science cannot disprove the claimed virgin birth of Jesus Christ. Yet, it can investigate the issue.

> *Either Jesus had a father or he didn't. The question is a scientific one, and scientific evidence, if any were available, would be used to settle it. The same is true of any miracle - and the deliberate and intentional creation of the Universe would have to have been the mother and father of all miracles. Either it happened or it didn't. It is a fact, one way or the other, and in our state of uncertainty we can put a probability on it - an estimate that may change as more information comes in.* (Dawkins 2006).

For the God hypothesis, he asserts that:

> *We cannot, of course, disprove God, just as we can't disprove Thor, fairies, leprechauns and the Flying Spaghetti Monster. But, like those other fantasies that we can't disprove, we can say that God is very very improbable.* (Dawkins 2006).

Dawkins holds that virtually all religious beliefs can be put under scientific scrutiny. Determination of the origin of the Universe in its present state is a task of science, not theological speculation. Branding the cosmological first-cause arguments about the presence a supreme entity as flabby arguments, he counters them with formulating one in the following style:

> *Some people are truthful; some other people are more truthful; and thus, there is a supreme being that is immeasurably truthful. And that is God. Now substitute 'dishonest' for 'truthful'. The same logic applies. So, God simultaneously is the most honest and the most dishonest being. Is that possible? If it is, is it a God a religious person would accept?*

Dawkins finds the God-of-gaps argument a spurious argument. He notes that how life began on Earth is a subject where science has not made much progress. But it is a subject of ongoing studies. Invoking God to fill this gap or the notion of Intelligent Design does not take us anywhere. But Dawkins differs from Collins on a major issue: Explaining why or how the Universe is fine tuned to generate and sustain life in some corner. To him, it is not a gap that science cannot, in principle, fulfill. Difficult it is, it has to be investigated by reason and evidence and not automatically assigned to a divine origin.

Dawkins is not impressed by the arguments from complexity and design either. Complex living entities did not arise from pure chance or the act of a supreme creator. Human organs like the eye and the brain did not emerge in an instant or over a short time period. They developed over millions of years of accumulation of simple changes (random mutations) that were beneficial from the viewpoint of survival. Given time, simplicity can create complexity and order, as explained by the Theory of Evolution. In his path-breaking work, *The Blind Watchmaker*, Dawkins makes an illuminating, evidence-based case for how the human eye evolved in a gradual process encompassing tens of millions of years.

> *Design is a workable explanation for organized complexity only in the short term. It is not an ultimate explanation, because designers themselves demand an explanation.*

> *Natural selection is so stunningly powerful and elegant, it not only explains the whole of life, it raises our consciousness and boosts our confidence in science's future ability to explain everything else.* (Dawkins 2006).

Ascribing the existence of physical laws and the narrow range of conditions that have made life on Earth possible to divine design is also an untenable proposition. Even if such conditions are present only in a planet circling one in a billion stars, in a Universe of billions of galaxies with billions of stars in each, there may be billions of planets with life-friendly conditions. And that is not considering non-carbon-based life forms, and the possibility of multiple Universes.

Dawkins is particularly irked by the notion of a personal God who chooses to answer or not answer prayers, reward or punish human beings, and affects changes through miracles like parting the sea. He calls that image of God a *'pernicious delusion'*.

While acknowledging that religion has contributed in important ways to human culture —language, literature, art, music, architecture—overall, he feels that religion has done more harm than good. Religious belief does not have a protective health benefit. It inculcates conformist, dogmatic attitudes in children from a young age, and inhibits the flowering of a questioning, scientific spirit. Religion is a divisive force as it promotes separate schools, discourages inter-faith marriage and rationalizes violence and terrorism. It is not a prerequisite for morality. Many passages in the scriptures glorify immoral conduct. Dawkins accuses Gould of invoking religion in an abstract sense, and not adequately addressing the historic and social realities—commendable and reprehensible—associated with different religions.

He dismisses those who invoke the words of eminent scientists like Albert Einstein to support religious belief. Einstein rejected the idea of a personal, interventionist God or a supernatural realm. He was a pantheist awestruck by the majesty of nature. For him, a spiritual sense of wonder was a prime driver of science.

Aware of the strong emotional attachment people have to religion, Dawkins does not support curbs on religious practice. But in school, children should be taught science, not superstition.

Victor Stenger, physicist, philosopher and popularizer of science, was as well an outspoken critic of religion, mysticism and paranormal phenomena (RPS 2022). Two of his well-known works, *The New Atheism: Taking a Stand for Science and Reason* and *God: The Failed Hypothesis: How Science Shows That God Does Not Exist*, reflect his clear disdain for NOMA. A compromise between science and pseudo-science weakens science.

> *Religion and science are like oil and water. They might co-exist, but they can never mix to produce a homogeneous medium. Religion and science are fundamentally incompatible.*

Victor J Stenger.

Convinced that religion exercises a deleterious influence on science research, business activities, politics and social policies, he favored control on interference of religion in public affairs. The key tenets of his thoughts: Religious beliefs, including the existence of God, are valid objects of scientific inquiry. There is no credible evidence for the doctrine of Intelligent Design. Universe and life would exist as they are whether a divine designer was present or not. The Universe is not fine-tuned to make life possible; life is fine-tuned to operate within the constraints imposed by the laws of physics. The laws of physics are not inconsistent with the idea that another Universe existed prior to the Big Bang. Science and human senses provide evidence beyond reasonable doubt that God does not exist. Free will and consciousness, if they exist, are potentially amenable to scientific explanation. The God-of-the-gaps form of reasoning is an escapist option that detracts from tackling daunting subjects.

Stenger took a keen interest in the origin of codes of morality and ethics. He observed that basically all religions propound an obligatory moral code said to be of divine origin. They imply the morality of the secularists, atheists and people who believe that humans descended from monkeys cannot be trusted.

With a focus on the United States, Stenger gives evidence from varied sources to argue to the contrary: That believers and non-believers adhere to similar moral standards. He indicates that religiosity is associated with higher rates of abusive and criminal conduct. The rates of child molestation, domestic violence are higher in conservative Christian families than in the non-conservative families, and religious people have higher incarceration rates than people with a secular disposition (Stenger 2006). The accuracy of some of his sources is, however, questionable. And, as he concedes, the evidence does not necessarily indicate a causal relationship between criminality and religiosity. Confounding factors like income, racism, joblessness, and bias against recording white collar crime need to be accounted for as well.

Yet, discrimination against women and child abuse are serious problems in the institutions of all major religions. Often the leaders of these religions adopted a defensive posture that hid the scope of the problem for decades. Victims got little assistance and the clerical perpetrators, but a slap on the wrist.

Sociological studies and media investigations generally show that criminality and violent conduct tend to be more common in religious nations and regions as compared to the more secular nations and regions (Baird and Gleeson 2017, Capps 1992, Chalabi 2015, Ellison et al. 2007, Fusch 2013, Gleeson and Baird 2017, McMaster 2020, Schwartz 2021, Zuckerman 2015). The proposition that religiosity reduces immoral conduct and secularism enhances it is not a tenable proposition.

Stenger argues that similar moral tenets—prohibition of killing, violence and stealing and valuing honesty, bravery, charity, altruism and compassion—in different social orders under different faith systems signifies a common need to maintain social stability and survival, not divine origination. He points out that such moral tenets had existed in societies that predated monotheistic organized religions. They have an evolutionary significance from both a biological and cultural standpoint and are seen in rudimentary forms among animals as well.

Stenger notes that major religious scriptures like the Bible and the Quran contain contradictory moral messages that historically have been used to promote respect for human rights as well as rationalize major violations of human rights. (RPS 2022). The modern codes of morality, law and justice resulted from human striving for a humane, democratic social order, and are not attributable to a divine being. Values and morals arise from 'our common humanity'.

We humans decide what is good by standards that have not been handed down by God. (Stenger 2006).

Despite their bold proclamations, in practice, the objectivity and moral tenets of the stalwarts of New Atheism—Richard Dawkins, Sam Harris, Christopher Hitchens and Victor Stenger—were shaped by the strong sociopolitical prejudices prevailing in the Western nations. Though he retracted some later, Dawkins made remarks that could only be interpreted as pandering to racism and patriarchy. And most importantly, in a patently Islamophobic fashion, they single out Islam as the main source of violence and human suffering in the world. In the name of the 'war on terrorism' they supported the brutal imperial aggression on Iraq and the Middles East in general, notwithstanding the fact that Saddam Hussein had no connections with the Twin Towers attacks in New York and had done much to fight fundamentalist movements on his own turf. To them, Islam, not religion as such, was the main enemy of 'civilization'. And curiously, in that respect, as Dawkins acknowledged, these champions of science and rationality shared the same platform as their declared enemies, the hardcore right wing Christian fundamentalists in the United States (RPS 2022). A more telling abandonment of 'our common humanity'—the total disregard for the immense suffering of the children and people of Iraq—would be hard to find.

Theologians for NOMA

The histories of Christianity and Islam disfavor the common perception that religion and science (faith and reason) have been perpetual antagonists (RPS 2022). The relationship between them has varied over time and place. There were times when institutionalized religion censored and persecuted rationalists and scientists who strayed beyond acceptable bounds and there were extended periods during which religious authorities sponsored academies and research in diverse scientific disciplines. Many scientists of the past and present were/are deeply religious as well.

The persecution of astronomers Giordano Bruno and Galileo Galilei is often cited to show the intrinsic hostility of the Catholic Church to science. But that is a simplistic picture. For centuries, the Church sponsored centers of learning where science was taught, and scientific research was carried out. Many famed scientists were practicing Catholics:

Since the Renaissance, Catholic scientists have been credited as fathers of a diverse range of scientific fields: Nicolaus Copernicus (1473-1543) pioneered heliocentrism, Jean-Baptiste Lamarck (1744-1829) prefigured the Theory of Evolution with Lamarckism, Friar Gregor Mendel (1822-1884) pioneered genetics, and Fr Georges Lemaltre (1894-1966) proposed the Big Bang cosmological model. The Jesuits have been particularly active, notably in astronomy.

Wikipedia (2022 - Science and the Catholic Church)

For nearly a century after it was formulated by Darwin, the Catholic Church maintained a discreet silence over the Theory of Evolution. And since the 1950s, several Papal encyclicals have proclaimed or implied that belief in evolution theory is not incompatible with Christian faith. In October 2014, Pope Francis extended the pronouncements by stating that the Big Bang and evolution reflect the work of God.

God is not a demiurge [demigod] or a magician, but the Creator who gives being to all entities. ... Evolution in nature is not opposed to the notion of Creation, because evolution presupposes the creation of beings that evolve.
Pope Francis (McKenna 2014).

153

Coming from the divisive American climate, Stephen Gould was surprised to encounter evolution-friendly attitudes among the Catholic clergy in Europe. They were perplexed by the debates over creation science and Intelligent Design raging in the US since they had been taught that a Christian could simultaneously accept Darwinian evolution and the idea of God as the ultimate power behind all phenomena. Accepting evolution did not entail surrendering the notion that the soul was created by God. The biology syllabuses of the Catholic schools in the US and elsewhere include the Theory of Evolution.

The Papal pronouncement on evolution was reinforced by the address of Cardinal Pietro Parolin, the Vatican Secretary of State, to a gathering of scientists, journalists, educators, politicians and theologians in 2020. The Cardinal urged science and religion to join hands to deal with the major problems—environmental despoliation, effects of modern communication technology on the youth, cynical rejection of authority and replacement of personal interaction by virtual contact, conflict between nations and cultures—facing humanity. The survival of life and humanity demanded an interdisciplinary approach that included not just varied scientific disciplines but also the wisdom derived from faith systems. Underscoring the unequal impact of these problems on nations and the importance of faith for motivating people, the Cardinal declared:

> *If we want to survive and if we want life on this planet to survive, then we still have to learn to assume a responsibility for our common home on the global level. At the same time, science by itself is not enough to resolve this problem.* Cardinal Pietro Parolin (Esteves 2020).

Neurotheology

Are religious or spiritual experiences—feeling of oneness with the divine and enlightenment—linked to activation of specific regions of the brain? Has evolution hardwired religion into the human brain? Does God communicate with humans through brain signals? These issues are among the concerns of a new discipline known as Neurotheology.

Neurologists explore such issues in several ways. One way uses a special helmet that injects weak magnetic fields in the temporal lobes of the brain. The aim is to see if it induces a sense of presence of divinity. Another way takes account of experiments linking hyper-religiosity, fainting episodes, and other behaviors to temporal lobe epilepsy. Neuroimaging investigates whether hyper-religiosity is just an emotional state or a religion specific state. Is there a 'God spot' in the brain? One study found that activating certain brain centers produced feelings like those found in intense devotional activity. It concluded that doctrinal religious practice generated emotive rewards. Another study claimed that altering brain signals generates transcendental states akin to the Buddhist experience of oneness with the Universe. Another approach used EEG scans of meditating subjects to find alteration in brain signals and whether such signals can generate perceptual clarity.

The third way employs psychoactive substances like psilocybin that stimulate the temporal lobe. These studies suggest that such substances produce feelings that resemble those found in devotional contemplation and mystical activity.

Neuro-theological studies covered rituals, meditation, chanting, praying, and speaking in tongues. Their subjects were nuns, Sikhs, Buddhists, Pentecostals, students, volunteers and elderly persons. Leading neurologists concluded that there is convincing evidence for associating neural activity with religiosity. Some add that religious activities produce long term positive changes in the brain. And these findings pertain to a variety of cultures and religious traditions.

The interpretation of these studies divides the field into two schools. One school holds that the results demonstrate a neuro-genetic basis for religiosity. Faith in the divine is an evolutionary trait that produces mentally soothing states and assists survival under conditions that induce helplessness and hopelessness. Organized religion persists because it confers mental and social benefits. The other school goes a step further to propose that the results point to the presence of the divine. Brain signals are the conduits through which God influences humans.

Neuro-theologians combine the religious and scientific visions of human life. The spiritual and biological aspects operate jointly and treating them as such will improve both physical health and mental well-being.

The validity of neuro-theological assertions has been critiqued on several grounds. Most studies in the field have had small sample sizes and small duration of follow up. Bias reducing techniques like control groups and double blinding have been rare. Some studies making strong claims were reported in conference proceedings but not published in peer reviewed journals. Follow up studies to replicate reported claims have mostly obtained disappointing results. Behaviors like intensive reading that may produce similar temporal lobe signals have not been given due attention. And, social and environmental factors associated with religiosity were not factored into the analysis.

That religious forms of behavior are genetically hard wired into the brain is a claim that rests on a shaky scientific foundation. And to extrapolate that it shows how God communicates with humans is to stretch the imagination. Basically, it is wishful thinking. How does the same type of hard wiring of the brain leads to diverse expressions of religiosity forms like Hinduism, Buddhism, Christianity, Islam and many more remains far from being resolved.

Overall

The above exploration of the relationship between religion and science leads us to posit four basic views on the nature of this relationship.

Scientific Incompatibility: Science doubts and inquires. Religion accepts via faith and scriptures; doubt is not permissible. The two are diametrically opposed perspectives.

Scientific NOMA: Science deals with facts; religion deals with values. They can live in harmony so long as they operate within their own domains. When they differ, science reigns.

Theological NOMA: Science deals with facts; religion deals with values. They can live in harmony so long as they operate within their own domains. But when they differ, the word of God reigns.

Theological Incompatibility: The word of God is the ultimate truth. Science is a useful tool, a gift from God. There are some questions science can never answer. The answers can be found only through faith and prayer. Assertions of science that contravene what has been revealed by God must be rejected outright.

In our exploration of where scientists and people of faith stand in relation to these four options, the focus thus far has been on specific personalities. It revealed that people engaged in routine scientific work as well as eminent scientists can be found all over the NOMA spectrum. Yet the question remains: How does the scientific community generally stand in relation to belief in God and NOMA?

Surveys in Western nations during the twentieth century found higher levels of disbelief among scientists than among the general public. About two-fifths of the scientists believed in a personal God, about two-fifths did not and the rest were agnostic. In the recent years, disbelief has been rising among scientists and the public, but the gap between them remains large. In the US around year 2010, four-fifths of the public believed in God but only one-third of the scientists did so; among the public, one-tenth did not believe in God but believed in a higher spiritual power but two-tenths of the scientists belonged to that group; and only 1 of 25 members of the public rejected belief in both God and a spiritual power while fully 4 out 10 scientists held negative beliefs towards both forms of deities. Belief in God varied by gender and discipline among the scientists but the variations were small. But the age effect was noticeable. Four out of 10 scientists of age between 18 to 34 did not believe in God while 3 out of 10 scientists aged 65 or more years held that opinion.

In most European nations, the level of disbelief is higher among both the public and the scientific community and the gap between them is not as large as in the US. But in the nations of the Global South the situation is reversed. Here the level of belief is higher among both the public and the scientific community and the gap between them is not as large as in the US.

An international survey of 22,000 physicists and biologists found that the large majority saw that science and religion could coexist in separate domains without conflict. Only a small minority of the non-religious scientists anywhere are actively hostile to religion. Many scientists who did not believe in a personal God nonetheless retained a sense of spirituality and appreciated the moral aspects of religious belief. The US National Academy of Science considers religion and science as two separate, legitimate realms of human experience. The doctrine of NOMA did not generate widespread opposition in the scientific circles.

4.13 BUDDHISM AND SCIENCE

Buddhism has garnered a special status among scientists and scholars of religion. Prominent scientists have cast it in a distinctly favorable light. Niels Bohr and JR Oppenheimer opined that it may answer the puzzling observations in quantum mechanics and explain the dual nature of light and Heisenberg's uncertainty principle. The agnostic mathematician and philosopher Bertrand Russell, who stridently critiqued monotheistic faiths, ranked Buddhism more favorably. But he held Confucianism in a higher regard.

> *Buddhism is a religion in the sense in which we understand the word. It has mystic doctrines and a way of salvation and a future life. It has a message to the world intended to cure the despair which it regards as natural to those who have no religious faith. It assumes an instinctive pessimism only to be cured by some gospel.* Bertrand Russell (IDR Labs 2021).

Elsewhere he opines:

> *Among present day religions Buddhism is best. The doctrines of Buddhism are profound, they are almost reasonable, and historically they have been the least harmful and the least cruel. ... Buddhism does not really purpose the truth; it appeals to sentiment and, ultimately, tries to persuade people to believe its doctrines which are based on subjective assumptions not objective evidence.* Bertrand Russell (IDR Labs 2021).

The moral values and social practices of Buddhism are more elevated than those of Christianity:

> *I think I should put Buddha and Socrates above* [Jesus Christ] *in those respects. ...* [Unlike Christianity], *Buddhism has never been a persecuting religion.* (IDR Labs 2021).

Russell's perception of the history of Buddhism is inaccurate (RPS 2022). In the past and at present, Buddhism has been enmeshed with political power, and has supported war, violence, and fascist persecution. Yet, his positive estimation of Buddhism influenced many secular and non-secular scientists.

Albert Einstein was not attracted to dogmatic religious traditions that invoke a personal, creator God who punishes and rewards humans. Yet, he often alluded to the cosmic religious sense, the sense of wonder at the magnificence of nature, a sense that was akin to the highest (mystical) experiences of traditional religions.

[The] *cosmic element is much stronger in Buddhism.* Albert Einstein (Wikipedia 2021 - Buddhism and Science).

Positive references from eminent luminaries raised the profile of Buddhism in the scientific world. After its adoption by the counterculture movement in the 1960s, Buddhism made inroads into the popular consciousness in the Western world. New monasteries and orders sprang up. Many converted to Buddhism. Today, many people visit Buddhist temples. Mindfulness and meditation training are a major industry. Buddhist books are bestsellers. Research into the psychological, behavioral and health effects of meditation is conducted on a large scale. Major corporate entities use meditation and mindfulness programs for their employees (RPS 2022).

Many scientists in the West sympathize with Buddhism. Some meditate and have converted to Buddhism. Immunologist Andrew Harman is not a rarity. The relationship between science and Buddhism is a key area inquiry. Among the conclusions drawn are: Both Buddhism and science discourage belief based on blind faith. Both stress unbiased (non-judgmental, impartial) investigations and deal with causation in a rational manner. Both require extended training and persistence and do not speculate about a spiritual domain but with life here and now, and not with a divine creator who intervenes in human affairs. Both represent areas of systematic knowledge about the nature of reality. Buddhism explores the inner reality (self and mind), science explores the external reality (physical and biological nature). Buddhism employs an experiential (meditative) approach, science uses the experimental, empirical approach. Buddhism is qualitative and contemplative, science is quantitative and rational. Buddhism promotes spiritual advancement, science promotes material advancement.

Going beyond NOMA, these views imply that Buddhism and science are not separate human endeavors but have much in common in terms of assumptions and method. Since science is blind about human matters like beauty, meaning, values, responsibility and humanness, it has much to learn from the Buddhist teachings on this front. The centuries of Buddhist wisdom on mental processes has effective techniques for healing mental afflictions and promoting desirable conduct. Overall, it is claimed that there is extensive room for a fruitful dialog between science and Buddhism.

Some scientists and Buddhist scholars claim that from a methodological perspective, Buddhism is a science. The teachings of the Buddha contain the essence of the modern scientific method.

> *Accept as completely true only that which is praised by the wise and which you test for yourself and know to be good for yourself and others.* Gautama Buddha (DK Publishers 2013; page 144).

A theoretical astrophysicist has declared the contemplative steps scientists adopt in dealing with evidence and synthesis is not just akin to what humans do in daily life but is also in line with how Buddhism promotes the acquisition of knowledge and wisdom. While humans approach the complex multidimensional reality by examining one facet at a time, ultimately, they need to be brought together in a holistic manner to form the true picture (Hut 2003).

Proposals to marry physics and science with Eastern mysticism gained steam in 1975 with the publication of *The Tao of Physics: An Exploration of the Parallels between Modern Physics and Eastern Mysticism*. Written by physicist Fritjof Capra, it has seen several editions and has been translated into 23 languages. Upon the explication of the basics of modern physics and various Asian mystical-cum-religious traditions, *The Tao of Physics* argues that all phenomena in the Universe—subatomic particles, stars and all in between—are interconnected. No entity in the Universe is static. Everything is perpetually in a state of flux. Entities exist in a dual form, yin and yang. The interaction between the forms propels change. The Theory of Relativity posits integration of space with time, and of energy with matter, while Quantum Mechanics reveals the interdependence of the observed and the observer, and that light acts in a wave-like form and a particulate form. Such principles of unity, dualism and dynamism have been known in the Asian mystical traditions—Hinduism, Taoism and Buddhism. Now they are being verified by modern science. The Asian mystical traditions and science include rational as well as intuitive modes of acquiring knowledge. Broader recognition of this fact will benefit science and human spiritual progress. Science and spirituality are not enemies. They need to cooperate to foster a world view that links external investigation, experiments and mathematics with introspection.

> *Science does not need mysticism and mysticism does not need science. But man needs both.* (Capra 1975).

Another acclaimed work is the 1983 book, *The Dancing Wu Li Masters: An Overview of the New Physics*, by spiritual teacher Gary Zukav. A best-seller, it has been praised for the simplicity and clarity of exposition of modern physics. Endorsing tenets similar to those proposed by *The Tao of Physics*, it declares that by questioning the existence of an objective reality and bringing psychology into physics, Western science is catching up with Eastern wisdom and recognizing that in essence humans are not just material but spiritual beings, integral to nature.

A recent work, *Buddhism & Science: Breaking New Ground*, goes further. Edited by AB Wallace with contributions from scientists, philosophers, Buddhist scholars and the Dalai Lama, it deals with physical and cognitive sciences. Not only does Buddhism provide an ethical corrective to social problems caused by science but also that major ideas in modern science are reflective of long known Buddhist principles. Take the Theory of Relativity. The Buddhist philosophy of the Middle Way—balancing between total self-denial and excess indulgence—is a relativist principle. And so is the Buddhist idea of emptiness that posits the universal relativity of nature. The famous equation $E = Mc^2$ is a confirmation of the Buddhist doctrine of the ever-changing, impermanent nature of things. The concept of non-locality (interdependence) employed in Quantum Theory is a basic Buddhist premise about the nature of life and reality. Cognitive sciences have much to learn from Buddhism. The new theories about the human nervous system and mind reflect precepts embodied in Buddhist meditation practice and philosophy. Emboldened by the erudition of the contributors, the editor declares:

> *Buddhism, like science, presents itself as a body of systematic knowledge, about the natural world, and it posits a wide array of testable hypothesis and theories concerning the nature of the human mind and its relation to the physical environment. These theories have allegedly[?] been tested and experientially confirmed numerous times over the past twenty-five hundred years, by means of duplicative meditative techniques.* (Wallace 2003).

Perhaps the most authoritative stand on the relationship between science and Buddhism emanates from the leading face of global Buddhism, Tenzin Gyatso, the 14th Dalai Lama. Convinced that it is essential to learn modern science, he fosters the teaching of science to Buddhist monks. He urges scientists to attend to the accumulated wisdom of Buddhism. Advocating the integration of the two worldviews, he has sponsored international conferences and talked at scientific meetings. Expounded in books and talks, his viewpoint is cogently synthesized in *The Universe in a Single Atom: The Convergence of Science and Spirituality* (Dalai Lama 2006).

> *My confidence in venturing into science lies in my basic belief that as in science, so in Buddhism, understanding the nature of reality is pursued by means of critical investigation.* (Dalai Lama 2006).

If a Buddhist precept is decisively contradicted by science, which should prevail? Unlike the leaders and thinkers of all major faith traditions, the Dalai Lama says that scientific evidence supersedes belief.

If scientific analysis were conclusively to demonstrate certain claims in Buddhism to be false, then we must accept the findings of science and abandon those claims. (Dalai Lama 2006).

A Critique

The voluminous literature promoting the complementarity of Buddhism and science has spawned sharp criticism from other scientists. *Buddhism and Science: A Guide for the Perplexed* by DS Lopez is a strongly critical work. We consider the main lines of criticism in this and similar texts. A number of these points were noted in RPS (2022).

Buddhist Exceptionalism: All the major Asian and non-Asian religious traditions have developed mystical practices and theories. One case is the Sufi school within Islam. They contain ideas similar to those expounded in Buddhism and can be interpreted in similar ways in relation to science. Buddhism appears to stand out only because of the exceptional attention accorded to it by Western scholars and media.

There is a basic query: What is Buddhism? As conceptualized in the West, it is an idealized, esoteric set of principles and practices taught by the Buddha which are bereft of social history. It is a Buddhism shaped by Western scholars in the 18th and 19th centuries whose knowledge of Asia and Buddhism was superficial. Seeking a faith without the philosophical and social limitations of Christianity, they overlooked the hundreds of incompatible subdivisions within Buddhism. Where differences are spotted, only those that are consistent with the ideas of the unity of science and Buddhism are stressed. It is not real life Buddhism.

The claim that Buddhism (or any religion) provides the antidote to the moral and psychological vacuum supposedly created by science is belied by history. Operating in conjunction with political and commercial power, Buddhist sanghas have been associated with war, violence, gross violations of human rights and genocide. The firm backing by Zen Buddhist monks and scholars of Japanese fascism and the enthusiastic participation of Buddhist monks in the ethnic cleansing of the Rohingya in Myanmar are but two aspects of that sordid history (RPS 2022).

To regard Buddhism as a non-doctrinaire evidence-based outlook is to ignore the multitude of Buddhist schools that contain rigid ideas about nature and reality that contradict observed facts. Some Buddhist schools reject the existence of an objective reality while other schools embrace it. Many contain patently flawed beliefs.

In 1938, the Tibetan intellectual Gendun Chopel wrote a newspaper article explaining to his compatriots that the world is round, rather than flat, chiding his fellow Tibetans for being the last Buddhists to deny the planet's true shape. He explained that the Buddha himself knew that the world is round but withheld this fact from his disciples because they would not have believed it. (Lopez 2021).

Theory of Evolution: The concept of *karma*—the present is caused by the past—is fundamental to Buddhism. Buddhist sacred texts do not mention evolution. But Buddhists implicitly or explicitly accept it at a general level. It is not viewed as inconsistent with *karma*. But for a key tenet, the story is different. Random mutations play a central role in biological evolution. Randomness, however, is difficult to reconcile with *karma*. What happens to a person is due to what he/she has done. Change is caused; it does not happen by chance.

From the Buddhist's perspective, the idea of these mutations being random events is deeply unsatisfying for a theory that purports to explain the origin of life. (Dalai Lama 2006).

159

Randomness has no room in the Buddhist ideas of rebirth into one of six realms. Gods, demigods, humans, animals, famished ghosts and relegation to hellish domains. Where you will land in the next cycle of life depends on your acts in the present one, not on chance.

In relation to random mutations, the Dalai Lama deviates from his stated view of the supremacy of science over traditional belief. He rejects random effects without adducing any evidence. That view contradicts the stand of Buddhism-oriented luminaries, himself included, on quantum theory. Probabilistic models and reasoning are central to quantum theory. Why accept randomness here and not in evolution?

According to the lore in key Buddhist texts, spiritual pre-humans evolved first, gradually took human form, adopted human habits, and were divided into males and females. Animals appeared after humans. The Dalai Lama, while accepting the Darwinian Theory of Evolution, neither mentions nor renounces these unscientific beliefs that are taught to and accepted by millions of Buddhists in Asia.

Buddhist Cosmology: The Big Bang theory posits a single time point of emergence and unidirectional development of the Universe. Supported by much observational data, it is the prevailing model in cosmology. But other models of the Universes have also been proposed. A firm scientific consensus on the issue has yet to emerge.

Buddhism holds that the Universe undergoes a four-phased process of emergence, existence, decline and destruction in a periodic manner. These phases are called *kalpas* (eons). Each *kalpa* has 20 sub-*kalpas* of similar length. The Universe is cyclical with no beginning or end. One cycle may not replicate the previous one. Moreover, there are cycles within cycles. The expansive canon of Buddhism contains variants of this general process with differing details about each phase. Their time scales, usually in the billions of Earth years, are also described.

According to one rendition, the first phase occurs after a primordial wind blows to create a new world upon the destruction of the previous one. Beings from a higher realm are born into one of the multiple lower realms. Human predecessors (demigods) who fly in air, do not need sustenance and live for thousands of years, emerge. Over time they lose their spirit-like properties, begin to eat food and gain weight. Humans as we know them appear. Division into males and females gives rise to sexual activity and presages emergence of social divisions, rulers and government. Meat is consumed. Vices—avarice, theft and violence—become more common. Life span reduces from thousands of years to less than a hundred years. (Wikipedia 2021 – Buddhist Cosmology). Other Buddhist traditions have somewhat different descriptions of this creation phase.

Buddhist cosmology has thirty-one realms of existence based on location, and the physical and mental features of the beings. Each realm has three sub-realms. Some realms are planar, and others are spherical. Some have deities, others have humans and yet others, animals. Different mental states permeate the realms. And the distances between these realms vary from tens of kilometers to hundreds of millions of kilometers. The position and motion of heavenly bodies, the sun and the moon, within some realms is also described.

One realm is the Earthly human realm in which a ring of mountains is surrounded by a vast ocean which in turn is encircled by a mountain ring. Four continents with distinct topographies, dimensions and natural features exist in this realm. For example, in one continent:

> *Aparagodaniya is ... located in the west, and is shaped like a circle with a circumference of about 7,500 yojanas. ... The tree of this continent is a giant Kadamba tree (Anthocephalus chinensis). The human inhabitants of this continent do not live in houses but sleep on the ground. Their main transportation is Bullock cart. They are about 24 feet (7.3 m) tall and they live for 500 years.* (Wikipedia 2021 – Buddhist Cosmology).

There are hellish realms in which the intense physical pain and mental anguish bestowed on their residents fall in a category that would shock even seasoned writers of horror novels (RPS 2022).

The Buddhist creation story and descriptions of the Universe are akin to the myths of virtually all religions—highly imaginative tales of morality and life. The far fetched layout of the Universe bereft of a scientific foundation hardly accords with the Big Bang theory. Only by overlooking a multiplicity of such perplexing details can the claim of Buddhism being a scientifically oriented religion be made to hold water.

Tradition has it that when the Buddha was asked by disciples about the expanse of the Universe and if it was finite or infinite, he first refrained from answering the question. He also discouraged people from delving into such matters as they were immaterial to understanding and resolving suffering. A total of fourteen similar queries were proscribed. Yet, Buddhist sages over the ages violated this edict and presented the physical and temporal nature of the Universe in fine detail. And even the Buddha ventured into them later on. The initial stand of the Buddha hardly promotes the scientific spirit of unrestrained inquiry. And when such an inquiry was undertaken the result was something that does not stand up to rigorous scrutiny.

Meditation and mindfulness have been promoted in many faith systems. The Sufi schools in Islam is one case. That they have moderate beneficial emotional, psychological, functional and health outcomes have been verified by scientific studies. They help people navigate through stressful times and life difficulties and cultivate resilience. But the claims about their benefits made by devotees abound with exaggerations. Novel forms of meditation and mindfulness practices have been developed for controlling blood pressure, inducing weight loss, dealing with substance abuse and harmful habits and other problems. Science learns from the past. It is a part of human scientific and cultural development and attests to the ongoing creativity of humans. On the other hand, corporate entities employ mindfulness and meditation to lower work tension, enhance productivity and serve the bottom line. As it promotes individual solutions to social problems, mindfulness and meditation serve to weaken collective organizing to affect social change—a blessing for the elite and the establishment.

Until some two centuries ago, most creative, inquiring activities occurred under the auspices of religious authorities. Education was the purview of priests and monks who spent years in mastering language, religious texts, and law as well as medical and scientific knowledge. It is of no surprise that much of the early discoveries in mathematics, science and technology were associated with religion. Some of these thinkers had far reaching insights that are pertinent to morality and science to this day. But there were also times when religion directly hindered the growth of scientific knowledge. When claiming that an idea from some religion has a scientific basis, these historical realities must be borne in mind. And it has to be recognized that often such claims, even when made by prominent scientists, are hand-waving arguments akin to clutching at straws.

Buddhism allegedly transcends the four modes of association between science and religion—Scientific Incompatibility, Scientific NOMA, Theological NOMA and Theological Incompatibility. Unlike other faith systems, it is supposed to have an integral, complimentary link with science. But that is a flawed point of view. In terms of method, primary tenets and major theories, science is as far from Buddhism as it is from any other religion. To posit a direct link between the Theory of Relativity or Quantum Theory and Buddhism does a disservice to both. Morally and factually, it is better to ascribe to SJ Gould's notion of non-overlapping domains, and let science and Buddhism adhere to their own domains.

4.14 WHICH RELIGION?

Historically, over 4,000 faith systems that include divine beings have been recorded. Yet, in discussions of religion and science, 'religion' often implicitly denotes a monotheist belief system. An equally important issue is: How do the varied faith systems relate to each other? During the colonial era in Africa, Western Christian missionaries implored the people to abandon 'pagan' beliefs and adopt the teachings of the Bible. The fundamentalist fringe of any religion regards its belief system to be superior to the rest. Some Muslims see Hindus as idol worshipers, some fundamentalist Christians declare Hindus as pagans and so on. Televangelists use adroit tactics to mystify their flock and extract large donations from them. Today it is considered impolite to rank religions. It is a recipe for acrimony and possible deadly consequences (RPS 2022).

The beliefs of one religion often contradict the beliefs of other religions. Their creation stories differ markedly, as, for example, between Islam and Hinduism. Mirza G Ahmad, the founder of the Ahmadiyya tradition wrote that Jesus Christ survived his crucifixion, went to Kashmir, India where he died at the age 120 years (RPS 2022). A grand, extant shrine in the area is said to be his tomb. To an Ahmadiyya this is a fact. Yet, Christians reject it outright.

Comparative Religion, a vast field of inquiry, systematically compares religious faiths in terms of historical origin, beliefs, rituals and rites. The diverse economic, political and social roles of varied religions are addressed. Yet, the empirical veracity of religions is rarely gauged along scientific lines. Using objective scientific criteria, can one say this religion is more true that that one? Or that this religion is authentic but that one is a cult?

The Family Federation for World Peace and Unification, popularly known as the **Unification Church**, is a case in point. It was founded by Sun Myung Moon in 1954 in South Korea. It is a Christian sect that claimed a global membership of 3 million in the 1980s. Now, after the death of the founder, the number has declined. The Church has two principal scriptural texts, the Bible and the *Divine Principle* written by SM Moon and one of his disciples. Reverend Moon was regarded by his faithful as a messiah who had a visitation from Jesus Christ, and was entrusted with completing the mission of saving humanity begun by Christ.

Official Emblem of the Family Federation for World Peace and Unification

The Unification Church has a number of doctrinal differences with main line Christian denominations. It questions the virgin birth of Christ, the nature of his divinity and ascribes to 'indemnity' as a key way of rescuing humanity from its spiritual abyss and establishing the Kingdom of God. The Church is famed for elaborate mass marriage and marriage renewal ceremonies and follows its own burial rites. Regarding the Christian cross as a divisive and elitist symbol, the Church decries its use. Aspects of Confucian beliefs are also blended within the theology of the Church.

Advocating harmony between religions, races and nations, the Church has worked with Muslim and Christian sects on social issues and sponsored several international interfaith meetings. Holding religion and science as complimentary endeavors, it has sponsored international meetings on issues like food scarcity, renewable energy and the environment. Yet, it has also promoted the pseudo-scientific doctrine of Intelligent Design as the explanation for the presence of life on Earth.

The Church has a major footprint in the educational, social services, business, political, communication, artistic and sporting realms. It runs several academies, sponsors academic organizations, and operates non-governmental organizations and peace foundations. Church owned companies control right-wing newspapers, publishing houses, press agencies and TV networks in the US, South Korea, Japan and South America. Its followers in both the US and Japan are thought to exceed 100,000. The Church and its members own automotive firms, computer companies, manufacturing concerns, a ship building firm, for-profit healthcare centers, hotels, and commercial real estate. The business arm of the Church is a major source of funds for the Church.

The basic planks of the Church's political program are unification of Korea, international peace and anti-communism. Accordingly, leaders of the Church are linked to influential personages and conservative causes across the globe. Japanese billionaires with fascist sympathies and far right politicians in the US are linked with the Church. The secular activities of the Church are justified as extensions of the drive to establish the Kingdom of God on Earth. Reverend Moon was convicted of tax evasion in the US. The Church has been accused of maintaining secretive ties with state intelligence agencies. And it has close ties with conservative churches and politicians in the US and Latin America, and anti-communist Islamic parties, and counts media celebrities, academics, and major politicians among its supporters.

Various Christian denominations have called Reverend Moon's teachings blasphemous. Main line Christian denominations see the Unification Church as a cult with heretical views. It brainwashes its members and induces them to make large donations. Yet, to his followers, he was a true parent and messianic savior who would return humanity to the Lord.

The Unification Church was brought into international limelight after the former prime minister of Japan, Abe Shinzo, was assassinated in July 2022. The killer was disgruntled by Abe's association with the Unification Church, which he accused of causing financial ruin to his family. In the subsequent investigation it came to light that about half of the members of parliament of the Liberal Democratic Party, the former ruling party in Japan, had some links with the Unification Church.

There are no objective criteria to determine what is a cult and what is an authentic religion. Who is to decide whether Reverend Moon or the Pope is a genuine Christian leader with a special spiritual status? The absence of such criteria implies that general discussions about the relationship between science and religion are incomplete without the inclusion of the relationship of science to specific religions. And that issue has to be pursued using objective, fair scientific criteria. General assertions like NOMA or perpetual antagonism will not take us far.

4.15 THOUGHTS

Religion and science emerged from a long evolutionary process in a conjoined fashion. Science was borne of practices like learning about food and water sources and predators, and religion was the basis for codes of behavior beneficial to the collective. While animals exhibit learning and morality at a rudimentary level, the relatively larger size and complexity of their brains and development of language enabled humans to accumulate a larger body of knowledge, form more complex codes of conduct and integrate them into a communal belief system. In the early days, empirical knowledge explained but a few things. The unexplained was ascribed to deities. Stories of human origin and worshiping and appealing to deities evolved. Fact and faith were intertwined. Shamans connected humans to gods.

As society became stratified, technology advanced, specialized labor emerged and writing developed, performing rites and rituals for the gods, exercising moral authority and acquisition of knowledge was monopolized by a priestly stratum operating under the authority of kings and emperors. The keepers of faith were the discoverers of the intricacies of nature. Religion and science were in the same hands.

Faith-based and fact-based mental constructs developed their own internal momenta deriving from varied social needs, mode of stratification and creative vagaries of the human mind. A tradition of secular, rational thought and empirical inquiry that viewed natural and social phenomena in a more organized fashion, and formulated theories about them evolved. Philosophers and scholars asked: What is life? Does it have a purpose or meaning? What is the shape of the Earth? How far is the sun from Earth? What happens when ice changes to water and water turns into steam? What are the functions of the different body organs? How can metals be purified? Can copper be turned into gold? How can we accurately measure distances, the flow of time, and speeds of moving objects?

Some queries reflected practical concerns; some were speculative. Both heralded a separation of faith-based and empirically-based ways of thought. Natural philosophy, an admixture of astronomy, physics, logic, biology and philosophy, emerged along with theology as an explanation of nature and the human condition. Yet, apart from a few places like ancient Greece and early Islamic societies, intellectual activity was still dominated by priests, monks and theologians. Even under the purview of religious authority, scientific disciplines, old and new, continued to develop right into the feudal era.

Revolts against Papal authority, the development of parliamentary rule in England, the French Revolution, the emergence of capitalist formations in Europe and the US, and European imperial ventures set in motion a qualitative break between science and religion. A convoluted process of experimentation, observation and analysis carried out by major scientists flowered the maturation of grand theories in physics, astronomy, chemistry, geology, biology, health as well as psychology and the social sciences.

Now the edifice of science stands on its own as a primary and integral part of the human condition vital for economic and material progress. Religion is no longer the main framework within which people understand the natural and social reality. Its role in this regards has receded in a dramatic fashion even as its emotive and cultural significance persists.

Together with art and culture, science and religion are the major pillars of society. Their relative importance differs in different parts of the world. In some nations, religion and scripture are not only the sources of moral authority but also the basis of the legal code.

Morality: Ethics and morality pertain not just to religion, but also the education and legal systems, the media and science. Basic ethical tenets like respect for the truth, intellectual integrity, open cooperation, and clear, full disclosure of study methods and findings are essential for the progress of science. But in practice scientists circumvent, in major and minor ways, these central tenets. Religious people and leaders also frequently betray the ethical principles they are meant to uphold. Neither religion nor science are guarantors of morality.

Endurance: A common perception is that religious ideas endure and scientific ideas change, sometimes drastically. It is a historically flawed perception. All religions have evolved over time, adopting new ideas and jettisoning old ones. Interpretation of holy texts has varied across different denominations and nations.

Falsifiability: Scientific propositions are testable and potentially falsifiable. Yet, there are exceptions. That an objective reality exists, a key premise of all science, cannot be put to test. Experiments to determine the speed of light always look at the journey of light signals from a point of origin to a destination and back to the origin, a two-way trip. Calculating the speed by dividing total distance by time taken assumes that light travels at the same speed in both directions. A one-way determination of the speed of light needs the presence of synchronized clocks at the origin and destination. Yet, ensuring that the clocks are synchronized requires knowledge of speed of light! And nothing can travel faster than light. It is an unresolved issue. And there are many more.

Limits to Knowledge: People of faith assert that the human mind lacks the capacity to comprehend God and His wisdom. Science asserts that no statement of fact is beyond scientific inquiry. Purely subjective emotions such as love and hate can be studied by sociologists, psychologists and neurologists. Yet, Quantum Mechanics has established fundamental limits to knowledge. It is not possible to know with exactitude both the position and momentum of a particle at one point in time. This is not a matter of measurement error. It is an intrinsic limit to simultaneous precision. No measuring instrument with zero error exists. In some fields, a small error drastically alters the conclusions.

Chaos theory shows that a minute alteration in the initial state of a system can significantly alter its development. The presence of intrinsic randomness bears upon the precision of the conclusions. The computation needed to exactly solve the formulas for interactive effects in a relatively simple system lie beyond any conceivable computer. These are insurmountable limitations.

Take a clinical trial to assess the utility of cognitive behavioral therapy for amelioration of compulsive behavior. One group gets educational literature only while the other gets the literature plus behavioral therapy. The trial does not find a significant difference between these approaches. Yet, it has been established in many medical, psychological and sociological studies that when people are enrolled in scientific investigation, it tends to alter their behavior in ways that reduce the outcome differences between the groups. Subjects may independently obtain the relevant information, get external assistance or otherwise modify their behavior. The magnitude of such an effect, the Hawthorne effect, is difficult to disentangle. The applicability of the study findings to the routine clinical setting, especially when the outcome difference is small, is then cast in doubt.

Religion and science both experience insurmountable limitations on what can be known. Religion resolves it through faith. Science confronts it by creative, enhanced resolve. For example, one cannot but be astonished by the precision with which the rest mass of a proton, a subatomic particle, is known. It is 1.67262×10^{-27} kg.

Free Will: Can a person have conscious, self-generated desires and act in ways based on her own free thoughts? Religion and science tackle this query via four avenues—determinism, soft determinism, soft libertarianism, and libertarianism. In the religious view, God is the determinant. In the scientific view, determinism arises from elemental forces—atomic, genetic, neurological, psychic or societal. In the former view, anything other than strict determinism confronts the Paradox of Free Will. If God fully knows the future, can there be free choice? Strict determinism and pure libertarianism are minorities in both domains. Believers ascribe moral deeds to God's grace and evil deeds to influences of the devil. Secular scholars posit causal forces acting in unison with free will and differ by the type and strength of the causal factors they emphasize.

Responsibility: A greater level of belief in free will—secular or religious—is associated with a greater emphasis on personal responsibility. Studies show that belief in free will tends to make people act in ways that are considerate of the possible consequences. On the other hand, thinking that life follows fate or forces beyond one's control induces indifference about whether one's act will harm oneself or others.

Believers in free will tend to judge acts of others with a harsh standard. If you commit a crime, then you deserve to be punished. If you killed, then you deserve the death penalty. On the other hand, those who hold that human behavior is largely governed by external circumstances stress the need to change social conditions to reduce the incidence of crime.

Existence of God: Religion asserts the existence of God, divine beings (gods, angels) and divine realms beyond the observable. Science says it is an assertion lacking credible evidence. The religious respond by saying that science has not provided proof for the non-existence of God. Scientists hold that the burden of proof for any proposition lies on those who advance it. If a person is charged with a crime, then the prosecution has to provide evidence that shows beyond a reasonable doubt that he did it. The non-existence of proof is not equivalent to a proof of existence. If a person does not have a valid alibi, it does not necessarily show that he was at the scene of the crime.

A variety of arguments for the existence of God have been advanced. These include the argument from the final cause, the argument from design and complexity, the fine-tuning argument, and arguments from morality. When interrogated in a thoroughgoing manner, such arguments are found wanting. The sole argument that holds up is that the existence of God is a matter of faith, not of logic or empirical inquiry.

Faith is the bird that feels the light
and sings when the dawn is still dark.
Rabindranath Tagore

165

To assert that God's logic is beyond human comprehension but then employ human logic to prove God's existence reeks of double standards. Flexible standards are a feature of fundamentalist thinking. The flat Earth believers utilize shaky evidence and dismiss the solid evidence pointing the other way. For them it is manufactured evidence.

To alter a theory when new evidence requires it is not a failing but a strong point. It indicates a capacity to grow. The existence of a supremely powerful God, life after death, miracles, and paranormal abilities are claims not consistent with known laws of science. Science does not dismiss them offhand but subjects them to the Sagan Standard, a standard first set forth by the mathematician, scientist, philosopher Simon Laplace in 1814:

Extraordinary claims demand extraordinary evidence.
Karl Sagan

Scientists reason in probabilities, not certainties. A multiplicity of factors may impinge upon a phenomenon. Evidence is subject to error. Probabilistic reasoning not only permits estimating the relative import of these factors but also accounts for random variation. It permits prediction of outcomes to a certain degree of certainty. A weather forecast model, for example, cannot tell you whether it will rain or not rain tomorrow. But upon processing the relevant data, it can say whether the chance of rain is as low as 10% or as high as 90%. And even then, the opposite of what it says may transpire. The scientific spirit is aptly summed by two distinguished voices in the field:

Science is uncertain. Theories are subject to revision; observations are open to a variety of interpretations, and scientists quarrel among themselves. This is disillusioning for those untrained in the scientific method, who thus turn to the rigid certainty of the Bible instead. There is something comfortable about a view that allows for no deviation and that spares you the painful necessity of having to think.

Isaac Asimov

I conclude that, while it is true that science cannot decide questions of value, that is because they cannot be intellectually decided at all, and lie outside the realm of truth and falsehood. Whatever knowledge is attainable, must be attained by scientific methods; and what science cannot discover, mankind cannot know.

Bertrand Russell

When dealing with competing theories, scientists normally gravitate towards the simpler one. This preference is called Occam's Razor. Science respects the right of persons to hold beliefs. If a person believes in angels, a scientist will shrug his shoulders. Suppose he says, 'My angel told me not to drive to work today. So, I did not. And you know what. There was a horrible crash on that freeway. Many people were seriously injured. I was saved by my angel'. The scientist may inquire, 'Why were those people not saved by an angel? What is special about you?' But when the man says, 'My daughter had fever. I prayed. My angel came and cured her. She is fine now'. Here the scientist will be perturbed. 'Are you not playing with her life? You can pray but you also need to send her to a doctor for diagnosis and treatment'. He will demand proof that prayers can cure fever. Beliefs that intrude in society in ways that may either harm or benefit people need good evidence and ethical justifications. Science cannot maintain a neutral stand towards such beliefs.

4.16 REFLECTIONS

Today, everyone uses the products of science and hardly anyone calls science a false doctrine. At the global level, more than 80% of the people follow religion. But people remain divided over whether science or religion is supreme. A tiny tale (source unknown) is revelatory:

Angels or Monkeys?

A five-year old girl asks: 'Mama, where did I come from?'
Mama: 'You were created by God and brought to us by angels'.

She tells Papa: 'Mama says angles brought me here'.
Papa: 'No, my dear. That is a fairy tale. We are descended, in a long line of evolution, from animals. Monkeys are our closest ancestors'.

She returns to Mama: 'You spoke a lie. Papa says we came from monkeys'.
Mama: 'We both spoke the truth, my love. He was talking about his side of the family and I was talking about my side'.

This quaint tale speaks to the popular misconceptions about the relationship of science to religion. Historically, science and religion have interacted in complimentary as well as conflicting ways. And the interactive pattern has varied across different belief systems.

Isaac Newton and Albert Einstein revolutionized our understanding of matter and the Universe and made invaluable contributions to the methods of science. Newton's formulation of the basic laws of motion was instrumental in cementing causal determinism as the dominant scientific philosophy. But they held divergent views on religion.

Newton was a devout Christian and erudite theologian who produced a mammoth body of work on Christianity, history of religion, interpretation of the Bible and related issues. His rejection of conceptualizing God as a Trinity and advocacy of an authentic interpretation of the Bible cost him a major academic position. He saw God not through the lens of revelation but of rational discourse and study of nature. Religion signified personal morality and piety. An advocate of religious tolerance and freedom, he also supported parliamentary democracy in an era when it rested on a fragile footing. He also defended academic freedom from monarchic intrusion.

Albert Einstein was critical of organized religion or scriptures. Eschewing religious rituals, he rejected the idea of a monotheistic, interventionist God. As a fan of Baruch Spinoza, his views blended agnosticism with pantheism. Like Newton, he subscribed to causal determinism and rejected that randomness was an inherent, basic property of nature. And he too took a progressive stand on societal issues. Opposing fascism, he rallied for nuclear disarmament and global peace. He was a socialist who spoke out against racism and discrimination. Regarding him a dangerous communist, the US FBI continually spied on him. Though ascribing to determinism, he was not a fatalist. His activism signified that humans can and should fight for freedom and institute social change.

Both these giants viewed their contributions to knowledge with a deep sense of humility.

I do not know what I may appear to the world, but to myself I seem to have been only like a boy playing on the seashore, and diverting myself in now and then finding a smoother pebble or a prettier shell than ordinary, whilst the great ocean of truth lay all undiscovered before me.

Isaac Newton

While the authenticity of this quote has been questioned, it does reflect his spirit. Einstein had a similar stand:

> *The child notes a definite plan in the arrangement of the books, a mysterious order, which it does not comprehend, but only dimly suspects. That, it seems to me, is the attitude of the human mind, even the greatest and most cultured, toward God. We see a Universe marvelously arranged, obeying certain laws, but we understand the laws only dimly.* Albert Einstein (Alpert 2019).

Einstein often invoked the idea of God in his public and private utterances; the above quote is not atypical. God denoted the 'marvelous' order of the Universe. His religion embraced spiritual contemplation of the majesty of nature and the human quest to decipher its mechanisms in the context of deeply felt humanism.

The views of Isaac Newton and Albert Einstein saddle the two strands of NOMA. Both firmly adhered to the method of science. But the former was more inclined towards religion while the latter leaned towards non-theistic spiritualism. Einstein also cautioned against scientific hubris.

> *We should be on our guard not to overestimate science and scientific methods when it is a question of human problems, and we should not assume that experts are the only ones who have the right to express themselves on questions affecting the organization of society.*
>
> Albert Einstein

Stephen Hawking, another titan of physics, designated the ideas of heaven and afterlife as fairy stories. Holding that the Universe was a product of spontaneous creation, he discarded the existence of a creator God. Yet, life and the Universe have meaning, and humans should exercise their curiosity and the methods of science to discover it. Being a firm atheist, he held that science and religion were conflicting, irreconcilable outlooks.

> *There is a fundamental difference between religion,*
> *which is based on authority, and science,*
> *which is based on observation and reason.*
> *Science will win because it works.*
>
> Stephen Hawking

Like the majority of scientists, Pope Francis and the Dalai Lama do not view religion and science in antagonistic terms. The latter sees them as conjoined entities.

> *Buddha's teachings are scientific methods to solve*
> *the problems of all living beings permanently.*
>
> The Dalai Lama

Science and religion are also to be compared on the plane of morality and values. Each religion has a code of conduct and some norms are common to all religions. No religion condones wanton killing. Most ascribe to the Golden Rule. Science too has its ethical standards such as respect for the truth, scientific integrity, cooperation, open disclosure of methods and data and avoidance of conflict of interest.

Morality primarily is an issue of practice. On that score, neither science nor religion can claim a loftier position. People of faith, priests, mullahs, monks and gurus and scientists have cheated, succumbed to greed and desire for fame, and violated the rights of people. Both were enmeshed with slavery, colonial rule, native people's genocide and economic injustice. Some scientists and religious leaders supported or stood on the sidelines of Nazism while brave scientists and religious leaders opposed the vile creed and paid a hefty price for it.

Morality has an evolutionary basis. But human morality largely derives from social and economic development of human society. It is an emergent entity within varied social formations that in the earlier times mostly functioned under the umbrella of religion. The course of human history has firmly embedded the ideas of compassion, decency, cooperation and the desire for freedom, dignity and social justice in the human mind. Today, a moral code based on such values is a universal, integral part of the human condition and is reflected in diverse ways in professional codes of conduct, national constitutions, national and international laws, and documents such as the Universal Declaration of Human Rights.

The actualization of noble moral and ethical tenets faces an uphill battle to this day. Neoliberal capitalism, the currently dominant mode of human society, is based on national and international exploitation, inequality, racial, ethnic and gender discrimination and militarism. Paying lip service to the ideas of freedom and justice, it serves the interests of the super-rich, large corporations and the politically powerful elites while neglecting or fueling the daunting challenges facing humanity today.

The fundamental challenge for people and leaders of faith and scientists and secularists does not pertain to the validity of the Big Bang theory or the Theory of Evolution. It relates to the challenge posed in 1931 by trade union activist Florence Reece and subsequently rendered into song by many artists:

Which side are you on?
Which side are you on?

Scientists and people of faith need to decide whether they will serve the elite or the broad humanity, whether they will partake in corporate greed or bravely fight the forces imposing misery and suffering on peoples everywhere. The words of the humanist philosopher Anthony C Graying are apropos:

> *The great moral questions of the present age are those about human rights, war, poverty, the vast disparities between rich and poor, the fact that somewhere in the third world a child dies every two and a half seconds because of starvation or remediable disease.*
>
> Anthony C Grayling

Both religion and science emerged from the human impulse to make sense of the world and deal with the uncertainties of life. United at the outset, they diverged methodologically and substantively over time and now form distinct arenas of the human condition, responding to distinct human needs. Yet, they still remain intertwined like an eternal knot on the conceptual and practical dimensions. The future of science and religion, indeed the future of humanity, depends on whether people of faith and people of science can creatively and harmoniously navigate the diverse strands of the complex knot to address these and related moral challenges.

CHAPTER 05: MATHEMATICS

The good Christian should beware of mathematicians.
Saint Augustine

*Mathematics is the language
in which God has written the Universe.*
Galileo Galilei

*From the intrinsic evidence of his creation,
the Great Architect of the Universe
now begins to appear as a pure mathematician.*
James Jeans

Where there is matter there is geometry.
Nicolaus Copernicus

Mathematics is not only real, but it is the only reality.
Martin Gardner

❖

MATHEMATICS, the undoubtedly least popular subject in school, evinces utterances like: I hate math; I hate algebra. Poor teaching plays a part. The austere, symbolic nature of the subject appeals but to a minority. Yet, modern life is not possible without mathematics. All activities relating to the production and distribution of essential and preferential goods at some level involve recourse to some tools of mathematics. All disciplines in the natural and social sciences use mathematical concepts and technique for investigation and inference. At times, these applications have an astounding level of accuracy. The formulas of the Theory of Relativity and Quantum Mechanics, for example, have verifiable predictions accurate to one part in a hundred billion. Art, music, literature, law and linguistics have some scope for quantitative analysis. If nothing else, their instruments and paraphernalia are made using tools that depend on mathematics. Its methods, content and extensive applications give mathematics a unique status among the sciences.

Mathematics is the queen of the sciences.
Carl Friedrich Gauss

Mathematics is the door and the key to the sciences.
Roger Bacon

Mathematics and religion are separate domains of human mental activity. But do they have anything in common, in substance and approach? How have they related to each other in human affairs over the course of history? Has religion fostered or blocked mathematics and has mathematics supported or negated religious ideas? Can a person of faith also be an aficionado of this austere subject? This chapter aims to address these and related questions.

5.1 NATURE AND ORIGINS

People see mathematics as a collection of numbers, figures, formulas and proofs. Yet, in essence, it is a logical exploration patterns within quantities (arithmetic), structures (algebra), space (geometry) and change (calculus or analysis). It is a deductive, not an experimental science. The idea of proof is central: Given a set of axioms, rules of deduction and already proven results, is proposition A true or false?

Mathematics has two sub-divisions—pure and applied. The former is an abstract field driven by internal logic while the latter applies the results and ideas from the former to technical and scientific domains. Thus, probability theory is a part of pure mathematics, and statistics has methods based on probability theory that are applied to physics, engineering, medicine, public health, agronomy, economics, actuarial science and psychology.

Globally, the number of people with a doctoral degree in mathematics, applied mathematics, statistics or biostatistics exceeds a hundred thousand, rivaling those with such academic qualifications in theology or religious studies. The number of users and teachers of quantitative subjects, in the tens of millions, far exceeds the number of priests, nuns and teachers of all the religions combined.

Origins

Number sense or the ability to distinguish one from many, is a survival trait that has been found in ants, bees, mice, wolves, birds and other animals. Are there too many foes? Are there enough allies around to embark on an attack? How many signs are there on the path to a food source? Making such distinctions, animals convey them by sound or behavioral signals. Some species make assessments beyond one and many.

Numerical cognition is vital to human survival. From the days of hunting and gathering, humans relied on accurate counts of objects and estimates of distance and time. That ability expanded qualitatively as agriculture, fishing and animal husbandry developed, better tools were made, dwellings and buildings became elaborate, trade expanded, and money circulation grew.

Estimation of resources and output, measurement of distance, area, volume and weight had to be accurate. States imposed taxes; merchants depended on credit; builders required good assessments of construction material. Large numbers were added, subtracted, multiplied and divided. Fractions and powers of numbers were formulated. Complex shapes—triangle, rectangle, cube, circle, sphere, pyramid—were deciphered. An accurate calendar to track seasons and days developed. Techniques for counting; measuring length, area, and volume, weighing, determining shape and measuring time developed in all ancient cultures. Some devices were rudimentary and some were elaborate.

The Ishango bone, made from the fibula of a baboon, is a counting device discovered in Central Africa. Possibly over 20,000 years old, it is regarded as an enumerating and multiplication tool. Some experts see it as a form of a six-month lunar calendar.

Ishango Bone

Quipu

The Quipu is an elaborately knotted branched pattern of fiber strings made in the South American Incas civilization. One use was to enumerate and systematically record the distribution of goods, structures and people. Different colors stood for different types of counts. The Quipu was also used for addition, subtraction and multiplication of large numbers.

Though integrated into the cultural, economic and religious edifice of the community, knowledge of mathematical techniques was restricted. A few basic techniques were commonly understood. The shamans and priests who the people consulted for physical illness or psychological distress held a monopoly in that domain. With their special knowledge, they facilitated the maintenance of the social and economic order as well as looked at the sky, assessed the seasons, and blessed the harvest and hunting parties.

Some 5,000 years ago, this body of knowledge congealed into a formal written discipline that had special symbols and rules to manipulate numbers and shapes. Mathematics was born, in varied forms, in Africa, Asia, South and North America and Europe. From early on, it benefited from cross-cultural interactions. An elaborate system of representing numbers, positive and negative, emerged. The number zero, devised in India, spread across continents, revolutionizing arithmetic. Word problems became formulae. Foundations of algebra were laid. Properties of geometric figures were ascertained. We note two examples from this multi-faceted process. Take a right triangle with height = a, base = b, and hypotenuse = c.

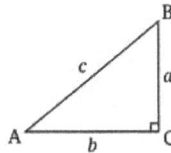

Right Triangle

For every such triangle, the square of the base and the square of the height total to the square of the hypotenuse. In symbolic terms, this is

$$a^2 + b^2 = c^2$$

All high school students now learn this formula. Called the Pythagorean Theorem after Pythagoras of Samos, Turkey, it was independently found in tropical Africa, China, Mexico, Central America, Iraq, Egypt and India. At the outset, the property was empirically ascertained for specific triangles and used in practical tasks like construction. The first general proof was given by Greek mathematicians including Pythagoras sometime before 500 BCE. The first systematic proof was devised by Euclid of Alexandria in his pioneering geometry text, *Elements*, around 300 BCE. *Elements* remained the main school geometry textbook until the early years of the 20th century.

Another geometrical fact every school child learns is that the ratio of the circumference of a circle to its diameter has a constant value. Denoted by the Greek letter π, it approximately equals 3.14. Its rough value was known in ancient Iran, Egypt, and other places. More

173

accurate estimates were derived by the Greek scientist and mathematician Archimedes around 250 BCE. Now we know that π is an unending decimal digit number.

$$\pi = 3.141592653589793238....$$

The Pythagorean Theorem and π, a tiny portion of this history, illustrate that what had emerged as a set of practical rules and empirical results detached itself from that mooring to be driven by an internal momentum based on curiosity and logic. Connection to natural science and practical application remained important. But they did not limit the field. Today, mathematics is a vast body of knowledge with many sub-disciplines. Professional mathematicians now are highly specialized scholars who know but only a small part of the gigantic edifice under which they function.

5.2 PYTHAGOREANISM

Our exposition of the relationship between mathematics and religion begins with a mathematically inspired faith system, Pythagoreanism, that emerged around 550 BCE. Deriving from the views and writings of Pythagoras and his followers, it was a mathematical school conjoined with a code of morality and conduct, and a system of supernatural beliefs.

Pythagoras of Samoa

Pythagoras enriched his mathematical knowledge from travels in Egypt, Iraq and possibly India. He and his followers are credited for a formal proof of the relation between the sides of a right-angled triangle. They also proved a series of results for triangles, parallelograms, polygons, circles, spheres and polyhedrons, and developed construction methods for geometric figures like pentagrams. Numbers were given geometric forms. Triangular numbers, for example, were visualized as dots in triangular formations.

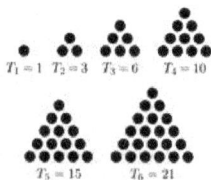

First Six Triangular Numbers

174

These numbers were recognized as cumulative sums of integers. Thus

$$T_5 = 1 + 2 + 3 + 4 + 5$$

and generally,

$$T_n = 1 + 2 + 3 + ... + n = n(n + 1)/2$$

Additionally, the Pythagoreans related the three different ways of averaging two numbers—arithmetic mean, geometric mean and harmonic mean—to each other. Initiating the quantitative analysis of music, they performed experiments to verify their ideas. Astronomy was a key area of interest. But their astronomical models did not always spring from observation. Thus, it was held that heavenly bodies including the Earth moved around a central fire, and the Earth was spherical because that was the most perfect shape of an object.

The Pythagoreans presaged algebra as an abstraction of arithmetic. The school is feted not just for the plethora of scientific and mathematical facts it unearthed but principally for its approach to mathematical analysis. The focus on axiomatic, deductive style led to the birth of pure mathematics. The linkage of mathematics with empirical investigation cemented the role of mathematics in science. And it signally influenced general Western philosophical thought.

> *Ultimately, Pythagoreanism has been a dynamic force on Western culture. It has creatively influenced philosophers, theologians, mathematicians and astronomers, as well as musicians, composers, poets and architects of the Middle Ages.* (BP 2021).

Pythagoras held strong beliefs about the nature of reality, humanity's role in the cosmos, and ways of living. Venerated by his followers as a humble, wise and pious sage, he was a broadly respected multi-talented scholar as well. His public orations drew large crowds. The educational reforms he advocated were ahead of his time. And his advice was often sought by the authorities.

Yet, he established a school that morphed into a secretive society, the entry into which was based on merit and character. Its beliefs, rules and rituals effectively converted it into a religion. The sayings of the founder attained a divine stature. Interestingly, Pythagoras lived around the same time as Confucius, the Buddha and Lao Tse. Learning mathematics and science from his travels, he also imbibed some of the religious lore of the visited lands.

Beliefs

The Pythagorean representation of reality resembled the Confucian principle of yin and yang: Limited and unlimited; odd and even; unity and multiplicity; left and right; male and female; rest and motion; straight and curved; light and dark; and good and evil. Reality was composed of opposing tendencies.

The binary essence of reality was buttressed by a key tenet: Numbers form the essence of reality. All phenomena are explicable as whole numbers or constructs based on whole numbers. Their defining dictum, placed on their school entrance, was:

All is Number

Numbers were linked to specific moral and intellectual features and events.

Number	Property
One	Reason
Two	First female number; opinion
Three	First male number; harmony
Four	Justice
Five	Marriage
Six	Creation

Source: Allen (1997)

The supremacy of whole numbers was an inviolable belief. Doubting it was a heresy punishable by expulsion or worse. Such rigidity was detrimental to the long run development of mathematics.

Consider a right-angled triangle with base and height equal to 1. The Pythagorean theorem tells us that the hypotenuse, c, is:

$$c^2 = 1^2 + 1^2 \quad \text{or} \quad c = \sqrt{2}$$

As shown in the diagram below, $\sqrt{2}$ is a measurable distance, which can be empirically approximated as 1.41. But what is its exact value? Can we find two integers x and y such that their ratio x/y is equal to $\sqrt{2}$? The answer to the second question is no. Further, written in decimal form, $\sqrt{2}$ has no end; the digits go on in a non-recurring pattern without limit. It is, as we say now, an irrational number. This fact, presumably known to some Pythagoreans, violated the key tenet that all was number and so was a closely guarded secret.

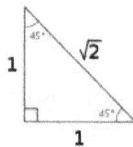

Unit Right Angled Triangle

Legend has it that Hippasus, the first Pythagorean to publicize the proof of the irrationality of $\sqrt{2}$, was expelled or drowned at sea by his compatriots. The other religious types of beliefs of the Pythagoreans were: Living beings have a soul. Humans are animals with advanced intellectual abilities. A soul may transmigrate into an animal or a human after death. Strict life rules and rituals purify the soul. And their burial rites reflected belief in reincarnation of the soul.

While Buddhists strive to attain spiritual liberation through right living and inner reflection, and Muslims seek ascendance into heaven via absolute faith in Allah and his prophet and abiding by the main tenets of Islam, the Pythagoreans prescribed liberation from the cycle of rebirth through correct living and devotion to rational thought and mathematics.

Practice

The Pythagoreans had strict dietary, clothing and behavioral codes. Holding that inflicting pain on animals was cruel, they were vegetarians. But they did not eat lentils. Valuing austerity, virtuous conduct and living in harmony with nature, they linked healthy bodies to healthy minds. Music was a path into the harmonies of the Universe and a means to calm the soul. With esoteric rituals and unique burial rites, they saw mathematical and philosophical pursuits as the basis for a moral life. Overall, they formed a politically conservative group focused on individual perfection that stressed respect for elders and the state.

Early Pythagoreans lived in exclusive communities. Life was governed by the Pythagorean code and cooperation between families. Knowledge and material goods were shared. Education was emphasized. Monogamy was the norm. These communities stood apart in one additional aspect: The role of women. In principle, men and women had equal status. Both had to abide by the same sexual code. Yet, for '*the sake of harmony,*' women had to submit to men. In spite of the burden of domestic work, their women had more freedom of movement and participated more in intellectual pursuits than women at that time elsewhere. It is also said that Pythagoras conducted special classes for female students.

Pythagoras Teaching Women

History records several eminent female Pythagoreans. Among them was Theano. Some sources have her married to Pythagoras while others claim she was his disciple. In any case, she was an intellectual in her own right, likely the first female philosopher-cum-mathematician of Europe. After the death of Pythagoras, she dedicated herself to spreading his ideas.

> *Beyond this work, it is possible that Theano wrote some of her own texts that were not published. In addition to some moral reflections, she also researched cosmology, medicine and mathematics. She was particularly interested in the theory of the golden ratio and regular polyhedrons, also known as Platonic solids.* (Timon 2018).

The distinct beliefs and unconventional ways of the exclusive Pythagorean communities generated hostility from their neighbors. Persecuted in many cities, some were killed, and their houses were set on fire. Pythagoras was also forced to flee. Eventually, Pythagorean communities suffered a steep decline, especially after the death of the founder.

But Pythagoreanism did not die. Philosophers continued to ascribe to its philosophy. The tradition split into two sects: one stressed mathematical pursuits and the lore of numbers while the other stressed the moral values, rituals and lifestyle. Neo-Pythagoreanism, an offshoot from the teachings of Pythagoras, continued to have a strong influence on mathematics, science and philosophy well into the modern era.

Pythagoreanism represented a major step in the transition from mystical, speculative thought to rational, scientific thought. Yet, it spawned its own brand of mysticism. Abstractly analyzing numbers and shapes, it heralded the emergence of pure mathematics. It made mathematics an intrinsically worthy pursuit. Yet, it mystified numbers and suppressed products of mathematical analysis—the incommensurable numbers. It utilized empirical observation to learn about nature but some of its beliefs strayed onto pure speculation.

> *Never before or since has mathematics played so large a role in life and religion as it did among the Pythagoreans.* (Boyer and Merzbach 1991).

Pythagoreanism illustrates the paradox ensuing from linking mathematics and science to religious beliefs. Thought it is an extreme case, in historical terms, it is not an anomaly. Mathematics and religion have impinged on each other in intriguingly diverse ways in many societies.

5.3 MATHEMATICS ACROSS CULTURES

Religion and mathematics have coexisted in a harmonious fashion in most societies and cultures. The foundation was laid in the ancient times when numeric and geometric skills were entangled with divine beliefs, shamanic praxis, art and play. We examine the intersection of religion and mathematics for several religious traditions.

+ Hinduism, Buddhism and Jainism +

The Indian sub-continent witnessed a major spurt of mathematical progress sometime before 1000 BCE. Commerce expanded as kingdoms took root. The Vedas were written as Hinduism developed into a unifying ideology of the land. Jainism and Buddhism, offshoots of Hinduism, emerged later in the course of social and economic developments. Each period brought newer mathematical techniques of value in measuring length, distance and weight, and construction, commerce and agriculture.

Yet, mathematics was also directly linked to religion. Mathematical ideas and skills were expressed through scriptures. The mathematics of that era is recorded in the Shulba Sutras, an adjunct text of the Vedas. Geometrical figures symbolize fire altars for deities. Cases of the Pythagorean theorem and geometric methods of solving linear and quadratic equations are recorded. Astonishingly, the value of $\sqrt{2}$ is given to 5 decimal digits of accuracy as 1.4142156. The motivation for determining $\sqrt{2}$ was to construct a devotional square altar twice the area of a given square altar. Important mathematical sutras were recorded by Baudhayana, Apastamba and Katyayanya who advised the faithful on ways of worship and maintained the altars and sacred fires in the temples.

The sutras contain methods to construct a cemetery altar in the form of an isosceles trapezium of precise dimensions, a circle of area equal to that of a square, and a complex tantric figure centered on nine interwoven isosceles triangles. Methods of reducing the multiplication of large numbers to a series of smaller multiplications employing algebraic identities also originates from those texts. The Vedas give methods of computing roots and have rules for combining numbers with exponents. Numbers as large as a trillion are expressed using powers of ten. Some numbers, like 108, are sacred in Hinduism as well as in Jainism and Buddhism.

Buddhism and Jainism considered the mastery of numerical skills essential for attaining a high spiritual status. Ancient Greek arithmetic rarely strayed beyond 10,000. But these faith systems expressed a passion for very large and very small numbers.

> *A 4th Century CE Sanskrit text reports Buddha enumerating numbers up to 10^{53}, as well as describing six more numbering systems over and above these,*

leading to a number equivalent to 10^{421}. Given that there are an estimated 10^{80} atoms in the whole Universe, this is as close to infinity as any in the ancient world came. It also describes a series of iterations in decreasing size, in order to demonstrate the size of an atom, which comes remarkably close to the actual size of a carbon atom (about 70 trillionths of a meter) (SOM 2021a).

Zero

Elementary school children now learn that zero is a number. But what to us is a natural idea took millennia to develop, and that, in just a few ancient cultures. Mayan arithmetic used it as a place holder in Quipu computations. But in ancient Greece, the idea of zero was disparaged. Aristotle and other Greek philosophers frowned upon it. The Pythagorean number lore had no room for it. The negativity towards zero persisted into the Christian era. For over a millennium, the Church equated God with the infinite and the devil with nullity. Zero was a satanic entity that confounded arithmetical operations. Any number could be divided by any other number, but not by zero. If you multiply by zero, everything vanishes. How can it be a number? Lacking zero, the West was stuck with the cumbersome Roman numeral system.

For reasons more cultural than economic, India was the birthplace of the modern idea of zero. The Vedic culture had no qualms with nothingness. It visualized the divinity as both existence and non-existence, all and none. In the Hindu, Buddhist and Jain lore, the deity was both infinite and null.

In the earliest age of the gods, existence was born from non-existence.
The Rig Veda (Seife 2000, page 66).

Fifth century Indian mathematicians developed a number system in which zero was not just a place holder but a number as well. Counting and arithmetical operations improved significantly. Further maturation occurred in the Islamic world after after 650 CE. Merchants brought back the idea of zero from India and Muslim mathematicians used it to devise the number system in use today. The English word zero originates from the Arabic word sifr.

Leonardo of Pisa (Fibonacci), the famed traveling Italian mathematician, returned with the Hindu-Arabic numeral system and the idea of zero from his trips to Muslim lands. His 1202 book *Liber Abaci* gave a clear exposition of its practical utility. Greeted with scorn, the new system was banned for a while. Yet, society was changing. Agriculture, commerce, transport and crafts expanded. Taxation and financial transactions involved large numbers for which the Roman numeral system was too cumbersome. It was time for change. Theological edicts gradually crumbled in the face of the needs of commerce. Computational drudgery was reduced and clarity was enhanced. Banking, commerce and accounting, indeed the development of capitalism, received a powerful boost. And all that from the elusive notion of nothing.

Japanese Temple Geometry

A distinctly novel blending of religion with mathematics emerged in Japan after the early 1600s. Till then, mathematics in Japan had largely derived from texts imported from China. Christian missionaries from Europe also introduced some mathematical ideas. Political, military and economic factors precipitated a major transformation in Japan after 1603. Apart from restricted trading, all linkages with the outside world were severed. Christian missionaries were expelled. The Bible was banned. Feudal parochialism declined as unified governance took hold. Social progress and education were accorded a high priority. For some two and half centuries, peace and stability prevailed. Cultural activities—art, poetry, book printing, music, public exchange of ideas—bloomed. Higher education expanded with an emphasis on science and mathematics.

Deprived of external links, science in Japan took on an autonomous trajectory. Based on the extant foundation, *Wasan* (Japanese mathematics) emerged. A booklet of Japanese abacus, with application to surveying and commerce was produced. Japanese mathematicians made discoveries in arithmetic, algebra and geometry that paralleled those in the West, and at times went further. Isomura Yoshinori devised original geometric problems and estimated a value of π correct to 10 digits. Seki Takakazu, the most distinguished Japanese mathematician of the era, formulated the theory of determinants, defined Bernoulli numbers and had an efficient method for finding roots of polynomial equations. Other mathematicians worked on infinite series and methods resembling those found in integral calculus.

Mathematics also flourished in the confines of Buddhist temples and Shinto shrines. Traditionally the visitors bore offerings and sacrificial material. But as the cost of such items rose, some devotees brought colorful, hand carved or printed, wooden tablets that were hung on the walls near the entrance or under the roof of the temple. At some point, general images were supplemented with images expressing mathematical problems. As the practice caught on, Japanese sacred mathematics was born.

These *Sangaku* tablets were of devotional origin. But their problems were secular. Most were geometry problems. Interspersed within was a smattering of arithmetic, algebra and logic problems. Anyone could hang a tablet. And many, young and old, of various social strata, did. Some tablets contained a solution, but others posed a challenge to visiting devotees. Mathematics was democratized.

A few mathematicians collected these problems and published them in books. Overall, many thousands of problems were placed on the tablets. Only about nine hundred have survived. Some require a basic knowledge of geometry—similar triangles, the theorem of Pythagoras—while others need advanced geometry. Some *Sangaku* problems tax modern-day experts. A few utilized complex theorems not known to mathematicians elsewhere. A basic *Sangaku* challenge is shown below.

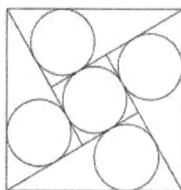

A Sangaku Problem

We have five equal circles of radius r contained in a square of side length 1. What is the value of r? A student with a sound understating of high school geometry can readily show that $r = (-1 + \sqrt{3})/4$.

As Japan resumed full contact with the West in the 1850s, this spiritual and democratic yet patently innovative mathematical tradition underwent a steep decline and died out. It has been brought to light again of recent, but no one places Sangaku tablets in temples any more.

Christianity

Christianity has related to mathematics with harmony, indifference or plain antipathy. It depended on what was at stake. Theological wrangling at times pitted one branch of mathematics against another. Here we give a brief look into that history.

Early Christian theologians and clergy were familiar with the works of Greek philosophers like Plato and Aristotle, and scientist-mathematicians like Euclid, Archimedes, Thales and the Pythagoreans. Among the diverse views on God and the cosmos of the Greek philosophers, some were declared compatible with the Christian doctrine, some incompatible, and some were in dispute. Some Grecian ideas became interwoven with the standpoint of the Catholic Church and other ideas, like the zero, were proscribed. The Grecian emphasis on logical reasoning was, however, imbibed by most theologians. And their unending disputes were adorned with finely parsed arguments over minutiae. Employing dense logic and esoteric terminology, they continually quibbled over interpretation of Biblical verses and other matters.

Yet, scholasticism was not all to no avail. In the centuries after the fall of the Roman Empire, when science and mathematics made little progress in Europe, the monks embraced rational analysis. Among other things, it was applied to predicting the cycles of the moon and constructing an accurate calendar to predict Easter, the day of birth of Jesus Christ. They also imbibed the works of Muslim mathematicians. Using Biblical references, numbers were assigned a spiritual significance. Seven was a sacred number as it was the sum of the numbers of the Trinity and the gospels. To the monks, seven represented the soul.

> *Medieval church mathematicians had inherited a canon of principles from pagan authors and needed to translate the meta-physics of antiquity into a contemplative format acceptable to Christian philosophy.* (Mulcare 2013).

When life took on a new direction as the feudal order and the stranglehold of the Vatican began to crumble across Europe, the logical and mathematical reasoning kept alive in monasteries and theological literature was inherited by a new breed of mathematicians and scientists who no longer functioned within the strict ambit of the church.

> *... Western science grew out of Christian theology. It is probably not an accident that modern science grew explosively in Christian Europe and left the rest of the world behind. A thousand years of theological disputes nurtured the habit of analytical thinking that could be applied to the analysis of natural phenomena. On the other hand, the close historical relations between theology and science have caused conflicts between science and Christianity that do not exist between science and other religions ...* Freeman Dyson (O'Connor and Robertson 2021).

While this quote ignores key social and economic factors, it makes a point. A further insight is provided by the views of some major Western scientists-cum-mathematicians on religion.

The Catholic Church expressed the Aristotelian geocentric view that the planets, the Sun and the stars move in circular orbits around a fixed Earth. The circle represents a perfect geometrical shape, embodying the divinity of God. The centrality of the Earth shows God's love of humanity. But it was a flawed model of the cosmos that was eventually overturned by the works of Nicolaus Copernicus, Giordano Bruno, Johannes Kepler, Galileo Galilei, Tycho Brahe and Isaac Newton. (The contributions of Islamic scientists are noted later).

Nicolaus Copernicus was a master of many disciplines. Fluent in several European languages, he advanced astronomy, mathematics, economics, medicine, classical studies and held important governmental positions. Upon accurately recording the motion of several planets, he concluded that they orbited the Sun, not the Earth. His circle of scientists concurred, and advised him to publish his findings.

But he was an ordained priest, often involved in ecclesiastical endeavors. Concerned by potential errors and deviation of his conclusions with the Church doctrine, he checked and rechecked his data. Each time, he reached the same conclusion. Yet, he hesitated for over a decade to publish. His ground-breaking ideas first appeared in *De Revolutionibus*, a book published just before his death, exercising a significant effect on astronomical research and science education.

Convinced that his propositions were sound empirically and mathematically, Copernicus felt that they revealed the grandeur of God and did not conflict with the scriptures.

> *Perhaps there will be prattlers who, although completely ignorant of mathematics, nevertheless take it upon themselves to pass judgment on mathematical questions, and on account of some passage in Scripture, badly distorted to their purpose, will dare to censure and assail what I have presented here.* Nicolaus Copernicus (O'Connor and Robertson 2021).

A man of high political standing, Copernicus had a cordial relationship with the Catholic authorities throughout his life. His friends included priests sympathetic to his work. When its circulation was low, his book only elicited indifference from the Vatican. But as some Protestant and Catholic dignitaries made hostile comments, and astronomers extended his ideas along lines that made the Church uncomfortable, it was banned. Announced in 1616, the ban remained in force until 1835.

Giordano Bruno, a Dominican friar, mathematician, astronomer, poet and occultist, was a central player in this saga. An innovative scientist familiar with the works of Muslim astronomers, he became a major proponent of the Copernican worldview. He also improved it by observing that the stars were suns with their own orbiting planets. For the church, it raised the specter of other Earth-like planets with life forms in the Universe. Was there more than one Jesus Christ? Did Jesus visit the other planets? Bruno also held other 'heretical' beliefs like reincarnation, considered the Earth a living entity with a soul, and disputed doctrines like the virginity of Mary. When told to recant, he refused. After years of being hounded, he was condemned by the Inquisition and burned at the stake in 1600.

Galileo Galilei, a creative astronomer, physicist, engineer, is regarded as a father of modern physics. His emphasis on accurate observation, carefully designed experiments, rational analysis, usage of mathematics, repeatability and prudent interpretation set a defining emulative standard for the world of science. Among the areas he probed were kinematics, optics, dynamics, material science, hydrology and cosmology. He improved the telescope and invented scientific instruments. Also excelling in poetry, music and painting, his multiple, striking discoveries earned him fame across Europe.

Galileo was a non-dogmatic Christian who decried literal interpretations of the Bible. The scriptures gave the final word on faith and morality, but not on laws governing natural phenomena. Pursuit of the truth via the method of science—reason and experiment—was a divinely sanctioned activity and mathematics was the language God had used to design the Universe.

> *It is surely harmful to souls to make it a heresy to believe what is proved.*
> Galileo Galilei

Galileo's accurate astronomical observations provided further support to the heliocentric model of the solar system. And he boldly and publicly promoted it. Being a man with connections in high religious and political circles, just a few theologians and scientists critiqued his ideas. The Vatican maintained a discrete silence at the outset. The Inquisitional ban on *De revolutionibus* in 1616 altered everything. Summoned to Rome, he was ordered by the Pope to cease teaching and advocating the heliocentric model. For the next one and a half decades he complied. But in 1632, he managed to get approval from the Papal censors to publish the *Dialogue Concerning the Two Chief World Systems*. Written in a debating format between two points of view, the book artfully presents the case for his stand on cosmological issues.

But approval was short lived. Irked by its terminology and implicit but unapologetic propagation of the heliocentric model, the Inquisition put him on trial for heresy. He was forced to recant and sentenced to house arrest. For the next seven years, until his death at the age of 77, he was a broken man. His scientific work ceased. Only one book explicating his ideas came forth. And it was published posthumously. The Vatican used his recantation in a shameless propaganda campaign against the heliocentric model. It took over two hundred years for the Catholic Church to concede the validity of this model and clear this magnificent man of science and mathematics.

Tycho Brahe, a Lutheran Danish noble, and a meticulous astronomer, scientist and designer of highly accurate scientific instruments, was another actor of influence in this saga. He proposed a model in which the Sun and the Moon circled the Earth and the other planets circled the Sun. While his model did not accord with the traditional Church doctrine, it at least rejected the Copernican model that was banned in 1616. Brahe strongly believed in the authority of the Bible, and the Church used his ideas to discredit the Copernican model.

Johannes Kepler, an astronomer, mathematician, and a student of Brahe significantly improved the Copernican model by proposing that planetary bodies move in elliptical orbits around the Sun. His mathematically precise three laws of planetary motion formed a stepping-stone for Newton's theory of gravitation. But like Copernicus, Kepler erred in not accepting Bruno's idea that the stars were not static objects a fixed distance away from the Sun but were Sun-like objects possibly with planetary systems.

Kepler was a devout Lutheran. He said that God had created humans in His own image so that they could study and understand the laws of nature He had devised. The Bible, moreover, was not a substitute for science:

> *To teach mankind about nature is not the purpose of Holy Scripture, which speaks to people about these matters in a human way in order to be understood by them and uses popular concepts.* Johannes Kepler (O'Connor and Robertson 2021).

Like Galileo, Kepler held that God had formulated the laws of nature in a mathematical form. Accordingly, the study of mathematics was divinely blessed undertaking.

> *Geometry existed before the creation; is co-eternal with the mind of God; is God himself ... Where there is matter there is geometry.* Johannes Kepler (O'Connor and Robertson 2021).

But feeling that he was usurping their prerogative to interpret the scriptures, the Lutheran authorities attacked his unconventional views.

Isaac Newton, a veritable genius of science and mathematics, leaped ahead of Copernicus, Bruno, Kepler, Galileo and Brahe to frame a set of simple, general principles from which ideas about the solar system can be derived or refuted. Appointed the Lucasian Professor of Mathematics at Cambridge University at the age of 25, his magnum opus, *The Principia*, presents the three laws of motion and the law of gravitation that revolutionized physics and cosmology. Kepler's laws of planetary motion were shown a specific instance of the general laws. He and GB Leibniz independently discovered calculus, which now is a mathematical pillar of all science, natural and social. He also studied the nature of light and made significant discoveries in optics.

Newton cemented the scientific method. Rejecting reliance on traditional authority, especially Aristotle, he preferred experiment and observation. Elegant laws and mathematical formulations held credence only to the degree they predicted phenomena with quantitative accuracy. Reproducibility was of utmost importance.

Newton was a man of the world. Besides serving as a president of the august Royal Society, he sat for Cambridge University in the Parliament. He garnered the royal knighthood and became the Master of the Royal Mint. He stood for religious tolerance, freedom and democracy at a time when such notions were an endangered species. And he decried monarchic intrusion in the academy. Venerated by scientists, scholars and the public, he was buried in Westminster Abbey in a ceremony befitting a monarch.

Yet, this towering scientist was, like Keppler, a devout but independent Christian. He rejected the key Christian idea of the godliness of Jesus, a position which almost jeopardized his academic career. Holding that modern Christianity had deviated from the faith of early believers, Newton studied ancient texts and wrote extensively on the interpretation of the Bible and Biblical events. He even predicted the second coming of Jesus. The ink he poured over theological matters far outstripped what he did over scientific issues.

For Newton, faith, morality and piety, not ritual, were the essence of religion. He held that the laws of nature revealed the magnificence of the mind of God and like Einstein, viewed his contributions to science with profound humility.

It is the perfection of God's work that they are all done with the greatest simplicity. He is the God of order and not of confusion. Isaac Newton

Newton also dabbled in alchemy, doing experiments and speculating about his findings. The striking thing about him is the lack of methodological consistency and coherence between his three areas of interest—science, religion and alchemy. In theological pursuits, he set aside his scientific dictum of empirical testing all hypotheses, and in alchemy studies, he abandoned quantitative validation. It is unthinkable that he would have put the following ideas in his works on projectile motion or the passage of light through a prism.

As a blind man has no idea of colors, so have we no idea of the manner by which the all-wise God perceives and understands all things.
Isaac Newton

The extreme level of cognitive dissonance manifested in a truly giant intellect speaks volume about the human need for psychic comfort, sympathy from fellow humans and effects of social conditioning. It also casts a distinct light on the potential for coexistence between rational and irrational beliefs, between science and religion, between reasoning in a mathematically precise way and proclivity to illogical speculation.

Islam

The Islamic empires of the Middle East, North Africa, Southern Europe, India, and Central and West Africa that flourished from the 8th century onward were generally characterized by significant progress in science and mathematics (RPS 2022). In a culture of interfaith tolerance, Muslim, Jewish and Christian scientists and mathematicians translated mathematical texts from Greece, Persia, India, and China, and blazed innovative trails in arithmetic, algebra, theory of equations, and geometry, extended methods of proving theorems, found properties of special numbers, devised the calculation of areas and volumes of complex figures, and investigated circle geometry, power series, binomial theorem, trigonometry, conic sections, combinatorics, solutions of algebraic equations and continued fractions. Promotion of the decimal notation, the idea of zero, and formal algebraic notation simplified the statement and solution of mathematical problems. Foundational queries about their discipline were broached. Later translations of their works into European languages was instrumental in the development of mathematics in Europe.

The African city state of Timbuktu was an outstanding instance of vibrant Islamic scholarship. From the 11th century on, it was the regional center of trade and transport for a wide variety of merchandise. Commerce produced prosperity and catalyzed education and scholarship. Institutions of higher learning housed libraries with holdings totaling over a million manuscripts. Only a few hundred thousand have survived. Among them are the mathematical works of Al-Samlali. Application to trade and industry was prioritized over pure mathematical pursuits.

Among the principal mathematical luminaries of the Islamic world (in no particular order) were Abu Al-Hasan, Ibrahim Ibn Sinan, Sharaf Al-Din, Muhammad Al-Khwarizmi, Muhammad Al-Karaji, Nasir Al-Din Al-Tusi, Omar Khayyam, Ibn Qunfudh, Ibn Al-Banna,

Abu Kamil, Abu Al-Khazin, Thabit ibn Qurrah, Abu Al-Kuhi, Ibn Al-Haytham, Abu Manṣur, Abu Al-Kindi, Abu Ibn Hussein, Al-Battani and Al-Buruni. These mathematicians also engaged in scientific, philosophical, literary, historical and poetic pursuits.

A key feature of the scientific and mathematical endeavors in the Islamic empires was: The mathematicians were esteemed personalities subsisting under the patronage of the Caliph, the Emir and wealthy merchants and based in institutions that housed hundreds of thousands of scholarly volumes. Applications to construction, technology, agriculture, military matters and travel were valued but no prohibition on pursuance of abstract mathematical issues existed. Pure and applied mathematics developed side by side.

> *Geometry enlightens the intellect and sets one's mind right. All its proofs are very clear and orderly. It is hardly possible for errors to enter into geometrical reasoning, because it is well arranged and orderly. Thus, the mind that constantly applies itself to geometry is not likely to fall into error. In this convenient way, the person who knows geometry acquires intelligence.*
> Ibn Khaldun

Mathematicians were respected for their contributions to the practice of Islamic faith. For example, they strove to determine accurate directions to Mecca for prayers, predict the phases of the moon and the sighting of the new moon for Eid and enable fair distribution of inheritance. Discord between religious authorities on the one hand, and science and mathematics on the other, of the form seen during the era of Inquisition, did not materialize in the Islamic nations. Some scientists and mathematicians experienced state repression. But it arose mostly from political differences and loyalties rather than from the nature of their work.

Islamic mathematicians-cum-scientists saw their secular and religious work as complementary endeavors. Passages from the Quran encouraging the pursuit of knowledge buttressed the view that faith and reason formed two sides of the same coin. Investigating natural phenomena and mathematical truth was a means to reveal the glory of Allah.

> *That fondness for science ... that affability and condescension which God shows to the learned, that promptitude with which he protects and supports them in the elucidation of obscurities and in the removal of difficulties, has encouraged me to compose a short work on calculating by al-jabr and al-muqabala, confining it to what is easiest and most useful in arithmetic.*
> Al-Khwarizmi

Some Islamic mathematicians, notably of Sufi inclinations, ruminated along agnostic lines. Yet they remained within the ambit of Islam, being more critical of the ossification of faith and emphasis on ritual rather than of Islam as such. Assigning religious significance to numbers became common. The number seven has significance as the number of times one goes around the Kaaba during the Hajj and the number of gates of hell. The number 786 denotes for Bismillah—In the name of Allah.

By convention, Islamic literature and art do not depict human or life symbols. Muslim artists focus their talents onto engraving stunning geometric patterns in the walls, doorways and structures of mosques and religious buildings. Mosques across the world—Saudi Arabia, Morocco, India, Spain, Pakistan, Syria, Nigeria, the US and beyond—display geometric design formations that are a delight to behold and reflect a deep understanding of spatial geometry and construction rules.

> *The marvels of Islamic patterns—the most recognizable visual expression of Islamic art and architecture—are not just a beautiful accident. The ancient practitioners of this craft used traditional methods of measurement to create dazzling geometric compositions, often based on the repetition of a single*

pattern. The results are magnificent in their beauty and awe-inspiring in their execution. (Broug 2019).

Quasi-crystal forms arise from interlocking simple blocks in patterns that do not repeat even if infinitely extended in any direction. First described in the West in the 1970s by Roger Penrose, the uncharacteristic features they exhibit are of both theoretical and practical significance in mathematics, crystallography, quantum mechanics, solar panel design, manufacture of material for insulation and cutting and bone repair. One instance is shown below.

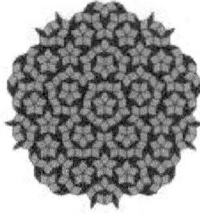

Penrose Tile

Yet, recent research has revealed the presence of quasi-crystalline design patterns in mosques and tomb sites in Iran and Uzbekistan that were built more than five hundred years ago. How these patterns were derived is a mystery. But they represent a level of geometric sophistication that can only arise from a sound understanding of basic geometry.

In addition to the mathematical import, Islamic architectural designs embody the basic tenets of Islamic philosophy, lore and mysticism as well. Like Japanese temple geometry, they are an exquisite blend of religion with mathematics.

++++

As the Islamic empires solidified into highly unequal and authoritarian social formations, as the ruling class indulged in opulence to the detriment of internally generated economic progress, and particularly as they fell into the clutches of Western imperialism, local scientific, mathematical progress and the academia suffered significantly. Muslim nations unduly emphasized rote memorization of the Quran in schools and placed science teaching on a lower rung. Rhetorically they rally against Western cultural influences but unreservedly embrace capitalist economic and cultural ideas and practices. Deemed an enemy, the West it also the major supplier of arms and backer of Islamic authoritarian regimes. Science and mathematics are not pilloried. In principle, they are praised. In practice, they are neglected over business studies, a manifestation of neoliberal dependency, authoritarianism and heeding to the interests of the wealthy upper stratum rather than the masses. The conundrum hobbles even major nations like Indonesia (RPS 2022).

Yet, despite the odds, path breaking mathematical work persists in some Islamic nations. Maryam Mirzakhani, Iranian mathematician who tragically died from breast cancer at the age of 40, was a recipient of the Fields Medal, seen as the Nobel Prize in mathematics, in 2014. Her remarkable contributions to complex geometry and dynamics propelled her to become the first woman, the first Iranian, and the first of the only two Muslim mathematicians to garner the high honor. Her example has inspired girls throughout the world to pursue a profession in which women are acutely underrepresented.

5.4 INVENTION OR DISCOVERY

The extensive utility of mathematics is beyond dispute. But why that is so continues to perplex scientists and philosophers.

> *How can it be that mathematics, being after all a product of human thought which is independent of experience, is so admirably appropriate to the objects of reality?*
>
> Albert Einstein

Take a steel kitchen knife. Iron exists in nature, but in a form that is not directly usable. The process through which it is converted into a knife—mining, thermal and chemical processing, smelting, etc.—was devised by human ingenuity over a long period of time. Iron is discovered, knives are invented. Photosynthesis in plants is an objective natural fact. It would occur even if humans did not exist. The electron microscope is an invention. Had it not been for humans, it would not be there.

The import of mathematics to modeling natural phenomena and resolution of practical problems has two facets: Techniques and ideas that emerged from endeavors to address specific scientific problems, and techniques and ideas that emerged from internal progress in mathematics but which later were found to have scientific import. In practice, utility often derives from a mixture of the two. Either way, the applicability of mathematics is nothing short of remarkable. To philosopher Eugene Wigner, the exactitude of mathematical models for natural phenomena appeared to be 'unreasonable'. In a pioneering paper, he laid out major instances of that property and opined:

> *The miracle of the appropriateness of the language of mathematics for the formulation of the laws of physics is a wonderful gift which we neither understand nor deserve.* (Wigner 1960).

The conundrum elicits a related query: Is mathematics invented or discovered? Scholars have probed this query for two millennia. But a consensus is far from being realized. There are three basic positions here:

Discovery: Mathematics subsists in a distinct conceptual domain ruled by the laws of logic. Mathematical truths have an objective existence independent of the human mind. Like other scientific truths, they await discovery. Mathematics is the language of science because it is embedded in the reality explored by science. Unlike other languages, and despite varied notation and details, it had uniform basic rules and concepts wherever it emerged. The theorem about the relationship between the sides of a right triangle was the same in Egypt, China, India and Greece. As a universal language, mathematics is the ideal way of communicating with extraterrestrial beings.

That mathematics has a separate existential reality is called the Platonic or idealist view. Mathematicians enter it by postulating axioms, applying the rules of logic and unraveling it bit-by-bit. Scientists utilize their constructs to explicate the natural reality. Discovery precedes application as mathematics is integral to the fabric of reality. At times, mathematical concepts that were deemed to have no connection to any natural process are later found to have critical significance in science. Examples include non-Euclidean geometry and the theory of groups.

Platonism envisions three dimensions of reality—material, psychological (awareness) and abstract (mathematical). Strict Platonism denies the reality of any but the abstract domain. The rest are illusions. Mathematics drives the Universe. For the Pythagorean creed, numbers are the true essence of reality.

Invention: Mathematics arose from human endeavors. Humans encountered basically similar challenges for survival and progress across the world. So the numerical tools they devised attained similar forms across time and space. The wide utility of mathematics to science and practice derives from the fact that over time humans fashioned it for specific needs. Numbers, geometric shapes and patterns of change are abstractions of real phenomena. That $2 \times 7 = 7 \times 2$ can be shown by totaling seven pairs of objects or two pairs of seven objects. The algebraic generalization a × b = b × a follows thereon. This is the constructivist (anti-Platonist or materialist) position. Primarily, no humans means no arithmetic, algebra, geometry or calculus.

Discovery and Invention: Basic mathematical truths were discovered in the course of historical endeavors. Mathematicians fashioned a language to deal with them and invented proofs of the truths. Properties of right triangles, for example, were initially found empirically. The number 2 is an abstraction of pairs of objects—bananas, sticks, lions. Mathematical entities resemble other human inventions. Numbers and geometry arise from human endeavors.

Thousands of different types of knives exist. But they all reflect the laws of material science and the basic purpose of a knife. Likewise, distinct geometries—Euclidean, hyperbolic and elliptical—have been developed and are applied in different contexts. Each has its own set of axioms. But they all have to conform to the basic rules of logic and cumulative axiomatic mode of argument. Similarly, several forms of algebra exist. Some permit commutative multiplication, some do not. You pick and choose.

A large domain of mathematics—statistics, financial mathematics, vector analysis, simulation methods—is linked to applications. Yet, mathematicians are not bound by physical reality. Employing distinct axioms and primary categories, they construct (invent) novel but internally consistent areas of mathematics. Mathematics is jointly driven by practice and internal logic. It is discovered and invented. At times, a branch of mathematics develops through cycles of theorizing and application. This is the realist position.

A variant of the realist position holds that mathematics is not as good a reflector of reality as is proclaimed. Many models have large margins of error, and clear criteria for choosing between different models do not exist. Models that provide a good fit to the observed data are found through balancing different forms of errors. In economics, mathematical models have generated global disasters.

Eminent scientists, mathematicians and philosophers take varied positions on whether mathematics is invented or discovered. The discovery school includes giants like GH Hardy, Roger Penrose, Kurt Godel and Martin Gardner. Giants like Albert Einstein, David Hilbert and Georg Cantor belonged to the invention school. Michael Atiyah, Gregory Chaitin, Philip J Davis, AK Dewdney, Reuben Hersh and A Renyi propose an admixture of the two perspectives. Physicist Max Tegmark sees mathematics as the prime propeller of nature but RH Nelson, a Professor of Public Policy, holds that mathematics operates in a God-like fashion. Surprisingly, on this issue, the contenders often deviate from the logical style of the mathematician, and quibble in ways akin to theological speculation.

Theological Connotations

The issue of discovery or invention has profound theological implications. If the former is correct, then it stands to reason that mathematics was created by a divine mind. Mathematicians have unraveled only a portion of God's majestic edifice. Since God's creations includes mathematics, then it is to be expected that the other things He created and the laws by which they are governed are amenable to the language of mathematics.

God is the supreme mathematician. By his grace, he has enabled humans to think in rational, logical and scientific ways. Some discerning human thinkers have been able to get a glimpse of His mathematical truths. Mathematics is God's blessing and we should be thankful to Him. GW Leibniz, the co-founder of calculus, opined that God is the perfect mathematician. Plato held that God communicates with humans through mathematics.

If mathematics was invented, then it previously did not exist, not even in a divine realm. If an all-knowing God exists, that is not possible. He knows all of mathematics, including the portion humans do not know. Designating it as a human invention is inconsistent with the existence of God. It is an atheistic stance.

The contention between discovery and invention of mathematics parallels the contention between determinism and free will. If an all knowing God exists, He determines all. Human free will does not exist. Mathematicians just uncover what God knows. The libertarian stand refutes such limitations. Mathematicians are free within the rules of their discipline to choose their areas of inquiry and prove new mathematical truths. The compatibilist stand is that while God knows all, humans have the freedom to explore and invent.

Belief in God essentially implies that mathematics is discovered, not invented. Its awesome edifice reflects the divinity of God. Its extensive applicability to life and science reflects the fact that God utilized mathematical tools for designing the cosmos. It is a further proof that God exists.

Theoretical physicist, co-founder of string theory and popular science writer Michio Kaku adopts the invention option. Important ideas in mathematics and even new fields in mathematics arose after scientists sought general theories for specific phenomena. Abstruse mathematical ideas that had emerged without link to science or practice were later found relevant for some scientific area. And new mathematical concepts emerged from the applications. Physics and mathematics have had a symbiotic developmental relationship at least from the days of Isaac Newton, if not earlier. Discovery and invention have gone hand in hand.

Deeply impressed by the elegant picture of the Universe drawn by modern science and mathematics, Kaku mentions God in his books and interviews. That the Universe is governed by a few basic laws, is regular and not a chaotic bundle of particles, shows simplicity within complexity, is symmetric, and can be described with mathematical equations make it look like the product of an intelligent force. He also refers to the equations of God. Given his stature, Kaku's references to God were taken imply that he was a theist. Fellow scientists were confounded and the media abounded with speculations.

Kaku clarified that his God was the God of Spinoza and Einstein. and not the personal God of the Bible.

> One god is a personal god, the god that you pray to, the god that smites the Philistines, the god that walks on water. That's the first god. But there's another god, and that's the god of Spinoza. That's the god of beauty, harmony, simplicity. Michio Kaku (Berman 2018).

The process of scientific discovery and new mathematical formulations were akin to reading the mind of God. But, in his view, God is the Universe and the Universe is God. And mathematics is the ideal language to read that mind.

The debate on whether mathematics is discovered or invented goes on. The theological implications of either are also being explored. Most recently it was catalyzed by astrophysicist Mario Livio book with the eye-catching title *Is God a Mathematician?* (Livio 2010). Yet, apart from new examples and terminology, the mathematicians, scientists and theologians involved in the exchanges have yet to contribute novel insights into the matter.

5.5 INFINITY

Say n is a large number. Then $n + 1$ is larger, implying that a largest number does not exist. The number horizon is unbounded. Yet, we feel that way out there, there is a humongous numerical entity larger than any conceivable entity. We call it infinity. Infinite entities are limitless. Infinity has a special symbol in mathematics.

Infinity

In high school, we learn that infinity is a 'number' larger than any number one can think of. If we ask what is ∞ + 1 or ∞ × 1, we are told it is ∞. It violates the rules of arithmetic. And we wonder: how that can be?

Beyond numbers, infinity is an attribute—unlimited size, extent, duration, ability or emotion. Infinity appears in science, psychology, art, literature, religion and mysticism. Physicists ponder: Is the Universe finite, infinite, or both? Religious scholars ruminate on the infinitude of God. Infinity has mesmerized towering intellects:

> *Nature is an infinite sphere of which the center is everywhere and the circumference nowhere.* Blaise Pascal

> *The known is finite, the unknown infinite.* Thomas Huxley

> *There are moments when all anxiety and stated toil are becalmed in the infinite leisure and repose of nature.* Henry David Thoreau

> *The power of imagination makes us infinite.* John Muir

> *You and I are essentially infinite choice-makers.* Deepak Chopra

> *Music fills the infinite between two souls.* Rabindranath Tagore

> *I give infinite thanks to God, who has been pleased to make me the first observer of marvelous things.* Galileo Galilei

> *Many are the names of God and infinite the forms through which He may be approached.* Ramakrishna

> *In Christ was united the human and the divine. His mission was to reconcile God to man, and man to God; to unite the finite with the infinite. Ellen G White*

> *We must accept finite disappointment, but never lose infinite hope.* Martin Luther King

(Source: www.brainyquote.com).

Rhetorical flourishes aside, infinity is an elusive, perplexing idea, hard to pin down concretely. As a famous mathematician opined:

> *No other question has ever moved so profoundly the spirit of man; no other idea has so fruitfully stimulated his intellect; yet no other concept stands in greater need of clarification than that of the infinite.* David Hilbert

What is infinity? Does it exist in nature or is it a purely theoretical construct? Is it an expression of divinity? Mathematical conceptualizations of infinity appeared in many cultures. A Jain mathematical text from the 4th century BCE India alludes to directional, spatial, temporal and numerical infinities and divides numbers into three categories: enumerable, innumerable, and infinite. Around the same time, the Grecian philosopher Zeno of Elena formulated four paradoxes based on infinite processes that continue to confound. One paradox went: A hare and a tortoise run at constant speeds on a straight line in the same direction. The tortoise is 10 meters ahead at the start. To catch up, the hare has first to reach where the tortoise was at the outset, and then to reach its new position, and then to reach its newer position, and on and on. They come closer and closer, yet a small gap always remains. Since an infinite number of steps are involved, the swift hare can never catch up with the sluggish tortoise. Some Chinese philosophers from the same era also formulated a paradox of this form. For Aristotle:

> [Something] *is infinite if, taking it quantity by quantity, we can always take something outside.* Aristotle (Blanc 2021).

He argued against the existence of actual (physical) infinity but allowed for potential (theoretical or mathematical) infinity. He postulated that Time has no beginning. Thereby infinity is a process rather an event or object.

Holding that the mathematical features of objects reflect their physical features, Islamic mathematician Avicenna extended the Aristotelian idea of infinity by forming a one-to-one correspondence between sets of objects. This method is a pillar of the modern concept of infinity. However, Avicenna alluded to the existence of infinite, non-ordered collections of non-corporeal entities like souls and angels as well.

Mathematicians across cultures used the idea of infinity and limits in an informal, utilitarian way to make novel discoveries. Differential and integral calculus at the outset deployed infinity in a casual manner, without a formal foundation. Placing the notion of infinity on a rigorous, logical footing was initiated in the 19th century by Georg Cantor. He divided numbers into two categories: Numbers denoting the size of a set (cardinal numbers) and numbers denoting the order of entities within a set (ordinal numbers). Ten boats pass through the gate of a canal and a particular boat is the tenth boat. It is an essential distinction for non-finite sets. Laying the basis for set theory, he invoked the idea of a one-to-one correspondence between sets to demonstrate patently counter-intuitive properties of infinite sets.

Consider a correspondence between the set of natural numbers and the set of even numbers:

$$1 \leftrightarrow 2; 2 \leftrightarrow 4; 3 \leftrightarrow 6; 4 \leftrightarrow 8; 5 \leftrightarrow 10; .... n \leftrightarrow 2n; .....$$

Thus, for every positive integer, there is a unique even positive integer and for every even positive integer, there is a unique positive integer. The two sets are of the same 'size' (cardinality). Yet, the latter is a part of the former. That cannot happen for finite sets. But for infinite sets, the cardinality of a proper subset can be the same as the cardinality of the entire set.

Cantor proved that the set of rational numbers, numbers of the form a/b (a and b integers), has the cardinality of the set of integers. The cardinality of the set of real numbers is identical to the cardinality of all numbers between 0 and 1. An infinitely long line is in that sense equivalent to a line of length 1! On the other hand, by ingeniously proving that the cardinality of the set of real numbers exceeds the cardinality integers, he demonstrated that there is an unlimited hierarchy of cardinals.

In sum, Cantor showed that starting from the integers, a hierarchy of levels of infinity exist. At any level, 'adding' or 'subtracting' a finite or some particular infinite quantity does not make a difference. If you remove the number 5 from the set of integers, you are still left with an infinite set of the same cardinality. Further, at any level of infinity, a proper subset of the whole can have the same cardinality as the whole. If you have an infinite supply of apples, and you can give away one every minute for a million years, the size of your pile will remain undiminished.

Yet, despite his rigorous approach, Cantor's intellectual edifice had flaws and required foundational modifications and expansion. Subsequent efforts of eminent mathematicians and logicians generated a generally robust but very complex modern theory of the infinite. Yet, contentious issues remain.

++++

The conception of divinity within religious traditions implies the idea of a being with supreme abilities. Whatever it is, the divine being possesses it at the extreme, infinite level. Divinity has no limits; it is infinite. If we say God [Allah, Brahman] has infinite goodness, it means that He is more good than any human and His goodness exceeds any possible level of goodness. God is not bound by space, time or power. He is ever present, everywhere. He knows everything, past, present and future.

> *The heavens, even the highest heaven, cannot contain thee* [God]. Bible Kings 8:27.

Theologians and religiously inclined mathematicians have pondered on the spiritual and mathematical versions of infinity. Viewing mathematics as a venue to fathom the handiwork of God, they befriend the two. Accepting the eternity of the soul, Christian and Islamic theologians posited that heaven has an infinite number of souls.

Ancient Indian mathematician Bhaskara grappled with the enigma of division by zero. For, if for any n,

$$n/0 = \infty \text{ then } \infty \times 0 = n \text{ for all } n$$

Lacking a rigorous axiomatic foundation, this formalist reasoning seemed to show that all numbers are equal, a patently false proposition. He thereby concluded that infinity is akin to an immutable, infinite deity. It indicates that truly understanding the infinitude of God is beyond the finite human mind.

The Christian notion of infinity evolved over time. Saint Augustine of Hippo, the most important of the early Catholic theologians, claimed that mathematics and mathematical infinity facilitate our understanding of God. God transcends the finite and the infinite. Subsuming time and space, God comprehends the incomprehensible, perceives without thought and counts without numbers. He transcends infinity. Thomas Aquinas, the 13th century Dominican friar who exerted a major influence on Christian thought, modified that formulation to stress the infinite character of God. Only God is absolutely infinite; all else is subordinate, perhaps relatively infinite. Rene Descartes and GW Leibniz argued that the infinity of God is reflected in the infinity of numbers yet is superior to it. Theological infinity supersedes mathematical infinity.

Cantor and Theology

From childhood to the end of his life, Georg Cantor was a devout Christian who saw his mathematical pursuits as a calling from God. He felt that the existence of transfinite numbers had been revealed to him. In the interest of the Church and humanity, he felt bound to promote his unconventional findings. He visualized three forms of infinity: mathematical, physical and spiritual (transcendental). The first two forms are subservient to the transcendental infinite.

Believing that God created the world from numbers, Cantor felt that the notion of infinity is inborn. God is above infinity and to God all infinities are finite. Calling God 'the Absolute infinite,' he denotes it with the capital Greek letter Ω. In his later years, Cantor dwelt deeply in theology and philosophy.

Doubting the hierarchy of infinities, leading mathematicians rejected his work on transfinite numbers. His academic career suffered and his attempts to publish his work faced steep obstacles. Though some Catholic scholars felt uneasy about his ideas, a good number supported his work. His faith in his mission sustained him. He corresponded with senior church dignitaries to explain that only God was perfect and absolutely infinite.

Ultimately, hostility from mathematical luminaries took its toll. His self-confidence and creativity waned, and his bipolar disorder intensified. Yet, his revolutionary ideas could not be suppressed. Towards the end of his life, he was celebrated but isolated. Sunk in poverty, he met a lonely death in a sanatorium.

Paradoxically, Cantor, whose ideas and methods constitute an essential pillar of the vast edifice of modern mathematics, was treated as a heretic by fellow mathematicians of his day but his work was psychically sustained by his religious beliefs and friendly religious scholars.

Cantor's influence persists in the theological arena. His ideas are utilized by some faith systems to buttress their vision of God. Paradoxical issues like the existence of evil, free will and omnipotence of God are explained away by arguing that the infinitude of God is beyond human comprehension. Many Christian websites give a reasonably accurate explanation of the mathematical ideas of Cantor and then make long leaps of faith to draw bold theological conclusions:

> *Infinity is one aspect of God's nature. He is limitless in power, knowledge, and majesty. We are finite and cannot fully comprehend the concept of infinity.* (Lisle 2017).

> *The infinite nature of numbers reminds us of our limited knowledge.* (Hannon 2010).

Today Cantor's legacy is claimed by mathematicians and theologians alike. For the former, it derives from the logically rigorous parts of his work, but for the latter, it is a matter of clinging to his pseudo-logical speculations.

Infinitesimal

Mathematical infinity comes in two forms—infinitely large and infinitely small, or infinitesimal. The latter is a numerical entity that assumes progressively small values. Its absolute value tends to zero while always remaining non-zero. The distance between the hare and the tortoise, as conceptualized in Zeno's paradox, is a case in point.

Mathematicians in India, Greece and Islamic empires partitioned lines into arbitrarily small lengths and used the idea of infinitely small lengths and areas to find areas and volumes of irregular figures. That was the basis of the method of exhaustion used by Archimedes and Eudoxus to approximate, among other things, the area of a circle.

The Iraqi scholar Ibn al-Haytham who lived around 1,000 CE extended the application of the infinitesimal. Though trained in Islamic jurisprudence, he was sufficiently disenchanted by ongoing theological conflicts to abandon legal practice and turn to mathematics and science. His erudite contributions to astronomy, physics, optics, medicine, psychology, and engineering as well as mathematics earned him region-wide fame. At one point, he was asked to advise on the construction of a dam on the Nile. His assessment was that the dam would not effectively regulate the water flow in the river. But that verdict so angered the Caliph that he was placed under house arrest.

In the ten years he spent under confinement, he churned out books on a variety of subjects. Later he taught science and mathematics in Cairo. In all, he wrote about 200 books, about half of which have survived. Many of his ideas predated what would transpire in Europe after the 12th century. He also enriched the methodology of science.

His contributions to mathematics related to proof by induction, number theory, analytic geometry, conic sections and summation series of powers of integers. Utilizing Archimedes' method of using infinitely small quantities, he proved additional results. Invocation of the infinitesimal in the Islamic world at this stage did not ruffle theological feathers. But in the Christian nations, it became a stage for a veritable theological war.

++++

From the 14th century on, seeds of social instability were being planted across Europe. The oppressive rural gentry and corrupt priestly orders faced challenges on several fronts. Emerging urban centers and expansion of transport and trade gave rise to new centers of economic and political power. Merchants, bankers, artisan guilds and peasants stood up for their own interests against landowners and kings. Peasant revolts spread; many fled to towns to escape utter misery. The excessive tribute exacted by local churches and the Vatican was no longer tolerable. Wars between nations were being fought.

As internal discord within the Catholic Church mounted, reformist voices grew in strength. Protestantism was on the rise. It was a time when new findings in science questioned doctrines deemed sacrosanct for centuries. The rulers and the Church mounted a fierce counter offensive. Here we are concerned with one arena of that conflict: mathematics.

The Society of Jesuits, founded in 1540 by Ignatius of Loyola with papal consent, was a major player in this war. Composed of men of nobility and learned scholars including mathematicians, scientists and astronomers, its aim was to halt the spread of theological 'chaos' and reassert the doctrinal authority of the Catholic Church. Its mission was a global one extending far beyond Europe into the Americas.

The Society attracted educated, disciplined persons who took monastic vows of simple living. Professing absolute loyalty to the Church leadership, their major goal was to vanquish the virus of Protestantism. With education a key instrument in that endeavor, the Order set up colleges and employed expert instructors. At the outset, theology, philosophy, humanities and languages were stressed. High standards, inclusion of secular subjects and rigorous curricula attracted students from non-Jesuit background as well. Many scientific luminaries were to emerge from Jesuit training.

Mathematics was not deemed a key subject during the first decade. But once Christopher Clavius assumed the professorship of mathematics at the premier Jesuit College in Rome, its status began to change. At that time, a Papal committee was tasked to construct a new calendar, a centuries-old concern for the Church. Clavius was a member. And his thorough, precise calculations were instrumental in the development of the Gregorian calendar, the modern calendar. A spectacular achievement, it buttressed Papal authority. Even Protestants and other adversaries of the Bishop of Rome had little option but to abide by it. Making Clavius a man of influence, the episode enhanced the prestige of mathematics in the Jesuit ranks.

> Rigorous, orderly and irresistible, mathematics was for Clavius the embodiment of the Jesuit program. By imposing truth and vanquishing error, it established fixed order and certainty in place of chaos and confusion. (Alexander 2014, page 65).

Recruiting master mathematicians as teachers, Calvius turned it into a core field of Jesuit training. He and fellow authors wrote new, rigorous textbooks. While arithmetic and algebra were included, geometry, as laid down in Euclid's *The Elements,* was the soul of Jesuit mathematics. Embodying certainty and perfection, it revealed the wisdom of God and His architecture for the Universe.

Official Seal of the Jesuits

Among the novel mathematical ideas being formulated in those days was the use of minute, indivisible entities in relation to matter, space and time. It was utilized by mathematicians and scientists to prove important results. Different types of infinite sequences and series were being evaluated. In one type of infinite series, the terms become smaller and smaller but do not reach zero. The new ideas led to computation of areas and volumes of complex entities. Basic ideas of differential and integral calculus, which profusely utilized infinitesimal quantities, were laid. Galileo opined that

> [A] *continuous line is composed of an infinite number of indivisible points separated by an infinite number of miniscule empty spaces.* Galileo (Alexander 2014, page 90).

Cavileri and Toricelli deployed infinitesimals to compute areas of complex figures. Isaac Newton and GW Leibniz applied these achievements to establish a coherent edifice for calculus, a new branch of mathematics.

The infinitesimal was a heuristic device, not an axiomatically generated entity. And it was a paradoxical entity. Lines contain an infinite number of points; planes, an infinite number of parallel lines; and a solid, an infinite number of parallel planes. If points have zero length, lines have zero area, and planes have zero volume, how come their sum is finite? Where does the infinitesimal fit into this scenario? Are infinitesimals actual numbers?

The Jesuits doubted whether the infinitesimal and the findings it brought forth were in accord with the Euclidean tradition. The matter was referred to the Revisor General, the organ with the final say on what was acceptable to the Jesuit Order and could be taught in its institutions. The rulings were respected across the Catholic lands. After years of prevarication, in 1632, it deliberated the proposition that the continuum was composed of infinitesimals or indivisible entities (atoms). The judgment was crystal clear:

> *We consider this proposition to be not only repugnant to the common doctrine of Aristotle but that it is by itself improbable, and ... is disapproved and forbidden in our society.* (Alexander 2014).

The ruling started an ardent theological campaign against the infinitesimal that was to last some two hundred years. Mathematicians promoting it were strongly denounced. Jesuit mathematicians who sympathized with the idea were socially isolated and humiliated. For many decades, legal and theological avenues were used to silence all pro-infinitesimal voices in Italy.

The ferocity of the campaign did not emanate from purely mathematical grounds. The Jesuits stood for order and social stability. Obeisance to established doctrine and authority was central. The infinitesimal was not rooted in Euclidean geometry, the divinely ordained branch of mathematics. It was mysterious entity linked to measuring change, a socially subversive idea.

> *To the Jesuits, tradition, resoluteness and authority seemed bound up with Euclid and Catholicism; chaos, confusion and paradoxes were associated with infinitesimals and the motley array of proliferating Protestant sects.* (Paulos 2014).

++++

Two decades after the verdict of the Revisor General, the dispute reached the English shores. The crusade against the infinitesimal there was led by the influential political philosopher Thomas Hobbes. A dabbler in subjects ranging from law, history, science, theology and mathematics, he is primarily famed for his 1651 book *The Leviathan*. Its basic message is that human society requires a strong central authority to avoid civil conflict and moral disorder. Else, people will become unruly and chaos will prevail. A supporter of absolute monarchy against the growing parliamentary authority, he opposed demands for academic freedom by the universities.

Hobbes, a mathematician of minor standing, highly valued geometry. It stood for order and stability. Among other things, he presented a method for finding a square whose area was equal to that of a given circle. It was soon seen that his proof was flawed, yet Hobbes clung to it until the end.

Though not a devout Christian, Hobbes concurred with the Jesuit stand on infinitesimals. The idea reeked of logical incoherence. Arguments invoking them seemed arbitrary. Instead of the tried-and-true axiomatic Euclidean framework, it used the questionable notion of induction. Infinitesimals discounted the finest intellectual tools of the human mind and by implication, put into question the basic ideas underpinning the social order. They were subversive entities.

The infinitesimal was advocated by John Wallis, a Presbyterian minister and Professor of Mathematics and Astronomy at Oxford University. A prodigious mental calculator and talented mathematician, he made novel contributions to algebra, analytic geometry, infinite series, integral calculus, and continued fractions. Besides pioneering a standard notation for powers in algebraic expressions, he introduced the symbol for infinity we use today. His ideas were partly derived from earlier era Islamic mathematicians. He also proved a remarkable infinite product for π.

The Wallis Formula For Pi

$$\frac{\pi}{2} = \frac{2}{1}\frac{2}{3}\frac{4}{3}\frac{4}{5}\frac{6}{5}\frac{6}{7}\cdots$$

Wallis Formula for Pi

A founder of the Royal Society, supporter of religious pluralism, John Wallis favored parliamentary rule. Among other things, he assisted the democratic side by using his mathematical skills to decipher coded messages.

Wallis and Hobbes stood on the opposite sides of the mathematical and political divides. Hobbes castigated Wallis for lacking Euclidean rigor while Wallis took apart the Hobbes' method of squaring the circle and defended the results obtained by the method of infinitesimals. It was an acrimonious, public battle with political overtones that raged for decades.

The growth and applications of calculus were unstoppable and critical for not just mathematics but science as well. The infinitesimal side prevailed but not in its original form. Infinitesimals have been replaced by the rigorous idea of limits. But they were a stepping-stone on the ascent to modern mathematics. But on the political front, the authoritarian tendencies favored by Hobbes are making a comeback on all the continents.

196

5.6 THEOLOGICAL MATHEMATICS

Mathematical concepts have been invoked to support as well as oppose religious ideas. The notion of infinity is but one example. We first take a general look at how religion has been disposed towards mathematics, then review other cases.

Mathematics: A Religious Endeavor

Our broad historical survey of mathematics under various cultures indicated that despite disagreements about specific ideas, mathematics and theology have generally maintained a cozy relationship. Religious scholars have opined that the logical rigor and austere elegance of mathematics reflect the divinity of God. He designed nature with mathematical tools and granted us the capacity to discover them. Buddhist, Hindu, Muslim and Christian scholars have enriched the discipline. Eminent mathematicians have been devout believers. Faith traditions generally have a harmonious, even celebratory, stand towards mathematics.

The Divine Ratio

Start a sequence with numbers 0 and 1. Add them to get a third number, 1. Add the last two (1 and 1) to get the fourth number, 2. Again add the last two (1 and 2) to get 3. The sequence of numbers generated by this process of adding the current and previous number, (0, 1, 1, 2, 3, 5, 8, 13, 21, 34, ...), is the Fibonacci sequence, after its famed advocate, Leonardo Fibonacci of Pisa. These numbers were known to Indian and Greek scholars of the ancient times and to Muslim mathematicians prior to 1200. Fibonacci, the conveyor of the Hindu-Arabic numeral system into Europe, reintroduced them to the region.

Arrange the Fibonacci numbers as sides of squares placed side by side, and form a spiral by drawing quarter circles within each square:

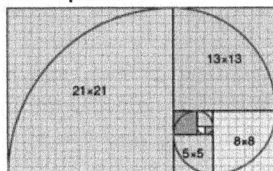

Fibonacci Spiral

Compare this spiral with the nautilus shell. The similarity is striking.

Nautilus Shell

Spiral shapes are ubiquitous in nature (molecular arrangements, flowers, animal physiology, galactic patterns), architecture (pyramids, temples, buildings), art and book design. The Nautilus seashell, together with a host of other seashells, is but one of the numerous cases of helical and spiral shapes found in the natural world. By optimizing space and energy needs, and functioning as efficient defense structures for some organisms, these shapes have evolved over eons.

Returning to Fibonacci numbers, omitting the first number, we compute the ratio of each number with its predecessor: 1/1 = 1.0; 2/1 = 2.0; 3/2 = 1.5; 5.3 = 1.67; 13/8 = 1.625; 21/13 = 1.615; 34/21 = 1.619, ... As we proceed, the ratio approaches a fixed number. Designated by the Greek letter φ (phi), it is

$$\varphi = (1 + \sqrt{5})/2 = 1.618033989....$$

φ is called the Golden Ratio, Golden Mean or Divine Ratio. Arising in many contexts, it was known to earlier era Indian, Greek and Muslim mathematicians. But only after *Liber Abaci* drew attention to the Fibonacci numbers was the connection between the two entities found. Not only is φ the limiting ratio of successive Fibonacci numbers but any Fibonacci can be directly computed with a simple formula based on φ.

Today, a huge, fascinating body of the properties of Fibonacci sequences and the Golden Ratio are known. There is a journal exclusively devoted to them and hundreds of papers showing the presence of Fibonacci like spirals in a wide variety of phenomena exist. Figures and shape incorporating the Divine Ratio has made inroads into varied human applications.

The Golden (Divine) Ratio has been embraced by several religious traditions. A paper on the website of the conservative Institute for Creation Research summarizes the history and properties of Fibonacci numbers and observes:

> [The Golden Spiral] *is visible in things as diverse as: hurricanes, spiral seeds, the cochlea of the human ear, ram's horn, sea-horse tail, growing fern leaves, DNA molecule, waves breaking on the beach, tornadoes, galaxies, the tail of a comet as it winds around the sun, whirlpools, seed patterns of sunflowers, daisies, dandelions, and in the construction of the ears of most mammals.* (Wilson 2002).

And that is not coincidental:

> [The] *Divine Proportion—existing in the smallest to the largest parts, in living and also in non-living things—reveals the awesome handiwork of God and His interest in beauty, function, and order. ... [The] correlation of the Fibonacci pattern to the periodic times of the planets is far more than just a chance arrangement. It is one more example of God's marvelous mathematical arrangement of His creation. The fact that it is not perfect reveals that although Adam's sin affected the whole creation (Romans 8:22), yet God in His goodness has not allowed sin to overcome all the marks of His great handiwork (Psalm 19:1).* (Willson 2002).

When the observed patterns and spirals deviate slightly from the expected patterns and spirals, the inaccuracies are ascribed to Adam's original sin, not God.

A posting on the website Islam Hashtag imputes Allah's majesty in relation to the Golden Ratio. The article details the existence of the ratio in several natural phenomena. Using geographical measures for the Islamic holy city of Mecca and patterns derived from Quranic verses, it declares that [the] *Golden Ratio proves that Islam is a true religion.* (Aafiya 2015).

Pi

Another entity that has captured both secular and theological imaginations is the ratio of the circumference of a circle to its diameter, π. It is a most useful and theoretically intriguing entity. Today, a veritable industry devoted to π exists. March 14th or 3/14 (in the US date notation) is Pi day, celebrated in creative and fun-filled ways every year in schools across the world. An opportunity to counter negative attitudes towards mathematics, it serves to attract children to the subject.

π is a transcendental number. Written in decimal digits, it never ends. For most practical applications, five digits of accuracy is ample. Some astronomical and space flight computations need 15 digits of π while an accurate determination of the fundamental constants of the Universe needs up to 32 digits of π.

$$\pi = 3.14159265358979323846264338327950288419716939937510 \ldots$$

Finding the digits of π is an ongoing mathematical quest. Efficient formulas, supercomputer power and clever programs have enabled determinations of more than 31 trillion digits of π. Written on a tape with each digit 1 millimeter in length, it would go to the moon and back over sixty times. Recitation of the digits of π from memory is a holy grail for mental marathoners. A few persons in Asia have publicly recited more than 60,000 digits of π. The current record is held by Akira Haraguchi of Japan. In 2019, he accurately recited 100,700 digits of π in an event lasting over 16 hours and 30 minutes. Fascination with this esoteric number extends into the religious domain.

Hinduism: Aryabhata, the legendary Indian mathematician astronomer who lived around 500, approximated the ratio of the circumference of a circle to its diameter as 3.1416, a five-digit level of accuracy. Learning in that era was virtually monopolized by the Brahmins who conducted religious rites and rituals and controlled the temples. It is claimed that a hymn to Lord Krishna contains, in an encrypted form, the value of π accurate to 31 decimal points. Hindu websites today resort to such amplified claims to declare that:

Science and spirituality both moved together in this land. (HASD 2014).

Srinivasa Ramanujan

The life and work of the early 20th century prodigy, Srinivasa Ramanujan, are apropos. At an early age he, on his own, rediscovered known mathematical results and formulated astounding expressions involving infinite series, infinite products, continued fractions, elliptic integrals and prime number theory that still intrigue mathematical stalwarts. Later he collaborated with two mathematical titans, GH Hardy and JE Littlewood, at the University of Cambridge. They jointly produced breakthroughs in several areas of mathematics. Ramanujan was inducted into the Royal Society and elected a fellow of the Trinity College,

Ramanujan became seriously ill within a few years, and had to return to India. After a partial recovery, and churning out more astounding results, he succumbed and died in 1920 at the tender age of 32 years. But his legacy endures. Regarded as a first-class genius, a good portion of his vast output awaits careful study. Enriching many branches of mathematics, his work has found varied applications in physics and other sciences.

Ramanujan and Hardy were contrasting figures in ways more than one. An eccentric persona enamored by cricket and linguistic flourishes, Hardy was a gregarious talker, admired instructor and sought-after mentor. Of left leaning political views, he opposed war and the warmongering common among fellow academics. And on top of that, he was an avowed atheist disposed to sardonic comments about God. With a devout Brahmin family, Ramanujan was steeped in Hindu traditions. A vegetarian, and socially conservative, he found the culture in England strange.

Hardy worshiped at the altar of rigor and step-by-step proof. The self-taught Ramanujan apparently had an intuitive method of deriving equations. Paper was costly at home. He could afford only a few notebooks and worked on a slate. As he went along, he wiped out the previous steps and wrote the final result in his notebooks. Hardy partly succeeded in reorienting him. Yet, with his overactive brain, he lacked the patience to make stops on the way to the destination. When the exasperated Hardy emphasized the indispensability of proof, Ramanujan is reported to have responded:

> [An] *equation has no meaning to me unless it expresses a thought of god.*
> (Aron 2016).

Positing a divine origin for his mathematical prowess, he held that the family goddess was revealing mathematical equations to him during sleep. It was but the joint love of pure mathematics that bonded the two vastly disparate personalities. It is not in doubt that Ramanujan was the leading genius of the partnership.

Buddhism: Japanese temple geometry was but one expression of the linkage between Buddhism and mathematics. However, a specific traditional link to π has not been recorded. The master pi reciter Haraguchi is a practicing Zen Buddhist. He believes that the spirit of the Buddha pervades all creation, including rotational motion, circles and π.

> *All things in this world, including ourselves, are aggregate sums of atoms, which are made up of rotating electrons. The ultimate history of mankind is moving toward a happy ending for people of all races. The Earth, the galaxy and the Universe all rotate. In other words, I think rotation is the absolute truth. So as long as I'm thinking about pi, I think I can live a life according to truth.* Akira Haraguchi (Bellos 2015).

Daily recitation of some 25,000 digits of π is akin to Buddhist meditation to him. The three-hour venture is a plunge into the ultimate reality. Bestowing mental tranquility, it reveals the meaning of life and tells him that humanity is marching towards a state of blissful harmony.

Islam: Mathematicians working during the heydays of the Islamic empires were generally practicing Muslims. Mathematical pursuits complemented faith. They were two venues for seeking the truth. Problems relating to circles were also addressed. Working around 1300, Al-Kashi determined π to 16 decimal points. Centuries earlier, the pioneering algebraist Al-Khwarizmi had proclaimed the Islamic stand on circle-related endeavors:

> *It is an approximation not a proof, and no one stands on the truth of this, and no one but Allah knows the true circumference of the circle, as the line is not straight and has no beginning and no end, we merely attempt to approximate and discover the root, but even the root has no definition as no one may know its exact value but Allah, and the best of these approximations that is to*

multiply the diameter by three and seventh as it is faster and simpler and only Allah might know its true value. Al-Khwarizmi (Abdelhamid 2021).

Islamic websites today take this to imply that any approximation we attain will not be the true value of π; only Allah knows the truth. Other sites claim that the numbers 3.14 and 2x3.14 = 6.28 are encoded in the verses of the Quran. They seem unperturbed that the computations of Al-Kashi that had given far more accurate values of π. Skeptical websites note that for any long book, scriptural or secular, computerized search and deployment of linguistic plasticity can yield numerical associations that are hard to explain in rational terms.

Christianity: Verse Kings 7 of the Bible details the construction of the house of Solomon. One passage reads:

> *And he made a molten sea, ten cubits from the one brim to the other: it was round all about, and his height was five cubits: and a line of thirty cubits did compass it round about.* The Bible, Kings 7:23.

This implies that π = 30/10 = 3. The passage has generated an extensive debate. Secularists note that some two hundred and fifty years before the birth of Christ, Archimedes had demonstrated that the value of π was less than 22/7 (3.1429) but more than 223/71 (3.1408). The appearance of a crude value like 3 in the Bible hardly supports divine origination.

Christian websites counter this claim in several ways. Passages of the Bible are to be interpreted in context, according to purpose. Rounding is a common practice. The level of rounding varies according to the task at hand. The single digit rounded value for π here is reflective of the Trinity of God. It is further noted that no distinction is made between the outer and inner rims of the vessel. Hence, the rounding is justified. More importantly,

> *Pi reveals The Lord's nature to us. He has everything surrounded (circumference), goes straight through the central point (diameter) and is infinite, boundless and limitless.*

> *The thing to keep in mind is to turn the wonder at pi itself into wonder at the Creator of all things–the one who understands what only baffles our comprehension. (*MCCH 2017).

Accordingly, some Christian websites give quite accurate descriptions of the history, applications and approximations for π, and take part in the Pi Day celebrations.

Religion and Counting

Religion and a counting system were present in neolithic and post-neolithic societies. The creative effort expended for exquisite shrines was matched by ingenious systems devised to represent and operate upon numbers. Ancient Sumerians developed a base-sixty number system which, via the Babylonians, was passed down to the modern era. It is still used for measuring time, angles and Earth coordinates. But now we use it with Arab numerals.

In the twelfth century Europe, crafts and commerce were expanding and town populations grew rapidly as well. Activities that entailed manipulation of large numbers, trade, manufacture, transport, banking and science, faced a major hurdle: The archaic, time-consuming Roman numeral system blessed by the Church. Despite the ecclesiastical opposition, a gradual transition to Indo-Arabic numerals was in the offing everywhere. The needs of capital outshone Papal preference.

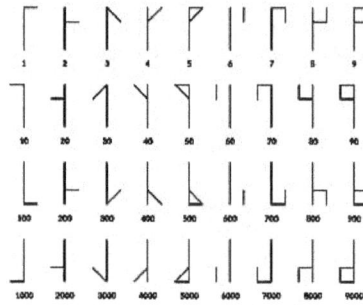

Cistercian Numerals

Some monastic orders felt they could serve both masters, secular and divine. Among them was the Cistercian monastic order which devised an elegant system of numerals in which any number could be represented by a single glyph devised in a systematic fashion. Cistercian numerals were more compact than Roman or Indo-Arabic numerals but could only cover numbers up to 9,999. Furthermore, they could be used for counting quantities (stock, pages, stock of goods) but not for arithmetic, accounting or banking. And they were useless for fractions, roots or complex mathematical operations. The chart above is demonstrative.

The Cistercian numerals never gained a broad following but survived within a small niche until recently. They are another example of the amiable relationship between the Christian church and mathematics. As the case of Jesuits shows, medieval monks and other theologians studied mathematics and engaged in innovative mathematical pursuits. And, many high ranking mathematicians like Cantor and Godel believed in God and produced proofs for His existence.

5.7 NUMEROLOGY

Numerology, assigning mystical significance to numbers and use of number-based rules for the conduct and foretelling of life events, has existed in cultures across the world. Holding that natural numbers were the essence of reality, the Pythagoreans assigned mystical properties to specific numbers. Numerology posits that numbers affect human characteristics and events.

Consider, for example, the full name of a person. Based on a scheme relating numbers to the letters of the alphabet and rules for manipulating numbers, we get the name number of that person. Similarly, birth date is used to generate the birth number. In some numerologic systems, name numbers indicate a person's demeanor and fate while birth numbers foretell talents and character. Many systems based on diverse alphabetical systems—Arabic, Chaldean, Chinese, English, Hebrew, Indian, Latin, Japanese—and using different sets of rules have been recorded. Our focus is on the association between numerology and the four major global religions.

Hinduism: Hinduism has an elaborate numerologic system derived from Vedic texts and traditions. It has two basic principles: Every number from 1 to 9 has an associated planetary deity. The deities influence—in positive, negative and interactive ways—the life, personality, character, intelligence, health and failings of an individual. A person's date of birth and name are reducible by special rules to three key numbers—birth number, destiny number and name number—that govern his or her life. For example, 1 is the number of the Planet Sun whose deity denotes '*strong individuality, masculinity, likes to be in control, authoritative, bright, freedom loving, intelligent, able-bodied, prefers luxuries and comforts of a good life,*' while 2 is the number of the Planet Moon whose deity denotes '*peace loving, soft, gentle, emotional, supportive, fluctuating, good looking, intuitive, attractive'*. (Jayaram 2021).

The number 3 has a special standing. It represents the triumvirate of the three supreme gods, Brahman, Vishnu and Shiva, who create, maintain and annihilate the Universe.

Decisions about important issues like date of marriage, naming a child, and time and date for holding a religious ceremony derive from numerologic advice. Business, work and property decisions utilize numerologic computations. Future projections also employ these divine numbers. Books, charts and websites for this purpose are abundant. Traditionally, a renown priest-cum-numerologist proficient in Sanskrit and the Vedic scriptures and with experience in conducting religious ceremonies is consulted.

Numerologic links with life and personal features are not cast in stone; there is considerable variation from source to source. Numerologists are expected to be honest, humble and virtuous. They should deal in a fair manner with their clients. In practice, numerologic gurus who worship at the altar of money and prestige are not uncommon.

Buddhism: Performing numerical calculations for life choices is unusual in any branch of Buddhism. However, Buddhist canons are awash in numerical listings of beliefs, practices and divinities. Examples include the Four Noble Truths, the Noble Eight fold Path, and the Six Realms of Existence. These canons refer to numbers larger than those found in any other religious tradition. For example, the Buddha apparently performed calculations whose result was a number with 421 zeroes.

Numerical lists facilitated memorization and transmission of the canons at a time when writing was rare. The numbers often had spiritual import. 1 signified unity, harmony and meditative focus; 2 signified, diversity and conflict between the divine and the mundane; 3 integrated harmony with divisiveness; 4 denoted perfection and balance; 7 alluded to the mystery of life and death; and so on. The Buddha took seven steps after his birth. The number 108 symbolizes the Buddha as a unique being. The number 3 is embossed in the *Tripitaka*, the three baskets of the Buddhist canon, and in the three-lions pillar of the emperor Ashoka, who adopted Buddhism as the formal religion of his empire and spread it across Asia.

As in Hinduism, considerable flexibility in relation to the divine significance of the hallowed numbers exists. For some Buddhists scholars, zero is the supreme number. Noting that the number 1 denotes spiritual unity, one scholar opines:

> [While] t*he Vedas yearn to return to the One, Buddhists seek to let go of the One so as to realize the only truly Buddhist number – zero.* (Sujato 2010).

Christianity: Numbers have a special place in the Christian lore. The venerated father, St. Augustine of Hippo, held that all phenomena had a numerical basis.

> *Numbers are the Universal language offered by the deity to humans as confirmation of the truth.* Saint Augustine (Harrison 2019).

The Bible, especially the Old Testament, has many passages that can be construed to imply divine import to numbers. Some examples:

> *Here is the secret meaning of the seven stars which you saw in my right hand and of the seven lamps of gold: the seven stars are the angels of the seven churches, and the seven lamps are the seven churches.* The Bible, Revelations of John, 1:20.

> *Lent lasts for 40 days and 40 nights; and Moses led the Israelites for 40 years in the wilderness. In Noah's time, rain fell for 40 days to flood the Earth. Goliath was defeated by David after he had challenged the Israelites*

for 40 days. Ascension Day comes 40 days after Easter. Such consistency can hardly be chance and must hold some meaning. (Harrison 2019).

The number 1, the generator of all integers, represents the supreme creator. In the New Testament we read:

One Lord, one faith, one baptism. The Bible, Ephesians 4:5.

The number 3, denoting the trinity of the Father, the Holy Ghost and the Son, has an elevated significance. The number 12 is linked to the 12 angels, the 12 stars, the 12 gates of heaven, and the 12 fruits in the tree of life. Before Judas departed to betray Jesus Christ, 13 people sat at the Last Supper table. Thirteen is thus an unlucky number, to avoided as much as possible. Some hotels in the USA do not have Room No. 13. And 666 is the number of the devil.

Early Christianity inherited the Judaic and Roman practices under which numbers and assigning numbers to the letters of the alphabet were used to ascribe meaning to life events and foretell the future. But after the Christian creed unified in the 4th century, numerologic divination was proscribed by the leading Church Fathers and codified into law.

The Catholic Church has historically believed in the significance of certain numbers, but ultimately rejected the systematic divination associated with numerology. (Gabriel 2017).

In force for nearly a millennium, the ban on divination through numbers fell apart as Christianity divided into denominations and sects following varied beliefs. In particular, the Greek Orthodox Church was favorably disposed towards numerology.

Despite the disapproval of numerologic divination by the mainstream Christian denominations, numerology has a fair following in the nations of the West. Christian numerologists who predict the date on which the world will end continue to ply their trade.

Islam: Numbers play an important symbolic role in Islam and are used to explore hidden meaning in the passages of the Quran. The number 4, linked to the four esteemed Caliphs, the maximum number of legal wives, and the four Sunni law schools, enshrines universal harmony. The number 5 is ensconced in the five pillars of Islam, the five daily prayers, and for Shia Muslims, the holy personages of Prophet Mohamed, his wife Fatima, Imam Ali and his sons, Imam Hassan and Imam Hussein.

The number 40 has a practical significance. The climax of the memorial after death is on the 40th day; some Sufi traditions have 40 days of seclusion and reflection; some religious feasts last for 40 days; and women are deemed impure for the first 40 days after childbirth. It also denotes the age of maturity. Muslims from the India and Pakistan regard 786 as a divine number representing the holy phrase: *In the name of Allah, the most merciful, the most compassionate.*

Some Muslim traditions use name numerology to depict the personality traits of individuals and match making for couples. Numerical juggling for divination and predicting the future together with astrological prediction and fortune telling are, however, strongly proscribed by the religious authorities. Only Allah can know what has not yet occurred.

Using astrology or numerology to determine the future is haram. The Prophet (peace and blessings be upon him) cursed people who practice fortune-telling. Therefore, the people should be discouraged from following this sinful practice. Mufti Ibrahim Desai (Esposito 2021).

++++

Historically the logically driven queen of sciences has provided ample room for pseudo-scientific ideas. Schimmel (1993) gives a survey of the numerologic lore. Most cultures had distinct systems for interpreting and manipulating numbers. The diversity reflects human creativity, not divine origination. Numerology has no empirical basis and no demonstrated predictive value. At best, it soothes the gullible mind.

Living on in the age of science, it is packaged with pseudo-science practices like astrology, tarot card divination, extra sensory perception, crystal healing, flower and scent remedies, occultism, miracles and magical rituals which still have a broad following even among well-educated populations. Some surveys indicate that three out of four people generally hold paranormal or esoteric beliefs of some kind. With the advent of the Internet and the so-called age of information, such beliefs and practices have bloomed.

Irrationality is indispensable for a system of inequality, economic injustice and violation of human dignity and rights—the global neoliberal capitalist order—to survive and prosper. Combating irrationality on the intellectual front has to be combined with combating irrationality on the social, economic and ecological fronts.

5.8 AXIOMATIC MATHEMATICS

All branches of mathematics—algebra, geometry, number theory, set theory, topology—rest on three foundational pillars: The rules of logic, primary entities and basic axioms (postulates). The sparse foundation then generates a complex edifice in a cumulative way. Each brick, attached by deductive reasoning (proof), is called a theorem. The rules of logic are called syllogisms. A simple example of a syllogism is:

> Major Premise: The angles of a triangle sum to two right angles.
> Minor Premise: Figure X is a triangle.
> Conclusion: The sum of the angles of X equals two right angles.

If the major or the minor premise is false, the conclusion is false. Now consider a variation.

> Major Premise: The angles of a triangle sum to two right angles.
> Minor Premise: Figure X is an equilateral triangle.
> Conclusion: The sum of the angles of X equals two right angles.

In this case, if the major premise is false, the conclusion is false but if the minor premise is false, the conclusion is not necessarily false because X may be some other type of triangle. Different types of syllogisms and formal ways of assessing their validity exist. They are the bedrock of mathematical inference.

High school Euclidean geometry uses points, lines, planes and solids as primary entities. Its five basic axioms are:

1. A straight line can be drawn from any point to another point.
2. A terminated line can be further produced indefinitely.
3. A circle can be drawn with any center and any radius.
4. All right angles are equal to one another.
5. Parallel lines never meet.

Using the basic rules of logic, Euclid derived a large body of elegant results from this simple foundation. Planar and solid Euclidean geometry is now a gigantic body of complex results that can fill hundreds of books. Euclid held that the five axioms were self-evident truths reflecting the structure of nature. Given the extensive utility of geometry in the sciences and all walks of life, it seems to be a plausible proposition.

Every system of thought has axioms. Axioms are sacrosanct. They are beliefs that provide meaning. Doubt an axiom and the entire field may be in jeopardy. What is derived from the axioms is testable, but the axioms are not.

Mathematical axioms are formulated with clarity and precision. A valid set of axioms must satisfy three key requirements: Completeness, consistency and non-redundancy. A set of axioms should be adequate for the desired purpose, not produce contradictions, and contain minimally necessary statements. Using the axioms, it should be possible to establish the truth or falsity of any statement in the field. What is true is provable and what is provable is true.

Non-Euclidean Geometry

The fifth postulate of Euclidean geometry about parallel lines has intrigued mathematicians for a long time. It seems too complex to be a postulate. Perhaps the fifth postulate can be derived from the first four. Over centuries, that task was tackled by numerous eminent mathematicians and many upstarts. But a flawless derivation did not emerge. This postulate can be formulated in several equivalent ways. One is:

> Given any straight line and a point not on it, there exists one and only one straight line which passes through that point and never intersects the first line, no matter how far they are extended.

By tweaking the fifth postulate, 19th century mathematicians took the bold step of querying the hallowed status of Euclidean geometry. Nikolai Lobachevsky, Janos Bolyai and Bernard Riemann fathered two branches of non-Euclidean geometry in which parallel lines either meet or diverge. In hyperbolic geometry, more than one line passing through the same point is parallel to a given line and in elliptic geometry, there is none. The figure below is illustrative.

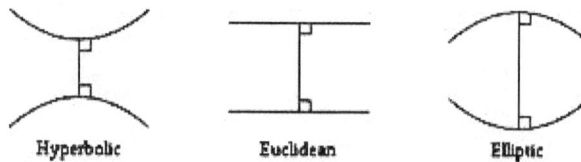

Hyperbolic   Euclidean   Elliptic

Three Geometries

Retaining the first four postulates of Euclidean geometry but modifying the fifth postulate generated geometries that were internally as consistent as Euclidean geometry but had counter-intuitive results. In hyperbolic geometry, the sum of the angles of a triangle is less than two right angles while in elliptic geometry, it is more than two right angles. While Euclidean geometry pertains to flat, planar surfaces, hyperbolic geometry applies to saddle-type curved surfaces and elliptic geometry, to spherical surfaces.

Non-Euclidean geometries brought forth far ranging challenges. A body of knowledge seen beyond dispute by mathematicians, scientists, philosophers and theologians was actually subject to doubt. If that could affect such a robust body of ideas, what of other fields of rational inquiry? Were they too not permeated with unfounded assumptions? Is all truth relative?

Mathematical truth differs from scientific truth. The former is impeccable; the latter is subject to change. The former depends on deductive reasoning, the latter on empirical validation. Thus, within the axioms of Euclidean geometry, the Pythagorean theorem for right triangles was true in the past, is true now and will remain true forever.

One strong basis of support for Euclidean geometry was its use in physics, cosmology, manufacture, architecture, construction and so on. The results shone with accuracy. Few doubted that Euclidean geometry reflected the actual nature of space. Its competitors, though logically sound, were pure mental constructs.

But that notion was shattered with the formulation of the General Theory of Relativity by Albert Einstein. The relativistic model of space, time and gravity distorts the structure of space, making it intrinsically curved. Time does not flow uniformly. For speeding entities it flows more slowly than for stationary entities. The relativistic effects make a major difference over interstellar distances and at speeds approaching the speed of light. For most real-life needs, the Euclidean model has adequate accuracy. But it does not represent reality. The axioms and results of Euclidean geometry, hitherto seen as self-evident and complete, are but an approximation of reality. Sense perceptions are not as dependable as we think they are.

++++

Does the axiomatic method constitute a firm, logically consistent foundation for mathematics? This issue has plagued mathematicians since Euclid constructed geometry from five key axioms. Can the whole of mathematics be generated from a few basic axioms? If so, it would be a stunning validation of the axiomatic approach. Following a program set by David Hilbert, Bertrand Russell and AN Whitehead embarked on the mammoth project of deriving arithmetic and the rest of mathematics from a sparse axiomatic foundation. They started from the idea of sets, as envisioned by Cantor.

A set is an unordered collection of objects. {Mango, Book, Mountain} is a set, and so is {5, 100, 3, 11}. { } is an empty set; {1, 2, 3, 4, ...} is an infinite set of natural numbers. Russell was aware that the notion of sets can give rise to logical paradoxes.

> Let L_1 be a list of entities, L_2 another list of entities, L_3 another list, and so on. Some lists list themselves while others do not. Let S be the master set of all lists that do not list themselves. Does it list itself? If it does, then it cannot be a list that does not list itself. But if it does not list itself, there is a larger list that does. The question is undecidable.

> A town has one barber who shaves those and only those who do not shave themselves. Who shaves the barber? If shaves himself, then he, the barber, cannot shave himself. If he does not shave himself, then, as the barber, he has to shave himself.

Such paradoxes are part of broad category of logical paradoxes associated with self-referential statements. The logically undecidable statements are either neither true nor false, or true and false at the same time. Russell and Whitehead reformulated the notion of sets to resolve the paradoxes of the naive set theory. Their three-volume opus, *Principia Mathematica*, attempts to axiomatically construct the entire edifice of mathematics from the basic principles of logic. That it took them more than a hundred pages to prove that '1 + 1 = 2' attests to the rigor of their work. Though some issues remained, it was felt that additional refinements would provide a consistent axiomatic foundation for mathematics.

But deep trouble lay ahead. In 1931, a short, monumental paper by Kurt Godel proved that within any formally consistent axiomatic system there will always be statements that can neither be proven nor be falsified. No system of axioms can form the basis for the whole of mathematics. An undecidable statement may be true yet cannot be proven. The link between truth and proof was broken.

This unnerving result derailed the attempts to formalize mathematics on an axiomatic basis. Reconstructing the axioms and creating a new version of the set theory could resolve some undecidable statements, but there is no way to remove all possible undecidable statements. The dream of a perfectly logical discipline lay asunder.

Nonetheless, the mathematical edifice kept on growing. Over a hundred thousand people with a doctoral degree in mathematically related fields across the world generate more than 150,000 peer-reviewed papers each year. Astounding breakthroughs are made; and previously unresolved problems are conquered. The undecidable statements are few and generally distant from the usual concerns of mathematicians. They do not disprove what has been rigorously proved. Godel or not, the formula for solving a quadratic equation with $a \neq 0$:

$$ax^2 + bx + c = 0 \quad \text{remains} \quad x = [-b \pm \sqrt{(b^2 - 4ac)}]/2a$$

The extensive diversity of developments in pure and applied mathematics amply justify the assertion that *'a new golden age'* has dawned for mathematics (Devlin 1988).

5.9 AXIOMATIC THEOLOGY

Religion and mathematics have generally been in a state of harmony, not discord. When there was conflict, as for instance when the Jesuit opposed infinitesimals, it was not opposition to mathematics *per se*, but a preference for one branch of the subject (Euclidean geometry) against another (calculus).

Theologians have a Platonist attitude: Mathematics exists objectively, outside the human mind. It is the essence of the physical reality because it was, like all else, created by the Divine Mind. God has given humans the ability to discover it. The study of mathematics is a spiritual quest.

> *The belief that we live in a divine Universe and partake in the study of the divine mind by studying mathematics and science has arguably been the longest running motivation for rational thought, from Pythagoras, through Newton, to many scientists today.* (Wilson 2014).

As noted for the Golden Ratio (φ), π and infinity, religious thinkers have embraced important mathematical developments and interpreted them as additional evidence to support their cause.

Consider the Fourier Transform (FT). The FT for any wave-like function over time or space is a complex valued integral that decomposes the wave into a series of basic components. Usually taught in graduate level courses, it is important for rapid computation in many areas of science, medicine and technology.

Vipassana, a meditation technique practiced in India more than two thousand years ago, is often taught in Buddhist medication training classes. A Buddhist scholar has recently argued that since objects in nature and the human essence have a wave like form, they too can be decomposed, in a manner akin to the FT, into their basic components.

> *I posit that the analogous technique to the Fourier Transform is Buddhist Vipassana meditation. It can be used to enable a meditator to engage in a systematic search for his sense of self.* (Manu 2008).

With due intellectual flexibility, a true believer can adapt the most intricate of scientific and mathematical ideas to support his/her beliefs, however unfounded.

Axioms of God

Religions have deployed the axiomatic approach and syllogism to support their beliefs. As every religion has primary beliefs on which an edifice of rites and rituals is constructed, that is to be expected. God or Allah created all that exists is a primary axiom for Christianity and Islam. The Bible represents the wisdom of God is a basic Christian axiom and the corresponding Islamic axiom is that the Quran is the final word of Allah.

Theologians and philosophers have employed axiomatic formalism to argue for the existence of a divine being, God. Among them are Aristotle, Al-Kindi, Al-Ghazali, Thomas Aquinas, and recently, William Craig. Using existential or moral premises, they derived the existence of a first cause, a necessary, perfect being, and the mover and designer of the Universe (Chapter 4).

Eminent mathematicians have joined the fray. Rene Descartes, the 17th century founder of analytic geometry, was one of the most influential figures in the development of modern rationalist philosophy. Ascribing to the binary doctrine of the mind and the body as distinct entities, and upon extended philosophical meditations, he drew the famed conclusion:

I think, therefore I am.
Rene Descartes

To him, it was beyond doubt that a self that thinks exists. On that foundation, he formulated an intricate, step-wise set of reasons to show that unlike other ideas, the idea of an infinite, immutable being was not conceived by himself and it was not inculcated into his mind by external influences. It was an idea innate to his state of being. The only possible source was the infinite being, God. And since God could not be a deceiver, it proved the existence of God. One summation of another of his arguments is:

1. *God is a being with all perfections.*
2. *Existence is a perfection.*
3. *Therefor God Exists.* (Look 2020).

Descartes' legacies in mathematics and philosophy persist. His arguments for the existence of God, however, were infused with conceptual and logical errors and have largely been discarded.

GW Leibniz

GW Leibniz, one of the founders of calculus, formulated a set of arguments in the same tradition. A multi-talented scholar, Leibniz innovated in mathematics and generated novel ideas in philosophy, law, physics, history and theology. He devised a mechanical calculator for basic arithmetical operations and devoted years to constructing a logically sound language for universal communication. Consistent with his later work on calculus, he held that change in nature occurred in a gradual, continuous way, not via discrete steps. Countering the emerging atomic theory, he proposed the matter was composed of non-corporeal monads.

Leibniz was a devout Protestant firmly attached to the doctrine of the Trinity. His doctoral dissertation used a geometric method applied to objects in motion to prove the existence of God. He wrote extensively on the subject, progressively refining and extending his arguments. Opposed to Cartesian dualism, he was particularly critical of the arguments presented by Descartes. By giving the body a separate state of being, dualism paved the way to materialism and atheism. To counter disbelief, it was important to show both the necessity and perfection of God. One scheme he gave ran:

1. *God is a being having all perfections. (Axiom)*
2. *A perfection is a simple and absolute property. (Axiom)*
3. *Existence is a perfection.*
4. *If existence is part of the essence of a thing, then it is a necessary being.*
5. *If it is possible for a necessary being to exist, then a necessary being does exist.*
6. *It is possible for a being to have all perfections.*
7. *Therefore, a necessary being (God) does exist.* (Look 2020).

Holding that all phenomena are interconnected, Leibniz asserted that God is a unique being. To him matter was secondary to mind and the mind was the soul. Leibniz was an unfazed social optimist. His primary social tenet, given in a five-step argument, was that God had created a world in which all entities operated in harmonious, divinely ordained ways. The mind and the body, though functioning under different rules, are matched in a perfect harmony. The ubiquity of harmony in nature and society implies that humans live in the best of possible worlds. Though problems still remain, fundamentally nothing can be better.

> *God has chosen the most perfect world, that is, the one which is at the same time the simplest in hypotheses and the richest in phenomena.* GW Leibniz (Look 2020).

The reformist philosopher Voltaire, a veritable thorn to the establishment, penned a short fictional account of Candide, a young man whose teacher has imbued him with a Leibniz-style perspective on life. But his blissful life is suddenly shattered as he is thrust into an uncertain world where he encounters hardships, untold suffering, and witnesses events like war, slavery, executions, cannibalism, and natural disasters. Gradually, it dawns upon him that the world is hardly the best of all possible worlds.

To Leibniz, the order and complexity of nature necessarily connoted the existence of a divine creator, the perfection of all perfections, a supremely benevolent being, a power above all conceivable powers, a power present everywhere, all the time. In a reflection of his divine perfection, God created an ideal world, a world that embodies his goodness. The existent evil is thereby necessary evil. Humans need the guidance of the divine to navigate in this world.

Leibniz's views reflected a conservative Christian stand in support of monarchical rule, a defense of the *status quo* against reformist philosophies. His theory of monads was ridiculed by scientists and philosophers. His axiomatic ways to prove the existence of God and a perfect social order are riddled with leaps of faith, dubious premises and flawed reasoning.

Nonetheless, he was a rare genius with an excellent command of law who easily mastered languages, taught himself advanced mathematics and physics, proposed measures for educational reform, strove to harmonize and unify conflicting branches of Christianity and develop a new language. Calculus, his crowning achievement, became a true gem in pure and applied mathematics and an indispensable tool for physics.

Godel and God

Kurt Godel, who more than anyone else punctured the mathematical dream of an almost divine state of certainty, spent a good portion of his latter life into seeking a mathematical basis for divinity. A Platonist who held that mathematics has an eternal, objective existence, he inferred that it reflected the mind of God.

A master of impeccable logical thinking who could spot the minutest of errors in reasoning, Kurt Goldel was a devout theist of Lutheran persuasion. Attached to Christian values, he also had a positive disposition towards Islam. But castigating existing religious institutions, he did not attend Church service but regularly read the Bible. A believer in a personal God, his outlook synchronized rationalist thought with spiritual tendencies. He reasoned that in a rational, meaningful Universe overseen by God, life after death is a logically necessary phenomena. No recourse to scriptures is needed.

> *Religions are, for the most part, bad—but religion is not.* (Wikipedia 2021 – Kurt Godel).

Ax. 1. $(P(\varphi) \wedge \Box \forall x(\varphi(x) \Rightarrow \psi(x))) \Rightarrow P(\psi)$

Ax. 2. $P(\neg\varphi) \Leftrightarrow \neg P(\varphi)$

Th. 1. $P(\varphi) \Rightarrow \Diamond \exists x \, \varphi(x)$

Df. 1. $G(x) \Leftrightarrow \forall\varphi(P(\varphi) \Rightarrow \varphi(x))$

Ax. 3. $P(G)$

Th. 2. $\Diamond \exists x \, G(x)$

Df. 2. $\varphi \text{ ess } x \Leftrightarrow \varphi(x) \wedge \forall\psi(\psi(x) \Rightarrow \Box \forall y(\varphi(y) \Rightarrow \psi(y)))$

Ax. 4. $P(\varphi) \Rightarrow \Box P(\varphi)$

Th. 3. $G(x) \Rightarrow G \text{ ess } x$

Df. 3. $E(x) \Leftrightarrow \forall\varphi(\varphi \text{ ess } x \Rightarrow \Box \exists y \, \varphi(y))$

Ax. 5. $P(E)$

Th. 4. $\Box \exists x \, G(x)$

Godel's Proof (Source: AW 2021).

Godel has to be credited with producing the most mathematically grounded attempt to prove the existence of God. Inspired by the earlier attempts of Leibniz and utilizing the modal logic he had devised for his mathematical endeavors, his proof strives to derive the existence of God through reason alone. His technical derivation is given above. In a simplified nutshell, this proof read:

1. God is perfect in all respects.
2. God is always good.
3. In principle, such a being is conceivable.
4. Hence, of logical necessity, God exists.

Years of training in symbolic logic are needed to follow Godel's formalism. He dispenses with empirical observation and relies on arguments from necessity, not contingency, to establish the actuality of an absolute divinity. References to miracles, design and symmetry in nature are superfluous. God exists because He is a logical possibility, and thus a necessity.

Humanity's beliefs about the divine emerged from struggles with nature and the play of social forces. Belief was fixed by scriptures and organized religion. Discarding the historical baggage, Godel proposed that mathematics suffices to establish divinity. While mathematics is incomplete and undecidable to a degree, no doubt about the existence of God is possible. Reason and faith are not strangers after all.

> *I believe that there is much more reason in religion, though not in the churches, than one commonly believes.* Kurt Godel (Vaughn 2021).

Some scholars hold Godel responsible, though inadvertently, for making mathematics a unique religion.

> *If a 'religion' is defined to be a system of ideas that contains unprovable statements, then Godel taught us that mathematics is not only a religion, it is the only religion that can prove itself to be one.* John Barrow

While his works in logic, mathematics, science and philosophy retain rock like solidity, Godel's theological work is at best viewed as an effusive curiosity. Like other logic-based arguments for God, they do not hold up to critical scrutiny. Primarily, his axioms are riddled with flaws. If you start with false premises, then no matter how sound your reasoning, your conclusion will be false. An eminent mathematician opines:

> [While] *Gödel's mathematical accomplishments were remarkable, his theological beliefs and statements were rather unoriginal.* (Davis 1998).

Godel's proof is a majestic monument to how readily rational thought can transmogrify into mysticism. Many brilliant mathematicians have attempted to deduce God's existence via mathematical arguments. But, like the alchemist's efforts to convert copper into gold, their efforts have turned to naught.

> [Rational] *or otherwise non-faith oriented attempts at explaining the existence of God have always and will always fall short; the only way to claim that God does indeed exist is through faith alone.* (Archuman 2017).

Despite all the past failures, efforts to establish divine presence in the Universe with intricate, abstruse, oft unfathomable philosophic, logical arguments persist.

A Certain Ambiguity

A Certain Ambiguity, a year 2007 book by G Suri and HS Bal compares, in a fictionalized but mathematically and historically grounded form, the role of axioms in mathematics and religion. The central character is Ravi Kapoor, a student at Stanford University majoring in Economics. The book begins with his attendance in a course on the mathematics and philosophy of infinity. Aimed at non-specialists, the course covers Zeno's paradoxes, Euclidean and non-Euclidean geometry, number theory, Cantor's work on infinity and Godel's incompleteness theorems.

Ravi's interest in mathematics was sparked at an early age by his grandfather, Vijay Kapoor, a famed mathematician in India. In the course of doing research for a paper for his class, Ravi is surprised to learn from past media reports that his grandfather had not only been a visiting scholar at the Morisette University in New Jersey, USA, but also that during that time, he had been incarcerated for a while on the charge of blasphemy.

One day, when returning from his office, Vijay comes upon a public park gathering wherein a preacher is saying that the Christian way of life is superior to life in 'heathen' nations that followed 'superstitious' Eastern religions like Hinduism. Vijay mounts the podium to shine a critical light on Christianity and alludes to Darwin's Theory of Evolution to counter the Biblical story of creation. Vijay speaks as a scientist who is sympathetic to the ideas of freedom and democracy.

His intervention angers the audience. He is heckled, arrested by the sheriff and jailed awaiting trial for violating the laws on blasphemy. The case could have international ramifications. Thus a prominent judge, Judge John Taylor, is assigned by the state governor to determine whether Ravi should stand trial or not. The year is 1917.

A large portion of the book is taken up by an extended prison-cell dialogue between Vijay and the Judge. Refusing to recant what he had said, Vijay tells the judge that his ideas stem from the exacting, deductive approach of mathematics. The Judge is lectured on Euclidean geometry, the notion of proof and certainty of geometrical theorems. Disturbed to consider that his Christian beliefs may not be as valid as these theorems, he pursues the matter. One day he comes upon a news story about an experiment that had been done to test Einstein's General Theory of Relativity. Einstein had triumphed over Newtonian physics. As his theory utilizes a non-Euclidean model, it shows that Euclidean geometry is not an accurate description of space.

The judge inquires of Ravi: If Euclidean geometry and its axioms are flawed, does it not mean that mathematics is not an impeccable science? Ravi concedes that non-Euclidean geometry stands on a ground that is as solid, from a deductive standpoint, as Euclidean geometry. Their axioms differ but both adhere to the same logical method. Just as Euclidean axioms can be scrutinized on rational grounds, theological axioms can also be tested in a similar fashion.

Plunged into an emotional abyss of doubt, Judge Taylor is rescued from his plight after a fortuitous sojourn at an African American church. Imbibing the spirit of communal solidarity and captivated by the emotive melodies, his faith is reinforced by the pastor:

> *There can be no proof. The acceptance of God can only come from faith. Faith is a starting point, and you can never prove a starting point because, well, it's a starting point.* (Suri and Bal, 2007, page 254.)

Vijay's certitude about Euclidean geometry also crumbles but his Platonic faith in mathematics and the axiomatic method remains:

> *I deeply believe that mathematics is not a mere chess game where arbitrary axioms devised by humans lead to arbitrary theorems. No, I do believe that mathematical truths have an external reality quite independent of the minds considering them. ... This truth is timeless and absolute.* (Suri and Bal, 2007, page 257).

Ultimately, both arrive at a vision that unifies their diverse outlooks:

> *The human experience is such that we yearn to find something that is lasting and true, something that speaks to our own hearts, and has meaning. Meaning, however, whatever its variety, seems to demand faith.* (Suri and Bal, 2007, page 257).

Observing that Vijay has softened his stand, the Judge recommends that he not be tried. Released from custody, he sets sail back to India the very next day. The duo remain lifelong friends, exchanging letters now and then. John Taylor visits India where they continue their discourse on faith, life and mathematics.

A Certain Ambiguity is a beautiful piece of prose, a gentle introduction to mathematical concepts beyond the high school level, and an illuminating foray into philosophy, religion, politics and human relationships. There is much else to the story that cannot be put down here. Yet, I take issue with the equivalency drawn between faith in God and faith in the methods of mathematics. While both have basic premises that are accepted without proof, the edifice of mathematics arises from a rigorous application of logic while the edifice of religion lacks a systematic construction. While both have integral uncertainties, mathematicians strive to resolve uncertainties with additional axioms while the religions accept them as reflective of the mind of God. And then there is the question of practice. Mathematics thrives on the basis of its extensive applications in science and technology while religion thrives because it provides emotional equipoise and meaning to life and existence.

++++

Attempts to place religion on an axiomatic foundation persist. Philosophers continue to produce apparently new forms of arguments and books with arcane terminology that befuddles the mind to same degree as Godel's proof. One more accessible axiomatic construction of Christianity systematically lays down the notions of faith, God, prayer, sin, afterlife, salvation, Jesus, the Holy Spirit, the Church and the Bible. Its first part runs:

> *The Axioms of Faith are a ladder. The starting point is complete religious and spiritual unbelief.*

Faith is AT LEAST a way to contextualize the human need for spirituality and find meaning in the face of mortality. EVEN IF this is all faith is, spiritual practice can be beneficial to cognition, emotional states, and culture.

God is AT LEAST the natural forces that created and sustain the Universe as experienced via a psycho-social model in human brains that naturally emerges from innate biases. EVEN IF that is a comprehensive definition for God, the pursuit of this personal, subjective experience can provide meaning, peace, and empathy for others.

Prayer is AT LEAST a form of meditation that encourages the development of healthy brain tissue, lowers stress, and can connect us to God. EVEN IF that is a comprehensive definition of prayer, the health and psychological benefits of prayer justify the discipline. (McHargue 2015).

Claims that religion can be derived from basic axioms with the same rigor as that of Euclidean geometry are heard. Supposedly definitive proofs for the existence of God are on offer too. But they are verbal gyration devoid of mathematical rigor and consistency.

5.10 A CONTESTED SEPARATION

Among the eminent mathematicians and scientists with a religious outlook were Aryabhata, Aristotle, Plato, Pythagoras, Al-Kindi, Al-Ghazali, Al-Kashi, Al-Khwarizmi, Nicolaus Copernicus, Johannes Kepler, Galileo Galilei, Rene Descartes, GW Leibniz, Isaac Newton, Georg Cantor, Kurt Godel and Srinivasa Ramanujan. To this list we add Nicholas of Cusa, Blaise Pascal, Johann Bernoulli, Jacob Bernoulli, Colin Maclaurin, Leonhard Euler, Maria Agnesi, Augustin Cauchy, Tycho Brahe, John Napier, Carl Friedrich Gauss, William R Hamilton, George Stokes and Bernhard Riemann. And more. A number were distinguished theologians who wrote major treatises relating to religion, and at times explained their mathematical findings in theological terms. And some also engaged in pastoral duties. And some produced mathematical, scientific and philosophic arguments to prove the existence of God.

These mathematicians made novel and monumental contributions to one or another fields of mathematics. GW Leibniz, Isaac Newton, Leonard Euler and CF Gauss fundamentally transformed multiple areas of the subject. Declaring their belief in God, they acknowledged His influence on their work. Consider a sampling of their statements extracted from McIntyre (2017):

All my discoveries have been made in answer to prayer. Isaac Newton

Although God is a perfect and free agent, unrestrained by any natural law, yet he did not want to pervert the order of nature that he set up. Tycho Brahe

I render infinite thanks to God for being so kind as to make me alone the first observer of marvels kept hidden in obscurity for all previous centuries. Galileo Galilei

Although the light of reason, however clear and evident it is, may seem to suggest something different to us, we should put our faith exclusively in divine authority rather than in our own judgment. Rene Descartes

Finally, two days ago, I succeeded not on account of my hard efforts, but by the grace of the Lord. Like a sudden flash of lightning, the riddle was solved. I am unable to say what was the conducting thread that connected what I previously knew with what made my success possible. CF Gauss

My intellect has never ceased to embrace Christianity with satisfactory and complete conviction. WR Hamilton

I know of NO sound conclusions of science that are opposed to the Christian religion. George Stokes

George Stokes was a 19th century polymath whose contributions to mathematics, physics and physiology reverberate to this day. A Lucasian Professor of Mathematics at University of Cambridge for over five decades, he also served as the president of the Royal Society and a conservative member of the British parliament. And he was as well a learned, influential Christian theologian who held high office in a Christian institute and a major Bible society. In particular, he campaigned against the Darwinian Theory of Evolution. Yet, he saw an essential harmony between science and religion.

We all admit that the book of Nature and the book of Revelation come alike from God, and that consequently there can be no real discrepancy between the two if rightly interpreted. The provisions of Science and Revelation are, for the most part, so distinct that there is little chance of collision. But if an apparent discrepancy should arise, we have no right on principle, to exclude either in favor of the other. For however firmly convinced we may be of the truth of revelation, we must admit our liability to err as to the extent or interpretation of what is revealed; and however strong the scientific evidence in favor of a theory may be, we must remember that we are dealing with evidence which, in its nature, is probable only, and it is conceivable that wider scientific knowledge might lead us to alter our opinion. George Stokes (Wikipedia 2021 -- Sir George Stokes, 1st Baronet).

If a scriptural view appears flawed, it is not because it is truly flawed but because our interpretation is flawed. Scientific theories can never be flawless as it is their nature to undergo improvement.

++++

Mathematicians were also shunned or persecuted by the Church. Yet, they did not see themselves as opponents of religion or non-believers. Making a distinction between hierarchically organized religion and the essence of faith, they held that it was their detractors who were at fault for neglecting true faith.

The Holy Scriptures can never lie or err, and its declarations are absolutely and inviolably true. I should have added only that, though Scripture cannot err, nevertheless some of its interpreters and expositors can sometimes err in various ways. Galileo Galilei (McIntyre 2017).

The mathematicians lived in eras saturated with religious ideas and rituals. The church was central to life. Loyalty to religion signified fidelity to the monarch and the ruling order. Often reliant on sponsorship from the nobility and the crown for their upkeep, it was unsurprising that they espoused the prevalent worldview.

A minority that held an ambivalent or agnostic stand on religion also existed. For instance, the 10th century Iranian physician, philosopher and logician Al-Razi opined:

> *Books on medicine, geometry, astronomy, and logic are more useful than the Bible and the Quran. The authors of these books have found the facts and the truths by their own intelligence, without the help of prophets.* Al-Razi (Davis 1998).

An oft cited exchange between Emperor Napoleon Bonaparte and the pioneering mathematician, statistician, engineer and philosopher Pierre-Simon Laplace is apropos. Though regarded as authentic, it is rendered in different forms. One version runs:

> Napoleon: *I don't find God in your book on astronomy.*
> Laplace: *Sire, I have no need for that hypothesis.* (Davis 1998).

Yet, according to a distinguished mathematician:

> *Practically every major theme of mathematics, its concepts, its methodology, its philosophy, have [historically] been linked in some way to theological concepts.* (Davis 1998).

A Gradual Separation

The vast expansion of education in the 20th century brought forth higher college level enrollments in courses in pure and applied mathematics. Mathematics and statistics made greater inroads in natural, medical and social sciences. Engineers studied advanced calculus, statistics and discrete mathematics. Commercial, manufacturing, advertising and state bodies, including the military, needed staff with mathematical and statistical skills. National and international mathematical and statistical associations were formed and hundreds of relevant journals came into being. Mathematicians and their institutions are funded by universities, governments, foundations and corporations. Only rarely do these bodies have a connection to religion. For such reasons and more, the historic bond between mathematics and religion has weakened over time.

Religion, to professional mathematicians, is no more an arena that takes up a major part of their public or private lives. At best, it is a weekend social event. They are engrossed in finding proofs of theorems or mathematical techniques for practical problems. The Platonist view that mathematics has an objective existence no longer has wide currency in their community. Even fewer assign the mathematical truths to a divine mind. The common view is that mathematical truths arise in specified axiomatic systems formulated by the human mind. Some truths may correspond to the physical reality and may be applied in the sciences; others will be abstract but logically true ideas.

Religious belief has experienced a steep decline in the industrialized nations in the recent years. In many, majorities do not take part in religious rites and rituals, and a quarter to a third of the adults consider themselves agnostics or atheists. That trend is more marked among scientists and mathematicians. They query religion from rational and moral perspectives. Seeing extensive suffering in the world, some wonder why an all loving God created such a world. The USA has higher levels of disbelief among all social segments compared to Europe but the trends are similar. In the poorer nations, the level of disbelief is much lower, yet here as well, the trend generally is similar.

A year 1914 survey of 1,000 American scientists—biologists, physicists and mathematicians—found that some 40% of the responders believed in a personal God and an almost equal proportion did not. In 1996, a similarly designed survey found that 40% believed in a personal God while 45% did not, a slight gain for non-belief. In both surveys, the rest saw themselves as agnostics or persons with no definite belief. Other surveys restricted to college professors found higher levels of belief in God. Interestingly, the instructors in the natural sciences and mathematics had higher levels of belief that those in social science, humanity or psychology.

Overall, religious belief in the US remained strong in the first decade of the 21st century. Surveys done in those years indicated that that while some 95% of the population believed in God or some form of a divine power, only about 50% of the scientists and mathematicians ascribed to such beliefs. Further, some 40% of the latter ejected any notion of divinity, only 4% of the public did so. A poll of the members of the US National Academy of Sciences found that some 15% of the mathematicians believed in a personal, creator God, only about 6% of the biologists did so. By 2020, however, the trend towards disbelief had grown among the public as well. By then, about a third had atheistic or agnostic beliefs.

Recent data show that about 40% of the British adults hold agnostic or atheistic beliefs. Among scientists, such beliefs prevail more. A year 2013 survey of the fellows of the Royal Society of London found that the greater majority rejected any idea of God or divinity. Belief in a personal God was not accepted by 87% of the respondents. Biologists were more atheistic than physical scientists. Yet, most scientists held that religion and science need not be in an overt state of conflict. No specific data for mathematicians were given. But it is safe to presume a trend similar to that in the US.

++++

Even as the mathematical world petitioned for a divorce from the faith-based world, religious institutions and theologians remained favorably disposed towards mathematics. The trend was especially marked in the USA.

Consider the higher education landscape. Over the past three decades, religious belief among college entrants in the US has declined significantly. In 1986, 10% of freshman college students had no religious affiliation and 85% attended religious service. The corresponding figures for 2016 were 31% and 69%. Further, of the about 5,300 colleges and universities in the USA, some 900, or about a sixth, have a religious affiliation. Of the latter, about half are linked with a Christian denomination. The student population at the religiously affiliated institutions ranges from several hundred to tens of thousands. And some of them feature high in the list of quality ranking of all the US colleges and universities.

Institutions with a religious affiliation often offer rigorous study programs in non-religious subjects and include programs that combine theology with subjects like management, social science, history, physics, chemistry, biology and mathematics. Some institutions offer a double major in theology and a secular subject. Independent academiis that combine theology with mathematics and science exist. The Center for Theology and the Natural Sciences in California is a case in point.

Mathematics is a prominent subject at Christian colleges. The faculty at some colleges include high ranked mathematicians. The Department of Mathematics and Computer Science at the Gordon University, a liberal arts institution in the Boston area, offers a distinctive BA degree program which links mathematical topics and faith.

> *Explore God's creation in its mathematical expression under-talented, experienced faculty whose expertise spans the spectrum from industrial applications to the frontiers of pure mathematics. You will gain an unmatched depth and breadth, including opportunities for collaborative research in statistics, operations research, and computational and pure mathematics. Our faculty are committed not only to teaching mathematics but also to helping you understand how it can inform life and faith.* (GU 2021).

The Emmaus Bible College in the state of Iowa has a Bachelor of Science degree program with a double major in applied mathematics and theology.

Applied Mathematics is a unique, four-year degree program that prepares graduates to impact the world for Jesus Christ through their capacity to solve real-world problems that require mathematical and reasoning skills. (EBC 2021).

Mathematics is a prominent field of study at the Thomas Aquinas College in Massachusetts, a reputable Catholic liberal arts college. Among the graduation requirements for a student is completion of 28 credits in mathematical subjects, which is much higher than at most colleges and universities in the US. According to the president of the college,

Mathematics serves to lay a strong foundation for the intellectual life in at least four ways: It frees the student from the grip of skepticism; It prepares the mind to think clearly and cogently, expanding the ability to know; It opens the mind in wonder to the beauty and order of God's Universe; It prepares the mind for a quantitative treatment of the natural world.

We are not gathering students from around the world to contemplate their navels or even to think fuzzy, sweet thoughts about Jesus. We do not aim at this caricature of a liberal arts student. We want to help young minds make a good beginning on the road to wisdom. This is very difficult and takes sharp and careful reasoning. A heavy mathematical foundation sets the bar high, but it also helps strengthen the mind for the heavy lifting that must be done. (Kelly 2021b).

Theologically blended mathematics is on offer at Christian elementary and high schools. The curriculum at the Abeka Academy in Florida focuses on the development of traditional mathematical skills but in a way that will orient the student towards divinity.

The students will find exactness, preciseness, and completeness in the subject matter of mathematics just as would be expected in God's world. ... Traditional math is Christian math, and it must be taught by traditional methods. A math lesson rightly taught is one more way that a Christian teacher can instill within students the principles of God's Word. (Abeka 2017).

A few Christian theologians decry the stress on mathematics in Christian education. In their view, it will divert the student away from true faith. But they constitute a small minority.

The Association of Christians in the Mathematical Sciences is a body of mathematicians and computer scientists who ascribe to the Christian faith. Formally founded in 1985, it holds well attended conferences every other year. A variety of high caliber papers dealing with the teaching, philosophy and history of mathematics, the linkage between theology and mathematics, and state of the art ideas in different fields of mathematics are presented. The refereed conference proceedings, freely available online, are a good source of innovative ideas for secular mathematicians and educators as well. Recognized mathematicians often attend.

Educational books form an active front for linking faith and mathematics. Aiming to harmonize Christian and scientific perspectives, demonstrate the consistency between faith and mathematics, they explore theological and mathematical truths, and cover basic and advanced ideas in mathematics. Some recent titles: *Mathematics Through the Eyes of Faith* (Bradley and Howell 2011); *Mathematics: Is God Silent?* (Nickel 2012); *Redeeming Mathematics: A God-Centered Approach* (Poythress 2015); and *Mathematics in a Postmodern Age: A Christian Perspective* (Howell and Bradley 2001).

Despite the ongoing secular trend in the world of mathematics, the affinity of religiously inclined people towards the subject indicates that a decisive divorce between mathematics and religion is unlikely to transpire any time soon.

5.11 ETHICS

Pure mathematicians subsist in a sphere detached from the hustle and bustle of reality. Yearning to prove theorems with steadfast logic, they value elegance and economy. The truth they desire has to sparkle with beauty.

> *A mathematician, like a painter or poet, is a maker of patterns. If his patterns are more permanent than theirs, it is because they are made with ideas.* (Hardy 1940).

The applied mathematicians and statisticians take pride in application of results from this rigorous intellectual edifice to problems in science, technology, health, economics and social science. Devoted to truth, beauty and social utility, and unlike lawyers or politicians, they are engaged in pursuits that apparently are free of adverse ethical connotations. Mathematicians are wont to project that their discipline is, as matter of principle, an ethical enterprise.

But that is an image devoid of substance. Mathematicians function in society, partake in social institutions and practices, and are influenced by prevailing values and attitudes. What they do and do not do as they practice their profession can and does have major ethical implications.

Prior to the 19th century, eminent mathematicians generally had the patronage of kings, emperors, nobility, the church or wealthy merchants. While this usually did not affect the kind of problems they tackled, they also undertook tasks for their patrons. Many had religious inclinations, brought out theological treatises and received material support from the church. As noted earlier, Isaac Newton, GW Leibniz and Rene Descartes were among them.

Mathematicians today depend on different patrons to fund their educational, teaching and research activities, and to pay the housing, utility, food and other bills. Their funds come from private and public universities, major corporations, foundations and governmental agencies, including the military. Some work at religiously affiliated academies. In all, hundreds of thousands of mathematicians, statisticians and quantitative analysts are employed in diverse sectors of the modern economy. In this section, our focus is mostly on the USA, and partly on the UK and Europe.

++++

Over 750,000 persons with a Bachelor's or higher degree in mathematics or statistics exist in the current US workforce. Over 50% just have the first degree. Most of them teach in schools, junior colleges and technical training institutes. About 30% are employed by federal, state and municipal governments. In 2020, the median wage for a statistician was $92,000, and for a mathematician, it was $111,000.

Mathematicians and statisticians with a doctoral degree constitute less than 10% of the total. They are the only ones who engage in research, and write in and can understand, articles in the technical journals in their fields. Teachers and statistical assistants mostly perform routine tasks. The largest employers of mathematicians and statisticians with a Master's or doctoral degree are the four-year colleges, universities, governmental agencies, computer-based corporations, research bodies, medicinal drug and device makers, and insurance companies.

Basic courses in mathematics and/or statistics is a requirement for high school students and virtually all those pursuing higher education in the sciences, social sciences, engineering, medicine, public health, business and marketing. Lawyers, journalists and attendees of theological seminaries attend courses in quantitative fields as well.

219

Mathematization of the social and natural sciences, the economy, social affairs and even private lives has assisted systematization and organization in complex settings. Without mathematics and statistics, modern life would not be possible. But it has also engendered two major social ailments:

Numerosis: The proliferation of quantification in all facets of society. Data, statistical analysis and mathematical models permeate all facets of the sciences, academy, business, social service and government. Quantification is synonymous with good science. Overuse of models detracts from conceptual understanding and interpretation of the phenomena under study. Concrete meaning is replaced by formulas and a numerical output.

Numeritis: Production of volumes of dubious quality data, and misuse and abuse of statistics are well documented in health and education research, psychology, economics and the social sciences. Numerous investigative reports, books and proclamations by scientific bodies on this matter have been issued. But the practice goes on. Commercial incentives and corporate intrusion in bio-medical, agricultural and other research areas result in conflicts of interest that bias the data and compromise the validity of the findings.

In the past four decades, there have been many well publicized scandals involving data manipulation and fabrication in the papers relating to medical, psychology and other research. A good proportion of these were published in highly reputed journals. Statistical and mathematical associations have responded by strengthening the codes of ethics for their profession. We give one example.

American Mathematical Society (AMS): The revised ethical guidelines of the AMS, issued in January 2019, require its members to abide by a series of ethical tenets: To adhere to high standards of integrity in research and presentation of their findings; To avoid plagiarism in any form, give credit to others when it is due, correct their published errors in a timely fashion and respect confidentiality; To be aware of their social responsibility and avoid discrimination on the basis of *race, gender, ethnicity, age, sexual orientation, religious belief, political belief, or disability*; To avoid conflict of interest in refereeing papers and taking part in decision making bodies; To promote free dissemination of information and avoid the excessive secrecy required by corporate and governmental bodies; Whistle blowers who disclose professional wrong doings should be protected and given a fair hearing; Mathematical departments should adhere to the required standards in their courses and granting degrees; Journal editors should strive to ensure timely, fair and high quality of refereeing of submitted papers and communicate in an adequate way with the authors. They should resist demands by external agencies for censorship. (AMS 2019).

Misuse and abuse of statistics can be intentional, due to poor training, or lack of due diligence. It produces flawed conclusions that may be harmful in terms of the progress of science or human welfare, or both. A large number of papers and books showing how bias and errors creep into study design, statistical analysis and interpretation exist. The types of research errors and misconduct include: biased sampling, biased framing of questions, exclusion of data, outliers and unfavorable findings, overemphasizing favorable findings, use of inappropriate analytic methods, partial reporting, outcome measures that exaggerate the importance of the intervention, data dredging, misleading charts and graphs, confusing statistical significance with practical significance, and spinning the basic message of the research.

Leading statistical and bio-statistical associations have guidelines that stress avoidance of such unethical practices. Yet, publication of papers in health, medical and other journals which incorporate these flaws continues, often in adroitly camouflaged ways.

The Pharmaceutical Industry

A large proportion of statisticians with a master's or doctoral degree work in the health sector. Dispersed among makers of medicinal drugs and medical devices, large hospitals, governmental health agencies and departments of medical statistics within universities, they constitute a major branch of their profession—the bio-statisticians.

In the pharmaceutical industry, they work on internal and collaborative pre-clinical and clinical drug studies, develop models for drug metabolism, and assist in the creation of marketing material for doctors and the general public. They as well play a critical role in the design, conduct, analysis and reporting of drug studies.

With an output of nearly $640 billion, the pharmaceutical industry forms one of the largest sectors of the US economy. Dominated by a few major corporations collectively called Big Pharma, it is among the most profitable sectors as well. The year 2015 average profit margin for the pharmaceutical firms was 17% while the margin for the top 500 non-pharmaceutical firms was 7%. High profits are generated by high drug prices. The industry claims that they have to cover the large cost of research and development. Yet, Big Pharma companies typically spend more on sales and marketing than on R&D. For example, Pfizer, recording an income of $51,750 million in 2019, spent $8,650 million on R&D and $14,350 million on marketing and sales. A portion of the R&D budget goes to fund university-based research to test their drugs.

The US and New Zealand are the only two countries in the world that permit direct advertising of drugs to the public. Besides TV and print media ads, drug companies spend a large sum on producing promotional material for physicians, surgeons and pharmacies, distributing free samples and hosting seminars and luncheons to promote their products. Senior medical experts receive fees, research funds and other perks from the companies. Lobbying politicians and funding their election campaigns is a time-honored device to create Big Pharma friendly laws and regulatory set-up. As a result, the budget and staff of the US Food and Drug Administration have remained grossly inadequate to ensured proper monitoring of the pharmaceutical and medical device industries.

The Big Pharma drug ads, often featuring celebrities, are a case in point. A 2014 investigation of the contents 168 television ads for medicinal drugs produced disturbing conclusions:

> *Of the most emphasized claims in prescription (n = 84) and nonprescription (n = 84) drug advertisements, 33% were objectively true, 57% were potentially misleading and 10% were false. ... Potentially misleading claims are prevalent throughout consumer-targeted prescription and nonprescription drug advertising on television.* (Faerber and Kreling 2014).

Flawed ads exaggerate the benefits and omit or minimize the side effects of the drugs. The relevant FDA unit is too small and can only spot a few misleading ads. Now and then, it nabs a culprit, and gives warning letters or imposes fines. Most ads pass under the radar. For Big Pharma, the fines are the cost of doing business. Yet, the fraudulent ads inflict extensive harm.

Opioids are a class of drugs that include the illegal drug heroin and legal medicinal drugs used for pain relief and inducing a relaxing effect. Three decades of extensive marketing by the manufacturers converted them into one of the most widely prescribed pain relief drugs in the US and other places. But they are also addictive substances that induce minor and major side effects including death. The statistics are shocking: OxyContin was introduced to the market in the mid-1990s as a miraculous, nonaddictive painkiller.

> *In 2017, approximately 12 million Americans misused opioids and more than 47,000 people died of opioid overdose.* (Jalali et al 2020).

The estimated fatalities over the twenty five years of medical usage exceed 1 million. And even more just suffered from side effects. The responsibility for the ghastly numbers is clear: Drug companies (deceptive promotion and paying kickbacks to doctors); the FDA (faulty oversight and erroneous labeling) and doctors (unjustified over prescribing and illicit distribution). In a spate of recent lawsuits, judges across the US found the manufacturers guilty of acts that led to rising addiction, overdose deaths and babies born with opioids in their bodies. In March 2021,

> *Purdue Pharma pleaded guilty on Tuesday to criminal charges that it misled the federal government about sales of its blockbuster painkiller OxyContin, the prescription opioid that helped fuel a national addiction crisis. ... The company agreed last month to plead guilty to criminal charges and face criminal and civil penalties of about $8.3 billion as part of the settlement with the Justice Department. ... The company's owners, members of the wealthy Sackler family, agreed to pay $225 million in civil penalties as part of the settlement but did not face criminal charges. The sales of OxyContin helped the Sacklers build a fortune estimated to be at least $13 billion, an amount that dwarfs the fine.* (Benner 2020).

The opioid episode is but one of hundreds of serious cases of misconduct by drug companies. Many studies have revealed that research done or sponsored by drug manufacturers is infused with pro-company bias. The flaws include selective usage of data, short periods of follow up, small sample sizes, partial reporting of side effects, and inappropriate statistical and clinical outcome measures. Drug company funded research tends to produce conclusions more favorable to the study drug than independent research. And medical researchers who sit on FDA committees are more prone to vote for drug approval if they have received funds from drug companies than those who have not.

The profit motive perverts modern medicine. The statisticians and quantitative analysts in the pay of for-profit medical entities are a party to that perversion. They handle and analyze the research data and play a key role in writing the reports, valid and flawed. They produce the misleading numbers, graphs and charts used for marketing to the doctors and the public. The statistics profession has failed to fully recognize and curtail such violations of bio-medical and statistical ethics.

Statisticians in the Pharmaceutical Industry brings together statisticians employed by pharmaceutical and allied firms and academic and research statisticians with an interest in pharmaceutical development. Founded in 1997, it promotes quality research and sound statistical practice in the industry. Besides publishing a journal and a newsletter, it holds discussion fora and conducts training classes and publishes a journal. Yet, it is hardly famed for taking pro-active or corrective measures to reign in the statistical abuses noted above.

National Security Agency

The US National Security Agency (NSA) is a high budget entity within the humongous American spying and covert action network of 17 intelligence agencies. Focusing on signals intelligence and cybersecurity, the NSA has a global presence and works in tandem with British, Canadian, Australian and New Zealand intelligence agencies. Its sophisticated machinery of eaves dropping can tap into the internet and telephone communications of over a billion people worldwide, track their movements, and capture their video streams and photos.

The NSA is the largest single employer of mathematicians with an advanced degree in the US, and perhaps the world. Besides publishing an in-house journal, it sponsors summer programs for students majoring in mathematics or statistics in several US universities. The mathematics and statistics departments in these institutions maintain close ties with the NSA.

The NSA areas of focus include discrete mathematics, cryptography, cyber-security, algorithms and related computer applications, optimization, facial recognition, and communication theory. Some of the senior experts in these fields work at or with the NSA.

The NSA also leads in the violation of the right to privacy of individuals on a global scale. In 2013, the courageous effort of whistle blower Edward Snowden revealed that NSA and its partners were *systematically monitoring ... emails, texts, phone and Skype calls, web browsing, bank transactions and location data of millions in many nations.*

They have tapped internet trunk cables, bugged charities and political leaders, conducted economic espionage, hacked cloud servers and disrupted lawful activist groups, all under the banner of national security. (Leinster 2014).

The NSA has made backdoor intrusions in financial and other networks to undermine credit card encryption. Like all US imperial projects that cost lives and disrupt livelihoods, these illegal violations of basic rights are justified in terms of prevention of terrorism. In truth, they are an adjunct to the practice of international state terrorism.

Yet these headline grabbing revelations of practices enabled by very advanced mathematical techniques have barely caused a ripple in the mathematical community. The NSA continues to send large delegations to and maintain well-funded publicity and recruiting booths at mathematical conferences. And their booths are among the most visited booths.

A few activist mathematicians and the Electronic Frontier Foundation have issued calls to fellow mathematicians and computer scientists to desist from working in NSA funded projects.

National mathematical societies can stop publishing the agencies' job adverts, refuse their money, or even expel members who work for agencies of mass surveillance. At the very least, we should acknowledge that these choices are ours to make. We are human beings first and mathematicians second, and if we do not like what the secret services are doing, we should not cooperate. (Leinster 2014a).

US mathematicians have shrugged off the Orwellian implications of the NSA mass surveillance and disruptive activities. Dissident voices flounder in the wilderness. The sharp minds engaged in proving esoteric theorems plow on with business as usual.

Financial Meltdown - 2008

The early years of the first decade of the 21st century appeared to climax the fundamental capitalist tenet: Capitalism is the best economic system. Stock markets in the industrialized economies recorded rapid, sustained growth. Personal wealth, property ownership and house prices expanded steeply. But economic inequality grew rapidly as well. That was justified under the claim that eventually everyone, including the peoples of the poor nations, would experience rising prosperity.

The two key features of this boom were rising rates of home ownership enabled by eased lending conditions by the banks and use of complex financial instruments called derivatives for stock trading. Issuing low interest, zero-down loans under eased the criteria for credit worthiness, the banks enticed millions to acquire homes. But the fine print in the contract enabled the lender to seize the property with ease in the case of a default. Most buyers overlooked the fine print. Derivatives bundle and sub-divide assets from different sources, which are sold to different buyers. The buyers create their own derivatives and resell them. In the end, the assets in a single home are owned by thousands of investors. Billions of dollars in shares, bonds, currency and derivatives change hands in the blink of an eye under computerized trading. The cascade of hundreds of millions of financial transactions create the impression of rapid economic growth. Investment agencies catering to the upper income class, the hedge funds, were the major force in the derivative trades. The economic growth in those years occurred primarily in the financial sector, not in the manufacture of real goods and essential services.

By 2007, the international financial system was trading derivatives valued at one quadrillion dollars per year. This is 10 times the total worth, adjusted for inflation, of all products made by the world's manufacturing industries over the last century. ... Derivatives created a booming global economy, but they also led to turbulent markets, the credit crunch, the near collapse of the banking system and the economic slump. (Stewart 2012).

The bubble burst in 2008. In the ensuing meltdown, major banks, hedge funds, insurance companies, public mortgage lenders and manufacturers— in North America and across Europe—faced bankruptcy. Individual wealth including retirement savings shrank. People were unable to pay credit card and mortgage debt. Hundreds of thousands lost their homes. Poverty and unemployment rose. The entire global economy was badly affected in the ensuing double dip recession. Only the massive bailouts by the central banks managed to reign in the spiraling catastrophe.

The responsibility for the chaos was multi-pronged. Unlimited greed fostered by ultra-consumerism, relaxed laws, poor state oversight, high risk lending, failure of the credit rating agencies and accounting firms to raise the alarm, infectious euphoria, ignoring history and reliance on patently unrealistic forecasts—all factored in the episode.

But this time a new culprit was present. Computerized trading calls for rapid decisions in risky settings. Financial institutions and major investors need strategies to optimize their assets and outdo competitors. This issue was tackled by several econometricians in the 1980s. By the early 1990s, a series of complex models and algorithms for high speed transactions had been devised. The initial successes of these methods in setting financial values and risks for derivatives gained the attention of the financial power houses. Economists, mathematicians and computer scientists found new, lucrative job openings. Top experts and esteemed dons, called the quants, were soon fine tuning complex trading models, developing more efficient ones, and programming them on computers to ensure that their firm was a step ahead of the rivals. A new field of specialization, Mathematics of Finance, was on offer at academies across the world. Two of the three creators of a commonly used model—the Black and Scholes equation—were honored with the 1997 Nobel Prize for Economics.

In no time, spectacular results were seen. Millionaires, even billionaires were created overnight as the stock market hit the roof. The rate of growth of the global economy created an impression that the process could not be reversed. A real period of unimpeded growth had finally begun; everyone will be rich; it is just a matter of time – that was the little-disputed mantra spread by the financial gurus, politicians and the mass media. Some enterprising mathematicians and scientists set up their own trading entities and hedge funds and prospered to the tune of hundreds of millions.

Financial mathematics is based on a key assumption. The market is rational and efficient. It may oscillate unpredictably and widely in the short run. But the pattern of long run oscillations will reinforce stability and progress. It is not an empirically invalid assumption. The relevant statistical models have an inbuilt though rare possibility of spiraling into catastrophic instability. A few scientists warned of the dangers of such models. Yet, their caution was ignored as young mathematicians and economists aspired to join the elite club of financial analysts and modelers. Even as signs of looming turbulence emerged, the experts and lay persons alike remained mesmerized in a get-rich-quickly process that essentially was a Ponzi scheme.

[Usually the Black and Scholes] *model performed very well, so as time passed and confidence grew, many bankers and traders forgot the model had limitations. They used the equation as a kind of talisman, a bit of mathematical magic to protect them against criticism if anything went wrong.* (Stewart 2012).

A common query posed in the aftermath of the 2008 financial crisis was: Did mathematical models and mathematicians share the blame? A majority of the mathematical and economics fraternity defensively said no. In their view, the known limitations of the models were ignored. Blindly chasing profits, financial managers, traders and experts at the scene had misapplied the models, placed inordinate bets and piled up excessive debt. Very low probability events implicit in the risk models were adjudged as zero probability events. It was a human, not a conceptual, mathematical failure.

Critics, including a few mathematicians, on the other hand, claimed that the models had basic flaws. The automated credit scoring risk models dispensed with traditional lender judgment and could not accurately assess the risk of mortgage loans. The models for derivative valuation and trading could not be tested empirically due to nature of the variables used. Their use was based on faith, not science. The models sidelined a critical factor, the not-always-rational human factor. According to a senior professor of physics at the Columbia University who at one time served as a managing director at Goldman Sachs:

> *To confuse the model with the world is to embrace a future disaster driven by the belief that humans obey mathematical rules. ...The models were more a tool of enthusiasm than a cause of the crisis.* Emanuel Derman (Lohr 2008).

A yet another line of criticism lambasted the mathematicians. They were branded immoral speculators or mercenaries working against the interests of the majority, even as proponents of mathematical terrorism.

> *University mathematicians have provided the finance industry with a steady stream of trained experts .... to accept models axiomatically, without skepticism.* David Steinsaltz (Harris 2015, page 107).

> *People assume that if they use higher mathematics and computer models, they are doing God's work. They are using the devil's work.* Investor Charlie Munger (Patterson 2009, page 295).

In the years since the crisis, a number of solutions to prevent a recurrence have been proposed. Besides legislative and managerial measures, they include developing better models, using chaos theory-based models to incorporate the human factor, eliminating or creating awareness of the blind spots in the models and ensuring prudent usage of the models. The models were value free; the key was to prevent their misuse. So continued the numerosis mantra.

Studies show that borrowers with poor math skills are more likely to be duped by deceptive marketing. Financial distress is negatively linked to numerical skill. Enhancement of mathematical literacy and introduction of financial computations in school curricula has thus been proposed an additional remedial measure.

In the couple of years after the financial crisis, economics received a heavy dose of criticism for using flawed assumptions and over reliance on models. Of all things, attention was turned on Karl Marx, the fervent critic of capitalism. Perhaps he was right. The capitalist economy is intrinsically unstable. The contradiction between the privatization of profit and socialization of risk makes cycles of boom and bust inevitable. It is a fundamental flaw that no model, however elaborate, can fix.

Soon such thoughts evaporated except among a tiny minority. Today, the field of financial mathematics continues to attract students. Career prospects are good. Well-known pure mathematicians are migrating into the field, in theory and practice. With stock markets booming once more, it is almost back to business as usual. The ethical codes of professional societies are placed on the back burner in a setting where single-minded profiteering and conflict of interest are but the norm.

Nazi Era Mathematics

225

Eugenic theories emerged in nations across Europe and North America during the 19th century. But it was in Britain that a systematic quantitative foundation for the doctrine was laid. The major scholars in the eugenics drive were at the same time among the principal founders of the modern science of statistics—Francis Galton, Karl Pearson and Ronald Fisher. (Chapter 3).

Practical implementation of the eugenic notion of racial purity reached its zenith in Nazi Germany. In particular, the rise and consolidation of Nazi power after the mid-1920s profoundly affected research institutions and universities in the nation. Departments of biology, psychology and medicine conducted research in eugenics, promoted the theory of racial purity, and collaborated in the Nazi scheme of sterilization, isolation and extermination. Many German eugenics research projects had linkages with the major American universities, foundations and research institutes (Chapter 3).

Prior to 1930, German universities housed many leading physicists and mathematicians. As the Nazis consolidated power in the 1930s, the situation changed drastically. Establishing dominance in the academy was a key priority. Take the Gottingen Mathematical Institute. Founded in the early years of the 18th century with the help of Carl F Gauss, one of the five greatest mathematicians of all time, and presence of Bernard Reimann, a leading light of non-Euclidean geometry, and the mathematics educator Felix Klein, the mathematics unit at the University of Gottingen was one of the premier centers of mathematics in the world.

In the initial decades of the 20th century, the University was home to first class mathematicians. The Mathematical Institute became an independent body under the directorship of Richard Courant in 1929. Its impressive lecture halls and an amply stocked library were complemented by ultra-modern facilities for staff, students and visitors. The Rockefeller Foundation had largely funded their construction. The Physics Institute housing eminent physicists like Max Born, James Frank and Werner Heisenberg stood adjacent to it.

As fascist Nazi ideas bloomed, the headship of physics and mathematics departments came in the hands of scholars who were sympathetic to or members of the Nazi party. At a few places, liberal scholars who did not express opposition to the Nazi ideology were tolerated. But mathematicians and physicists of Jewish ancestry or with dissident or communist views were marginalized, demoted or forced to flee abroad. Some were interred in the concentration camps while a few took their own lives. Given that about a third of the professorial positions in Germany in those days were held by mathematicians of Jewish ancestry, the impact was Earth-shaking. We first note a few mathematical victims of Nazism.

The Victims

Richard Courant was a Jewish German mathematician who gained fame for important contributions to mathematical analysis, calculus of variation, differential equations and mathematical physics. He co-wrote a popular book on mathematics and authored several university level textbooks on mathematics. The first director of the Gottingen Mathematical Institute, he was also a notable figure in the left leaning German Social Democratic Party. Nazi opposition to his politics led to his dismissal from the Institute and departure from Germany in 1933.

Hermann Weyl was a non-Jewish German mathematician who took up a distinguished chair at the Gottingen Mathematical Institute in 1930. In 1933 he succeeded Courant as the director. During his career, he developed novel results in several branches of pure mathematics. He also worked on theoretical physics and wrote philosophical tracts. He is now seen as one of the most influential mathematicians of the 20th century.

His tenure as the Institute director was short lived. Nazi youth gangs were increasingly harassing Jews and political dissidents. And his wife was Jewish. Fearful of their future, he resigned and departed with his family to assume an academic position in the US. Like Baruch Spinoza and Albert Einstein, Weyl distanced himself from organized religion and the idea of a personal God and instead embraced the pantheistic attribution of spirituality to the magnificence of nature.

Emmy Noether

Emmy Noether, an outstanding abstract algebraist and discoverer of results with important applications in physics, is one of the most distinguished woman mathematicians of all time. In the early 1900s, German universities barred women from joining the faculty. Despite writing a stellar doctoral thesis, for the first seven years, she could only do unpaid research at the Mathematical Institute of Erlangen. An invitation from David Hilbert to join the Gottingen Mathematical Institute prompted her to move there. But some faculty members were opposed to a woman assuming a teaching position. Thus, for four years she taught courses assigned to Hilbert. In 1919, she secured the position of a lecturer—the first woman in Germany to hold the position.

The hurdles she had faced did not prevent her from generating novel ideas and proofs that earned her an international reputation. After the Nazi decree ejecting all Jewish teachers was announced in 1933, she took up a university post in the US. Unfortunately, she died from cancer just two years later.

Edmund Landau was a wealthy Jewish German mathematician who specialized in number theory and complex analysis. Appointed to a chair at the University of Gottingen in 1909, he did not ascribe to Judaism but was a dedicated Zionist. In the 1920s, he spent some time in Palestine, helping to establish a mathematical institute at the Hebrew University of Jerusalem. In early 1933, he was told by the Nazi Minister of Culture to desist from lecturing at the Institute. His courses were conducted by a mathematician who was a member of the secret police. When he tried to lecture, he was hounded by fanatic Nazi students and the secret police. Dejected, he moved to Berlin and died from cardiac complications in 1938.

Felix Bernstein was a Jewish German mathematician at the Gottingen Mathematical Institute who made contribution to set theory and statistical analysis of blood group inheritance. For a while, he was active in politics and a senior figure in a political party that was a rival to the Nazis. His politics and ethnic background both contributed to his dismissal from the Institute in 1933.

The Bystanders

David Hilbert was a German mathematician whose prodigious forays into the foundations and edifice of mathematics and foresight into its future made him one of the undisputed doyens of mathematics. In 1900, he set 23 challenging problems many of which baffled mathematicians until quite recently. They also opened up a vast area of research in the field. A senior professor of mathematics at the University of Gottingen from 1895 to 1934 and editor of the then most prestigious mathematical journal for nearly four decades, some of his 76 doctoral students later became prominent mathematicians. Raised as a Protestant, he later became an agnostic. According to him,

> [Mathematical] *truth was independent of the existence of God or other a priori assumptions.* (Wikipedia 2021- David Hilbert).

Hilbert was a political liberal who refrained from political association with the Nazi party. Favoring the inclusion of women into the academy, he gave practical support to women mathematicians. Emmy Noether was one of his proteges. The Nazi era purge began after he had formally retired. But, he used his influence to assist some mathematicians who had been victimized by the Nazis. He died in 1943.

The Enablers

Helmut Hasse, a ranking mathematician and professor of mathematics at Gottingen Mathematical Institute, assumed the directorship of the Institute in 1934. A die-hard Nazi and anti-Semite, he was one of the signatories of a vow of allegiance by German professors and high school teachers to Adolf Hitler. During WW II, he worked with the German navy on the ballistic missile program.

Paul Teichmuller, a gifted mathematician who made significant contributions to complex analysis and differential geometry, was an ardent Nazi. A member of the Nazi party and its paramilitary wing, he was in the front line of the crusade for the dismissal of Jewish mathematicians like Richard Courant and Edmund Landau. Apart from taking part in military action during WW II, he also did cryptographic work for the German military high command. He died on the battle front in 1943.

Ludwig Bieberbach was a prominent mathematician with strong fascistic and anti-Semitic leanings. As a member of the Nazi party and its storm troopers, he was deeply engaged in the Nazi crusade against Jewish mathematicians. In the process, he betrayed the colleagues with whom he had collaborated earlier, facilitating their removal and arrest by the secret police.

Karl T Vahlen, an Austrian born mathematician who specialized in number theory and applied mathematics, was a professor of mathematics at the Humboldt University of Berlin. A member of the Nazi party from its founding days and a high-ranking officer of the secret police, he was deeply engaged in the campaign against non-Aryan and dissident mathematicians.

Together with Bieberbach, Vahlen injected racism into mathematics by promoting the idea of an indigenous German approach to the subject under the label of *Deutsche Mathematik* and founded a journal of that names. Like the *Deutsche Physik* movement set up by senior pro-Nazi physicists, its aim was to counter the allegedly Jewish dominance of mathematics. Vahlen also served in the leadership of the Kaiser Wilhelm Society for the Advancement of Science and served as president of the Prussian Academy of Sciences. He attained a high rank in the secret police and was active in the war effort. Captured at the end of WW II, he died in custody in Czechoslovakia.

Hellmuth Kneser was a mathematician who made novel additions to group theory and topology. He was also an anti-Semite, and a member of the Nazi party and the Nazi storm troopers.

> *May God grant German science a unitary, powerful and continued political position.* Wikipedia (2021 -- Hellmuth Kneser).

Wilhelm Suss, a mathematics professor at the University of Freiburg and an editor of *Deutsche Mathematik*, was a member of the Nazi party, its storm troopers and a senior member of the National Socialist German Lecturers League.

++++

The Nazi onslaught against Jews, Roma people, political dissidents, the disabled and homosexuals was conducted methodically. Once in power in 1933, the regime initiated a national census which recorded not just sex, age, location but also ethnicity. An accurate determination of ethnicity was needed. Later that was needed in areas of Europe occupied by the German forces. Census data had to be reconciled with community, church and state records. It was a mammoth task requiring an efficient and elaborate system of keeping and processing data.

For this task, the Nazis obtained assistance from IBM, a major US corporation that was a world leader in data management and processing. James Watson, the profit seeking IBM CEO, had scant moral qualms about working with the Nazis. A German subsidiary of IBM, whose management was ardently pro-Nazi, was established. In no time, Germany became the second most important market for IBM technology after the US. Its punch cards and machines were utilized for the census and the military and the railways, and for managing occupied areas and the extensive concentration camp system. About 1.5 billion IBM punch cards were sold annually in Germany alone.

> [IBM's] *Hollerith system was used to identify, sort, assign, and transport millions in Europe during the Holocaust, particularly in the death camps.* (Wikipedia 2021 -- IBM and the Holocaust).

Watson and his senior aides regular visited Germany between 1933 and 1943. IBM set up branches across Germany and occupied Europe and provided training to Nazi personnel. In 1937, Watson received a medal from Hitler. Even after the US entered the war and working with the Nazis was forbidden by US law, IBM continued to work covertly with them. Senior officials of its German subsidiary were arrested after the war but Watson and IBM USA escaped legal scrutiny. Their dastardly role was, until recently, erased from history.

The elaborate data collection and processing system set up under the Nazis could not have functioned efficiently without involvement of numerous statisticians, demographers and statistical assistants in Germany and the US.

The Rank and File

The bulk of the mathematical community in Nazi Germany, researchers, academics, school teachers and university students, knew what was afoot but chose to look away or engage in activities to which the Nazi regime would not object. They just wanted to be left alone so that they could get along with their professional work.

The German mathematical society expelled non-Aryan members. Some students in the mathematics departments were engaged in pro-Nazi activism. Some mathematical faculty joined the Nazi party and the National Socialist German Lecturers League. About a fourth of the academic staff were members of the League. The majority came from the humanities, but the leadership was in the hands of physicians and medical faculty. Its aim was to make university courses and research conform to the Nazi philosophy and priorities. It also exercised a strong influence on university appointments. But since many of the leaders were loud mouths with scant academic credentials, the League did not significantly alter the university curricula.

A few of the mathematicians who were supporters of the Nazi were disturbed by the ensuing deleterious impact on the quality of mathematics. A few of them later moderated their pro-Nazi stance. Hellmuth Kneser and Wilhelm Suss were among them.

The Opponents

Well known German mathematicians who fervently opposed the Nazis were a distinct rarity. **Emil J Gumbel**, a professor of mathematical statistics at the University of Heidelberg and a pioneer of the statistical theory of extreme values, was a communist and an outspoken critic of the Nazi regime. In 1932, he cosigned, along with more than thirty artists, scientists and authors, a public appeal to defeat Nazi electoral candidates in the upcoming elections. Forced into exile by the authorities, he first moved to France and then to the US, where he continued his anti-Nazi writing and activism.

> *A courageous man, Gumbel spoke out passionately against the Nazis and came to symbolize a 'one-man party' at the center of controversy in German academia. His intellectual and moral vigor never waned, and despite his significant scientific contributions, it is his legacy of political ideology that endures for later generations to learn from.* (Brenner 2002).

Florence Nightingale

The birthday of Florence Nightingale is now celebrated as the International Nurses Day. The Florence Nightingale Medal is the highest international award for a nurse. These distinctions are in recognition of her unique role as a pioneer of modern nursing.

Of wealthy parentage and home-schooled by her father in Latin, Italian, Greek, history, writing, mathematics and philosophy, Florence's life underwent a major transformation after she observed the compassionate care given to indigent patients by a priest. The incident, occurring when she was 31, inspired her to train as a nurse. After working as a clinical nurse for a year, she was assigned to train and lead a contingent of over 50 nurses serving the British armed forces in the Crimean War. Her devotion to the wounded and ill soldiers as well as her tireless efforts to improve nursing and medical care during the war formed the basis of her fame.

The innovative measures she championed during the war and later included: Better hospital layout, administration and ward management, improved hygiene, ventilation and sanitation for the wards, better nutrition for patients, sufficient medical supplies, high standard of care for the wounded, infection control via hand washing, washed towels and linen, and attention to cleanliness by medical staff. Promoting adequate, systematic training for nurses and work schedules that reduced overwork, she aimed to professionalize nursing and secure higher status for nurses. Instrumental in the establishment of the first secular nursing school in the world, she authored a simply written text, *Notes on Nursing*, that came to form the basis for the curricula of future nursing schools.

Her innovative measures drastically reduced death rates in hospital wards on the war front and in general medical facilities. Nightingale was a dedicated social reformer and a champion of women's rights. Though from the upper stratum of British society and politically well connected, she promoted improved health care for all sections of the British society, greater participation of women in the work force and removal of laws that needlessly punished poor women. She was involved in improving food delivery and health services in British colonial India and betterment of hospital conditions in Turkey as well.

> *By the time Nightingale left Turkey after the war ended in July 1856, the hospitals were well-run and efficient, with mortality rates no greater than civilian hospitals in England, and Nightingale had earned a reputation as an icon of Victorian women.* (Rehmeyer 2008).

Florence Nightingale in a Field Hospital

The Crimean War killed 500,000 Russians, 100,000 Frenchmen, 50,000 Turks and 25,000 Britons. Besides Nightingale, other reform-minded health workers involved in the War were Mary Seacole, a British Jamaican-born nurse, Russian nursing assistant Daria Mikhailova and Russian surgeon Nikolai Pirogov. More than three quarters of the war fatalities were caused by infectious diseases, not combat. The efforts of these reformists no doubt reduced the death rates.

Nightingale's advocacy of beter nursing care and social reforms was guided by numbers. Attracted to mathematics from childhood, she received private tuition in the subject when she was twenty. With a fascination for statistics, she systematically compiled large datasets relating to public health and social welfare. During her nursing work in the Crimean War and later social reform campaigns, she became remarkably adept at presenting complex and voluminous data in the form of easily understandable charts and graphs. Her innovative diagrams attest to her status as a major founder of graphical statistics. Her graphs that show seasonal trends and regional patterns in cause-related mortality and morbidity facilitate uncovering the associations between health and other factors.

Graphs are a key tool for conveying the message to policy makers and the public who generally disdain numbers conveyed in traditional ways. Working with eminent statisticians, she gained the ears of senior policy makers, prompted the appointment of a Royal Commission and wrote an 830-page report that among other things examined death and disease rates among hospitalized patients throughout England. Her efforts had far-reaching consequence on public health and hospital hygiene in England.

In 1959, she was inducted into the Royal Statistical Society, the first woman to gain the honor, and fifteen years later, she was made an honorary member of the American Statistical Association. She viewed statistics as a central facet of modern life, a basic requirement for health and social reform.

> *Statistics is the most important science in the whole world: for upon it depends the practical application of every other science and of every art: the one science essential to all political and social administration, all education, all organization based on experience, for it only gives results of our experience.* Florence Nightingale

Nightingale was a devout Christian who interpreted the mystical experiences she underwent as invocations from God to live a life of public service. But she looked askance at organized religion. The Church of England had worsened the lot of the poor. Religious health workers often were driven by selfish motives. After extensive studies of Christianity and Eastern religions, she penned a veritable theological opus, *Suggestions for Thought*, and numerous religious books. Religion combined faith in God with altruistic service to humanity. God was a merciful, not punitive, being. Respectful of Eastern and other religions, she opposed divisiveness among Christians and hostility based on religion.

Nurses and other professionals should combine spirituality, faith and technical skills to render the best possible service. Faith complemented statistics.

To understand God's thoughts we must study statistics, for these are the measure of his purpose. Florence Nightingale

Her efforts to improve public health and the lot of the poor were a religious calling. The traditional religions focus on afterlife and personal salvation was not true religion.

Mankind must make heaven before we can 'go to heaven' (as the phrase is), in this world as in any other. Florence Nightingale

The 'kingdom of heaven is within,' indeed, but we must also create one without, because we are intended to act upon our circumstances. Florence Nightingale

Florence Nightingale uniquely blended professionalism, mathematical skills, compassion and devotion to social reform with a nondenominational, multi-faith, spiritualist outlook.

Yet, one point must be made. By improving the conditions in British military hospitals, she strengthened the capability of a military machine bent on subduing foes and establishing British imperial presence in the Middle East and beyond. The Crimean War was a precursor to more deadly wars that were fought with the improved versions of weapons and tactics used in this war. British imperialism was on the verge of rampaging in Asia and Africa with little concern for the lives of the indigenous peoples. Nightingale's work was morally compromised in that respect.

++++

Our comparative sojourn into the ethical engagements of mathematics and mathematicians underscores a major point: The claim that mathematics and statistics are ethically neutral entities is a fictional claim. It does not hold water at the conceptual, personal or social level. Like religion, these disciplines have served beneficial and harmful purposes. And that feature is not simply a historic curiosity but also a major concern in modern society.

5.12 PASCAL'S WAGER

It is a dark, cloudy morning. I have a series of errands to run. Should I carry an umbrella? If I do and it rains, I will be protected. If I do not and it rains, I will get soaked. Yet if I do and it does not rain, I will have an unneeded burden to carry around. But if I do not and it does not rain, there will be nothing to complain about.

Life brims with uncertainties. What we choose when faced with one can have good or bad consequences, minor or serious. We weigh risk against benefit and consider the chance that the event in question will occur. My decision to carry an umbrella will differ if the forecast says that the chance of rain is 80% or 20%.

Blaise Pascal, a towering 17th century mathematician, physicist and inventor, pioneered a quantitative approach to rational choice based on probability theory. Pascal is also famed for his contributions to philosophy and theology.

His most well-known theological text, the *Pensees*, was written after he had an intense religious experience at the age of 31. It applies probabilistic reasoning to two basic questions: Should we believe in God? Should we live a virtuous life? His resolution of these questions is called Pascal's Wager.

The two key premises of this wager are: (i) You either believe God exists or you believe He does not exist. There is no third option. (ii) Existence or non-existence of God cannot be

232

established by rational discourse. Belief or disbelief in God carry risks and rewards. Weighing them, he argued that the best choice is to believe in God.

> *God is or He is not. Let us weigh the gain and the loss in selecting 'God is.' If you win, you win all. If you lose, you lose nothing. Therefore, bet unhesitatingly that He is.* Blaise Pascal (Wikipedia 2021 – Pascal's Wager).

Such an argument for believing in a divine power had been proposed by Hindu, Greek and Muslim scholars. Jafar al-Sadiq, the 6th Imam of Shia Muslims and an influential jurist, philosopher and Sufi mystic is said to have formulated one such argument during an exchange with a non-believer.

> *If what you say is correct – and it is not – then we will both succeed. But if what I say is correct – and it is – then I will succeed, and you will be destroyed.* Imam Jafar al-Sadiq (Wikipedia 2021 – Pascal's Wager).

Pascal went further than his predecessors by formalizing his argument in terms of placing a bet, considering the probability of existence of God, quantifying the rewards and costs, and computing the expected value for the choice made.

Probabilities are assigned values from 0.0 to 1.0, where 0.0 zero denotes an event that never occurs and 1.0, an even that always occurs. Consider an event E. You are to choose between its occurrence (E) or non-occurrence (not-E). Suppose the probability that E will occur is p and thus that it will not occur is $1 - p$. If E occurs and you had selected it, your benefit is B but if you had not selected it, your loss will be equivalent to -B. If E does not occur and you had not selected it, the gain is C but if you had selected it, your loss will be equivalent to -C. Under this scenario, your expected gain (expected value) for choosing E is:

$$EG(E) = p \times B + (1 - p)(-B)$$

And your expected gain for choosing not-E is:

$$EG(not\text{-}E) = p \times (-C) + (1 - p)(C)$$

Suppose E stands for existence of God. If E is true and you express belief in him, you get eternal salvation (a gain of B gold bars) but if you do not express belief in him, you get eternal damnation (a loss of B gold bars). But if E is false and you do express belief, you have minor inconvenience costing you C gold bars but if you do not express belief, it saves C gold bars.

Let us assign some numerical values. Suppose the probability that God exists, p, is 0.01, eternal salvation (B) is valued at 10,000,000 gold bars and the cost of expressing belief (C) is 200 gold bars. If you express belief in God, your expected gain will be

$$0.01 \times 10,000,000 + 0.99 \times (-200) = 99,802 \text{ gold bars}$$

If you express disbelief in God, your expected gain (loss) will be

$$0.01 \times (-10,000,000) + 0.99 \times 200 = -99,802 \text{ gold bars}$$

Hence, even if the probability of the existence of God is small, believing in Him will yield a substantial reward while not believing in Him will entail a large loss. Whatever the truth, you are better off in believing in God and living your life accordingly. It is better to be safe than sorry.

++++

Pascal's Wager has been critiqued on several grounds. First, it does not guide actual religious belief. Religion is not a unitary entity. The God of a Hindu, the God of a Muslim or the God of a traditional African religion is not the God of the Christian. Each demands vastly different beliefs and practices. Pascal resolved the dilemma by adopting a prejudiced, racist position. Anything other than Catholicism was a pagan or faulty religion. God will only save a select few; the others (non-Christians) will be condemned to hell whatever their belief.

The wager fosters a gain driven attitude towards God. A true devotee will not see belief in God equivalent to placing bets. God rewards those who love Him unconditionally, not opportunists. Pascal assumed that good morality and belief in God are necessarily linked. That is also not supported by the historical record. (RPS 2022).

Pascal was cognizant of uncertainty in other aspects of life. His wager has been linked to the Precautionary Principle: It is prudent to adopt a cautionary approach to the acts, inventions and policies which can cause great harm. It is best to leave an ample margin of safety in the design of vehicles, airplanes, bridges, dams and medical devices. Two engine airplanes are so designed that if one engine fails, they can fly on a single engine. It is evident in nature. Humans have two lungs, two eyes, two ears and two kidneys, but can survive on one of each.

The precautionary principle has been advocated by leading climatologists in a bid to spur complacent global policy makers to initiate firm climate control policies. The cost of not doing it may prove catastrophic for all life on Earth.

5.13 REFLECTIONS

The conceptual, ethical and societal ramifications of religion, mathematics and their interactions are multifarious.

Conceptual Connections

Did the theorem of Pythagoras exist before humans came to know it or is it purely a product of the human mind? In general, is mathematics discovered or invented? The Platonist school holds that mathematical truths exist in a mental Universe independent of humans. Via deductive reasoning, mathematicians lay bare what hitherto was hidden. The invention school holds that mathematics is a creation of the human mind. Using non-contradictory axioms and the basic rules of logic, humans have cumulatively built a magnificent, multifaceted edifice of theorems and formulas.

> *The science of pure mathematics ... may claim to be the most original creation of the human spirit.* Alfred N Whitehead

Both positions have theological implications. It is argued that ideas do not arise by themselves. Ideas need a mind. The objective existence of the complex edifice of mathematics points to the existence of a divine mind.

> *In the pure mathematics we contemplate absolute truths which existed in the divine mind before the morning stars sang together, and which will continue to exist there when the last of their radiant host shall have fallen from heaven.* Edward Everett

The invention school first notes that animals can distinguish one from many and distinguish shapes in very rudimentary ways. With that evolutionary background, mathematical concepts

234

like numbers and triangles originated from what humans observed in nature. Initially, they just served a practical purpose. But as societies progressed, humans created an abstract conceptual edifice that slowly congealed into what we now know as mathematics. In the same fashion as they wrote poetry, painted, sang and played music, they indulged in the mathematical arts. The difference was that unlike other creations of the human mind, mathematics was based on strict rules of logic, and was in part driven by the need to solve practical problems.

How do these two viewpoints impinge upon the astonishing applicability of mathematics to science and technology? If mathematics exists in a Universe of its own, then this Universe of ideas is possibly in harmony with the material Universe and the applicability of the former in the latter is consequential upon that fact. This is more plausible if both the material and the mental Universes were created by the same divine being. God formulated the laws of nature in a mathematical form. Mathematics is a divine quest, a mode of logically driven spiritual worship.

> *Geometry existed before the creation. It is co-eternal with the mind of God. It is God himself.* Johannes Kepler (Davis 1998).

The inventions school notes that the greater bulk of mathematics is in the domain of pure mathematics. Yet, formulation of mathematical ideas to resolve concrete issues is a continuing exercise. The applicability of some parts of mathematics to science and technology, however surprising, is essentially not a remarkable thing.

> *It can be shown that a mathematical web of some kind can be woven about any Universe containing several objects. The fact that our Universe lends itself to mathematical treatment is not a fact of any great philosophical significance.* Bertrand Russell

The vision of mathematics as a purely human practice does not entail recourse to the existence of a divine mind as an explanatory factor for its utility.

++++

The formulation non-Euclidean geometries had unnerving implications. If a shining star like Euclidean geometry can have logically consistent alternatives, then other parts of mathematics may likely have sound alternatives. Einstein's theory of relativity posited a non-Euclidean structure of space-time. This implied that the issue had major practical implications as well. To the theologians who designated God as the supreme mathematician, the quandary was: What does God prefer, Euclidean or non-Euclidean geometry?

Determination that any mathematical system based on a non-trivial, consistent set of axioms has statements that can be neither proven or dis-proven by Kurt Godel posed a dilemma for mathematicians and the theological community. The former rationalizes that such statements are a rarity. But the latter wonders how a discipline that reflects the mind of God can have such a basic limitation? Others opine that Godel showed that humans cannot truly fathom the mind of God. Their discernment of mathematics is uncertain. God is beyond logic, beyond certainty and uncertainty. Mathematics as formulated by humans is but a poor version of divine mathematics. Perhaps it was this form of thinking that propelled Godel to form a modal logic-based proof of God but not realize that his proof of God stood on shaky ground while his proof on the incompleteness and inconsistency of well-defined axiomatic systems was impeccable.

The paradoxes revealed by Bertrand Russell and Kurt Godel and solid proofs of impossibility of certain mathematical tasks appeared to abnegate the notion of a being with supreme powers.

> *Can God create a stone that He cannot lift?*

Mathematicians have proved that it is impossible to find a general formula based on basic algebraic operations for solving equations degree 5 and higher. Can God find such a formula?

For each query, if He can, then there is something He cannot do. And if He cannot, then He is not God. As a distinguished mathematician satirized:

While the other sciences search for the rules that God has chosen for this Universe, we mathematicians search for the rules that even God has to obey. Serre JP (Tu 2013).

Mathematical axiomatic systems are formed in a systematic, rigorous fashion. Faith-based axioms lack those features. Axiomatic arguments for the existence of God pertain to religion in an abstract manner. But religions are many. The religions based on the Vedas, the Tripitaka, the Bible, and the Quran and their concepts of divinity are as far apart as the Earth is from the Sun. Their followers trust their holy books as a matter of faith, not from logical reasoning based on a set of axioms. To place religion and math on an equivalent plane just because both have axioms is a simplistic, perfunctory exercise.

Another difference with faith and religious beliefs is that nobody has yet be condemned to eternal hell, unspeakable suffering or the like by not accepting the rules of Euclidean Geometry, ZFC or Peano's Axioms...in fact, nobody has even been put to death by that, although the other way around has happened! (Antonio 2012).

++++

For a long time in history, mathematics was intertwined with religion. But now that bond has been shattered. Each exists within its own domain and has its own conceptual foundation. Using mathematical ideas to shore up religious belief is as puerile as using religious ideas to prove mathematical theorems. While many mathematicians are devout believers, none integrates topology, vector analysis or set theory with his or her religious beliefs.

Besides truth, mathematicians revere elegance and brevity. They see beauty in what they create. Thus, by now over 350 different proofs for the Pythagorean theorem have been devised. Examining them elicits a sense of marvel. The sheer elegance of some proofs is breathtaking. In the following two figures, the areas of their white portions are evidently equal. Just a glance proves the theorem. Yet, it is an as rigorous a proof of the theorem as any.

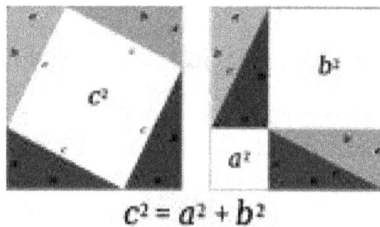

$$c^2 = a^2 + b^2$$

An Elegant Proof

Mathematicians harbor mystical or spiritual feelings towards the beauty of the edifice they live in. Deist spirituality does not, for scientists and mathematicians, translate into the belief in a divine, all powerful creator God. As a veritable intellectual giant put it:

236

I am a deeply religious nonbeliever - this is a somewhat new kind of religion.
Albert Einstein

Ethical Connotations

Besides elegance and truth, applicability and good ethical standing are said to be key salutary features of mathematics. Even if pure mathematicians insist that mathematics ought to be pursued for its own sake, there is no doubt that without mathematics modern life would not be as it is. By improving humanity's lot and reducing suffering, mathematics is an ethically laudable discipline. Even though they claim that the contents of mathematics are ethically neutral, mathematicians feel that they are among the good guys.

Yet, as shown earlier, the obverse side of this idealized portrait exists as well. Throughout history, prominent mathematicians functioned under the tutelage of emperors, wealthy merchants, landlords and powerful religious institutions. Our sojourn into the practice of mathematics showed mathematicians have worked for pharmaceutical corporations, big banks, security agencies and fascist, genocidal regimes.

During WW II, not just German mathematicians but mathematicians in the most powerful nations—USA, Britain, France, the USSR—were involved in the war effort. They worked on development of ballistic missiles, coding and decoding secret communication, radar technology, military logistics, operations research, computer development, etc. The linkage between mathematics and the state was further enhanced during the Cold War. Nuclear weapons development involved many mathematicians.

Many mathematicians and statisticians today are engaged in research and operational aspects for essential human endeavors. But the scope of their discipline has widened considerably. We have now data mining, efficient search methods, artificial intelligence, robotics, satellite operations, military and civilian drones, preference generating algorithms, super computers, cybersecurity, stock market decision making, predictive policing, environmental science, and so on that use mathematical techniques. Major firms in electronics, production of computers and cell phones, marketing, banking, insurance, medical, chip making, energy, industrial farming, air, sea and land transportation, manufacture of weapons—Apple, Microsoft, Google, Amazon, Ford, IBM, Bank of America, Shell, Exxon, Boeing, Walmart and their counterparts in other nations—need mathematical, statistical and computer science tools at some level. Many either hire or contract out experts in such fields to optimize their operation, expansion and profits.

> *The most powerful weapon in business today is the alliance between the mathematical smarts of machines and the imaginative human intellect of great leaders. Together they make the mathematical corporation, the business model of the future.* (Sullivan and Zutavern 2017).

Profit motive and state interest often collide with human interest. Disregard for the environment, public health and global peace and harmony, reckless and irrational consumerism, national and international inequities, hunger and malnutrition ensue from reckless, unethical corporate and state endeavors. While mathematicians, statisticians and computer scientists and their societies express ethical concern related to research integrity and publication, they are generally silent on the societal implications of what they do or what is done with the methods they have developed. Many mathematicians, statisticians and computer scientists are directly embroiled in activities with negative ethical implications. They serve corporate, state and imperial interests to buttress an unsavory *status quo*. Their colleagues feign neutrality or bury their heads in the sand. Only a few brave souls raise ethical, human rights concerns. Even fewer worry that mathematically buttressed neoliberal capitalism is sending humanity into an environmental abyss, profound economic instability, vast inequality, racism, religious and political extremism and unbridled authoritarianism.

One began to hear it said that World War I was the chemists' war, World War II was the physicists' war, World War III (may it never come) will be the mathematicians' war. (Davis and Hersh 1999).

5.14 CONCLUSION

Religion and mathematics are products of the human quest to tackle the challenges of existence in a risk laden environment and give meaning to nature and life. Conjoined at the outset, mathematics gradually began to carve out a space of its own. Yet, it continued to retain tangible links to the world of the divine. Until about 200 years ago, eminent mathematicians often had religious beliefs. Some of them professed axiomatic proofs for the existence of God.

Today, the historic conceptual and personal bond between mathematics and religion has largely been severed. Fewer than 1 in 5 mathematicians believes in a personal, creator God. More express belief in spirituality of some type. Gurus, priests, imams and theologians, on the other hand, often present mathematics as a discipline that complements their beliefs.

Having faith in mathematics is similar to having faith in religion: it is to believe in something intangible which yet is tangible in the minds of believers. (Vasak 2020).

Mathematics expresses values that reflect the cosmos, including orderliness, balance, harmony, logic, and abstract beauty. Deepak Chopra

[Mathematics] *opens the mind to the wonders of God's creation. This is true even of disembodied numbers, figures, and solids, but it is especially true in astronomy.* Brian T Kelly, Thomas Aquinas College (Kelly 2021b).

It is argued that mathematics and religion have an axiomatic orientation, and the presence of mathematical entities like π and the Golden Ratio in nature as well as the fact that nature follows mathematically formulated laws point to the presence of a divine power. Srinivasa Ramanujan, a genius of the highest order, said that his formulas were revealed to him in dreams by a goddess.

Yet, all attempts to enjoin religion with mathematics into a coherent whole have floundered. None of the axiomatic or mathematically inspired proofs for the existence of God, including those formulated by eminent mathematicians, have the robustness of the proof of the Pythagorean Theorem or the irrationality of $\sqrt{2}$.

To mathematize theology or theologize mathematics is to discredit both religion and mathematics. They are distinct human endeavors. While taking pride in their distinct contributions to human culture and flourishing, they need to maintain a respectful, conceptual distance between each other.

Mathematicians should concede that mathematical reasoning cannot either prove or disprove the existence of God while believers in religion need to refrain from selective appropriation of mathematical ideas to buttress their faith. The wisdom of a prominent Sufi mathematician is apropos:

Faith is an oasis in the heart
which will never be reached
by the caravan of thinking.
Khalil Gibran

++++

238

Ethics is the fundamental concern. Religion and mathematics have functioned on both sides of the ethical coin. Both have served noble human interests, and both have partaken in gross abuse of human rights. None can boast a spotless ethical record. And that duality of ethics persists to this day.

Francis Su, a past president of the American Mathematical Association, has declared that mathematics not only enables play, beauty and truth, but also justice and love. But for promoting justice and love, mathematicians need to put in more effort than done thus far. They need to champion racial and gender inclusiveness and stand up for social justice and human rights.

> *I feel that our mathematics community can do better; we can become more just. I see a lot of ways in which we can do better and become more virtuous as a community.* Francis Su (Harnett 2017).

One of the most prominent defenders of human rights of our times had a similar vision.

> *Those who wish to change the world should have the best possible understanding of the world, including what is revealed by the sciences, some of which they might be able to use for their purposes. That's why workers' education, including science and mathematics, has commonly been a concern of left intellectuals.* Noam Chomsky

The teaching of mathematics and statistics, from elementary to university level needs to be augmented with accurate renditions of the history of the subject, narrations of the lives of eminent mathematicians, and depiction of ethical and unethical roles played with the usage of these disciplines. Mathematical utility has to respect the code of ethics enshrined in The Universal Declaration of Human Rights. Many proposals to formulate the teaching of mathematics and statistics in conjunction with concerns for social justice and human rights exist. But public and private educational institutions have hardly paid attention to these proposals.

Mathematics and statistics are magnificent aspects of human culture. Like religion, they have ennobled human culture in diverse ways and are worthy of attention without external concerns. Yet, ultimately, they will earn their keep not on account of the stunning intellectual character of their output but primarily for their role in the resolution of the daunting problems facing humanity. The sooner they and their organizations wake up to this reality, the better will it be for their professions and humanity. Their skills must be mobilized to consciously serve global existential and ecological concerns. Collaboration with like-minded believers and spiritualists in this noble task is also an essential requirement.

CHAPTER 06: RUMINATIONS

Because we all share this planet Earth,
we have to learn to live in harmony and peace
with each other and with nature.
This is not just a dream, but a necessity.
Dalai Lama

From religion comes a man's purpose;
from science, his power to achieve it.
William Henry Bragg

God is always invented to explain those things
that you do not understand.
Richard P Feynman

W E NOW revisit eugenics and climate change to further explore their religious ramifications.

6.1 A DIVERSION

Anecdotes involving religion and mathematics are an amusing point of departure.

Commenting on the dry, pedantic manner of teaching mathematics generally seen in schools, a popular American television personality opined:

As long as algebra is taught in school,
there will be prayer in school.
Cokie Roberts

Two algebraic powers confront a thorny theological issue:

X^2 : *Are you religious?*
X^3 : *I believe in the existence of powers higher than us.*

A skeptic propounds devilish numerology:

If 666 is an evil number,
then 25.8059758 is the root of all evil.

A creative Biblical scholar discovers the algebra embedded in the Bible:

Biblical Algebra
Subtract all your fears
Divide your blessings with others
Multiply your good deeds
Add Jesus to your life
Equals a wonderful life

The famed spiritualist Radhanath Swami posited a new brand of arithmetic:

The more we take the less we have.
The more we give the more we have.
This is spiritual mathematics.

Leonhard Euler, a genius whose prodigious mathematical output elicits wonder to this day, offered a Proof by Intimidation for the existence of God. Serving as a court mathematician for the Russian empress Catherine the Great, he was asked to debate a famed atheistic philosopher, Denis Diderot. As Diderot poured out his ungodly tirade, the empress was disturbed. Sensing her unease, Euler solemnly stated:

Sir, $(a + b^n) = x$. *Therefore, God exists. Reply.*

The astounded Diderot beat a hasty, speechless retreat. Whether this episode ever took place is in doubt. But it is reminiscent of the arcane terminology found in the theological discourse about existence of God.

6.2 NEW EUGENICS

Eugenics is an elitist and supremacist doctrine. It has two basic premises: Human society should be organized on biological grounds and the simple and complex mental and physical abilities and characteristic of humans are passed on from parents to offspring. Besides disease risk, height or eye color, intelligence, educational attainment, anti-social behaviors, wealth and criminality also have a genetic basis. It declares that if the transmission of the associated 'bad' genes is not controlled, the quality of human stock will diminish over time.

These assertions stemmed from baseless assumptions and pseudo-scientific investigations. Yet, eugenics became a popular idea in the nations of the West between 1850 and the end of WW II. It was embraced by secular and religious people, liberals and conservatives, scientists as well as literary figures. It not only led to large scale forced sterilizations but also to the Nazi gas chambers (Chapter 3).

Today hardly any scientist dares to explicitly adorn the garb of eugenics. But eugenics still flourishes under the rubric modern genetics. Since the deciphering of the human genome, a literal explosion of genetic studies have concluded that a wide variety of human physical, psychological, health and social characteristics, simple and complex, have a genetic basis. Depending on the characteristic, some geneticists assert that genes are the primary, if not the sole, determinants. Some ascribe environmental and social factors to be as important while some hold that after the effects of other factors are considered, genes at best have a minor role. Over the years, and for a wide variety of life outcomes, greater import has been assigned to genetic factors.

Let us set aside the validity of these studies and focus on their implications. Consider neurofibromatosis, a debilitating, multi-symptom childhood disease in which tumors crop up across the nervous system. One form of the disease is caused by a mutation in a gene called *NF1*. If the parent has the disorder, the chance that his or her child will have it is 50%. Treatment options exist but are of limited value. Prevention options include parental screening, voluntary sterilization, and abortion. Since the newborn may not have the mutation, should the prospective parents take a chance? Is there a societal interest in preventing the birth of a child with this disease? Should sterilization or abortion be enforced by law? Eugenics once again raises its head.

Thalassemia is a spectrum of diseases in which the red blood cells have no or too little of hemoglobin, the molecule that transports oxygen in the blood. Those with severe forms of thalassemia appear pale and suffer from extreme fatigue, shortness of breath, heart arrhythmia and other ailments. Without specialized care and regular blood transfusions, they experience a poor quality of life and have a short life expectancy.

The malady poses ethical concerns for individuals and society. Take a family whose child has thalassemia. Driven by love and ethics, parents will do their best for the child. But what they manage to achieve greatly depends on their economic status, the setup of the health care system and where they live. Children with thalassemia fare better if they live in rich nations, have affluent parents or access to a universal, public health care system. Under favorable conditions and modern healthcare, babies with thalassemia have a 50% chance of reaching the age of fifty. But an afflicted child of a poor family in East Africa, where health care is skewed by the neoliberal system, has a slim chance of a decent existence and survival. As for diseases in general, economics triumphs genetics. Those with the means can minimize the impact of skewed genes. Else genetic determinism portends a miserable life.

If you have thalassemia or you know that you are an asymptomatic carrier, should you procreate? If you are pregnant and it is found that your unborn child is at a high risk of acquiring thalassemia, should you abort?

These queries were explored in a recently study of 67,089 intending-to-marry couples in Tehran, Iran. After being screened for major thalassemia, the at risk couples had the option to forego marriage, sterilize or marry with a view to raise a family. Women who were already pregnant had the choice of aborting the fetus. Considering the costs for the government, insurance agencies and families and estimating the number of cases prevented, the authors employed sophisticated statistical analysis to conclude:

> *Screening is a long-term value for money intervention that is highly cost effective and its long-term clinical and economic benefits outweigh those of managing thalassaemia major patients.* (Esmaeilzadeh et al. 2022).

Society and family will benefit, financially, if children with thalassaemia are not allowed come into this world. Is that conclusion acceptable to people of faith and secularists?

++++

Modern medicine, genetics and technologies have spawned scenarios that carry perplexing moral dilemmas for individuals, families, healthcare providers and the state. Going beyond screening for major maladies, they enable control of the genetic makeup of offspring with genetic engineering and embryo selection. The underlying rationale is the classic eugenic rationale: To improve the quality of human life. Called Liberal Eugenics or New Eugenics, its ethical justification rests on the claim that no coercion or state intervention is involved. Parents, the prime authority for deciding what is good for their children, informed and of their own free will, make the choice. It is good for the family and, ultimately, good for a society based on individual autonomy, freedom and equality.

Perhaps that assertion is plausible in the case lethal diseases in the newborn. But what of characteristics like eye color, skin lightness, intelligence, educational attainment, athletic prowess and musical ability? Modern genetic and medical technologies will soon make it possible for prospective parents to select babies with a variety of features other than gender. The technical possibility of editing genes to produce 'designer babies' exists, though it is not legal anywhere. It has been attempted once, though not with success, by a rogue scientist. In making the choices, what will differentiate cultural prejudice and cosmetic embellishment from health reasons? Will this set the stage for a revival of the Galtonian program of protecting the 'good' germ plasm? Should the medical system enable parents to request the insertion or deletion of genetic markers in their offspring that are said to be associated with such features? Will that not eventually lead to one mostly wealthy social stratum with 'superior' features and another mostly poor with 'inferior' features? Where does one draw the line?

Say, one set of genes has been linked to criminality or anti-social behaviors and another, to low educational attainment. Should the state mandate embryonic deletion of these genes in the same way as it makes education and vaccination for children compulsory? Should insurance companies be permitted to set their rates based on genetic risk?

As with classic eugenics, new eugenics operates under the existence of sharp inequalities within and between nations and ethnic groups. Genetics will be adduced to explain why some children go to bed hungry while the delicious food on the plates of others goes into the dustbin? Why is Africa poor and Europe rich? Why are there, relatively and in absolute terms, so few African American mathematicians in the US? Is it because they have 'bad' genes? History tells us that these are not just theoretical concerns. By diverting attention from the real causes, they impact the lives of billions.

Sex Ratio at Birth

In most nations, for every 1,000 female births, there are between 1,030 to 1,070 male births. Over the past two decades, the nations with the highest ratios of boys to girls have been Azerbaijan (1,150), China (1,150), Armenia (1,140), Vietnam (1,110), Albania (1,110) and India (1,100). High male-biased sex ratio at birth is generally attributed to a cultural and economic preference for boys, use of ultra sound for fetal gender detection, and availability and access to abortion services. The male bias is compounded by higher mortality among young girls caused by malnutrition, neglect and, in some cases, female infanticide.

In China, the one-child policy was the main contributor to the male-child bias. In the former socialist states, Azerbaijan, Armenia, Vietnam and Albania, male-child bias was low or absent in the socialist days but rose once state support for families and the elderly was drastically reduced under capitalism.

In India, sons are generally preferred because they perpetuate the family name and take care of elderly parents. Daughters entail dowry payment and would eventually be a part of the husband's family. The major contribution that young girls provide to family welfare by cooking, washing and other chores, collecting water and firewood, looking after babies and the cost saving due to denial of education are ignored. Pregnancy termination based on gender was banned in India in 1994 but the practice persisted at many illegal clinics. Male bias at birth rose in the 1970s after abortion was legalized and ultrasound screening became available, reaching the peak of 1,112 to 1,000 around 2010. Since then, mainly as a result of official campaigns and expanded free education, it appears to have declined to 1,081 to 1,000. But independent observers dispute these numbers, alleging, with good reason, that the BJP government tends to put a rosy spin on statistics.

According to the 2011 census, the sex ratio at birth in India varied by religion. It was highest among Sikhs (1,210), followed by Hindus (1,120), Muslims (1,090) and Christians (1,050). Since then, a significant decline has occurred for all religious groups and, by 2021, inter-religion variation was not as pronounced. The ratio for Christians, at 1,030, was quite low by global standards. Besides religion, other factors like economic security, education and urbanization affect the birth ratio. The residual effect of religion after taking their effects into account has not been estimated.

High male to female ratio at birth is a sign of gender based negative eugenics enabled by modern medical technology, ultrasound and abortion. And it reveals the persistence of discrimination of women in society.

++++

We compare the views of a Christian and a Muslim scholar on modern genetics and some controversial medical technologies.

A Christian Perspective

A paper by JR Nelson (Nelson 1988) overviews the Christian perspective on modern genetics.

Genetics and genetic engineering carry risks and benefits for humanity. Since the 1970s, several Christian groups have accordingly organized meetings and issued policy statements about genetic technology. Underscoring the sanctity of human life, the risk of applying genetic techniques must be weighed against the benefit for the patient. Use of genetics to enhance human abilities is fraught with eugenic dangers and may cause immoral treatment of some people. Equity and justice require that the harm and benefit of genetic technologies must be evenly spread across all strata in society. Faith groups should neither exaggerate nor minimize the primary tenet that human life is sacred and should work to ensure that genetics research and application are subject to strict public safeguards.

Theologians from varied Christian traditions hold a spectrum of positions around these issues. At one end are the hard line Catholics and Protestants who reason from Biblical verses to repudiate medical interventions that interfere with God's design. Some reject contraception and reproduction through external sperm and ova; only natural procreation is acceptable. Some reject organ donation, blood transfusion, surgery and vaccination. Genetic modification, in particular, is sinful.

These conservatives are countered by the pragmatists who accept medical interventions in which the benefit outweighs harm, both for the subject and social groups. They condone stem cell research, withdrawing life support for those in irreversible coma, and genetic engineering.

A prudent course dispenses with the two extremes, upholds God's power and design for the Universe and humanity and, at same time, critically heeds to the findings of scientific and genetic research.

> *Judaism and Christianity .... affirm that God desires the enhancement and fulfillment of each human life and the integrity of families and the wider human community. These values correlate with advances in science and technology that preserve life -- whether by electronic diagnosis, prenatal and neonatal therapy, nutrition, pharmacology or organ transplants.* (Nelson 1988).

The Bible also teaches that we are God's stewards for our own lives and for the environment. Scientific data, both medical and ecological, warn us of the limited resources we have for supporting the human race. We have to ensure that genetic technology helps to preserve life and the environment, not destroy them.

The Biblical vision of humans projects them along three dimensions. First, they have a physical being, including a complex, multi-functional brain. Second, with the brain as the foundation, humans have a mind, an entity that cannot be reduced to chemical reactions and neural signals within the nervous system. The mind enables humans to exercise free will and have a sense responsibility and morality. And transcending the brain and the mind is the unique, immortal soul, the true seat of individuality and connection to God, the divine, supreme being.

A binary, disconnected depiction of humans is flawed in spiritual terms and it fails to provide a valid understanding of human behavior. It also negates free will and responsibility. The three-way model integrating the body, the mind and the spirit is the best way to visualize the findings of science, including genetics, and design methods treating human ailments and avoiding simplistic reduction of complex human characteristics to one gene or a few genes.

Nonetheless, genetics and theology complement each other.

245

[Genetics] and theology provide different kinds of data, in different dimensions of cognition, which are ultimately complementary. Genetic science is opening new vistas for understanding, but it will remain insufficient without the insights of faith and theology. (Nelson 1988).

Another Christian Perspective

With about four million adherents dispersed over some 10,000 congregations, the Evangelical Lutheran Church in America (ELCA) is one of the largest Christian denomination in the United States. In 2008, it convened a task force of experts in theology and science to formulate the church's assessment of and policy towards genetic technologies.

The task force's report, ECLA (2008), addressed varied issues relating to genetics in health, agriculture and animal husbandry. Packed with scientific facts, it evoked the Biblical premise that humans are charged with exercising creative stewardship to protect life and nature. People of faith must be aware of the potential for extensive benefit as well as serious harm posed by genetic tools. Research and application of genetics must be done under strict guidelines that heed to social justice, respect for community culture, individual freedom and protection of persons.

Scriptures do not mention genetics yet contain indications for Christians on how to deal with it. Science is God's gift and should not be misused. Genetic research and application must not fall succumb to the human proclivity to sin. Working with integrity, researchers should avoid exaggeration, and cautiously apply the findings. The precautionary principle and the Golden Rule are apropos in genetic investigations. In light of the interdependence of modern existence, undue haste and neglect of the social and global context in the use of genetics can generate significant harm. Respecting diverse cultures and communities, science should protect the biosphere and the interests of the future generations. These considerations also apply to reproductive cloning, *in vitro* fertilization and human embryonic stem cells and other techniques in this domain.

Current economic and social inequalities and concentration of power within a small elite increase the possibility of an unjust distribution of benefit and harm of genetics (and other fruits of science). Christians should empower themselves to operate under Biblical and secular values in the challenging environment. They should reach out to peoples of other faiths and transform the church into a public venue for discussion of vital issues:

The confidence to act in the face of both the promise and the peril arise from trust in God, fed and informed by the cross and resurrection; the necessity to deliberate, act and evaluate genetic developments on their merits leads to the need for a common understanding of values and directives, that is, an ethical framework. (ECLA 2008).

A Muslim Perspective

For an overview of the Islamic perspective on modern reproductive and genetic technologies, we turn to a paper by H Hathout (Hathout 2006).

Islam is a continuum of past monotheistic faiths and shares many of their ethical and spiritual tenets. Special features demarcate humans from lower animals: a sense of right and wrong, and responsibility, free will and a desire for knowledge.

Humans are biology plus something else that is unique to the species. We are spiritual beings even though housed in a biological container. (Hathout 2006).

Humans should grasp the basic elements of science like atoms, molecules, cells, general biology and the human reproductive process. The principles of Islamic jurisprudence are a guide for complex questions posed by modern genetics and technologies like *in vitro* fertilization, genetic engineering, cloning, stem cell research and abortion. The relevant Islamic tenets include:

> • *the choice of the lesser of two harms if both cannot be avoided* • *necessities overrule prohibitions* • *avoiding harm takes priority over bringing good* • *public interest overrules private interest.* (Hathout 2006).

Applying these principles together with what the scriptures say, Islam prohibits abortion because it halts a process that will generate a human being. According to Prophet Muhammad, the spirit penetrates the embryo on the 120th day. Hence punishment for abortion is less sever if done before this time. If continued pregnancy will endanger the life of the mother or the newborn will have severe abnormalities, then the fetus can be aborted.

In vitro fertilization is acceptable provided it just involves a married couple. The use of sperm or ova from others and surrogacy infringe upon the sanctity of marriage and are not allowed. Genetic engineering interferes with God's plan for life and is not sanctioned by the Quran. As a matter of principle, it is disallowed. But if the technology is needed on medical grounds, its use is permissible.

Reproduction among animals and humans requires sexual relation between males and females. By making asexual reproduction possible, cloning violates the Quran and is thus proscribed. The high rate of fetal wastage in cloning and its potential to undermine social relations make it even more undesirable. Cloning is allowed for research but is absolutely banned for procreation.

Stem cells are intermediate cellular forms that have not become specialized tissue cells. As such, they have the potentiality to turn into tissue cells for varied organs. For example, a stem cell may be directed to become a heart muscle cell and used to repair a diseased heart. Stem cells are harvested from varied sources like the placenta, the umbilical cord and surplus cells from *in vitro* fertilization.

While stem cell research has garnered much controversy in other cultures, it is not that controversial in Islam. Elsewhere, the issue has been politicized and the dictum of sanctity of human life has been misapplied. If adequate safeguards are observed, Islam permits stem cell research.

God appointed humans as the stewards of His creations. The Islamic view on modern medical technologies has to bear this tenet in mind. He has granted them free will that enables them to make sound decisions, live balanced lives in harmony with nature and fellow humans. Today, humans have fallen into excess materialist consumption, selfishness and are influenced by racism, elitism and militarism, all of which negate the religious principles of peace and compassion. Under the present system, medical research generates profit and benefit for a few but leaves the rest of humanity behind. Medical care should be based on the principle of egalitarianism.

> *People of conscience, especially those of faith, should join forces to create a counter wave against this rampant, selfish, materialistic, utilitarian philosophy in favor of a human attitude guided by compassion and love and human togetherness.* (Hathout 2006).

Most Muslim scholars would generally concur with this perspective on modern genetics and medical technologies provided above.

Hindu and Buddhist Extremism

Hinduism and Buddhism are considered tolerant, compassionate faiths that do not harbor inhumane doctrines like ethnic cleansing and eugenics. But history shows that they too have generated extremist fringe groups that have stood for expulsion, if not extermination, of minority cultural groups.

The current ruling party of India, Bharatiya Janata Party (BJP), operates under the umbrella of two doctrinal Hindu organizations, Rashtriya Swayamsevak Sangh (RSS) and Sangh Parivar (SP). Both organizations espouse Hindutva, a social vision that sees India as a pure Hindu nation cleansed of external cultural contamination, especially Islam and Christianity. In a nation with multiple ethnic groups speaking over 700 languages, they seek a homogenized nation with Hindi as the main language. Founded in the colonial era, the pioneering leaders of the RSS and SP were sympathetic to the Nazi genocidal ventures, stressing the alleged common Aryan origins of Indians and the German-Nordic stock. RSS is a militant organization with members in the hundreds of thousands and over fifty thousand branches spread across India. Given training in Hindu philosophy and practice, its members also learn martial arts, fighting with staves, and handling firearms: A modern version of the Nazi Brown Shirts.

Islam and Muslims are their prime targets. The denigration of Muslims, about 15% of the population, has been rising since the BJP came to power. Senior officials and major Hindu priests have demanded strict control, if not expulsion, of Muslims, before they allegedly overrun the Hindu nation. Their supposed high rate of procreation must be curbed. Laws calling their citizenship into question have been passed.

The latest incidence involves T Raja Singh, a BJP leader and ex-legislator famed for his fiery rhetoric against Muslims. In August 2022, he was detained by the police for making remarks perceived to be insulting Prophet Muhammad and catalyzing tensions between Hindu and Muslim communities. Though suspended from the BJP, he remains a popular figure among Hindu nationalists. A cheering crowd garlanded him after he was released on bail. Three months earlier, he was suspended for hate speech against Muslims. Though the BJP publicly distances itself from such rhetoric, it depends on anti-Muslim verbiage to maintain its political constituency. As a Muslim leader in Delhi put it:

> *Many BJP leaders target Muslims the way Singh does. Hate and othering of Muslims is a major plank of the Hindu nationalist party's political strategy in the Hindu-majority country.* Zafarul-Islam Khan (Rahman 2022).

The pernicious doctrine of Hindutva has permeated the Indian diaspora in the United States. While most immigrant communities were alienated by former president Trump's odious, deceitful castigation of immigrants, he was applauded by many Hindu residents in the US. They also welcomed the close political ties between Trump and Prime Minister Modi, raised funds for Trump and organized events featuring movie celebrities to highlight their common political philosophy. Both leaders branded Islam as a violent faith that encourages terrorism. If the history of communal violence in India is anything to go by, the BJP's internal and external policies may unleash a major eugenic crusade directed against Muslims in India (RPS 2022).

++++

The saga of the Rohingya people in Myanmar represents another gruesome facet of modern external eugenics. Just before 2017, they numbered about 1,300,000. Today only about 600,000 remain in the country. The pogroms mounted by the civilian and military governments and assisted by local militias led by Buddhist monks have razed hundreds of Rohingya villages to the ground, killed or maimed thousands and forced about 700,000 to flee to Bangladesh and other nations. Those that remain are mostly confined in internal refugee camps in miserable conditions. Hitherto, they were getting some assistance from aid agencies, but now that has been blocked by the military government. Children are hungry and malnourished. Health and education services are inaccessible. Living quarters are congested ramshackle sheds providing little protection from rain and summer heat. Neglect also prevails in the camps in Bangladesh where the conditions are slightly better.

In a country that has been their home for centuries, they are now foreigners without any rights. It is illegal to mention the name of their ethnic group. Abroad, they are torn between having a bona fide refugee status or a localized status that gives partial rights to work. Most want to return home but, like the Palestinians ejected from their homeland, have little chance they ever will. The government of Myanmar faces charges of genocide in a UN court, but the final verdict is yet to be delivered. (RPS 2022).

Both the past civilian government and the current military regime denigrate the Rohingya and have launched deadly assaults on them. Most people of this Buddhist majority nation support the anti-Rohingya crusade. While thousands of Buddhist nuns and monks are vigorously fighting the military dictatorship, as far as the Rohingya are concerned, they stand shoulder to shoulder with the soldiers. Opposition to the policy against the Rohingyas from Buddhist circles is a rare entity here.

++++

In theory, Hinduism and Buddhism advocate universal compassion. But the cases of Muslims in India and the Rohingya show that they can be party to vile forms of modern external eugenics. Eugenics remains alive not just in the context of utilization of modern genetic and reproductive technologies but also with respect to classic ethnic cleansing and genocidal campaigns. Most of the discussions on modern eugenics omit mentioning the latter forms. As in the past, religion is involved in such forms of eugenics. Vast economic inequalities between nations and ethnic groups in the context of neoliberal policies that usually carry a historic baggage of discrimination and impoverishment, also constitute a potent form of systemic (external) eugenics.

Industrial Greenhouse

6.3 CLIMATE CHANGE AND RELIGION

The climate on planet Earth—temperature, rainfall, humidity, solar radiation, wind speed and atmospheric pressure—vary within ranges that make existence of life possible. Too little rainfall generates drought and famine. Thousands of animals and humans perish. If it is too hot or too cold, life is also endangered. The average surface temperature on our planet varies around 15° C in a pattern that enables a humongous diversity of flora, fauna and microbes to flourish on land, sea, air and under the ground. Some life forms endure under very extreme conditions.

A green house is a transparent structure designed to maintain its internal temperature and humidity at levels ideal for cultivation of fruits, vegetables and flowers. They vary from small to giant industrial facilities. Like a blanket, they trap radiant heat and maintain stable internal conditions conducive to farming via a feedback loop. This process is called the greenhouse effect.

Life on Earth is protected by a greenhouse effect. A complex process of absorbing and emitting radiant energy enabled by the presence of certain gases within the atmosphere maintains climatic conditions suitable for life. Though present in low concentrations, a major change in their levels affects Earth's energy balance and causes significant climatic changes. The main greenhouse gases are water vapor, carbon dioxide, methane, nitrous oxide and ozone.

Remove carbon dioxide, and the terrestrial greenhouse effect would collapse. Without carbon dioxide, Earth's surface would be some 33°C (59°F) cooler. (NASA 2022).

Developments in agriculture, industry, transportation and consumption over the past two centuries have entailed increasingly higher usage of fossil fuels for energy requirements. As a result, the concentrations of the greenhouse gases has reached levels which have the potential to drastically affect the existence of life forms on Earth. The level of carbon dioxide, the main greenhouse gas, increased from 313 parts per million to 400 parts per million. As a result, the Earth's climate has altered significantly. Years 2016, 2019 and 2020 were the hottest three years on record. Eons old weather patterns have been disrupted. Rainfall seasons on which farmers plan their activities have changed. Glaciers, snow caps, and polar ice caps are melting rapidly, sea levels are rising, and coast lines are being redesigned. Small island nations face submersion. There are more serious wild fires year by year, and global food supply is threatened. Catastrophic events like hurricanes, drought and extreme weather variation are increasingly probable. Portending major losses of species and biodiversity, the process threatens all life, and particularly the billions of humans in the poor nations. If not controlled, it may become a run-away mutually reinforcing cycle that may wipe out most of the life on Earth.

These conclusions of climatologists, oceanographers and other scientists have been derived from thousands of meticulous studies published in leading science journals. Despite disputes over details, the consensus on the principal mechanism and effects of global climate change is more solid than on most major issues.

The Intergovernmental Panel on Climate Change (IPCC) is the UN body charged with investigating climate change and formulating policies to control it. Bringing together nearly four thousand experts on climate and related matters, it issues regular reports on the state of the global climate and makes projections for the future. The basic message is unequivocal: Climatic conditions worldwide are becoming graver by the year. Serious steps are needed to curb fossil fuel usage and reduce emissions of green house gases. Between 1880 to 1980, the Earth's temperature rose by 0.08° C per decade, but from 1981 onward, it rose at the rate of 0.18° C per decade. The average global temperature in 2022 is projected to be around 1.15° C higher than during the second half of the nineteenth century and projections are that if greenhouse gas emissions are not lowered, the mean global temperature will rise by 1.5° C by 2050 and about 3° C by 2300.

Global warming of such a magnitude will entail major heat waves, excessive rainfall and flooding in some areas and severe drought and famine elsewhere, more frequent hurricanes and storms, and major soil erosion. Water sources will dry out, glaciers and snow caps will recede faster, and water supply will dwindle at an alarming rate. Plant and animal diversity will plummet. Increasing oceanic acidity will subject marine life and coral reefs to extinction. Reduced availability of water, food and essential resources will lead to social conflict and war as well as massive population migrations. If immediate drastic measures are not taken globally, the tipping point threatening human civilization is not too far away.

These are the projections of a cool-headed body of knowledgeable scientists. There is no doubt that climate change is one of the main existential concerns facing humanity.

This is the year for action. Countries need to commit to net zero emissions by 2050. They need to act now to protect people against the disastrous effects of climate change. UN Secretary General Antonio Guterres (Al-Jazeera 2021).

The Conference of the Parties to the United Nations Framework Convention on Climate Change, designated as COP, is the largest annual official gathering on climate change. Attended by heads of state, ministers, business leaders, mayors, activists, lobbyists, scientists and faith leaders, it discusses the state of the global climate, evaluates the implementation of earlier resolutions and issues the latest UN-endorsed policy on climate change. The 27th COP was held in November 2022 in Egypt. We look at the successes and failures of the COP process larter. For now, we focus on how religious organizations interpret and deal with climate change.

++++

Polls show that 6 of 10 Christians, including conservative evangelicals, and 8 of 10 of people of other faiths in the US hold that responding to climate change is an urgent Congressional requirement. Many faith leaders are raising the issue of climate justice through advocacy directed at the Congress. More than 3,400 faith leaders recently signed a new statement by Interfaith Power & Light in support of prudent investments in climate protection infrastructure. The letter states:

As leaders from many diverse faith traditions, we are united in our call for a bold economic recovery and infrastructure package that creates family and community sustaining jobs while caring for our climate and our neighbors. It is the moral responsibility of our nation, and our sacred task as people of faith, to protect our ecosystems, work for environmental justice and public health, and address the climate crisis. (Graves-Fitzsimmons and Siddiqi 2021).

Buddhism

Buddhist philosophy has two central tenets: Interdependence (nothing exists in isolation) and the ubiquity of dukkha (suffering, disaffection, imbalance). Yet, it is not a pessimistic doctrine because it also provides a path for the alleviation of disaffection through the Four Noble Truths: Dukkha is real. It has a cause. The cause can be uncovered. And dukkha can be transcended by addressing the cause. Climate change is dukkha affecting humanity, life and the ecosystem. Everyone is affected by climate change. Nothing is beyond it. Hence, humanity must heed the call of the scientists and leaders of goodwill to live in a responsible manner. Discarding selfishness and craving, and guarding against disinformation, they must embrace universal compassion, science and simple lifestyles.

The four noble truths provide a framework for diagnosing our current situation and formulating appropriate guidelines—because the threats and disasters we face ultimately stem from the human mind, and therefore require profound changes within our minds. (OES 2015).

The Dalai Lama claims that Buddhism is quintessentially an eco-friendly religion and that were the Buddha alive, he would be green. From the start of the 2010's, he has spoken out for measures to protect the global environment. Invoking the general scientific consensus, he urges global leaders to take urgent steps to avoid the devastating effects of climate change that place millions of lives at risk. He notes that Nepal, where he was born, is seriously affected by climate change. Calling protecting the environment a common human duty, he draws attention to the rapid rate of species extinction, the serious threat to Pacific Island nations of sea level rise, and water shortages that endanger lives across the globe. Many scientists praise his stand.

Thich Nhat Hanh, the late distinguished Buddhist sage, stressed the integral bond between life and the environment. Besides ecological spoilage across the globe, humanity is afflicted with persistent racism, rising economic inequalities and a devastating pandemic. United action is an urgent necessity.

> *Our way of living our life and planning our future has led us into this situation. And now we need to look deeply to find a way out, not only as individuals, but as a collective, a species.* (Hanh 2022).

According to a major Buddhist organization, by increasing our carbon-footprint:

> *Collectively, we are violating the first precept—"do not harm living beings" —on the largest possible scale.* (OES 2015).

Buddhist scholars have called for an overhaul of human consumption patterns and reorientation of the profit-driven economy. They stress usage of renewable energy for meeting energy needs in diverse activities. Protection of rain forests and regeneration of damaged ecosystems are a priority too.

Hanh and other Buddhist luminaries recommend the Zen Buddhist practice of meditation as a powerful tool for building resolve, enhancing resilience and cultivating compassion in the face of the multiplicity of grave dangers we face.

The Time to Act is Now: A Buddhist Declaration on Climate Change is a 2015 declaration on climate change endorsed by the Dalai Lama, Thich Nhat Hanh, and 215 Buddhist scholars, luminaries and senior monks of varied Buddhist traditions from 13 nations. Subsequently it was signed by 2,354 members of One Earth Sangha, an international Buddhist organization. And when it was opened to the public on the Sangha's website, it gathered more than a million signatures of Buddhists and their sympathizers from all around the globe (OES 2015). The declaration was presented to the annual COP as the official Buddhist stand on climate change.

To date, effective action on climate change has been blocked by powerful vested interests. However, the disconcerting gap between words and deeds from quarters that favor strong action to control it does not help either. As with other religions, practical Buddhist actions on climate change do not reflect the fine values embodied in the declarations of their leaders.

> *There are some merits to these Buddhist outlooks regarding climate change, but representatives at COP26 did not always mention these views' shortcomings. For instance, in 2020, Yale University's Center for Environmental Law and Policy ranked 180 countries in terms of positive ecological performance. The average primarily Buddhist nation ranked 102 out of 180, with Burma's finishing next to last at number 179. While diverse factors shape such rankings, these evaluations still make it difficult to accept some of the eco-friendly praises that Buddhism has received.* (CUP 2022).

Hinduism

Hinduism views humanity, nature and the global environment as components of an interconnected whole. Protecting what has been created by the gods is a basic responsibility and central to dharma (divine obligation). Acting in ways that despoil nature affects karma (destiny) and hinders salvation from the cycle of rebirth (samsara). The river Ganges, Mother Ganga, is a goddess that protects humans and the Earth. Its water is holy and is used in religious rituals and for ablution to cleanse a person's sins.

According to the *Hindu Declarations on Climate Change* issued in December 2009 and endorsed by many Hindu dignitaries:

> *Knowing that the Divine is present everywhere and in all things, Hindus strive to do no harm. We hold a deep reverence for life and an awareness that the great forces of nature—the Earth, the water, the fire, the air and space—as well as all the various orders of life, including plants and trees, forests and animals, are bound to each other within life's cosmic web.* (IEF 2022).

The Green Temples Guide is a document issued to promote good environment management for Hindu holy sites where at times millions of worshipers gather for special events.

Of the over 1.3 billion people in India, about 80% are Hindu. And the latter form just more than 90% of the global Hindu population. The topic of Hinduism and climate change is basically equivalent to that of India and climate change. Presently, India is ruled by the BJP party under Prime Minister Narendra Modi. The BJP policies are based on Hindutva, a fundamentalist Hindu doctrine. The aim of Hindutva is to make India a great and pure Hindu nation. Accordingly, the climate change policies of the BJP government are said to be inspired by the Vedas and other ancient Hindu texts.

Environmental management and climate change are critical issues for India. Of the 25 most polluted cities in the world, nine are in India, with the capital city Delhi ranking at number 2. Heat waves and noxious gases cause many deaths in urban India. Indian cities also face serious water shortage on a regular basis. For rural women and girls, ensuring minimal availability of water for the family is a daily back-breaking undertaking (RPS 2022). Overall, less than half of the nation's population has access to safe, piped water.

India suffers from a serious lack of adequate sanitation facilities, ranking the worst in the region. Despite recent progress, only a third of the population has access to good sanitation and more than 20% still defecate on open ground. Rows of people crouching early in the morning on their pots on the street is an ugly but common sight in many Indian cities. Communicable diseases take a heavy toll, and about half the children are poorly nourished.

Prime Minister Modi made a surprising pledge during his election campaign: *Toilets before temples*. His government has plans to avail safe water to each household by 2024, enhance the national electricity supply, and embark on a large scale toilet building program. Urban pollution is to be reduced by income tax rebates for the purchase of electrical vehicles, and the carbon footprint to be lowered through usage of renewable sources for 40% of the energy needs by 2030, especially through production of affordable solar panels.

But words and actions do not rhyme. The populist BJP government serves three masters, the plutocrats, the Hindu fundamentalists and the public, effectively in that order. Its neoliberal policies often are not in consonance with sound, long-term environmental management in the public interest.

> *India's water crisis is often attributed to lack of government planning, increased corporate privatization, industrial and human waste and government corruption.* (Snyder 2022).

Hindutva is a divisive, hate-generating ideology. It pits the majority Hindus against Muslims and Christians, and upper caste Hindus against the Dalits. It can hardly provide a basis for sound climate change policies. To quote an analyst:

(i) Hindutva advances spurious science, thereby structurally impairing prospects for combating climate change; (ii) The advancement of Hindutva culture wars distracts attention from pressing issues, including climate change; (iii) Given Hindutva's authoritarian quality, it presents a bigger black box compared with transparent, inclusive, accountable democracy; (iv) The tensions that Hindutva nurtures, regionally and internationally, weakens the capacity for concerted action against climate change. (Chandra 2021).

Under the BJP, increased investments in mining accelerates deforestation and semi-legal construction impedes natural water flow. Dams and conservation channels are managed poorly. and flood control is ineffective.

There is considerable opposition to the BJP within India. Many Hindus stand in unity with non-Hindus. Yet, the dominant outlook among the Hindus is pro-Hindutva and not reflective of the fine declarations on climate change issued by Hindu organizations and the BJP.

Islam

According to scholars of Islam, the Quran and the scriptural texts collectively known as Hadith, enjoin Muslims to cultivate respect for nature and all life. They represent the magnificent creations of Allah. Humans must protect what has been mercifully bestowed upon them by Allah. Living in gentle harmony with nature, they should be mindful of animals, plants and trees, use water in a prudent way, and guard the environment to enable the generations to come to live in safety and comfort.

> *There is not an animal that lives on the Earth, nor a being that flies on its wings, but they form communities like you. Nothing have we omitted from the Book, and they all shall be gathered to their Lord in the end.* Quran 6:38 (Zafar 2021).

Several major Islamic organizations convened an international symposium on climate change in 2015 and issued the *Islamic Declaration on Global Climate Change*. Urging Muslims to recognize the damage humans have caused through materialistic, profit-driven pursuits, the Declaration implores them to act in consonance with their faith and redress the effects of climate change. Its recommendations are consistent with the guidelines developed by the UN. Because the Declaration poses climate matters as faith-based issues, Muslims are generally receptive to its proposals.

Such environment friendly invocations are particularly apropos because many nations with large Muslim populations are seriously affected by climate change. The effects are expected to magnify over time. The nations of the Middle East, where Muslims comprise the majority, experience extreme heatwaves each year. In the not-that-distant a future, many may become bereft of flora and fauna. Temperature rise, combined with lower rainfall, is progressively making Turkey a seriously water-stressed nation. Storms and flooding may displace more than 15% of the population of Bangladesh.

Yet, Muslim nations are doing little to control climate change. Even the call for meat-free meals often falls on deaf ears.

> *However, despite their vulnerability, many Muslim countries are contributing to the problem. Indonesia, the most populous Muslim-majority country in the world, is the world's fifth-largest emitter of greenhouse gases, and is doing little to curb emissions. Bangladesh and Pakistan are the two most polluted countries in the world, but have taken no serious measures to*

address pollution. Inaction in the Muslim world persists despite a declaration by Muslim countries in 2015 to play an active role in combating climate change. (Ozdemir 2020).

In the nations of the West, awareness of climate change is generally high among the various religious groups, including Muslims. Many of them demand urgent action. In a poll conducted in 2021 in the US, some four out five Muslims sided with other faith groups to demand urgent congressional action on climate change. But in the Muslim nations of the Global South, where most of the Muslims live, neither climate change awareness nor activism is politically noticeable.

Saudi Arabia, the nation where Islam was born, houses the two most holy cities in Islam. Over a million Muslims visit the nation for the annual Hajj pilgrimage. In this absolute monarchy, Sunni Islam is the official religion. The national motto of Saudi Arabia is

Lā ʾilāha ʾillā Llāh, Muḥammadur rasūlu Llāh
There is no god but God; Muhammad is the messenger of God.

Ascribing to the orthodox Wahhabi tendency in Islam, its national emblem is

The Saudi Arabian National Emblem

With over 90% of the population Muslim, Islam permeates all facets of Saudi society and polity. Citizens and non-citizens are expected to abide by Islamic restrictions, and penalties for violations are severe. Public beheading is a part of life here. With vast reserves of oil and gas, Saudi Arabia is the largest exporter of oil in the world. Its highly conservative regime is staunchly allied with the West. Despite a very poor human rights record, it imports billions of dollars worth of sophisticated weapons and garners advanced military training from the West on a regular basis.

The mostly desert nation is marked by very hot days and very cold nights. Summer temperatures reach up to 54° C. Yet, it has made remarkable progress in agricultural development in the past fifty years. It is not only self-sufficient in basic food items but also exports a variety of farm and dairy products. Modern technology has enabled achievement of high productivity in agriculture and dairy industry. Famed for specialty dates, the desert nation has nearly 20 million olive trees.

The Saudi government projects itself as an environmentally forward-looking state committed to sound environmental policies for the Middle Eastern region. In 2021, it launched the Saudi Green Initiative and the Middle East Green with plans to restore biodiversity, plant ten billion trees, generate forests, create parks and improve the soil on 40 million hectares of land. The key aim is to sequester a large volume of carbon dioxide from the atmosphere and control global warming. Launching these laudable initiatives, the ruler of the nation declared:

255

As a leading global oil producer, we are fully aware of our responsibility in advancing the fight against the climate crisis, and that just as we played a leading role in stabilizing energy markets during the oil and gas era, we will work to lead the coming green era. Crown Prince Mohammed bin Salman (Radwan 2022).

But there is a much less-publicized, not-so-green angle to the story that relates to Aramco, the state-owned Saudi oil and gas corporation. With annual sales at nearly US$350 billion and valued around US$1.7 trillion, it is the wealthiest commercial entity in the world. And its record on climate change is nothing to be proud off. From 1965 to 2017, Aramco emitted nearly 60 million tons of carbon dioxide into the atmosphere, accounting for nearly 5% of the global total. By most estimates, it was the first or second top corporate global greenhouse gas emitter over this period. Though it plans to invest US$1.5 billion in renewable energy projects, it also expects to develop new oil fields and increase oil output by 7% over the next 15 years.

In line with their financial interests, Saudi Arabia and Aramco are parties to campaigns that undermine sound international agreements on climate change. Under the effective leadership of the US government, they deploy their financial muscle to lobby against, and engage in disinformation drives to side-track the international efforts at reductions in greenhouse gas emission.

Green washing is a time-honored strategy used by the polluting corporations. They invest in some clean energy initiatives, support a few sound climate change policies, and re-brand themselves as socially responsible entities while at the same time expand fossil fuel emitting activities, fund climate change disputing think tanks and adroitly dispute the reports and recommendations of the IPCC. In tandem with energy giants like Exxon-Mobil and Shell, Aramco has proven to be a master of green washing.

The UK advertising regulator received over 60 complaints about Aramco's advertising in which it claimed to be "powering a more sustainable future" in April 2020. (CE 2022).

In stark contrast to its current splashy green washing campaign, Saudi Arabia has played a quiet yet powerful role in thwarting proactive climate policy at United Nations conferences and U.S. domestic policy battles alike. (Fang and Lerner 2019).

The reality is undeniable. The deleterious effects of climate change such as heat waves and blinding sand storms are appearing more frequently and intensely in Saudi Arabia. Yet, the nation remains an obstructionist force against sound global climate change policies at the global level. On the domestic front, however, it has taken noteworthy initiatives to protect itself from the calamities expected from the rise in average global temperature. The glaring contradiction between local and global priorities reflects the neoliberal profit-driven mindset of the ruling class.

++++

The common theme of the declarations on climate change issued by Islamic organizations is that Allah bestowed upon humans the responsibility of protecting His domain on planet Earth. Muslims need to abide by scientific proclamations on climate change matters and adjust their conduct accordingly. There is no contradiction between science and Islam. That is the message of the scriptures.

Muslims, especially in the Western nations, who tend to be better informed, generally accept this message. In theory it is accepted by the governments of Muslim-majority nations. But their practices place them in the ranks of climate change deniers. A few good initiatives based on privatization and neoliberal policies hardly amount to sound implementation of the recommendations of the UN bodies on climate change. For Islamic nations, profit triumphs over religion, as it does for Hindu and Buddhist nations. The primary hurdle thwarting effective policies on climate change is not religion but neoliberal capitalism.

Christianity

On climate change, Christianity stands apart from Hinduism, Buddhism and Islam on two fronts. Fewer Christians than people of the other three faiths deem climate change a priority, and there are sharper divisions among Christians about whether human induced climate change is real or not. These assertions are more valid for the United States. In one nationwide year 2021 poll, while some 8 out of ten Hindus, Buddhists and Muslims held that climate change should be *'a top or an important priority for Congress,'* only 6 out of ten Christians did so. Another, more detailed poll conducted around the same time had similar findings. It noted that the support level was lowest among white evangelical Christians (5 out of 10). (Graves-Fitzsimmons and Siddiqi 2021). Yet, it is encouraging that in a nation that is the headquarters of climate change deniers, the support level in any faith group did not fall below 50%.

The views of Christian groups on climate change span the spectrum from hard-core denial that climate change is human-induced to virtually full acceptance of the international scientific consensus. All these tendencies cite the Bible as the authority for their stand. We give a few examples.

Jehovah's Witnesses, a Christian denomination with several beliefs that are not supported by mainline Christianity and about 9 million members, holds that the day the Earth will be cleansed of evil-doers is not far away. It also believes that the conglomeration of problems faced by humans can only be resolved through the establishment of the kingdom of God on Earth. Invoking the Biblical prophecy that human action will one day bring the Earth to ruin, it sees climate change as one instance of the prophecy. As the Bible says, many humans will act in irresponsible, greedy, materialistic and destructive ways and continue to do so in the face of mounting problems. It cites the general unwillingness of people to modify their life style and the lack of genuine resolve and cooperation among governments as confirmation of the words of the Bible.

Yet, Jehovah's Witnesses almost fully accept the diagnosis, prognosis and prescriptions for climate change provided by the IPCC, UN and WHO. And in the face of the impending calamities, it maintains an optimistic stand.

> *The Bible reveals that Jehovah, a God, our Creator, is committed to our planet and those who inhabit it.* (JW 2022).

Accordingly, Jehovah will reverse the damage caused by wicked humans to our planet, control extreme weather patterns, and teach people to take good care of the global environment.

Christianity.com, a mainline Protestant outlet, presents climate change in terms similar to those noted above. An opinion piece on the outlet notes that climate change is not mentioned in the Bible. But the holy book predicts that the Earth will become hotter and hotter, portending a final destruction. Christians are urged to heed the concomitants of climate change like extreme heatwaves, melting glaciers and icecaps, altered timings of seasons and rise in sea levels.

> *Scientists have studied the planet and the surrounding atmosphere, which is why we should respect their work and hear what they have to say. Science attests to God, as God is the One who invented science.* (Becker 2021).

Christians need to be cognizant of their God given responsibility to cherish and protect His Earthly creations. Denying climate change is fruitless. Instead, they should play their part by recycling, using efficient light bulbs, reducing usage of fossil fuels and similar climate friendly devices.

The Bible predicts in some detail that the Earth and the heavens will undergo a process of total destruction. But after that, God will recreate everything anew.

> *The New Heaven and New Earth will be a place of perfection, happiness, and joy in the Lord. There will be no more fear, crying, or pain.* (Becker 2021).

Christians should be mindful of their role as the stewards of planet Earth, and act prudently in a knowledgeable fashion.

Pope Francis, the global Catholic leader, is the most outspoken and progressive religious dignitary on the subject. Repeatedly stressing the warnings of scientists and climatologists, he has urged people to act with the urgency attending to matters of life and death. At the Earth Day Summit in 2021, the Pope said:

> *We need to keep moving forward and we know that one doesn't come out of a crisis the same way one entered. We come out either better or worse. Our concern is to see that the environment is cleaner, purer, and preserved. We must take care of nature so that it takes care of us.* Pope Francis (Graves-Fitzsimmons and Siddiqi 2021).

Noting the policy level lethargy, American Catholic organizations came together to urge the US President and Congress to allocate more funds for the problem, participate seriously in UN deliberations and abide by the international accords.

Catholic Charities of Southwest Kansas, a campus-based US Catholic organization, recently issued a statement that reflects the general stand of the Catholic church on climate change. It begins by reiterating the key facts about climate change, namely increased temperatures, higher sea levels, melting of glaciers, severe droughts, more deaths from malaria, diarrhea, poor nutrition and heatwaves. The poor everywhere, who bear the least responsibility for global warming, disproportionately bear the brunt of climate change. Students are urged to become climate activists in their personal and public lives.

> *Be energy efficient. Calculate your carbon footprint. Adjust the thermostat. Adjust your driving routine. Go meatless on Fridays. Recycle More. Use less water. Petition policy makers. Write to your local newspaper. Form a study or action group at church. Start a "Wash and Dry" team in your workplace. Turn off your office. Get involved on campus.* (CCSWK 2022).

Answers in Genesis, a think tank reflecting conservative Christian views, has a somewhat differing perspective on climate change. Writing on its website, Andrew Foley decries the frequent alarmist media headlines on climatic events. Instead of being unduly fearful, he urges Christians to seek guidance from the Word of God. While firmly believing that *God is ultimately in control*, they should not adopt a lackadaisical or dismissive attitude or carry on with business as usual. They are obligated to protect God's creation to the best of their ability.

[Throughout] *Scripture, we see the balance of man's responsibility and God's sovereignty. They aren't mutually exclusive; they go together. Despite God's ultimate sovereignty, we have an obligation—given to us by our Creator in the Garden of Eden (Genesis 1:28) and never revoked—to have dominion over God's creation as his stewards, made in his image.* (Foley 2020).

Christians should eschew fear, recall the Biblical promise that the pattern of seasons will persist, and assist the poor and the needy. His concluding words, however, betray another agenda, and are music to the ears of political right wingers and big corporate polluters:

[The] *ultimate answer to environmental issues isn't more laws, bigger fines, or more power to the government. The answer is the gospel of Jesus Christ!* (Foley 2020).

Got Questions Ministries is, according to its website, a *Christian, Protestant, evangelical, theologically conservative, and non-denominational* organization dedicated to providing valid Bible-based answers to queries about religion. (GQ 2022). To the question: *How should a Christian view climate change?*, it has a succinct, forthright response. It begins by noting that scientists have shifted from talking about 'global warming' to talking about 'climate change'. They are not sure about what is really going on. Their terminology reflects the use of scare tactics to gain public attention.

While some scientists are doing honest research, others are driven by a fixed mindset. Some scientists have an eye towards financial gain. Hence, Christians should view climate change:

skeptically and critically, but at the same time honestly and respectfully. (GQ 2022).

What does that mean in concrete terms?

Is there anything wrong with going green? No, of course not. Is trying to reduce your carbon footprint a good thing? Probably so. Are renewable energy sources worth pursuing? Of course. Are any of these things to be the primary focus of followers of Jesus Christ? Absolutely not! (GQ 2022).

The National Association of Evangelicals (NAE) is a US-based association of over 40 Protestant denominations representing more than 45,000 local churches. Members regard the Bible as *'the inspired, the only infallible, authoritative Word of God'*. Hitherto, they were among the faith groups least likely take climate change seriously, let along engage in activism to ameliorate it. Many evangelicals reject science and invoke theories alluding to elite conspiracies about climate change.

But that stand changed in August 2020 when the NAE issued a detailed report that accepted the scientific consensus on climate change, and called it a global emergency needing urgent action. It noted that while the Bible does not directly relate to scientific research, the Bible has moral principles like caring for God's creation, loving one's neighbors and being generous that should guide Christian action during environmental and other crises.

Drawing attention to increasing deforestation, air pollution, child health and melting ice sheets, the report says that poor families are the most likely to be adversely impacted by climate change.

Creation, although groaning under the fall, is still intended to bless us. However, for too many in this world, the beach isn't about sunscreen and body surfing but is a daily reminder of rising tides and failed fishing.

Instead of a gulp of fresh air from a lush forest, too many children take a deep breath only to gasp with the toxic air that has irritated their lungs. NAE President Walter Kim (Jenkins 2022).

According to the report's main author, evangelical youth are increasingly perturbed by the climate crisis and resorting to social activism in consonance with Christian teachings. He sees that as a most hopeful sign.

Pastor Michael F Chandler, writing in the *Victorville Daily Press* addresses the key question: What does the Bible say about climate change? He begins by noting that God is supreme, majestic, compassionate and powerful, higher than all. Delighting in his splendor and grace, humans should worship him and be thankful. He invokes a Christian song relating to climate change that featured at the 2021 UN Climate Change Conference in Glasgow, Scotland. The first part is:

The Climate is Changing by Carolyn Winfrey Gillette
The climate is changing! Creation cries out! Your people face flooding and fire and drought. We see the great heat waves and storms at their worst. We pray for the poor, Lord — for they suffer first. We pray for the animals here in our midst who cannot defend their own right to exist. We pray for the mountains and forests and seas that bear the harsh footprint of our human greed. (Chandler 2021).

To Chandler, the song is blasphemous because it places God's creation over God.

Friends, the current language of climate change is incompatible with Biblical teaching. It is altogether godless and ascribes to little man far more power than invested him by the Creator. (Chandler 2021).

God and the Bible over humans and science, that is his clear message.

Christian Enquiry Agency, a British Christian organization, has a different message. Observing that climate change is a global emergency, its November 2022 statement posits that unsustainable and wasteful usage of Earth's resources as the primary cause. Christians have abdicated their spiritual duty to be the stewards of planet Earth. Though the situation is grave, the Bible provides room for hope. Upon a detailed exploration of Biblical passages, it is declared:

Christian faith and teaching enables reflection on the current climate and environmental challenges. Scientists and campaigners, even those who do not profess a faith, have commented that we need a new sense of spirituality to reform our relationship with our planet's ecosystem. The climate crisis is multi-dimensional and consequently we will find a range of themes in Biblical teaching that appear pertinent to our time. Christians have not always been alert to the ecological crisis but as we look on God's creation today, its beauty and its damage, we find the bible speaking to us in fresh and different ways. (CEA 2022).

In a marked contrast to the declarations seen earlier, the CEA emphasizes global systemic change. The dominant economic model that uses oil, gas and coal for energy needs to be replaced by one that promotes renewable energy sources, recognizes the major responsibility of the industrialized nations for the problem of climate change and allocates sufficient resources for the poor nations to address climate issues and, at the same time, improve the living standards of their people.

The CEA credits most UK churches for divesting their funds away from oil, gas and coal companies and demanding stronger action from the UK government. The churches are lauded for taking part in the Eco-church program and mounting climate awareness programs.

> *A growing environmental awareness supported by our knowledge of God's love for the whole of his creation enables us to engage with the difficult challenge of lifestyles. The path may be long but we each know that we are not in this alone.* (CEA 2022).

Pope Francis sums up the general Christian perspective on climate change:

> *We are stewards, not masters of our Earth. Each of us has a personal responsibility to care for the precious gift of God's creation.* Pope Francis (CCSWK 2022).

++++

The overall message of this chapter is that on an issue as fundamental for humanity as climate change, science and religion can amicably work together. Despite having different justifications for their actions, religious people, leaders and organizations have little compunction with abiding by the science-based measures proposed by the UN and international conferences. Protecting the creations of God (Allah, Brahman) is a primary spiritual obligation, sanctioned by their holy scriptures.

While most people of faith ascribe to this viewpoint, an intransigent minority of conservatives deems climate change analysis of scientists as an exaggeration or fraudulent, deriving from ulterior motives. This minority is allied with major energy corporations, right wing think tanks, super-wealthy individuals and right wing politicians. Their influence extends far beyond their numbers.

The problem with religion based denial of climate change is not theology. It is neoliberalism and the extremist religious forces that are wedded to the neoliberal forces in society.

6.4 EMERGENT VISIONS

The multiplicity of economic, social, health-related and environmental crises of the modern era has generated new visions of the relationship of humanity to nature and the relationship of human communities to each other. Implicit to each vision is a notion of how religion relates to science. Many of these visions are enhanced reformulations of earlier ideas. But the context is new, and the scholars have the liberty to draw upon the failures and successes of the older visions. Prior to synthesizing their essence, we present a few articulations of these visions. First, a definition:

> *The **Anthropocene Epoch** is an unofficial unit of geologic time, used to describe the most recent period in Earth's history when human activity started to have a significant impact on the planet's climate and ecosystems.* (NGL 2023).

Earth scientists generally accept to date 1950, the year of significant post WW II acceleration of global economic activity, as the start of the Anthropocene Epoch. Within a few decades, this epoch witnessed the dominance of the consumerist neoliberal economic system across the globe. It is also an era of major increases in global usage of fossil fuels. The British and US engineered overthrow of the democratically elected, reformist government of Mohammad

Mosaddegh in Iran in 1953 constitutes a major signpost of that era. We are the masters of the global energy resources: that was the unmistakable imperial message.

Karen Armstrong, an eminent authority on world religions, presents an integral, sacral vision of the relationship between humans and nature, and between science and religion (Armstrong 2022). According to her, over the course of evolution, humans developed participatory reciprocity with nature. They were dependent on nature but were also a part of it. Societal and natural realities were inextricably intertwined. Nature was a living entity permeated by an '*immanent sacred force*,' an entity to be respected and revered.

This sense of a solid bond with nature weakened with the development of science and technology. The Anthropocene Epoch has transformed it in a basal manner. Urban living and all-pervasive technology have drastically altered the global landscape and biosphere. Psychically and physically alienated from nature in a progressive fashion, people are losing their ability to appreciate its intrinsic beauty. The rapid pace of change has generated a mindset in which the external world appears like a 'simulated' reality, an entity as amenable to alteration as a computer image. No longer sacred, nature is a mechanical, external entity, a resource, a commodity. Dominating nature is the key to human progress.

Traditional religions embodied the idea that humans were integral to nature. There was no artificial separation between the two. The rituals of these religions were endeavors to restore the disruption of harmony with nature generated by human activity. They cultivated a higher sense of consciousness and restored the awareness of physical and psychic immersion into nature. The shamans were the conduits through whom people entered such a state of consciousness.

Armstrong stresses that human perception of nature underwent a foundational change during what is known as the Axial Age, roughly from 800 to 200 BCE. Large cities within powerful empires emerged during this period in different parts of the world. It was an era in which diverse, expansive forms of systematic world outlooks emerged.

> *In Greece, we saw the likes of Socrates, Archimedes, Hippocrates, Plato, and Aristotle. In the Middle East, we got the Jewish prophets like Isiah and Ezekiel, as well as Zoroaster in Persia. In India, we got the Buddha and the writing of the Hindu Upanishads. Whilst in China, Confucianism and Daoism came into their own, as well as the famous Sun Tzu.* (Thompson 2021).

Despite their differences, these outlooks stressed thinking and reflection, and the need to question authority. Autonomy and rights of individuals and communities were emphasized. People had the right to question the power of the emperors and kings. The concepts of truth, common sense, morality, justice and of philosophy as a venue for exploring the essence of reality were borne of the Axial Age. The foundation for modern science was laid. But at this stage, science was both a tool for exploiting nature and a means for reflective, spiritual interactions with nature.

The Axial Age outlook was affirmed and weakened as monotheistic religions, especially Christianity and Islam, secured a large base. Nature remains a sacred entity, but one that was created by God or Allah. And humans are designated as the superintendents of nature. They must respect it, but are no longer integral to it in the traditional sense of the term. They rule over it on behalf of the almighty.

Armstrong posits this domineering vision of nature as a primary affliction of modern society. It is not enough to see major problems like climate change to be due to inappropriate human meddling with nature, problems that can be resolved by applications of science and technology. No, the essence of how humans relate to nature has to be changed in a way that restores the sacrality of nature. The arrogant attitude towards nature, that we can alter it according to our needs, has to be discarded. The integral psychic bond between humanity and nature must be revived and reinforced. And it has to be done without sacrificing the key gains of the Axial Age like rational and critical reflection, codes of morality, and universal justice. A higher form of consciousness, both mystical and rational, has to be cultivated.

> *We think nothing of driving our cars to shops when we could easily walk or use public transport. We seem to regard flying as a human right, even though we know it pollutes the atmosphere.* (Armstrong 2022, page 166).

Armstrong sees the meditative practices found in diverse religions—Kabbalism, Islam, Christianity, Confucianism, Daoism and Hinduism—as aids for restoring the ancient sense of sacred integrality and love with nature. Humans should create social structures to propagate such feelings. A new set of values and moral code is apropos. '*Haughty individualism*' must be replaced by compassion, justice and equity. The Golden Rule—Do not do unto others what you would not have done to yourself; and do unto others what you would like to be done to you—which is central to most religions, must guide human conduct and official policies. Our planet is a fragile entity. We need to uphold moderation, balance, tolerance, respect for human dignity and worthiness of all living beings as central life tenets.

> *Awe at the natural world and unity between humanity and nature are beautifully expressed in the image of human beings forming a trinity with heaven and Earth.* (Armstrong 2022, page 133).

Individualism, the sense of personal entitlement, can be overcome with spiritual immersion within oneself. Meditation and yoga are not just mental exercises but a means of attaining a state of repose, wholeness and compassion.

> *Prayer cannot be a private union with a supernatural God; it must include our fellow humans and 'the things' of nature.* (Armstrong 2022, page 171).

Albert Einstein believed with Baruch Spinoza that God is nature and nature is God. That pantheistic vision regards the elegant laws of nature with awe and accords them a spiritual status. Armstrong augments it by invoking the integrality and sacrality of nature. In combination with love for fellow humans, a blend of these visions symbolizes the scientific spirituality and philosophy for the modern era. It neither dismisses traditional religions and nor does it make a foundational compromise with it. It favors a setting whereby religion and science can coexist peacefully without either feeling that it has compromised a basic aspect of its tenets and values.

Mary Evelyn Tucker, an acknowledged expert on ecology and religion, holds that the Universe is a single interconnected entity marked by a progressive rise of more complex entities from less complex entities. Though a universal process, formation of complexity does not necessarily induce hierarchy. Humanity is the most complex entity, living or non-living, but it does not have a divine right to rule over everything less complex. The religious and science-driven visions that posit humans at the top of the hierarchy reflect human hubris. But in the former case, it is tempered by God's ultimate authority. The rapid progress of technology in the Anthropocene Epoch has strengthened that hubris and inflicted extensive harm upon the environment.

Adopting a cooperative vision of the relationship between humans and nature, Tucker implores humans to realize that they are a part of a unitary reality. They must change their sense of identity and, as in ancient times, inculcate the sacral integrality with nature into the core of their consciousness. A sustainable future demands recognition of the '*evolutionary unity of the Universe, and humans [being] a part of it*'. (Moore and Tucker 2015). And scientific endeavors need to foster, not endanger, that unity.

The Penultimate Curiosity, a collaboration between scientist **Andrew Briggs** and art historian **Roger Wagner**, provides a fresh perspective on how science has been linked to religion (Wagner and Briggs 2016). Starting from the observation that curiosity is a primary trait acquired through evolution, they focus on two human curiosities: religion and science. The former asks why; the latter asks how. Religion is the ultimate curiosity, and science, the penultimate curiosity that swims behind the currents of the first curiosity. A rich array of case studies illustrate that this relationship is integral to and beneficial for both religion and science.

Religion and science are both attempts to explain the world. Each uses its own distinct method and rationality. Each has its social function. None can dismiss the other in an outright manner. They are the two sides of the same coin. And they feature common queries: Who are we? Where did we come from? As humans formed settled civilizations, they expanded intellectual endeavors to understand nature, a process that gave rise to elaborate, institutionalized religion and science. The two endeavors were often in the same hands at the outset.

The thesis of a permanent conflict between science and religion, as argued by Bertrand Russell and Francis Crick, is fallacious. The conflict appears as such only upon superficial examination. In reality, science and religion have bonded to each other in a '*mutually enriching relationship*'. And despite the divisive role of a few hot-button issues like creationism and the Theory of Evolution, indications are that the primarily harmonious relationship between religion and science will endure.

Contrary to earlier European prejudices, humanity is a single race descended from an ancestral African mitochondrial Eve. This is not just a matter of science. Rejecting a common human ancestry was used to support slavery and racism. There were scientists and theologians on both sides of the fence. Non-European ancient societies had far more complex ideas about God and religion than acknowledged by the early European scholars.

The take home message of the *Penultimate Curiosity* is that humans have to foster a harmonious relationship between science and religion and, at the same time, promote the unity of humanity.

In a path-breaking work, *The Social Instinct*, **Nichola Raihani**, an expert on evolution and behavior, makes a persuasive case that natural evolution has as much been driven by cooperation as by competition, both between and within species. Interdependence and cooperation are both critical for fostering survival and progress of species and communities. Cooperation also assists the selection of beneficial genes. By harming the environment, it may also play a regressive role in the broader context. For a sustainable, egalitarian future on this planet, humans must harness their competitive and cooperative spirits in a balanced, morally justified and science-driven manner. (Raihani 2021).

Amitav Ghosh is an award-winning author of grounded and exquisitely framed historical fiction, and poignantly elegant non-fiction books about major problems facing humanity today. His latest treatise, *The Nutmeg's Curse*, is a cogent rendition of climate change that covers critical social and economic topics usually avoided even by prominent scholars.

Building upon the environmental themes addressed in his earlier works, and organizing around the social history of the spice nutmeg, Ghosh makes a persuasive case that the environmental crises of today are a product of the development of European capitalism and colonialism.

In the sixteenth century, the seed and the covering of the nutmeg were coveted by Europeans traders. European powers competed to control the source of this highly profitable commodity. Through their ventures in Asia, the Dutch came to learn that nutmeg was cultivated only on the Banda Islands of Indonesia. *The Nutmeg's Curse* vividly describes the gruesome and duplicitous history of the Dutch attempts to bring the area under its control. To the colonists, only the profitability of the venture counted. The harm they inflicted upon the indigenous population and the environment were secondary, They even sent their forces to other islands where their competitors were trying to grow nutmeg and hacked down the nutmeg trees.

The nutmeg illustrates the central facet of modern civilization: extraction and consumption under a market-driven economy, an economy enabled by an ever-expanding use of fossil fuels. Over two centuries of this economic model has brought the planet to the verge of an existential environmental crisis. Yet, humanity is largely unperturbed.

> *Much, if not most of humanity today lives as once colonialists did—viewing the Earth as an inert entity that exists primarily to be exploited and profited from, with the aid of technology and science.* (Ghosh 2021).

The traditional, precolonial economies were driven by a conservationist ethos that strove to preserve a balanced relationship with the environment. Striking natural objects and resources were viewed in an animated light. Rivers, thus, were carriers of ancestral spirits. Embodying that outlook, shamans and religious movements were often in the forefront of the struggles against the colonialists. Modernity has discarded the sacrality of nature. It can be harnessed as we see fit. Arising problems can be fixed by technology. Observing that the carbon dioxide levels in the Earth's atmosphere are reaching 'unmanageable' levels, Ghosh sees this mentality as the 'collective delusion' of capitalism. He also emphasizes that what is seen as progress continues to be accompanied by injustice, hunger and ill-health in many nations.

Ghosh is almost unique among the authors concerned with climate change to draw attention to the large volumes of military driven emissions of greenhouse gases. A comprehensive assessment reveals that the scale of such emissions is stupendous. Considering the US military only:

> *The [US Department of Defense (DOD)] is the single largest consumer of energy in the US, and in fact, the world's single largest institutional consumer of petroleum. Since 2001, the DOD has consistently consumed between 77 and 80 percent of all US government energy consumption. ... Indeed, the DOD is the world's largest institutional user of petroleum and correspondingly, the single largest institutional producer of greenhouse gases (GHG) in the world.* (Crawford 2019).

For the military, reduction of greenhouse gases is not a primary issue. As one senior military officer stated:

> *We are in the business of protecting the nation, not the environment.* (Ghosh 2021, page 127).

Strategic military planning regards climate change as an inevitable aspect of the future landscape, a 'threat multiplier' that will generate its own forms of conflicts and crises, and calls for adequate budgeting to deal with such threats. Many nations are increasingly deploying the military to assist with the rescue and relief efforts when dealing with floods, hurricanes and wildfires. Civilian agencies that used to be at the forefront are displaced by the army. Environmental crises thereby engender militarization and increasing militarization engenders long term environmental degradation.

Ghosh warns that in this security driven political landscape, major powers are not ignoring climate change. They are addressing it, but hardly as directed by the UN and the IPCC.

> *[It] is a grave error to imagine that the world is not preparing for the disrupted planet of the future. It is just that it is not preparing by taking mitigatory measures or by reducing emissions; instead, it is preparing for a new struggle for geopolitical dominance.* (Ghosh 2021, page 129).

Amitav Ghosh presents one of the most lucid and comprehensive analyses of climate change. He categorically declares that the roots of the problem are the extractive, profit driven neoliberal system, the mentality of reckless consumption that goes along with it, and the imperial ventures by the dominant nations the frame current international relations.

While Ghosh presents an evidence-based case, his writings underscore the importance of spirituality of the precolonial communities that regarded nature as a sacred entity. Not a pessimist either, he is heartened by the numerous national and international movements that have sprung up to deal with local and broader environmental problems. And he is particularly aware of the important role played by progressive religious and spiritual leaders in this effort.

> *Pope Francis speaks directly to more than a billion people, and has already done more perhaps, to awaken the world to the planetary crisis than any other person on Earth.* (Ghosh 2021, page 244).

According to **Eduardo Gudynas**, the Director of the Latin American Center for Social Ecology, over a century of expansive mining, agriculture, forestry and construction, mostly export oriented, has generated *'an avalanche of environmental, territorial, and social problems'* in Latin America (Gudynas 2018). These extractivist activities have induced major social tensions and armed conflict, affected treatment of marginal and poor communities and respect for human rights, and slowed down the dismantling of military dictatorships and the instituting of democratic reforms.

This period witnessed outbreaks of discourse involving many social actors about governmental and commercial environmental policies. Gudynas attends to a neglected aspect of this process: Many religious institutions and indigenous traditions have been party to these tensions, conflicts and debates. Environmental activism has been linked with the struggles for human rights, and indigenous leaders and priests have been involved in both. Conservative Christian groups have backed the forces engaged in violations of human rights and environmental degradation. For example, in 2008:

> *[a] small number of people [appealed] to religion* in defense *of extractivism, as in Colombia in the conference series 'Christianity and Mining,' which, with the participation of government officials and businessmen, promoted the mining industry, downplayed its impact, and cited the story of Genesis to legitimate it.* (Gudynas 2018).

The case of the former President Jair Bolsonaro and his fundamentalist Protestant supporters is apropos. Yet, Christian churches have also sided with indigenous communities to support the protection of the environment, human rights and the right of the local people to improve their living conditions free from corporate domination. Many progressive environmental and human rights activists had religious roots and the words of Pope Francis have inspired and reinforced environmental and social activism. One example is the formation of the Churches and Mining Network that seeks to address the adverse impact of mining on the ecosystem and communities.

But there is a fundamental limitation affecting the philosophy underlying progressive activism, secular or ecclesiastical: It is too anthropocentric. Humans should care for nature. Yet as the stewards of God's creation, they stand above it. It is a philosophy that in practice has legitimized reformist policies of liberal and pseudo-socialist governments in Latin America. Progressive leaders like Evo Morales of Ecuador mouthed respect for indigenous values and peoples, yet focused on mining and other extractive activities that eroded indigenous cultures. Emphasis was on expanded welfare programs for the poor, not the right of the people to determine their own future. In the process, environmental activism became detached from social justice activism.

A radically different outlook based on the idea of *Buen Vivir* that is prevalent among indigenous communities is needed. *Buen Vivir* is a biocentric perspective that accords value to nonhuman nature independent of its value to humans. Nature is under the purview of the all-powerful goddess Pacha Mama (Mother Earth) who has designated humans not as overseers of nature but as an integral part of it. Nature has rights too. A popular interpretation of *Buen Vivir* is having a good life while respecting the rights of nature.

Since *Buen Vivir* doen not posit humans as the central elements of nature, it is at variance with the Catholic and Protestant outlooks on nature. And respecting the rights of nature is not possible under the profit-driven consumerist neoliberal framework. In its most radical form, *Buen Vivir* incorporates a solid critique of global capitalism. Though akin to Liberation Theology in many aspects, in the equal standing it accords to nature, it transcends Liberation Theology.

A New Synthesis

Karen Armstrong, Mary E Tucker, Andrew Briggs and Roger Wagner, Nichola Raihani, Amitav Ghosh and Eduardo Gudynas present refreshing, evidence-based visions on the relationship between humanity and nature with implications for the linkage between science and religion. Joined by a host of other scholars, climate change is a central issue in their discourse.

We synthesize several key principles from this discourse. One, humans are integral to nature. Nature is sacral, to be harvested in a harmonious fashion. Two, cooperation, not competition, must be the primary driving force for existence and progress. But cooperation must follow principles of universal justice and equity. Three, science is essential for probing complex natural and social realities and discover ways of development that will sustain a balanced, bonded existence with nature. It is not just a tool to tame nature. Four, nature and society constitute a complex, intertwined wholes, and acquisition of knowledge about them is best done using an interdisciplinary, Systems Theory approach. Five, humans need to cultivate a spiritual sense of oneness with nature by meditative reflection as well as by living in moderation and exercising compassion for all life. Six, continuous growth is not an integral facet of the economy. Owning five pairs of shoes and desiring more while some have none is a cruel absurdity that should be derided. Children should be taught to live in a responsible, simple manner.

The new vision posits that natural and social evolution reflect the growth of complexity in the Universe. The Second Law of Thermodynamics declares that the Universe is moving towards a state of high entropy, a measure of disorder. Randomness and extinction, not a complex, orderly existence is what the future ultimately portends for all living beings. The new vision modifies this forecast by positing that the spread of general disorder is accompanied by emergence of order, regulation and complex entities in some niches of the Universe. The solar system, life on Earth, animals and modern society exemplify such complexities. But progress does not always occur in a linear fashion. Reversals and destructive tendencies are common.

Harmony with nature also entails a non-antagonistic linkage between science and spirituality (or religion). Each fulfills human needs in its own way; each serves to satisfy human curiosity; each can foster universal cooperation and compassion; and, as the case climate change shows, each can play important roles in addressing the major problems facing humanity.

Yet, there is a significant lacuna in these visions. They primarily examine the relationship between humanity and nature, but not societal divisions. Humanity is not a unitary biological entity. It comprises people divided into strata, classes, races and nations. They exist within social formations that do not accord them equivalent relations with nature. The strata have differential impact on nature. This lacuna has been tackled through the Marxian idea of alienation.

Alienation is essentially experiencing the world and oneself passively, receptively, as the subject separated from the object. (Fromm 1961).

Humans relate to nature through manual and mental productive labor. It forms the foundation of existence, enables them to exercise control, and enables culture and creativity to flourish. Modern society has changed that bond with nature by converting labor into a commodity not controlled by the worker. The group that owns means of production controls the labor process. Modern society has two groups, a minority that decides and the majority that follows the decisions made by them or their representatives. This primarily economic division inexorably spills into governance, politics, education and culture. The benefits of labor mainly accrue to the minority. Despite the existence of institutions of freedom and democracy, politics, media and social institutions largely serve minority interests.

Capitalism fragments life into desired and imposed segments, home and work. In one arena people have relative autonomy; in another they must follow orders. All objects and activities become commodities. Capitalism inhibits humans from self-realization and entraps them within the ethos of consumptive individualism. Their ability to relate to nature and fellow humans in an independent manner is sharply compromised. Even if they want to, they are unable to treat nature as a sacral entity. Most people are so deeply immersed in the capitalist mode of life that it requires an incredible effort on their part to refrain from climate damaging activities and products. Capitalism alienates humans on the mental domain by restricting their ideas and visions within boundaries that do not critique its basic structure and power relations.

> *Consciousness, which man has from his species, is transformed through alienation so that species life becomes only a means for him.* Karl Marx (Fromm 1961).

In the Marxian perspective, restoring the sacral, integral bonding with nature and compassionate existence with fellow humans requires not just good science-based policies but also replacement of the rule of capital with an egalitarian, just, non-imperial and genuinely democratic society. In the group of progressive scholars addressing climate change, Amitav Ghosh and Eduardo Gudynas are among the few that come close to the Marxian holistic vision.

As a plethora of historical examples attest, religion can serve as a force for fostering alienation and domination as well as a force for enlightenment and liberation. In relation to an issue as grave as climate change, religious leaders and followers have appeared on both sides of the fence. But our cursory examination has shown that the vast majority regard protection of the environment as a religious duty and firmly support the relevant UN and IPCC policies. It is an inspirational, promising sign that despite their fundamentally different conceptual philosophies, science and religion may maintain an amicable coexistence at the practical, societal level for the benefit of humanity.

Even as they are deeply entangled in a life based on science and technology, people relate to religion and spirituality in emotive ways not available through science. Science expands our mental horizons and quenches our desire to make sense of the world around us. Yet, religion also has an explanatory appeal. And it creates a distinctive sense of closeness with people and nature, gives comfort and solace in times of stress and loss, and provides explanations that give an overall meaning to life. Indications are that science and religion will continue to coexist in the foreseeable future, and do so like an intertwined and complex eternal knot.

6.5 A ROAD TO NOWHERE

Human personality varies extensively. Some people just tend to their own needs and some also tend to the needs others or the community. Some focus mainly on material needs; some value spiritual needs. One person is egoistic; another, a prosocial individual. When the former types of traits dominate, the person is self-centered. Else, she is self-transcendent, that is, a person who sees beyond herself. Psychologists have formed several numeric scales to measure the degree of self-transcendence in people.

Molecular biologist Dean Hamer uses this concept to draw far reaching conclusions about genes and religion. Using the self-transcendence scale devised by Robert Cloninger, and data from a broad range of subjects enrolled in different studies, he asserts that spirituality in humans has a strong genetic basis. His book *The God Gene: How Faith is Hardwired into Our Genes* claims that the genes affecting the levels of monoamines in the central nervous system are related to self-transcendence. Monoamines are a class of molecules that are involved in the transmission of signals between neurons, and their levels in the brain affect a person's mental status in varied ways. Hamer holds that the expression of the genes controlling the production of a monoamine transporter called VMAT2 is linked to self-transcendence. Spirituality, a major component of transcendence, has a genetic basis.

His analysis indicates that gender, race, ethnicity, age, intelligence or anxiety level do not significantly affect self-transcendence. And his conclusion has a caveat. Spirituality as measured by a self-transcendence scale is not associated with membership of, or participation in, organized religion. One can be spiritual but not religious.

> *[Feelings] of spirituality are a matter of emotions rather than intellect. …. It is our genetic makeup that helps to determine how spiritual we are. We do not know God, we feel him.* (Hamer 2004, page 139).

Hamer's book has been critiqued by scientists and theologians on several grounds. One, the title misleads. Formulated to attract publicity, it asserts the existence of a God gene. But that claim belies Hamer's delineation of spirituality from religion. Actually, he argues that spirituality, not religiosity, is hardwired into our brains. While the former is, at least in part, hereditary, religiosity is linked to learning from family and others, culture, rituals and social transmission of ideas. No linkage between genes and membership in a specific religion exists.

Two, Hamer does not demarcate statistical significance from practical significance. While his reasoned analysis of the association of self-transcendence to VMAT2 levels yields significant p-values, it explains only 1% of the variation in self-transcendence scores. Thus, 99% of the variation is due to other genes, the possible interactions between genes and the environment, and social factors. Depiction of VMAT2 as the God gene is unwarranted by his own analysis. Even if such a gene exists, it would have no bearing on the existence or otherwise of God. Though aware of such subtleties, Hamer makes bold claims to present too simple a picture of a complex situation. The cogently written book abounds with speculative propositions that have not been published in a peer reviewed journal.

Three, Hamer's conclusions reflect hardcore reductionist determinism. They are at variance with the Emergence Theory approach that is gaining ground in many scientific disciplines. His conclusions also imply the neglect of personal responsibility. If you have the God gene, you will tend towards altruism, concern for others, and socially responsible conduct. But if you do not have it, what you do or do not do, is not your fault. Your genes are to blame.

Four, vexed at the idea of a God gene, Christian theologians echo the science-based criticisms leveled at Hamer's book, Seventh Day Adventist scholars Peter Landless and Zeno Charles-Marcel say that the idea of a God gene developed through evolution contradicts the Biblical message that all humans have a soul. The soul is not reducible to a single gene or self-transcendence. It cannot be that some have more soul and some have less. But whatever the truth, Christians should draw solace from the fact that:

> *The wonder of genetics itself reveals the handiwork of a wise, powerful, and gracious Creator in whom we live and move and have our very being (see Acts 17:28). We're not to be fearful or deceived. God invites us to know Him personally, and Jesus, whom He sent to give us abundant and eternal life.* (Landless and Charles-Marcel 2021).

Theoretical physicist, Anglican priest and ranking theologian John Polkinghorne does not accept the idea of God gene either:

269

The idea of a God gene goes against all my personal theological convictions. You can't cut faith down to the lowest common denominator of genetic survival. It shows the poverty of reductionist thinking. John Polkinghorne (Wikipedia 2023 – God Gene).

To date, not a single research paper supporting the hypothesis of a God gene has appeared in a peer-reviewed science journal. Like many headline-grabbing papers that have claimed a genetic basis for complex human behaviors, that hypothesis has fallen by the wayside. Interestingly, Hamer's earlier claim that homosexuality is genetically determined has also been discredited.

++++

The God gene episode imparts a key lesson. When science or religion intrudes into the other's territory in ways that violate sound principles of inquiry and make exaggerated claims, it will generally be to the detriment of both.

Consider the question: Can God or prayer change fate (destiny)? Christian sources generally say yes. God is merciful. If you pray sincerely, God may make a bad situation good. Muslim sources say no. God has predestined everything. His plan cannot be altered. If you tackle the question from a strictly logical angle, then you land in a conundrum. If God changes fate, then it was not fate. If he cannot, then he is not an almighty being with the power to do everything. People of faith will reject this argument saying human logic cannot fathom the divine mind. To pursue this issue is a futile endeavor both for religion and science.

The questions 'Is there a God gene?' and 'Is belief in God located in a specific part of the brain?' fall in a similar category. Theologians dismiss that possibility. It is a flawed reductionist, materialistic portrayal of the soul and spirituality. Many geneticists and neurologist reject simplistic reductionist inquiries and seek tenable explanations from the Emergence Theory (Systems Theory) approach. The emergence approach holds that belief in God emanates from multiple sources—personal, familial, educational, cultural, historical and, perhaps to a degree, genetic. Media like sensation; it sells. And some scientists are inclined towards it. Consider two headlines: 'Scientist says there is a God gene' and 'Scientist identifies a gene that partly inculcates a sense of spirituality'. No newspaper will carry the latter, more accurate headline.

Science and religion will not progress through simplistic, populist ventures. It is a road to nowhere. Science and religion are intertwined like the strands of an eternal knot. Some strands are conceptual, and some are practical. Some strands are edifying and enlightening. Some are morally repugnant. Some produce a logical muddle. None of these strands can be wished away. Each has a legitimate cause to be a topic of intellectual discourse. But at this stage of human history, instead of belaboring the fruitless conflict generating ideas, humans should focus on how science and religion can fruitfully coexist and function at the practical, societal level in a morally uplifting manner. The case of climate change vividly shows that they can. The important issue for both people of faith and secularists is to create a viable framework and logistics for the noble endeavor to jointly serve humanity and the planetary ecosystem.

CHAPTER 07: FINALE

God is compatible with science
because God invented every atom, molecule, and chromosome.
You can respect science and still be a believer.
Vivian Becker

Religion and science are the two wings upon which
man's intelligence can soar into the heights,
with which the human soul can progress.
Abdul Baha

You can't convince a believer of anything;
for their belief is not based on evidence,
it's based on a deep-seated need to believe.
Carl Sagan

The difference between science and religion
is that the former wishes to get rid of mysteries
whereas the latter worships them.
Sidney Hook

OUR SCRIPTURAL, practical and historical exploration of Hinduism, Buddhism, Christianity, Islam and several minor faiths together with secularism in *Religion, Politics and Society*, RPS (2022), concludes that no faith system is more exalted than other faith systems, morality can exist in the absence of religious belief, and at the practical level, religion and secularism can coexist peacefully. In this book, we observe that while religion and science differ in terms of their basic philosophy and approach, they need not be socially antagonistic forces. Sharing humanistic goals, they can work together to undo the damage wrought on people and our planet by profit-driven neoliberal capitalism and hyper-consumerism.

In RPS (2022), we outline the steps necessary to promote genuine harmony between all faith systems. These are universal acceptance of a declaration of spiritual equality; focus on common ethical principles such as the Golden Rule, honesty, and non-violence; separation of religion and state; science-based public education; freedom, within bounds of common decency, to choose what to wear; principled tolerance of faith systems other than one's own; acceptance of conversion, sexual preference and gender equality; and regular joint celebrations and inter-faith prayers.

We also identify factors that in their own ways hinder or promote religiosity. Group level factors include the level of societal stability and general prosperity, economic inequality, power and wealth of religious institutions, migration and demographics, ability of faith leaders to act as role models and syncretism. At the personal level, religious faith depends on religion acting as a moral compass, an instrument for coping in times of suffering and loss, a life regulator, a pillar for inner spiritual quest and a source of meaning for things mysterious and unknown. The social cost of disbelief is generally a strong deterrent against shedding one's faith.

Projections are that in the prosperous nations, mostly in the West, disbelief will continue to rise, yet an influential conservative Christian minority will also make its presence felt beyond its numerical strength. Many of those discarding traditional religions will turn to new and old modes of spirituality and alternative faiths.

In the nations of the Global South, high levels religiosity will persist, even among the new middle class. Secularism will continue to be a precarious tendency while syncretism will be common in the nations of Africa and Asia. China will remain a predominantly secular nation but many people will culturally adhere to Confucian, Taoist or Buddhist rites and rituals. And within a couple of decades, Islam will overtake Christianity as numerically largest religion in the world.

While women have made major strides in work, education and health status in many nations, most faith systems lag far behind on gender balance and decision-making equality. Critically, neither religion nor secularism can, on its own, remedy economic inequality, authoritarianism, racism, terrorism, environmental degradation, imperial militarism and hyper-nationalism. Only the combined force of common people, religious and secular, stretching across nations and firmly committed to social and economic justice, promises the hope of attaining a humane, tolerant social order.

Nothing in the present book leads us to question these findings. With its focus on the relationship between religion, on the one hand, and science and mathematics, on the other, on both the conceptual and practical domains, this book has cast a distinct light on the role of religion and science in society and catalyzed additional concerns. We discuss them in these concluding pages.

7.1 SCIENCE, FAITH & CONSCIOUSNESS

Religions project themselves as conceptually unifying visions of life and reality that are of divine origin. Adorning the mantle of (spiritual) holism, they castigate reductionism and materialism as atheistic, sinful tendencies. Only the belief in an overarching divine being or God can provide true meaning to life and account for the existence of consciousness and the soul.

Hitherto, science has meticulously applied reductionism as the key way to the truths of nature and society. It has likened holism to the effusive notion of a vital force (elan vital). Holism cannot explain the difference between life and non-life. Holistic notions are unscientific because they are not testable.

Reductionism is a dirty word in religious circles and holism has a likewise reputation in science. Alternative health modalities are deemed holistic while mainstream medicine is called reductionist. In the book of science, reducing complex phenomena into simpler processes and investigating one at a time is the practical and accurate approach to understanding them. The knowledge gained from these micro studies provides a valid picture of the whole. Holism says that reductionism ignores the interrelationships between the units.

We argue that both these approaches are flawed (Chapter 2). In actuality, science never adheres to hard core reductionism. Cell biology examines cellular components and also the interactions between them. Psychologists do not view a person's characteristics separately but also how they relate to each other. We also hold that while spiritualistic or liberal holistic critiques often embody a germ of the truth, what they offer as the alternative is a mishmash of techniques, a few valid but most misleading, with little indication of distinguishing between them.

Venturing beyond hard core reductionism, modern science has begun to integrate secular forms of holism into its philosophy and methodology. Under the label of emergence (Systems Theory, complexity theory), some science disciplines now study the units of a system and their interactions along with examining the behavior of the system as a whole. The major tenets of Emergence Theory are noted in Chapter 2: Existence of levels within all systems. Higher levels are more complex than lower levels. At each level, look at properties of its units and their interactions. And examine the properties of the system as a whole.

Consciousness

A query that generates acute conflicts between holism and reductionism, and thereby between science and religion is: What is consciousness? In Chapter 4, we describe the meaning of 'consciousness' in general terms and note the three basic approaches for explaining its nature: the theistic, pantheistic and scientific approaches. The scientific approach has two camps, the neuro-deterministic (reductionist) camp and the Emergence Theory (Systems Theory, holistic) camp. Building upon that material, we now review some recent scientific findings and theological perspectives in this area.

Spiritual holism claims that science cannot explain consciousness because it is more than a biological or psychological phenomenon. It is a unique human property that derives from the soul. The scientific standpoint first notes that some lower animals exhibit rudimentary forms of self-awareness and morality, which are integral to consciousness. It is an evolutionary feature. Neurologists have related specific areas of the human brain to emotions, the sense of identity and so on, that form a part of consciousness. The holy grail of this project is to locate the primary area in the brain that generates consciousness.

These experiments produced diverse but amazing claims (Chapter 4). Some say that consciousness and free will do not exist. Others claim to have accurately located brain areas relating to different elements of consciousness. But many of these studies have methodological and interpretive flaws and their claims are found wanting upon critical scrutiny.

Philosopher Thomas Nagel is regarded as the most important advocate of a non-reductionist scientific approach to consciousness. He asserts that:

> *an organism has conscious mental states if and only if there is something that it is like to be that organism—something it is like for the organism.*
> (Nagel 1974).

Consciousness is self-awareness. It is an eminently subjective experience of the organism that none but itself can truly have. Humans may imagine how a bee is conscious, but cannot reproduce the bee's awareness of itself. An entity beyond reductionism, it cannot be explained by the laws of physics. In Nagel's book, belief in God does not yield a rational explanation for consciousness because life has no essential, intrinsic meaning.

A new school of neuroscience transcends reductionism by postulating that consciousness is a whole brain property. Two leading non-reductionist theories are the Global Neuronal Workspace Theory and the Integrated Information Theory. The former posits consciousness as a facet of the mind that emerges from concurrent, non-linear firing of interconnected neurons from different areas of the brain. The latter depicts the mind and consciousness as emergent phenomena, a property of the total bio-system that constitutes the human being. Emotions and awareness cannot be pinned down to specific locations in the brain.

A further step along this direction rejects Cartesian mind-body dualism and integrates the body and brain (matter) with mind (emotions, sense of identity) to explain consciousness. Mind emerges from both the complex neural networks in the brain and the person's interactions with her surroundings and memory. Depending on the emotion or perception involved, some areas of the brain play a greater role than others, but none constitute the sole correlate of consciousness.

> *Being conscious does not just mean having awareness of the outside world. It means being aware of one's self within one's surroundings. The way we experience our body is central to how we perceive our self.* (Costandi 2022b).

More controversially, this line of thinking imputes that not only do creatures like bees possess a modicum of consciousness but also that sufficiently complex robots may also become conscious in some fashion.

Neuroscientist Bobby Azarian expounds the Emergence Theory approach to phenomena in general and consciousness in particular in a cogent fashion. He begins with the Second Law of Thermodynamics, a well established law that states that the Universe is inexorably marching towards a state of randomness (disorder). Azarian does not deny its validity but faults scientists for looking at it in a one-sided fashion and overlooking its obverse side, the continual emergence of self-perpetuating islands of complexity in the Universe. Drawing cases from many areas of science, he labels the realization that there is a built-in tendency of complexity to beget higher complexity as a paradigm shift. It is as important as the transition from Newtonian physics to relativistic physics. Many non-theist scientists of high standing implicitly recognize this tendency, which is what underlies their almost mystical vision of the Universe.

Consciousness is best explicated by the Systems Theory approach. Under the definition used by Nagel, consciousness is the outcome of *'the continual creation of organization and information in nature'* .

> *Given enough time, recursive self-organization will inevitably produce all kinds of wondrous emergent phenomena with surprising new properties like agency and mind.* (Azarian 2022, page 187).

Philosopher, mathematician Rene Descartes is famed for posing the question: *Do I exist?* The fact the question was being asked meant there was a conscious entity asking it. He concluded: *I think, therefore I am.* Yet, beyond this, he saw no way to prove whether the other things he perceives are real or illusory. As a devout Catholic, he resolved this conundrum by separating the body, including the brain, from the mind, and declaring the mind to be equivalent to the soul. Azarian accepts the initial line of reasoning but rejects the step into the religious domain.

Evidence from neurology and anthropology undermines the mind-soul link because animals also possess forms of consciousness. In place of determinism, strict body-mind duality and reductive theories that emanate from the Cartesian view, Azarian moves to an information theoretic formulation of consciousness like the ones noted earlier. Consciousness is an emergent phenomenon, a product of the highly organized matter that the body and brain represent.

The existing systems models of consciousness need extensive development. And however refined, it is unlikely that they will cogently address some key paradoxes. If consciousness is integrated information, then perhaps information is all that exists. The material world is an illusion. Perhaps the Universe is just a simulation, as some philosophers and computer scientists claim. Only the 'I' that thinks exists, all else is a figment of its imagination. If somebody pricks it with a pin, it says both the act of pricking and the feeling of pain exist in its mind. Even death is an illusion. No sound way to persuade a hard-core (philosophical) idealist to abandon that position exists. In practice, however and like hard-core determinism, it is a position that is almost impossible to sustain.

> *Those who believe that our agency is an illusion contradict themselves at just about every waking moment. Ultimately, they end having to accept that the choices they are making in life are real at some level-otherwise every waking moment would be filled with cognitive dissonance.* (Azarian 2022, page 192).

The brain enables us to concentrate, acquire and process information, store and recall events, and make decisions. It enables us to think. It has been traditionally accepted by experts that the evolution of the brain is guided towards development of such cognitive skills. But Lisa F Barrett, a highly cited neuroscientist presents a strong challenge to this idea. In a path-breaking synthesis of brain function, she notes that the brain enables us to have rational thoughts, express emotions and empathy, engage in imaginative, creative activities, and remember past events. But the brain did not evolve primarily to exercise these skills. Boldly asserting that '*your brain's most important job is not thinking,*' she declares that:

> *Your brain's most important job is to control your body ... by predicting energy needs before they arise, so that you can efficiently make worthwhile movements and survive. Your brain continuously invests your energy in the hopes of earning a good return such as food, shelter, affection or physical protection, so that you can perform nature's most vital task: passing on your genes to the next generation.* (Barrett 2021, page 10).

The brain is a physical organ, the mind is a mental entity. The former has some 128 billion interconnected neurons continually firing in a biochemical substrate of transmitters and modulators. Over 500 trillion neuron-to-neuron linkages functioning in a complex network that is organized across millions of interlinked hubs give rise to mental phenomena. A dynamic, plastic organ, the brain evolves from birth to old age, and acquires, consolidates and sheds neural links and properties all the time. The multi-purpose nature of many neurons enables it to function effectively with minimal energy needs.

The brain and its mental superstructure, the mind, are dynamic products of the natural and social environment. They also impinge upon nature, family, and the broader society through their sensory and motor capabilities.

> *The brain does not exist in isolation; it is one part of a complex and dynamic system that also includes the body and the environment.* Costandi (Smith 2022).

The modern Systems (Emergence) Theory approach for understanding the mind and consciousness is moving towards a synthesis that implies the replacement of the Cartesian dictum, I think therefore I am, with a more expansive one:

> *I sense therefore I am.* (Smith 2022).

The modern synthesis recognizes that collective and societal factors—culture, work, race, ethnicity, education, economic status, nationality, and history—are a part of the systemic model of the evolution and functioning of the mind and consciousness. Integrating these factors in a satisfactory manner into the model for consciousness has a long way to go, affected as it is by multiple societal biases. In particular, the Marxian notions of alienation and class consciousness have yet to receive the critical attention they rightly deserve in this field.

Faith and Consciousness

Religious traditions regard consciousness in a hierarchical, almost Systems Theory form. But their system is a top-down entity presided by a supreme, divine entity. Consciousness emanates from a realm that is beyond science. Beyond the human form of consciousness lies spiritual consciousness and beyond it, consciousness attained upon unity with God. With the potential for sacred feelings embedded in the brain, God has granted humans the power to attain true emancipation and awareness.

> *As we abide with Christ, our thoughts, emotions, and desires are transformed. A greater self-awareness leads to greater God-awareness, and a greater God-awareness leads to greater self-awareness. Only by abiding with Christ and living in community are we able to understand ourselves and God.* (Cheong 2020).

The diverse religious views on consciousness, described in Chapter 4, noted that Neurotheology examines the relationship between faith, on the one hand, and the brain and the mind, on the other from a scientific perspective. The discipline covers many faiths, psychology and neurology. Now we augment the earlier material by presenting the views of a Christian and a Muslim scholar.

Christian: After posing the question 'Is consciousness a product of the brain or/and a divine act of God?', theologian Mark Pretorious presents a systematic, scientifically informed Christian perspective on the subject. Quoting major scientists, he reinforces the view that a generally accepted, adequate definition of consciousness does not exist. While covering things like emotions and self-awareness, the concept is too subjective for scientific analysis. The science-based definitions lack conceptual solidity (Pretorious 2016).

Consciousness apparently arises from some forms of brain activity and has developed in an evolutionary manner. The details are controversial, and scientists remain deeply divided. Evidence that some animals exhibit conduct that seems conscious only shows that what they possess is inferior to human consciousness. The evolutionary process, in any case, is guided by God.

The forms of consciousness studied by science must be delineated from spiritual consciousness. The latter is a supernatural, altered sense of perception that has a different origin.

> *A spiritual awakening is therefore not a direct product of evolution but a consequence of the evolution of the brain and consciousness. Thus, God works in people by using the faculties of an evolved brain to bring about supernatural spiritual experiences.* (Pretorious 2016).

He encourages Christians to study neuroscience and psychology because it will help them understand the relationship between the mind and the brain, appreciate the value and limitations of science, and facilitate their spiritual endeavors. It will also help them appreciate the God-given freedom of choice in their lives (free will) and exercise it in an ethical way.

Islamic: Muslims generally accept the precept that 'Whosoever knows himself knows his Lord'. Consciousness is a human property that connects the brain to the spirit. Animals are not conscious. From that vantage point, Islamic scholar Nevzat Tahan presents an in-depth Muslim perspective on consciousness (Tarhan 2022).

Plants are not conscious. Animals possess rudimentary forms of awareness directed towards meeting basic needs. Humans transcend them by engaging in purposeful activity, thoughtful decisions and forming abstract ideas. Human consciousness is a developmental entity. The basic sense of awareness possessed by babies develops towards the abstract thinking of adults. And, '*along with being a biopsychosocial being, man is also a spiritual being*'.

Tahan draws on biology, psychology, physics and quantum mechanics to formulate an Islamic theory of consciousness. Consciousness spiritually elevates humans over all living beings. It is a gift from Allah that bridges the gap between the mind and the body.

> *Men view themselves and events with three different eyes called the eye of the mind, the eye of the heart and the eye of the spirit.* (Tarhan 2022).

Tahan's views would not draw opposition from Muslims, though some will prefer that they be rendered in a gender-neutral form.

A group of six scientists led by Nicola L Bragazzi focused on the spiritual aspects consciousness from an Islamic perspective (Bragazzi 2018). Reviewing varied multidisciplinary studies, they aimed to summarize the link between mental well-being and religiosity. A key idea they used is 'higher consciousness,' a mental state ascribed to gurus, monk and ascetics who mediate intensely, eat sparsely and live simply. It inculcates a deep sense of calmness and acceptance of worldly realities. This well-researched, scientifically coherent paper prudently draws a conclusion that has been drawn with respect to other faiths:

> *Available scientific evidences seem to suggest that Islam and Sufism (prayer and meditation) can contribute to the achievement of higher consciousness states.* (Bragazzi et al. 2018).

Meditation enables attainment of a state of consciousness beyond the regular forms of awareness.

++++

Faith-based perspectives on the mind and consciousness generally accept the diversity of the findings of modern science, though not always in a consistent fashion. The ideas of free will and responsibility are often integrated within these perspectives. They stress the health and mental benefits of attaining the higher levels of consciousness through focused religious practice. And they come with a major caveat: Science is acceptable, but it can never grasp spiritual consciousness and God-awareness. It is an unbridgeable gap.

7.2 THE FUTURE OF RELIGIOSITY

Global trends indicate that in the Western and some other affluent nations, religiosity is declining while varied shades of secularism are gaining a larger following. This trend is more marked among the youth. Though the US lags behind Europe, it too has experienced a marked decline in church attendance in the past two decades. But mainstream religion is often displaced by a spiritual doctrine. The number of avowed atheists is rising, yet they remain a small proportion of the total population. Scientists are more likely not to believe in God than non-scientists.

In the Global South, however, adherence to traditional religions, Christianity, Islam, Hinduism and Buddhism, remains virtually stable. It is so not only among the impoverished majority but also among the middle class whose numbers have sharply increased. Adherence to a main religion combined with attachment to an old or new folk religion is common. Secularists and atheists in Asia, Africa and South America remain a distinctly tiny, endangered segment of the population.

Overall, it is estimated that Islam will overtake Christianity as the dominant global religion in the next three decades, and the global level of religiosity will basically remain stable.

These trends, however, need to be viewed with several qualifications. One is provided by modern genetic studies that look at the relative effects of genes and the environment on people's attitudes towards, among other things, religion. The key finding is that though the social environment exercises the major effect, the effect of genes cannot be discounted. There is an ingrained baseline proclivity towards religiosity among humans. But that religiosity is not linked to a specific religion. It is mostly linked to cultural, historical factors and connotes some form of spiritualism.

Studies claiming that religiosity has a genetic basis, that a 'God-gene' exists and that religiosity is passed on from parents to offspring are complemented by neurological investigations that posit specific brains locations as centers for spirituality. Pointing to a biological basis for belief, these studies favor the long-term persistence of religiosity. But these reductionist studies have not held up to critical scrutiny. The connection of genes to the brain and mind are modeled as complex, dynamic systems interacting with the natural and social environment. The conflation of religiosity with spirituality is a common shortfall of these studies.

The second qualification is that even where a marked decline of religiosity has occurred, especially in adherence to mainline religions, a core and not-that-small group of conservative believers remains stable. Generally, it is connected to right wing political parties and think tanks, and exercises a disproportionately large influence on the political, educational and legal systems.

The third qualification relates to the non-religious benefits of religiosity, an issue covered in RPS (2022). Recently completed long-term, large sample studies conducted in many nations consistently reinforce earlier findings that regular church, temple or mosque attendance, praying, communal faith-based activities and community service are positively linked to improved outcomes in mental health, chronic diseases and longer life span.

Rather than ascribing such findings to divine intervention, scientists note the well-known stress-reducing, coping-related effects of living within a supportive and caring community. Religious rites and rituals stabilize life and lower anxiety levels when faced with work, family and other problems. God is merciful and He will come to your help. Have faith, be thankful and do your best. Faith calms the mind and enhances optimism.

> *Besides encouraging social connection, religion can help people to cultivate positive emotions that are good for our mental and physical well being, such as gratitude and awe.* (Robson 2022a).

A 17-year follow up study of 102 people with HIV infection found that the survival rate among those who prayed for known others was twice as high as that for those who did not. The conclusion was robust to adjustment for confounding factors like medication adherence, initial viral load and drug abuse. (Ironson and Ahmad 2022). Faith institutions provide a venue to combat loneliness and social isolation. Faith gives one's life purpose and meaning. It motivates one to engage with other believers.

Scientists explored a possible confusion of cause and effect in these studies. Perhaps, it is the healthy and mentally well-adjusted who are able to participate more in religious activities. Smoking, drinking, drug abuse, unsafe sex and other risky behaviors are generally less prevalent among the religious than among the non-religious. Moderation, not faith, is perhaps the key to good health. Yet, the health benefits of faith persist even after allowing for such factors. But some religious beliefs induce pessimism and have deleterious effects on health, mental stability and life span. But they are not as widespread as the beneficial effects of faith.

But it goes beyond religion. Non-religious people obtain health related and other benefits by practicing loving-kindness meditation, cultivating empathy and gratitude, engaging in voluntary and charitable activities, taking nature walks, reducing self-isolation, exploring the neighborhood, following a routine, often expressing gratitude, and reducing consumerist tendencies. If done regularly and with commitment, similar positive effects follow. The key hurdle is that the non-religious rarely have an established, organized community of like-minded persons within which to undertake such activities on a regular basis.

Religious institutions have centuries of history behind them. Since the days they were associated with power, they have amassed extensive resources and manpower to run their affairs. They have a diverse package of cultural practices, literature, music, art, dance, dress codes for clergy and lay people plus organized activities like child care, religious education, discussion circles, homeless feeding programs, charity donation collection events and so on under one roof. The non-religious rarely have anything equivalent.

> *The power of religion is that it gives you this package of ingredients that are pre-made and organized for you.* Professor Patty Van Cappellen (Robson 2022a).

In the prevalent neoliberal environment, where public service programs get a declining portion of the budget, where the ethic of individualism is promoted broadly, and where hyper-consumerism dominates, the religious have a venue to escape, at least partially, from that dehumanizing setting and find meaning in ways that soothe their spirits. In that regard, neoliberalism acts as a force to entrench religiosity.

Religiosity is a complex entity with biological, social, historic and cultural roots, and needs to be analyzed in that comprehensive framework. The alternative formulations noted above, path-breaking as they are, have mostly neglected a potent force in this equation, neoliberalism.

7.3 CLIMATE ACTIVISM AND RELIGION

In this section, we paint three diverse portraits of the interaction between religion and science by focusing on the ideas and activities of Chico Mendez and James Hansen, two stalwart environmental activists, and the Chipko Movement, a brave group of women environmental activists.

Chico Mendez

Covering 6 million square kilometers, the mammoth Amazon rain forest has been a major stabilizer of the Earth's climate. Its nearly 400 billion trees daily store significant amounts of carbon dioxide from the atmosphere and release large quantities of water vapor. With one in ten species located within its confines, it presents a magnificent spectacle of biodiversity. Of its 47 million residents, about three million are indigenous peoples. About 60% of the forest lies within Brazil; the rest is spread out among the adjacent countries.

The Amazonian indigenous people and the resources of the area have been under ferocious assault since the 1960s from loggers, miners, rubber harvesters, road builders, livestock breeders and oil companies. New settlements, economic activities and tree felling have destroyed 18% of the forest. The consequential increase in wildfires and reduction of the foliage has compromised the function of the forest as a protector of the planet from global warming.

Walking Tree - Amazon Rain Forest

The rain forest is a major arena of conflict between profit-hungry predators and the local people and activists. The former deploy armed thugs to launch violent assaults on those who dare to oppose them. Many local and foreign activists have lost their lives in this conflict.

One of them was Chico Mendez. Son of a rubber trapper, Mendez grew up in a poor family. Required to assist his father in trapping work from a young age, he never had the chance to attend school. It was only when he was 18 that he began to learn to read and write. Yet, he grew up with an acute perception of the natural and social environment, and a remarkable ability to articulate what he perceived.

His family was one of the tens of thousands of families of rubber trappers left in misery with broken promises after their studious efforts to meet America's demand for rubber during WW II. Mendez embraced the struggle for their rights and rose to the helm of the national union of rubber tappers. His dedication, clear articulation of the rights of workers, peasants and indigenous groups, and his astute ability to link, in a sustainable way, the rights of people to improve their lives with protecting the environment elevated him into a prominent international environmental activist. And his innovative, non-violent campaigns earned him the title, the Gandhi of the Amazon.

> As well as lobbying successfully to end international financing for Amazon clearance, [Mendez] organized the rubber tappers in non-violent resistance. Men, women and children would form human barricades known as "empates" to prevent the bulldozers from tearing down trees. (Rocha 2013).

But it also earned him the wrath of powerful ranchers and landowners in the area. He was constantly subject to death threats. In 1988, despite a modicum of police protection, he was slain by a bullet fired by a distraught rancher while resting at his modest home.

Yet, his legacy endures. Now a Patron of the Environment in Brazil, he is also an iconic figure for the global environmental movement. Major institutes and national parks bear his name and his project to combine economic growth with improving the lives of the poor in an environmentally sustainable fashion is being implemented in some areas of his nation. His death raised the global profile of the environmental movement and inspired many young people to join the cause.

Green Anacoda - Amazon Rain Forest

Like the majority of Brazilians, Mendez grew up a Roman Catholic. Priests from the Liberation Theology movement that was spreading across central and south America informed his activism. Besides leading the trapper's union, he was instrumental in founding a branch of the Worker's Party in his area.

Much of the higher leadership of the Catholic clergy in Brazil was allied with the wealthy landowners and right wing political forces. But a significant portion of the lower clergy and a few bishops sympathized with plight of the poor, landless majority (RPS 2022). On the international front, the progressive vision articulated by Pope John Paul II to be stewards for God's creation affected many Catholic priests in Latin America. Thus, many of them began to support local and global actions to protect the environment.

In the month after the assassination of Chico Mendez, the United States Conference of Catholic Bishops issued a statement calling him a courageous and dedicated trade union leader and environmentalist who understood the critical importance of the Amazon rain forest to the global environment. As the Secretary General of the Conference said:

I pray that the sacrificial death of Francisco [Chico] *Mendez will encourage many to take up where he left off and that it will focus renewed attention on the ever-growing urgency of safe-guarding the integrity of creation.* Monsignor Daniel F Hoye (USCCB 1989).

Of recent, the stance of the Catholic Church on the protection of the Amazon forest and controlling climate change has been galvanized by Pope Francis. In 2019, he convened a special synod of Catholic bishops from nine nations in the Amazon River basin to reflect on the religious, environmental and social issues concerning the area. It drew major international attention due to the significance accorded by the Pope to the rights and dignity of indigenous people, the need to respect for their beliefs and traditions, and protection of the Amazon rain forest. During his visit to Brazil, he interacted closely with the people and leaders of indigenous groups who hailed him as their champion. He put his words into action by attending an indigenous worship ceremony held near some sacred sites in the Vatican Gardens. It was an unprecedented act. Conservative Catholics and evangelicals excoriated him for participating in an '*act of idolatrous worship*' and desecration of holy places. But he was not deterred.

In the run up to the synod, the Brazilian government under the authoritarian, far right presidency of Jair Bolsonaro indicated that the synod constituted an act of foreign interference in the internal affairs of Brazil. He also initiated surveillance of the organizing meetings by the military and security agencies. The President was a firm ally of the business community in Brazil, particularly the agricultural, ranching and mining firms. Echoing pseudo-scientific doubts on global climate change, he began his term by relaxing many regulations restricting commercial activities in the Amazon rain forest. In his opinion, the data on increased forest fires due to expanding commercial activities were lies.

Richard Salles, the first Secretary of the Environment in the Bolsonaro administration took a further swipe at the environmental movement by brazenly distorting the legacy of Chico Mendez. Branding him an opportunistic profiteer, he alleged that Mendez was not as heroic a person as he was being portrayed:

> *On the part of environmentalists, more connected to the left, there is an aggrandizement of Chico Mendez. The people who are in agribusiness and are from his region say Chico Mendez was not all that.* Richard Salles (Lima 2019).

Jair Bolsonaro ran on a populist platform of fighting corruption and 'communist' influence in Brazil, enhancing Christian values and fostering economic growth through private initiative. But his real agenda was to implement firm neoliberal policies to benefit people of wealth and major corporations. As the inequality and misery generating results of his policies became clear to the public, his popularity plummeted. He was unseated in 2022.

Chico Mendez, 1988

281

And evangelicals formed the firmest base of support for Bolsonaro. But on the issue of the environment and climate change, many of them parted company with him. The dictum about the Christian responsibility to protect God's creation made generally conservative groups like the Environmental Evangelical Network to adopt a progressive stand on climate change. Chico Mendez, to them, was a 'Christian eco-warrior'.

The activism of Chico Mendez was not driven by theological concerns. But he was involved in church initiated efforts to bring education to rural communities and many local churches supported his environmental drives. His general outlook was molded by his close association with liberation theology proponents. And by calling him a steward of God's creation, major local and international religious organizations supported his activism. Mendez had a humanistic, internationalist outlook:

> At first I thought I was fighting to save rubber trees, then I thought I was fighting to save the Amazon rain forest. Now I realize I am fighting for humanity.

James E Hansen

James E Hansen, once the director of the Goddard Institute for Space Studies at the US National Aeronautics and Space Administration (NASA), is one of the world's top authorities on climate science. His pioneering research in the 1980s firmly supported three key ideas. One, human activity is mainly responsible for the buildup of greenhouse gases in Earth's atmosphere. Two, the planet's surface is heating up at an alarming rate. Three, if the emission of carbon dioxide is not reduced, catastrophic climate change will occur across the globe. The solidity of his research and his clear, remarkable testimony before the US Congress in 1988 stimulated extensive research on climate change and catalyzed policy makers to take it more seriously. He was one of the major voices behind the establishment of the authoritative UN-based Inter-governmental Panel on Climate Change (IPCC). If a single person has to be credited for altering the global discourse on climate change, it is James Hansen.

Hansen's projections on the rate and pattern of global warming have been amply verified. The deleterious effects of climate change are now manifested across the world. Most governments have policies to deal with it and there exists a major international mechanism to address it. Yet, as noted earlier, official action lags far behind what is required.

After his congressional testimony, Hansen increasingly realized that it is not sufficient to perform scientific studies, write reports and present them to policy makers. Powerful corporate and state vested interests are fighting tooth-and-nail to distort the scientific consensus relating to climate change, raise dubious doubts about the IPCC reports, spread disinformation and obstruct even the moderate decisions made at the global climate change conferences.

Since climate change is an urgent, existential problem for life on Earth, scientists must go beyond talk and unite with citizens to exert pressure on the lethargic authorities. In that spirit, Hansen began his public activism when he was at NASA. His congressional testimony was just a start. As the George W Bush White House began to subvert climate science, Hansen accused his employer, NASA, of succumbing to political pressure and altering his public statements on the issue. Other public agencies saw their reports being edited. Many scientists acquiesced, but Hansen did not. Later, he accused the Obama administration of feigning to heed climate scientists but adopting pro-business, counter-productive policies. He asserted that politics was subverting science and both the Republican Party and the Democratic Party were to blame.

Because of their role in spreading disinformation about climate change, he called for the arrest and trial of the executives of major energy firms on charges of crimes against humanity and nature. Together with over a hundred scientist, he urged scientists and scientific organizations to eschew sponsorship from the firms peddling climate related disinformation.

The oil company [Exxon] *knew about climate risks and factored them into planning as early as the 1970s – but continued to fund organizations rejecting established science well into the 2000s.* (Goldenberg 2016).

To counter the powerful economic and political interests that are egregiously distorting science, Hansen mounted street action. In 2011, he and over 1,250 activists were arrested for protesting the permit for the Keystone pipeline extension. He was handcuffed over the same issue in 2013. His firm opposition to the pipeline earned him a sharp public rebuke from the Minister of Natural Resources of Canada, Joe Oliver. The Minister accused him of double-talk, exaggerating the risks of the pipeline and peddling nonsense. In February 2021, he appealed to the UK Prime Minister Boris Johnson to cease governmental support for fossil fuels and put a brake on the machinations of corrupt *'special financial interests'*. Specifically, he called for disbanding the plans to open a new coalmine and cessation of UK governmental funding for *'fossil fuel projects'* abroad. Not doing so would only show a *'contemptuous disregard of the future of young people and nature'*. (Harvey 2021). Hansen is not afraid to call a spade a spade.

You just have to say what you think is right.
James Hansen

James Hansen Under Arrest

His keen scientific research and vibrant activism has earned James Hansen the title of *'the god-father of climate change,'* and he has received awards and honors for his work. A courageous man of integrity, he is admired by many scientists and non-scientists. Yet, some scientists and media personalities blame him for injecting politics into science and overplaying the dangers of global warming. The facts are the opposite. Hansen's aim is to raise evidence-based awareness of climate change, and expose and remove corporate interference and right wing political distortions in the process of dealing with climate change.

Hansen's climate science papers have come under sharp but distorted attacks from so-called climate skeptics who are associated with major right wing think tanks that receive hefty funds from energy corporations and billionaires. With conservative media, front groups with innocuous names and influential state and federal politicians amplifying their voices, the climate deniers, whose credibility is next to none within the scientific world, find a large public audience and are able to hold back effective, prudent policies on climate change.

The assaults on climate science proceed in a multi-pronged way. Marshaling a few articulate contrarian experts, minor errors in science papers are magnified, reputed climate scientists are misquoted, and emails are hacked to present private conversations as peer-reviewed science. A host of dirty tricks are deployed to tarnish the reputation of the IPPC and prominent climatologists, and create scandals without foundation. Climate science is derided as *'junk science'*. Mocking authentic scientists like James Hansen is a favored ploy. For example, because he felt that a bad agreement was worse than no agreement, Hansen had called for the cancellation of the year 2011 Climate Summit. While Hansen had adopted that stand to stress the criticality of global warming, the contrarians twisted his words to imply that 'the Father of Climate Change' was doubting the actuality of climate change. As one of the prominent right wing lobbyist framed the message:

> *James Hansen! James Hansen said this conference must fail! So if anyone asks you this week, How can you be against this? say, We stand shoulder to shoulder with NASA's James Hansen!* Marc Morano (Richardson 2010).

By the time such crude efforts are exposed, the damage is done. They sully the atmosphere at major conferences on climate change, produce squabbles among policy makers and seriously compromise the targets for emission of greenhouse gases. Only the executives of the energy corporations are left smiling. In the early 1990s, about 80% of the Americans felt that global warming was a threat. A decade later, after the corporate funded propaganda had inundated the media, the figure went down to 60%, and only 35% thought that it was caused by humans. Similar tactics were deployed by tobacco giants to hold back effective policies to control smoking and marketing of tobacco products for nearly five decades.

Recent studies indicate that much larger and more rapid reductions in the use of fossil fuels are required in order to prevent a climatic tipping point that would spark an irreversible, self-reinforcing process threatening humans and all life on Earth, especially in the poor nations. We are reaching a precipice, yet distorted perceptions about climate change were until recently commonplace among the American public and across the nations.

James Hansen is a stalwart scientist propelled by a deeply felt concern for humanity and the planetary ecosystem. He holds that the present generation has a fundamental obligation to safeguard the environment for the future generations. He does not have a religious agenda. Yet he sees people faith who are inspired by scriptures as allies in the struggle to tame climate change.

In 2006, along with scientific luminaries like biologist EO Wilson, botanist Peter Raven and epidemiologist Rita Colwell, he attended a gathering organized by the Harvard Center for Health and the Global Environment, and the National Association of Evangelicals (NAE). The joint statement issued at the end noted key areas of agreement between the secular and religious delegations. One, that climate change was mainly caused by humans.

> *We agree that our home, the Earth, which comes to us as that inexpressibly beautiful* and *mysterious gift that sustains our very lives, is seriously imperiled by human behavior.* (Sandell and Blakemore 2007).

Two, steps should be taken to protect the over a billion poor, suffering people from the vagaries of climate change. As a pastor attending the gathering said:

> *Unless we care for the vulnerable, we are not representing Jesus well.* Pastor Joel Hunter (Sandell and Blakemore 2007).

Acknowledging the importance of the IPCC, the overall message called for '*a new moral awakening*'. Drawing inspiration from science and scriptural values, people of varied backgrounds need to join hands to prevent calamitous climate change. More than 45,000 churches with 30 million members are associated with the NAE. It is also allied with the New Evangelical Partnership for the Common Good which aims to educate church attendees about climate change and involve them in personal and public efforts to control it.

Yet, under the umbrella of as the Cornwall Alliance, some major evangelical leaders remain opposed to these initiatives. A progressive theologian sums up the message of the Alliance:

> God is sovereign over creation and therefore humans can do no permanent damage. God entrusted the Earth to human dominion and we should not be afraid of economic development or other uses of human creativity. God established government for very limited purposes such as providing for the common defense–government should not intervene much in the workings of a free market economy. The Republican Party has taken a skeptical posture toward climate and we support that posture and that party. The media is overplaying climate change worries, at the behest of scientists who cannot be trusted anyway; it may all be a conspiracy to limit our personal and business freedoms and tax us even more. The environmental movement is secular/pagan and has always been a threat to American liberties and has always been anti-business and exaggerated environmental problems. David Gushee (Curry and Gushee 2010).

Basically, it is a pro-corporate neoliberal agenda cloaked in theological garb.

Chipko Movement

Rural communities maintain a stable existence by living in harmony with nature. Forests are valued for food for humans and livestock, firewood, weaving and building material, and medicinal herbs. Trees are sacred, not to be abused. The Bishnoi community in India follows a Hindu creed in which trees are especially venerated. The 29 principles laid down by their holy sage, Guru Jambhoji, include 'Do not cut green trees, save the environment' and 'Do not eat meat, always remain purely vegetarian'.

The forestry policies instituted in colonial India restricted villagers from using local forest and other resources according to their needs. Logging companies, foreign and Indian, however, obtained permits to fell large numbers of trees to fulfill the demands of the colonial economy. For the rural folk, it was a burden added on top of that imposed by the rajahs and large landowners.

The wood was needed as fuel for making lime to be used for building a new palace. The Khejri (*Prosopis cineraria*) is a robust plant that can withstand very hot, windy and low-rainfall conditions. Humans and animals consume the Khjeri bean; its bark has medicinal uses. Hindu communities across India regard it a holy tree, featuring it in special religious festivals. For the Bishnoi it thus has a double significance.

Khejri Tree (Prosopis cineraria)

It was in this context that in 1730, the Maharajah of Jodhpur authorized the felling of Khejri trees growing near a Bishnoi village. As the trees fell, Amrita Devi, a Bishnoi woman, was deeply incensed. It was a sacrilegious act. So she and her three daughters began to literally hug the trees to stop the felling. The tree cutters offered her a bribe but she refused. When a contingent of soldiers arrived, she told them that she would rather die than allow a tree to be felled. And when the soldiers killed her and her daughters, hundreds of villagers joined in, hugging the trees to save them. A massacre ensued. Three hundred and sixty three villagers (294 men and 69 women) were mowed down. When the Maharajah learned about the killings, he put a stop to the tree felling.

The Bishnoi villagers were the first modern champions of the environment, and originators of the tactic of clinging to trees in order to protect them from axes and saws. And they have for long been a source of inspiration for activists in India and beyond.

Laws restricting access to forestry resources introduced in the colonial were continued after Independence. State forest departments readily granted logging companies permits to harvest hundreds of trees while the local residents had a difficult time obtaining a permit to cut down a few trees. Commercial logging caused soil erosion and landslides in mountainous regions. Dissatisfied with how they were being treated, many villages mounted protests against deforestation in the 1960s. Followers of Mahatma Gandhi and Communist Party members backed the protesters. The peaceful demonstrators often faced violent police attacks.

One of the affected locations was a hilly area near the border with Tibet. For ages, women in the area had depended on nearby forests to satisfy home needs. When the government granted permits for a major deforestation program while restricting forest access by the locals, it made their lives even more arduous. Dissatisfaction rose, and the Chipko movement was born.

Chipko Tree Huggers

The name Chipko, a Hindi word that means 'to cling,' signified that the movement was using non-violent tactics similar to those of the Bishnoi more than two centuries earlier. The determined rural people, students and Chipko leaders were able to force the loggers to back down. Yet, it was hardly the end.

> [The] *tempo was upped in 1974 when the Reni Forest, in what was then the state of Uttar Pradesh near the Tibet border, was auctioned to supply timber to the Symonds sporting goods company to make cricket bats and other sporting equipment.* (Kilvert 2022).

Chipko leaders pledged to firmly oppose the tree felling. Applying the Gandhian tactic of non-violent civil action to environmental activism, they managed to stop the company from executing the permit. The Chipko agenda was later broadened to projects like opposing dam construction and mining, protecting water sources, and upholding the interests of marginalized communities.

Tree hugging, a tactic of Indian origin, has been creatively adopted by forest protection activists in many places. The effort to save the giant ancient redwood trees in California saw dedicated activists risking their lives to halt the chainsaw bearing lumberjacks from practicing their trade. Many lived atop the trees for years. Environmental groups in New Zealand and Australia were also influenced by the Chipko activists. Within India, they stimulated a host of environmental drives.

> *In recent years, the* [Chipko] *movement not only inspired numerous people to work on practical programs of water management, energy conservation, afforestation, and recycling, but also encouraged scholars to start studying issues of environmental degradation and methods of conservation in the Himalayas and throughout India.* (DC 2016).

The UN Environmental Program credited the Chipko for catalyzing:

> *a socioeconomic revolution by winning control of their forest resources from the hands of a distant bureaucracy which is only concerned with the selling of forestland for making urban-oriented products.* (DC 2016).

Environmentalists Sunderlal Bahuguna and Chandi Prasad Bhatt were the main leaders of the Chipko movement. Like Gandhi, Bahuguna organized, petitioned, fasted, marched and endured jail. Despite police attacks, they persevered. Most of the co-leaders and individuals at forefront of the Chipko drives were women. Among them were Gaura Devi, Suraksha Devi, Sudesha Devi, Bachni Devi and Chandi Bhatt and Virushka Devi. Admittedly they pioneered what is now called eco-feminism. Forming chains of women tree huggers, they chanted, sang and tied sacred ribbons around the trees. One of their songs went:

287

> *What do the forests bear?*
> *Soil, water and pure air.*
> (FORE 2022).

Gaura Devi played a prominent role in this drive to protect the people's heritage. Her group of defiant women were able to stop the tree felling by a sporting goods company. After a committee of inquiry report, the state government imposed a ten-year moratorium on tree cutting in the area.

> *A milestone was the mobilization of women in March 1974 by Gaura Devi, a woman elder who confronted gun-wielding loggers saying "Brothers! This forest is the source of our livelihood. If you destroy it, the mountain will come tumbling down onto our village. This forest nurtures us like a mother; you will only be able to use your axes on it if you shoot me first." The Chipko movement is noted for extensive participation by women and women leaders such as Gaura Devi. Women advocate for forests as self-renewing life-support systems rather than economic "resources," fusing their practical expertise with scientific knowledge.* FORE (2022).

The Chipko movements and its offshoots achieved many things. At the practical level, their work led to several ten to fifteen year logging moratoriums issued by the government of India. Tree felling declined across India. Forestry output in eight hilly regions of India went down from 62,000 cubic meters in 1971 to 40,000 cubic meters in 1981 (Mitra 1993). Raising environmental awareness across India, it revived Gandhian tactics and encouraged women to participate in environmental and civil struggles. It brought home the point that economic development need not be antithetical to environmental protection. But it should be a people-oriented, sustainable mode of development. It underscored the importance of enabling ordinary people to control their own local resources and protect their livelihood. And it taught social activists to form broad coalitions. As the nuclear physicist and biodiversity expert Dr Vandana Shiva expresses it:

> *[Chipko] taught me about self-organising.*
> *Chipko taught me the value of self-help and solidarity.*
> *Chipko taught me ecology and biodiversity.*
> *Finally, Chipko taught me humility.*
> Vandana Shiva (Kejriwal and Vora 2019).

The Western pro-corporate media has turned the term 'tree hugger' into a term of derision. It is now taken to denote activists who care more about trees than about people. But that is an artificial distinction. In India and elsewhere, tree huggers have been devoted to protecting forests and the environment as well as to improving the lives of ordinary people. The two objectives are inseparable.

Leaders of the Chipko movement were given many national and international awards. The government of India now commemorates the legacy of the Bishnoi who died in 1730 with the annual Amrita Devi Bishnoi Wildlife Protection Award. It recognizes the efforts of individuals and organizations for their service to wildlife conservation.

Yet, in the long run, the Chipko movement suffered the same fate as many well meaning social movements. The reforms made by the government helped in the short run, but ended up transferring the initiative to the state bureaucracy. Its leaders became more focused on conservation than on the right of the people to self-determination. Instead of engaging with the grassroots, they spent more time lobbying politicians and bureaucrats. And the movement did not fuse its goals with a broad social, political and economic agenda. Reformist ideas dominated and broadly inclusive struggle for change was neglected.

> *What this means is that a movement that could have given the world its most*
> *powerful green party with village self-governance at its heart, fell apart.*
> (Mitra 1993).

The irony is that now in India, even the authoritarian, religiously extremist BJP government, which is wedded to a pro-corporate, neoliberal program, claims to uphold the legacies of the Bishnoi and the Chipko. And it accuses its opponents of paying lip service to environmental protection while ignoring the interests of the people. A greater inversion of fact with fiction is hard to imagine,

The Chipko movement signifies an integral blend of a religious tradition with secular, science-based activism to protect nature and human rights.

++++

The work of Chico Mendez and James Hanson together with the history and initiatives of the Chipko movement underscore a key point of importance to this book. Despite the presence of conservative theological obstructionists, people and institutions with religious and secular motivations can work side by side in progressive, fact-driven environmental protection programs like the control of climate change. Each side has its own rationale but their methods and goals can converge. This message portends a promising future for interfaith tolerance and harmony.

7.4 SCIENTISTS AND RELIGION

Scientists view religion along a diverse spectrum ranging from compatibility to shaky or firm tolerance and onto outright hostility (Chapter 4). In order to further illuminate this issue, we augment the previously noted views of eminent scientist with the views of Charles Darwin, Stephen Hawking, Charles A Coulson, Francis Collins, Jim Al-Khalili and Michio Kaku.

Charles Darwin

Conservative people of faith depict Charles Darwin as a prime embodiment of evil. His view that random mutations and natural selection suffices to explain how living beings, including humans, evolved is seriously blasphemous for many Christians. Yet Darwin felt that the *Origin of Species* was not a profane work and would be acceptable to people of faith. There was '*no reason why the disciples of* [religion and science] *should attack each other with bitterness*'. (Spencer 2009). Instead of hostile confrontation, a gradual process of education and free exchange of ideas was needed. Only that would enable a wider, more solid appreciation of scientific ideas including the Theory of Evolution.

Darwin's ideas on religion evolved as he explored the nature of evolution. Eventually, he rejected the idea of a personal God, or the validity of the scriptures and miracles. A key reason he held back the publication of his *magnum opus* was to not perturb the feelings of his devout wife or inflame public sentiment. He felt it odd that God would allow extensive pain and suffering among humans. Darwin suffered deep anguish at the death of his beloved daughter Annie. It was the final straw that alienated him from the Christian notion of a loving God. Yet, he never categorically rejected God. There was some meaning behind the existence of the Universe. And he was aware that some questions were beyond science:

The mystery of the beginning of all things is insoluble by us; and I for one must be content to remain an agnostic.

Charles Darwin

He did not discount the possibility that God had created primordial life forms that had the capacity to evolve along the lines he had discovered. And, he cautioned against unwarranted anthropomorphism.

Man in his arrogance thinks himself a great work, worthy the interposition of a great deity. More humble and I believe true to consider him created from animals.

Charles Darwin

Appreciating the impact of religion in fostering good behavior among people. he maintained amicable relations with people of faith in his vicinity and beyond. Corresponding extensively with theologians on primary issues, he also replied to people who inquired about his beliefs.

Though a decent, tolerant person, Darwin was ingrained with the prejudices of his times. Unduly influenced by Malthusian ideas on population growth, he ascribed the same importance to competition in natural and societal development. He believed that men were intellectually superior to women, without evidence to support his belief. But he did not believe in racial inequalities. His research had amassed a massive body of facts showing that humanity is a single species. Important physical and behavioral features do not have a racial basis and that the idea of race is not scientifically tenable. At one point, he talked of '*civilized*' races replacing '*savage*' races but subsequently he envisaged global compassion and cooperation as the way forward for humanity.

As man advances in civilization, and small tribes are united into larger communities, the simplest reason would tell each individual that he ought to extend his social instincts and sympathies to all members of the same nation, though personally unknown to him. This point being once reached, there is only an artificial barrier to prevent his sympathies extending to the men of all nations and races. Charles Darwin (Rutherford 2021).

Primarily, Charles Darwin was a humanist and a dedicated scientist not bound by dogma and for whom truth was what could be proved. Evidence counts but faith should not be discounted.

Love of science, rather than hatred of religion, was what powered [Charles Darwin]. (Spencer 2009).

Stephen Hawking

Afflicted with a progressive, incurable motor neuron disease at the age of 21, Stephen Hawking surmounted incredible odds to become a theoretical physicist of the highest distinction occupying one of the most prestigious academic chairs in science. Over the course of half-a-century marked by extensive paralysis and loss of speech, he made prodigious theoretical and observational discoveries in general relativity, quantum gravity and cosmology. Elected into elite scientific societies, he garnered esteemed scientific and civilian awards. He penned several popular books on science and a series of fictional works illustrating key ideas in science for children. *A Brief History of Time* captured the public imagination across the world.

Hawking generally held progressive positions on social and political issues. Supporting universal health care, universal basic income, international economic cooperation and strict measures to control climate change, he opposed nuclear weapons, the war on Iraq and repression of the people of Palestine. But he was not optimistic about the future of humanity and life on Earth. The current trends were too dangerous.

Hawking's conceptualizations of time and the origin of the Universe led him to conclude that there is no need invoke a god to explain the natural world. Divine entities and realms are figments of human imagination. He was a firm atheist and a humanist who held that science and religion are incompatible. The existence of God is improbable. Denoting the idea of life after death as a *'fairy story,'* he doubted whether the Universe had been created by a supreme being. The brain was not driven by a soul. Basically, it was a computer.

Though life and the Universe are imbued with meaning, it is not of divine origin. Humans should strive to grasp the essence of existence, *'the grand design of the Universe,'* through science.

> *God is the name people give to the reason we are here. But I think that reason is the laws of physics rather than someone with whom one can have a personal relationship.* Stephen Hawking (Ducharme 2018).

Stephen Hawking

Unlike the proponents of New Atheism, Hawking did not castigate religious faith or people of faith. Nor did he single out Islam as the principal foe of peace and rationalism. He acknowledged that religion has a positive function since some people find comfort and meaning in religious beliefs and rituals.

Conventional wisdom holds that matter is composed of elementary particles like protons, neutrons, electrons and quarks that interact under the push and pull of the weak force, the strong force and gravity. The String Theory replaces that model by postulating that matter at the most basic level is composed of vibrating strings. The elementary particles arise from interacting strings. The M-theory takes it further and depicts the Universe as an eleven-dimensional entity where the three basic forces and the strings are unified into a single model. To Hawking, the M-theory was the holy grail of physics, the grand unified theory that Einstein could not find. Hawking felt that our Universe exists within a conglomeration of infinitely many parallel Universes, the multiverse.

The String Theory, the M-Theory and the multiverse construct are not based on experimental evidence. They are consistent and comprehensive models based on complex mathematical formulations. Other than that, they are as elusive as the idea of vital force underlying life. For Hawking, an overarching physical model of the Universe obviated the notion of a supreme being. Reality is fully explicable by science. God is superfluous. His designation of the multiverse as a plausible alternative to God, however, derived from tenuous arguments. Mathematical formulation displaced evidence. It was akin to proposing a new religion based on arcane formulas.

Charles A Coulson and Francis Collins

Charles A Coulson, an applied mathematician and chemist, held professorships in theoretical physics, mathematics, theoretical chemistry at two major British universities during his career. His pioneering research in quantum chemistry, his prolific scientific output, mentoring of a multitude of graduate students, and authorship of distinctive books in science showed that he was a man of multiple talents. On top of that, he was a devoted Christian, and an influential scholar of religion who delivered occasional sermons.

Coulson viewed Christianity and science as complementary traditions. The vocal conflict between them was caused by misrepresentations of the adherents of both traditions. But that ought not be. The Bible points towards a meaningful, ordered Universe, a *'grand purpose'* to existence, and science investigates and discovers its majestic patterns. Noting that many prominent scientists had been devout Christians, he critiqued modern scientists for overlooking that history and positing an inherent conflict between science and religion. A reconciliation between them, he felt, would strengthen both and also benefit society.

Coulson was a trenchant opponent of writers of faith who said they had found scientific facts and theories in the scriptures. Chiding those who pointed to what science had not been able to explain as proof of the existence of God, he berated them for adopting a futile position. To lean on the claim that human logic cannot discern the mind of God only drives religion into a narrower corner whenever a new scientific discovery occurs.

> *The Christian God, at any rate, is not God of the gaps, though we have often been tempted to make Him so.* Charles Coulson (Kalthoff 2015).

Christians should recognize that science is an expanding body of knowledge, and should celebrate its findings as demonstrations of the *'order and beauty'* of nature, as created by God. The God-of-the-gaps argument may appeal in the short run but ultimately, it undermines religious belief. And people of science should not seek God within their discoveries but rather become better scientists. God is not hiding in the crevices of nature. He is represented by *'the whole of nature'*.

In Coulson's perspective, religion is *'the total response of man to all his environment'*. As a part of this, science is *'an essentially religious activity'*.

> *The scientist who has no use for religion, and the Christian who has no use for science, are each condemned for the narrowness and poverty of their view.* Charles Coulson (Kalthoff 2015).

As an astute scientist and a discerning Christian, he urged a respectful, rational interaction between science and religion. Lacking that, both the appreciation of the *'transcendental element in life'* and the *'confidence in the wholeness of life'* would be diminished.

A left-leaning pacifist who eschewed wartime enlistment by taking on extra teaching duties, Coulson was also involved in a number of local and international charitable activities.

> *In 1962-68 he was a Member of the Central Committee of the World Council of Churches, and during this period he became very active over Third World problems. He was Chairman of Oxfam (1965-71) but also made great efforts to provide help in education for developing countries, by often lecturing there, by bringing over students and by offering advice.* (Altmann and Bowen 2021).

A modest, unassuming man of a lighthearted disposition who always adorned a simple attire, Coulson abounded with energy and concern for those around him. His persona aptly reflected his Christian ethic. If one may critique him, it is to say that his conceptualization of religion was largely confined to Christianity and his assessment of the social functions of religion and science did not fully bring out some important but not-that-inspiring facets of their history like slavery, genocide and colonialism.

The eminent geneticist Francis Collins is a devout Christian who is not, to the consternation of fellow scientists, averse to forthrightly proclaiming his faith. Envisioning science and religion as compatible domains, he upbraids doctrinaire Christians who reject the Theory of Evolution and peddle pseudo-scientific creationism. The evidence for Darwin's theory is '*absolutely overwhelming*'. Yet, that does not constitute evidence against the existence of God. On the contrary, it is but an indicator of the majestic plan of God, a plan that covers the Big Bang, the precise tuning of the constants of the Universe, the usefulness of mathematics to describe the properties of matter and energy and other natural phenomena as well as an elegant code of morality. On these issues, Collins and Coulson held largely concurrent views.

Asserting that two out of five scientists are comfortable holding a belief in a personal God along with their scientific ideas, Collins explicates Him thus:

> *God, who is not limited in space or time, created this Universe 13.7 billion years ago with its parameters precisely tuned to allow the development of complexity over long periods of time. That plan included the mechanism of evolution to create this marvelous diversity of living things on our planet and to include ourselves, human beings. Evolution, in the fullness of time, prepared these big-brained creatures, but that's probably not all we are from the perspective of a believer.* Francis Collins (Resentiel 2009).

Taking the example of the multiverse theory, Collins points out that science is not just cold facts. Critical parts of science derive from supposition and faith. A sharp disjuncture between science and religion is unwarranted. He disdains attribution of as yet scientifically unexplained phenomena to God. God is not a filler of gaps. Yet, he holds that there are some gaps in our knowledge, like what preceded the Big Bang, that science can never fill. They are gaps only God can fill.

Jim Al-Khalili and Michio Kaku

Jim Al-Khalili, an award-winning theoretical physicist, author of popular books on science, and broadcaster of science programs, is a self-declared atheist. His father was a Shia Muslim, and mother, an evangelical Christian. But as he grew up, he did not embrace either faith.

> *My study of science, trying to find a non-theistic answer to the questions of how the Universe is and our place in it, did probably push me away from religious faith.* Jim Al-Khalili (PC 2015).

His quest for meaning propelled him towards humanistic atheism and, for two years, he served as the president of the British Humanist Association. With his amicable and humble persona, Al-Khalili does not disparage religious belief but says that his engagement with theoretical physics provides him with reasonable answers to basic questions about origin and function posed by religion.

As to unexplained issues like the fine-tuning of the Universe and the presence of exquisite complexity in nature, he says attributing them to God just leads to an unending backward process. Complexity cannot be explained via relying on a higher level of complexity. Al-Khalili faults physicists who fall back on notions like the multiverse theory to account for the complexity of nature. That is not a scientific position. Why does mathematics appear to be the language of nature? He remains baffled by this question.

> *Al-Khalili is humble enough to admit that science will not necessarily furnish humanity with all the answers, either now or in the future.* (PC 2015).

Al-Khalili has no qualms in saying that he envies people of faith. Having found an acceptable meaning of life, they are at peace with themselves. He swims in uncertain waters. But, in the spirit of science, he remains '*at peace with not knowing*'.

Michio Kaku, a distinguished theoretical physicist who co-founded a major branch of string theory, is an outstanding public intellectual. Author of a number of popular books on science and science fiction, he often appears on television and radio programs, and in print media. Blending the personae of James Hanson and Jim Al-Khalili, he produces shows on issues relating to modern science and critical societal matters like the environment, climate change, militarism, nuclear weapons, and the abuse of science by corporations and governments. In what is a distinct rarity for a senior scientist, his weekly radio program *Exploration* appears on the leftist Pacifica Foundation's flagship station, WBAI, in New York. He has also worked closely with Peace Action, an organization campaigning for nuclear disarmament and global peace. He portrays science in cautionary terms:

> *We have to realize that science is a double-edged sword. One edge of the sword can cut against poverty, illness, disease and give us more democracies, and democracies never war with other democracies, but the other side of the sword could give us nuclear proliferation, bio-germs and even forces of darkness.*

Michio Kaku

Michio Kaku grew up in surroundings infused with Shinto and Buddhist culture. With a high IQ, he began performing remarkable scientific experiments on his own from his school days. Awed by the beauty of nature, he began wondering about the origin of the Universe, meaning and purpose life, and the existence of a creator. But instead of turning to religion, he immersed himself in theoretical physics, hoping to glimpse into these mysteries therein.

Like Albert Einstein, Kaku does not believe in God, and surely, not a personal God. And like Einstein, he does not desist from positively invoking God in his public utterances. He is a pantheist. His God is the God of Baruch Spinoza. The fantastic complexity, order and symmetry found in nature and the elegant design of natural phenomena as captured by mathematical models of nature to him elicit a spiritual aroma. Nature is God; God is nature. Kaku's oft use of spiritual terminology has caused confusion about his belief or lack of belief in God. He clarifies:

> *When scientists use the word God, they usually mean the God of Order.* Michio Kaku (Berman 2018).

His secular philosophy enjoined with his commitment to social justice, peace and protection of the environment makes Kaku a humanistic pantheist. Despite the current load of major problems, Kaku is hopeful about humanity's ability to overcome them through social reorganization and prudent use of science. And importantly, Kaku does not see a basic cause for a conflict between science and religion:

> *They can be in harmony, but only if rational people on both sides engage in honest debate.* Michio Kaku (Kershaw 2012).

++++

The overall picture of scientists' views on religion emerges from a set of three studies of major scientists done in 1914, 1933 and 1998. Conducted with similar methodologies, they indicated an initial trend towards disbelief followed by a leveling off. In 1914, 28% of the scientists believed in a personal God. In 1933, the figure declined to 15%, and by 1998, it had reached 7%. Correspondingly, in 1914, 53% of the scientists had no belief in a personal God. In 1933, it rose to 68%, and by 1998, it reached 72%. Yet, about a fifth remained agnostic all along. The percentages of agnostics in 1914, 1933 and 1998 were 21%, 17% and 21%, respectively. (Larson and Witham 1998). A limitation of these studies was that they were confined to the US. But it is likely that the trends in Western Europe were similar.

The geographical lacuna was bridged for the first time by a pioneering survey that enrolled 22,525 biologists and physicists from eight nations—France, Hong Kong, India, Italy, Taiwan, Turkey, the UK, and the USA. The subjects varied from graduate students to senior academics and researchers in elite and non-elite universities and institutions. The eight nations differed in terms of their scientific infrastructure and overall level of religiosity (Ecklund et al. 2016, Ecklund and Johnson 2019). Six key conclusions were drawn:

One: Formal identification with a religion among scientists varies significantly between nations. In France, it is 30%, but in Italy, Turkey and Taiwan, it is above 50%. In India, it reaches nearly 95%. Yet, formal identification does not reflect religious belief or practice. It often reflects culture and custom.

Two: The level of religiosity, as indicated by belief and practice, is, in most places, lower among scientists than in the general population. One exception is Turkey, where regular attendance at religious services among scientists is similar to the general population. In the UK and the US, it is a half, and in France, it is a third of that among the general population. The overall nonattendance rates by scientists for religious service in these nations are: US (60%), UK (66%), France (81%), Turkey (40%) and Italy (44%).

Three: In the US and the UK, 10% of scientists do not doubt the existence of God. In India, the figure rises to 25% in India and about 65% in Turkey. Apart from Turkey, the majority of the scientists in the other nations are non-religious.

Four: Yet, in all the eight nations, scientists generally do not think science is in conflict with religion. Most of the biologists and physicists surveyed view them as distinct traditions operating in separate, non-antagonistic domains. Even scientists in the nations of the Global South with a high degree of public religiosity, science and religion are not seen as foes by scientists.

Five: Many scientists who have discarded the belief in God retain a sense of spirituality emanating from their work or through practicing meditation, yoga or another such activity. Secular scientists generally have no compunction in taking part in religious festivities like Christmas, Diwali and Eid or attending religious services with family. They freely partake in the general culture.

Six: But there is one major anomaly. On the public expressions of antagonism between science and religion, the US is an almost one-of-a-kind entity. In the European Christian majority nations, people of faith generally accept science, and their scientists and secularists are more tolerant of religion than in the US. The disproportionate influence of evangelical Christianity in politics, the political party duopoly and the tendency of the media to highlight wedge issues and cultural differences rather than class differences make the US a fertile ground for raising the profile of anti-science bigots who reduce science to things like the Theory of Evolution and stem cell research.

A year 2015 survey of Americans indicated that about 60% felt that science often conflicted with religion. But when questioned about their own beliefs and science, only 30% said the two were in conflict. It also varied according to level of religiosity. Of those who took part in religious functions at least once a week, 75% visualized a conflict between science and religion but among those who hardly or never did, only 50% indicated such a conflict. There were two major sources of conflict: evolution and creation of the Universe.

> *Two-thirds of Americans (65%) believe that humans evolved over time. About a third of U.S. adults (35%) say that humans evolved through natural processes, while about a quarter (24%) say that human evolution has occurred with the guidance of 'a supreme being'. .... About a third of adults (31%) say that humans did not evolve but have always existed in their present form; white evangelical Protestants (60%) are more likely than those in other major religious groups to hold this view.* (Funk and Masci 2015).

Yet, on key practical matters like climate change and genetically modified foods, religious belief or lack thereof had little effect on their views. About half of the respondents agreed that religious institutions should participate in important science related issues and the other half did not. Evangelicals were more likely to support such participation. Recent trends indicate that the nature and level of religiosity in the US is converging towards that in Western Europe. The religious right remains a political force to contend with, but socially it is becoming more and more isolated.

7.5 A PRUDENT TRUCE

Science depends on investigation and analysis. It is both a method of acquiring knowledge and a varied body of knowledge. Combining inductive and deductive reasoning with empirical evidence, it tests hypotheses and formulates theories. It has no absolute truths. Practically reliable theories are also subject to doubt. Ever afflicted with measurement, methodological and conceptual errors, it is always evolving. By now, it boasts a massive base of reliable theoretical and practical knowledge about all areas of nature and human affairs. Yet many primary and secondary issues remain unresolved. The debates in genetics, neurology, climate science, cosmology, mathematics and other areas exemplify areas of uncertainty and contention.

Religion, in essence, is a fixed body of ideas about existence, meaning, life, creation and the future that come along with a code of morality. Religious ideas and rituals are enshrined in scriptures deemed to be of divine origin. In religion, doubt is heretical and absolute faith, a virtue.

Fundamentally, science and religion are as distinct as oil and water. But, like most general assertions, it is subject to qualifications. Some faith traditions allow doubt and debate. But they form a minority. On the other hand, some aspects of science at times become ossified into dogma whose doubters are treated like heretics.

There are issues religion has no capacity to explain; there are issues science lacks the wherewithal to explain. The scriptures cannot disclose the molecular structure of water, unravel the process of global warming or identify the causes of heart disease. Science cannot disclose the meaning of life. Science assumes that the basic laws of nature do not change over time, but has no way of proving that. It is '*an unwritten article of faith*'. (Hossenfelder 2022). Is mathematics just a useful tool for understanding nature or is it embodied in nature or is it all that is? Is there an objective reality or is it all in the mind? Is mathematics an integral element of nature? Science is unable to resolve such conundrums. One scientist goes further:

There are quite a few areas where the foundations of physics blur into religion, but physicists don't notice because they're not paying attention. (Hossenfelder 2022).

One writer likens the relationship between science and religion as that between reason and emotion. Both are integral to human life. The former is *'a rational exploration of the natural world'* while the latter is an *'emotional communion with the supernatural'*. (Arand 2022). And both are processes, though driven by distinct methods. Science and religion both are ethical activities since they pledge to explore major questions in life with integrity and an open mind:

> *Just as a scientist who put certain preferred conclusions above the scientific method could be said to have abandoned the project of science, so too someone who confuses faith as an existential stance with faith as a mere body of beliefs, has ceased to be truly religious.* (Arand 2022).

Despite the mostly dissimilar natures of science and religion, their adherents can embrace both without psychic unease. Many scientists are religious and many people of faith accept science without reservation. But there are subgroups within each camp that express stark hostility to the other camp. However, they usually are in a minority.

The fundamental issue does not lie on the conceptual dimension. Rather, it lies on the practical dimension. And here, both religion and science have served noble as well as ugly societal ends and institutions. Both have served the causes of justice, protection of human dignity and enlightenment. And when allied with wealth and political power, major and minor religions together with science have been allied with violence, slavery, colonialism, racism, economic exploitation, eugenics, and suppression of human and women's rights. Religions have condoned militarism while science has enabled the most barbaric forms of militarism.

In these times of grave maladies that threaten the existence of life and humans on this planet, the need for science and religion to pool their resources to build a common movement is more critical than ever before. Both have to detach themselves from the interests and forces that are driving the flora and fauna of the globe towards oblivion. And both have to agree upon a common framework to ensure minimal friction in that endeavor. There is no other rational, moral choice:

> *[We] must either learn to live together as brothers [and sisters] or we are all going to perish together as fools.*
> Martin Luther King

The first book in this series, RPS (2022), presents a framework for the promotion of harmony and joint engagement between different faith traditions, and between religions and the varied secular traditions. Its components are: Declaration of spiritual equality, demarcation of common ethical principles, separation of religion and state, demarcation of the roles of religion and science in education, freedom of dress, fostering principled tolerance, acceptance of faith-based conversions, promotion of gender equality, non-discrimination over sexual preference, and regular joint celebrations and inter-faith prayers.

Taking that recipe as the foundation, below we note six additional items for a harmonious, productive social engagement between science and religion.

Mutual Respect: All major scientific organizations and religious institutions shall be signatories to a global pact accepting the fundamental rights of science and religion to exist and freely propagate themselves. Every human being shall be free to accept or reject any scientific proposition or religious belief. Like all other freedoms, these freedoms may be exercised only to the extent that they do not directly harm other human beings.

Critical Engagement: Accepting the rights of existence of science and religion shall not imply automatic acceptance of ideas. Scientists shall be free to criticize religion and people of faith shall be free to criticize science. The normal rules of libel and defamation shall apply. No religious authority may issue edicts that advocate violence against scientists and dissenters, and no science organization shall advocate the banning of religion. Either party shall be free to peacefully protest or mount civil action against the other party. Peaceful dialogue between science and religion on a regular basis should be encouraged and allotted public funding.

Public Policy: State policies on health, education, the economy and the environment shall emanate from a genuinely democratic process and be based on the state-of-the-art science and technology, not scriptures. Religious people and organizations shall have the freedom to participate in these decisions.

Public Education: All children should get free schooling up to the high school level and shall be required to learn the major science disciplines as determined by the state educational authority and science bodies. Their contents and methods shall be free of religious influence. Public schools may not teach the beliefs and rituals of a specific religion. But comparative religion as a subject may be taught, provided it is not biased towards a particular religion.

Human Flourishing: The people and organizations of science and religion shall pledge to conduct themselves according to the highest standards of morality. They shall firmly desist from succumbing to racism, ethnic exclusion, extreme nationalism and gender discrimination, and actively promote within and between nation economic, educational and health-wise equality, international peace and disarmament, and honest cooperation on important global issues like poverty, hunger, militarism, climate change and economic inequality.

Planetary Welfare: In the light of the imminent danger of reaching the tipping point for catastrophic climate change and the deceptive policies of corporations and governments that are undermining the key goal of preventing a 1.5 degrees rise in the average planetary surface temperature, all people of faith and science should consider it a matter of urgency to mount strident civil action to force the local and international authorities to immediately take the actions recommended by the UN and the IPCC.

++++

We end this book with the words of two persons of eminence, a theologian and a scientist.

Dietrich Bonhoeffer was an influential German Lutheran priest and theologian during the dark days of Nazi rule. Unlike the majority of the German clergy and the bulk of the German scientists and mathematicians of that era, he did not acquiesce, or take part, in the evil unfolding before his eyes. Instead, he spoke up and condemned Nazi policies. Consequently, he was sent to a concentration camp and hanged. His compelling words remain as relevant today.

Silence in the face of evil is itself evil:
God will not hold us guiltless.
Not to speak is to speak.
Not to act is to act.

Dietrich Bonhoeffer

Albert Einstein was not a just a great theoretical physicist but also a humanist and a socialist who denounced the Nazi rule in his homeland, supported the struggle for racial equality in the US, decried the McCarthy era suppression of basic rights, and campaigned for nuclear disarmament. His exalted vision of an all-embracing global family denotes the essential foundational tenet for the code of morality of the future.

Our task must be to free ourselves
by widening our circle of compassion
to embrace all living creatures and
the whole of nature and its beauty.
Albert Einstein

The fundamental issue people of faith, secularists and scientists must confront is not to acquiesce with injustice, inequality, discrimination and militarism. That is more critical than whether or not one accepts that the world was created in six days. On the latter front, people with diverse beliefs can coexist peacefully. But significant disagreement on the former front is a recipe for disaster for humanity and all life on Earth.

SOURCES

Nothing is infallible.
Nothing is binding forever.
Everything is subject to inquiry and examination.
BR Ambedkar

From religion comes a man's purpose;
from science, his power to achieve it.
William Henry Bragg

1. INTRODUCTION

Section 1.0: HH 2022; Trefil and Hazen 1998; Wikipedia (2022 - Albert Einstein, Louis Pasteur, Michael Faraday, Rosalind Franklin).
Section 1.1:
Section 1.2: Aga Khan III 1954; Gandhi 1929, 1983.
Section 1.3:
Section 1.4: Boyett 2016; DK Publisher 2013; Hirji 2022; Wikipedia (2022 – Endless Knot).

2. EMERGENCE

Section 2.0: NWE Contributors 2018; Rae 2013; Rothman 2001; Verschuuren 2017; Wikipedia (2022 – Holism, Reductionism).
Section 2.1: Anderson 2014; Angell 2005; Astbury 2001; Burton et al. 2021; Groopman 2005; Jaber et al. 2021; Schmidt 2017; Spizziri 1990; Weinrib 2019.
Section 2.2: Aaen-Stockdale 2012; Alimohammadi et al. 2013; Amaro 2015; Awaad et al. 2021; Barni and Mahdany 2017; Bastian 2018, 2019; BT 2019; Chan 2016; Dhonden 2000; Draper and Uhl 2019; Editors 2019; Egnor and Gallagher 2022a, 2022b; Evans et al. 2020; FH 2020; GQ 2022a, 2022b; Hamdy 2009; Hawramani 2018; Helminski 2017, 2022; Holt 2017; HSS UK 2022; Heuertz 2009; IB 2018; James 2007; Koch 2014; Llewellin 2022; Marc 2018; Miri 2022; Murray 2017; Nanda 2016; Preston et al. 2013; Promtha et al. 2021; Sider et al. 2019; Skarbek 2021; Slingerland 2018; van Bohemen 2010; Wikipedia (2022 – Holism, Reductionism, Vitalism).
Section 2.3: Clarke 2017; Cozzarelli 2020; Garnham 2018; Hari 2022; Kangal 2020; Levine et al. 1987; Mouzelis 1980; Thompson 2011a, 2011b, 2011c, 2011d, 2011e, 2011f, 2011g, 2011h.
Section 2.4: Aaen-Stockdale 2012; Al-Khalili 2009; Bais 2010; Bar-Yam 1999; Clark et al. 2018; Clayton and Davies 2006; Coyne 2013; Evans et al. 2020; Frank 2021a, 2021b; Geiser 2022; Glyn 2013; Handelman 2015; Harper 2019; Horgan 2022; Laughlin 2005; Nintil 2015; NWE Contributors 2018; Oberlander 2022; Oreskes 2019; Palmer 2021; Preston et al. 2013; Siegel 2022; Slingerland 2018; Tse 2015; TSL Editors 2022; Voosholz and Gabriel 2021; Wagner 2009; *Wan 2011;* Wattles 2002; Wikipedia (2022 – Emergence, Emergentism, Scientific Method, Vitalism); Wintjen 2022.

Section 2.5: Bais 2010; Carey 2012; Harden 2021; Hercher 2018; Klug et al. 2019*;* Laughlin 2005; Mandal 2020; Meloni 2016; Mukherjee 2017; Nintil 2015; NWE Contributors 2018; Palmer 2021; Weber 2019; Wikipedia (2022 – Genetics).

Section 2.6: Ball 2022; Barret 2015; Edwards 2015; O'Grady 2022; Trefill and Hazen 1998; Wikipedia (2022 – Brain, Mind, Nervous System).

Section 2.7: Ahlstrom 2001; Ann 2019; Bang 2012; Baumeister 2013; BS 2021; Burlew 2022; Cave 2016; Critchlow 2020; Editorial 2012; Ellis 2020; Gagnon 1981; Guttridge 2012; Harris 2012; Hoekstra and Robinson 2022; Hurd et al. 2018; List, Caruso and Clark 2019; Miksha 2016; Narain 2014; Palmer 2010; PB 2023; Pipkin 2019; Rios 2007; Timpe 2021; Tse 2015; Wikipedia (2021 – Free Will).

Section 2.8: Davies 2022; Davis 2022; Harvey 2007; Marable 2015; Milanovic 2021; Patnaik and Patnaik 2019; Stiglitz 2019; Thier 2020; Yablon 2020.

Section 2.9: Bais 2010; Bhattacherjee 2012; Trefil and Hazen 1998; Wagner and Briggs 2016; Wikipedia (2022 – Scientific Method, Social Sciences).

3. EUGENICS

Section 3.0: Churchill 2001; Encyclopedia 2020; Wikipedia (2022 – Eugenics).

Section 3.1: Bianculli 2017; Bruce 2016; Cain 2020; Darwin 1871; Delzell and Poliak 2013; Encyclopedia 2020; English 2016; Foster 1998; Galton 1873; Media 2020; Pearson 1901, 1912; Wagenmakers 2018; Wikipedia (2021 -- Charles Darwin, Eugenics, Francis Galton, Karl Pearson, Thomas R Malthus).

Section 3.2: Citizendium 2020; Kevles 1985; Mandal 2020; Wikipedia (2021 – Eugenics, Introduction to Genetics).

Section 3.3: Baker 2015; Black 2003; Cavanaugh-O'Keefe 2020; Cohen 2016; Frazier 2019; Kurbegovic 2014; Samaan 2020; Wikipedia (2021 – Charles B Davenport, David Starr Jordan, Eugenics, Irving Fisher, Oliver Wendell Holmes, Jr., Oliver Wendell Holmes, Sr., Theodore Lothrop Stoddard).

Section 3.4: Baker 2015; Bjorkman and Widmalm 2011; Black 2003; Carter 2017; Cavanaugh-O'Keefe 2020; Cohen 2016; Donahue 2016; Frazier 2019; Freedland 2019; Gerais 2017; Grue 2010; Ings 2016; Kevles 1985; Kurbegovic 2014; Lombardo 2016; Pernick 1997, 2002; Samaan 2020; Wikipedia (2021 – Anti-miscegenation Laws in the United States, Charles B Davenport, Compulsory Sterilization, David Starr Jordan, David Herbert Lawrence, Eugenics, George Bernard Shaw, Introduction to Genetics, Irving Fisher, Mary de Garmo, Oliver Wendell Holmes, Jr., Oliver Wendell Holmes, Sr., Theodore Lothrop Stoddard).

Section 3.5: Black 2003, 2004, 2012; Blakemore 2019; Cavanaugh-O'Keefe 2020; Chase 2017; Cochran 2020; Devolder 2015; Fitzpatrick and Moses 2018; Friedman 2020; Geiderman 2002; Grue 2010; HE 2020; JVL 2020; Katznelson 2017; Kernan 2020; Klein 2011; Levine 2019; Lombardo 2020; Miller 2020; Moya-Smith 2017; MPS 2020; Murphy 1935; Nock 1941; OC 2016; Ray 2020; Richardson 2019; Stahnisch 2013; USHMM 2020; von Lupke-Schwarz 2013; Walker 2015; Waller 2012; Weindling et al. 2016; Whitman 2017; Wikipedia (2021 -- Adolf Hitler, Akton T4, Anti-communist Mass Killings, Auschwitz Concentration Camp, Ernst Rudin, Extermination Camps, Harry H Laughlin, Joseph Mengele, Law for the Prevention of Genetically Diseased Offspring, List of Prisoners of Dachau, Nazi Concentration Camp, Nazi Eugenics, Nazi Human Experimentation, The Holocaust, The Nuremberg Laws); Wolff 2020.

Section 3.6: Baker 2015; Berenbaum 2020; Bianculli 2017; Brown 2007; Bruce 2016; Burton 2020; Cain 2020; Churchill 2001; Clayton 2020; Darwin 1871; Delzell and Poliak 2013;

Edwards 2019; Ehle 1998; English 2016; Gerais 2017; Gerhart 1996; Guha 2019; Hari 2010; Helfand 2020; Jeffries 2012; Kurbegovic 2014; Lee 2020; Luff 2020; Media 2020; Mukerjee 2010; Pearson 1901, 1912; Pernick 2002; Polya 2012; Wagenmakers 2018; Wikipedia (2021 – Alexis Carrel, Eugenics, Nazi Eugenics, Racism in the Work of Charles Dickens, The Herero and Nama genocide); Younge 2002.

Section 3.7: Allen 2000; Fleury 2015; Frazier 2019; Freedland 2019; Gibbons 2020; History Editors 2020; Horvath 2020a; Krome-Lukens 2009; Kurbegovic 2014; Lombardo 2019; Lyster 2014; Majfud 2020; Mena 2017; Nuriddin 2017; Oveyssi 2015; Sanger 2011; Singleton 2014; Tonn 2017; Wikipedia (2021 – Feminist Eugenics, Gertrude Davenport, Lothrop Stoddard, Margaret Sanger, WEB Du Bois, Women's Rights).

Section 3.8: Adam 2003; Ali 2007; Berg 2013; Berzin 2020; Cama 2012; Chaudhury 2018; Chazan 2018; Chowdhury 2018; Eig 2017; Epstein 1999; Hemmer 2017; Johnson 2011; Kaminski 2010; Malik 2019; Motadel 2017; Redmond 2017; Sachedina 1995; Sharma 2018; Shtrauchler 2017; Thangaraj 2017; Victoria 2013, 2015; Victoria and Muneo 2014; Wark 2010; Wikipedia (2021 -- Armenian Genocide, Daisetsu Teitaro Suzuki, Hindu Mahasabha, Kurds in Turkey, Rashtriya Swayamsevak Sangh, Relations between Nazi Germany and the Arab World, Religion in America, Takenaka Shogen); Wilkes 2020.

Section 3.9: AES 1926; Alomia 2010; AP 2020; Appleman 2018; Baker 2014; Ball 2021; Barnes 1949; BBC 2021; Bergman 2011; Billinger 2014; Blaich 1993, 1996; Brian 2016; Burleigh 1994; Chesterton 1922; Clark 2017; Dhingra 2018; Draper 2019; Durst 2017; EA Editors 2021; Encyclopedia 2020; Farley 2019; Fiedler 2013; Garver and Garver 1992; GQ 2021b; Keel 2019; Kellner 2005; Kurbegovic 2013; Lombardo 2016; LT 2020; MacKinnon 2021; McClay 2010; Merricks 2012; Naicker 2010; Nelson 2015; O'brien 2020; OR 1924; Pearson 1930; Phillips 2018; Quartey 2019; Rogers 2021; Rosen 2020; Schroder 2003; Selden 2021; Shudy 2012; Smith 2004; Spenkuch and Tillmann 2017; Stern 2014; Toth 2005; UMC 2016; USHMM 2021; Wiggam 1924; Wikipedia (2021 – Augustus Hopkins Strong, Catholic Church and Nazi Germany, Confessing Church, Emanuel Hirsch, Gilbert Keith Chesterton, Jehovah's Witnesses, John Harvey Kellogg, Matthew (3:10, 7:17, 7:18), Paul Althaus, Religion in Germany, Religious Views of Adolf Hitler, The Galton Society).

Section 3.10: Adams 1989; Agarwal 2020; Allen 2018; Angus 2009; Bergman 2001; Blacker 1951; Brezis and Young 2003; Citizendium 2020; Crew et al. 1939; deJong-Lambert 2017; Evans 2018; Follet 2019; Foster 1998; Frazier 2019; Gould 1981; Haldane 1934; Hogben 1948; Hughes 2020; Krementsov 2011, 2018; Lienhard 2021; Lucassen 2010; Luff 2020; Maisey 2020; Meloni 2016; Milmo 2011; Nuriddin 2017; NYT 1975; Pontecorvo and Pallardy 2020; Rejon 2018; Rosenthal 2019; Rudling 2014, 2015; Slorach 2020; Tabery 2011; Trotsky 1935; Weber 2019; Weisman 1970; Wikipedia (2021 – Alexander Sergeevich Serebrovsky, Eugenics Manifesto, Franz Boas, Health Care in Russia, Hermann Muller, Jack Lindsay, JBS Haldane, John Desmond Bernal, Lancelot Thomas Hogben, Lysenkoism); Williams 2010.

Section 3.11: Adams 2021; Brignell 2010a, 2010b; Durst 2017; Duster 2003; Foster 1998; Freedland 2012; Gordon 2012; Hirji 2019; Kellner 2005; Lombardo 2011; Oreskes 2019; von Lupke-Schwarz 2013; von Sponeck 2006.

4. SCIENCE

Section 4.1: Angell 2005; DeAngelis 2000, 2006; Editorial 2004; Hirji 2009; SC 2021; Wikipedia (2021 – Science); Wilson 2009; Zinn 1980, 1990.

Section 4.2: Anderson 2008; Anderson 2020; BBC 2022; Cowie 2019; Craig 1999, 2000; Dawkins 2015, Denigris 2015; Duffy 2013; Duigman 2021; Fallon 2021; Galadari 2019;

Gross and Wright 2017; Hogan 2019; Kandi 2013; Koch 2009, 2018; Lewis 2018, 2019, 2021; McLuhan 2019; Mowe 2011; Nelson 2017; O'Leary 2017; Pretorius 2016; Randi 1982, 1995; Rawlette 2020; Reed 2012; Romero and Perez 2012; Schwartz 2017; Schwartz and Simon 2007; Tharoor 2014; Wikipedia (2021 – Big Bang Theory, Kalam Cosmological Argument, List of New Religious Movements, Religious Interpretations of the Big Bang Theory, Scholarly Approaches to Mysticism, The Mind–Body Problem, Unification Movement); Wikiversity (2021 – Does God Exist?); Wolchover 2017; Wright 2018.

Section 4.3: Garrido 2013; Jyotiraditya 2018; McLeod 2019; Musser 2011; Narain 2014; Rovelli 2022; Sengupta 2016; Stevenson 2019; Trefil and Hazen 1998.

Section 4.4: Anonymous 2020; Cave 2016; Chilton 2018; Doorn 2019; Editorial 2012; Ellis 2020; Leasure 2019; McLeod 2019; Narain 2014; Park et al. 2020; Pipkin 2019; Priest 2018; Sproul 2021; Timpe 2021; Wikipedia (2021 – Free Will).

Section 4.5: Ahlstrom 2001; Anonymous 2020; Betuel 2020; Carey 2012; Cave 2016; Chabris et al. 2012; Chilton 2018; Comfort 2018; Conan et al. 2010; Corbyn 2022; Critchlow 2020; Doorn 2019; Ebstein et al. 2015; Editorial 2012; Ioannidis 2005; Koenig et al. 2005; Lewis 2020; Lurie and Lurie 2020; Mehta 2014; Meloni 2016; Mitchell 2019; Mukherjee 2016; Mullin 2020; Narain 2014; Pipkin 2019; Plomin 2019; Resnik and Vorhaus 2006; Savage et al. 2018; Scharping 2018; Sullivan 2019; The Onion 1999; Tielbeek et al. 2017; Wikipedia (2021 – Free Will); Wintour 2013; Ziliak and McCloskey 2008; Zwart 2014.

Section 4.6: Afsaruddin 2005; Ahlstrom 2001; Al-Jubouri 2004; Ananthaswamy 2012; Anne 2016; Anonymous 2020; ANU 2021; Arnold 2020; Beaumont 2019; Betuel 2020; Bhat 2021; Bitesize 2021; Brass et al. 2017; Furstenberg and Mele 2019; Carey 2012; Carrier 2018; Cave 2016; Chabris et al. 2012; Cherry 2020; Chilton 2018; Choate 2013; Choy et al. 2022; Clayton 2018; Colmez and Schneps 2019; Comfort 2018; Conan et al. 2010; Critchlow 2020; Davis 2020; Diodati 2020; Doorn 2019; Ebstein et al. 2015; Editorial 2012; Ellis 2020; EPFL 2020; Evans 2018; Federman 2010; Fraser 2015; Garrido 2013; Gier and Kjellberg 2004; Gholipour 2019a, 2019b; Goetz 2014; Gooding 2018; Graffin and Provine 2007; Graham-Leigh 2018; Heflick 2010; Ioannidis 2005; JW 2021; Jyotiraditya 2018; Koenig et al. 2005; Kulkarni 2020; Leasure 2019; Levine 2021; Lewis 2020; Libet et al. 1983; Libet 2007; Lott 2013; Lurie and Lurie 2020; McLeod 2019; Mehta 2014; Menaker 2012; Mitchell 2019; Mukherjee 2016; Mullin 2020; Musser 2011; Narain 2014; NTU 2022; O'Brien 2020, 2021; OT 2015; Park et al. 2020; Parrott 2017; Petkov 2019; Pipkin 2019; Plomin 2019; Repetti 2015; Saini 2020; Savage et al. 2018; Sayadaw 2019; Scharping 2018; Sengupta 2016; Sennett 1971; Sproul 2021; Stafford 2013; Sullivan 2019; The Onion 1999; Tielbeek et al. 2017; Timpe 2021; Wikipedia (2021 – Free Will, Neuroscience of Free Will); Woods 2001; Ziliak and McCloskey 2008; Zwart 2014.

Section 4.7: Baars 2003; Bar-Yam 1999; BS 2021; Budson et al. 2022; Castillou 2022; Choy et al. 2022; Corbyn 2022; DiGravio 2022; Farnsworth 2020; Frank 2021a, 2021b; Geiser 2022; GL 2022; Ghose 2012a, 2012b; Horgan 2022; Jaeger et al. 2022; Khadka et al. 2014; King 2013; Laughlin 2005; NTU 2022; Oberlander 2022; Rios 2007; Siegel 2022; Seth 2007; Suttie 2013; Vithoulkas and Muresanu 2014; Wikipedia (2022 – Mind); Wintjen 2022; Wintour 2013; Wolchover 2017.

Section 4.8: ACLU 2021; ACS 2020; Anne 2019; Berard 2022; Dennett 2015; Dennett and Caruso 2018; SD 2012; Vonasch et al. 2017; Wertz et al. 2018; Wikipedia (2021 – Neurolaw).

Section 4.9: Al-Jazeera 2021; Cave 2016; Gagnon 1981; Hirschfeld 2009; Neri 2018; Oreskes 2019; Oreskes and Conway 2010; Palmer 2010.

Section 4.10: Al-Jazeera 2021; Augustyn 2022; Cave 2016; Crossman 2020; Gagnon 1981; Hirschfeld 2009; Larsson 2011; Marcuse 1991; Marx 1852, 1959; Marx and Engels 1846; Palmer 2010; Neri 2018; Oreskes 2019; Oreskes and Conway 2010; Palmer 2010; Spirkin 1983, 1990; Wikipedia (2021 – Alexander Spirkin, False Consciousness, Marx's Theory of Alienation); Wills 2021.

Section 4.11: Afsaruddin 2005; Ahlstrom 2001; Al-Jubouri 2004; Ananthaswamy 2012; Anne 2016; Anonymous 2020; ANU 2021; Arnold 2020; Beaumont 2019; Betuel 2020; Bhat 2021; Bitesize 2021; Brass, Brown and Behrmann 2017; Furstenberg and Mele 2019; Carey 2012; Carrier 2018; Cave 2016; Chabris et al. 2012; Cherry 2020; Chilton 2018; Choate 2013; Clayton 2018; Colmez and Schneps 2019; Comfort 2018; Conan et al. 2010; Critchlow 2020; Davis 2020; Diodati 2020; Doorn 2019; Ebstein et al. 2015; Editorial 2012; Ellis 2020; EPFL 2020; Evans 2018; Federman 2010; Fraser 2015; Garrido 2013; Gier and Kjellberg 2004; Gholipour 2019a, 2019b; Goetz 2014; Gooding 2018; Graffin and Provine 2007; Graham-Leigh 2018; Heflick 2010; Ioannidis 2005; JW 2021; Jyotiraditya 2018; Koenig et al. 2005; Kulkarni 2020; Leasure 2019; Levine 2021; Lewis 2020; Libet et al. 1983; Libet 2007; Lott 2013; Lurie and Lurie 2020; McLeod 2019; Mehta 2014; Menaker 2012; Mitchell 2019; Mukherjee 2016; Mullin 2020; Musser 2011; Narain 2014; O'Brien 2020, 2021; OT 2015; Park et al. 2020; Parrott 2017; Petkov 2019; Pipkin 2019; Plomin 2019; PLOS 2020; Priest 2018; Repetti 2015; Saini 2020; Savage et al. 2018; Sayadaw 2019; Scharping 2018; Sengupta 2016; Sennett 1971; Sproul 2021; Stafford 2013; Sullivan 2019; The Onion 1999; Tielbeek et al. 2017; Timpe 2021; Wikipedia (2021 – Free Will, Free Will in Theology, Predestination in Islam); Woods 2001; Ziliak and McCloskey 2008; Zwart 2014.

Section 4.12: Aaen-Stockdale 2012; Alpert 2019; Anonymous 2003; Asher 2012; Assad 2021; Austin 1998; Baird and Gleeson 2017; Bardon 2019; Bentley 2018; Blackford 2016; BL Editors 2019; Bloom 2016; Brice 2021; Cain 2021; Capps 1992; Chalabi 2015; Collins 2007; Dawkins 2006, 2008; Ecklund 2010; Ecklund et al. 2016; Ecklund and Scheitle 2017; Ecklund and Johnson 2019; Ellis 2007; Ellison et al. 2007; Esteves 2020; Fuchs 2013; Funk and Masci 2015; Gleeson and Baird 2017; Gajilan 2007; Gould 1997, 1999; Inch 2014; Jones and Leicht 2016; Johns 1999; Johnson 2005; KR 2001; Lavers 2001; Lawless 2021; Lee 2018; Liu et al. 2009; Masci 2009; McKenna 2014; McMaster 2020; Morales 2009; Newberg 2010; Newberg, D'Aquili and Rause 2002; Nsar 2006; Omundson 2020; Orr 1999; Paulson 2006; Ricker 2007; Rhodes 1999; Ridley 1998; Salleh 2018; Schuler 2016; Schwartz 2021; Skatssoon 2006; Stankorb 2021; Stenger 2006, 2007, 2011; Stenger et al. 2015; Tan 2019; Ward 2008; Whitehead 1925; Williams 2006; Wikipedia (2021 -- Alfred North Whitehead, Anthropic Principle, Evolution and the Catholic Church, Francis Collins, Francis Crick, James Watson, Neuroscience of Religion; Non-Overlapping Magisteria, Stephen Jay Gould, Victor J Stenger); Youra 2020; Zuckerman 2015.

Section 4.13: Ames 2003; Barash 2014; Capra 2010; CBC 2020; Dalai Lama 2006; Donis 2021; Finkelstein 2003; Hut 2003; IDR Labs 2021; Impey 2020; Johnson 2005; Klinghoffer 2012; Lopez 2010, 2021; Mansfield 2003; Merali 2018; Wallace (editor) 2003; Wallace 2003; Westmoreland 2019; Wikipedia (2021 – Buddhism and Evolution, Buddhism and Science, Buddhist Cosmology, Relationship between Religion and Science, The Unanswered Questions); Zukav 1983.

Section 4.14: Khalil 2022; Snodgrass 2022; Wikipedia (2021 – Comparative Religion, Unification Church).

Section 4.15:

Section 4.16: Alpert 2019; Ducharme 2018; Gewertz 2007; Harrison 2018; Jammer 1999; Jones 2011.

5. MATHEMATICS

Section 5.1: Ascher 2002; Beckmann 1967; Caldwell 2021; Euclid and Heath 2017; Gerdes 1994, 2007; Goldman 2012; Hom 2015; Joseph 2000; Katz 2019; Lakshmi 2017; Nieder 2020; SOM 2021a; Ward 2018; Wikipedia (2021 – Mathematics).

Section 5.2: Allen 1997; Boyer and Merzbach 1991; BP 2021; Clegg 2009; Ferencik 2017; Huffman 2019; Pomeroy 2013; Timon 2018; Wikipedia (2021 – Hippasus, Pythagoras, Pythagoreanism).

Section 5.3: Alexander 2014; Boltz 1983; Castelvecchi 2020; Dunham 2007; Dutta 2002; FSTC 2013; Harrison 2018; HC Editors 2019; Hidetoshi and Rothman 2008; Hosking 2017; IM 2021; Jarus 2013; Jnana 2021; Jones 2011; Joseph 2000; Kaplan 1999; Lakshmi 2017; Martinez 2018; Mastin 2020; Mulcare 2013; O'Connor and Robertson 2021; Seife 2000; Solis 2013; SOM 2021a; Strevens 2020; Sutton 2007; Tate 2010; Ward 2018; Wigner 1960; Wikipedia (2021 -- Galileo Galilei, Giordano Bruno, Indian Mathematics, Isaac Newton, Johannes Kepler, Maryam Mirzakhani, Nicolaus Copernicus, Sangaku, Shulba Sutras, Tycho Brahe).

Section 5.4: Abbot 2013; Atiyah 1995; Berman 2018; David 2017; Dewdney 1999; Freiberger 2008; Howell 2005; Kaku 2022; Kaufman 2009; Kuhn 2010; Lamb 2011a, 2011b; Livio 2010, 2015; Nelson 2017, 2021; Otis 2020a; Pedigo 2019, 2020; Renyi 1967; Tyson 2011; Wigner 1960; Wikipedia (2022 - Michio Kaku, The Unreasonable Effectiveness of Mathematics in the Natural Sciences).

Section 5.5: Alexander 2014; Blanc 2021; Campbell 2016; Csillag 2008; CSM 1985; Faena 2021; Freiberger and Thomas 2013); Glutsyuk 2014; Grabiner 2014; Hannon 2010; Hosch 2016; Humphrey 2018; Jongsma 2005; Kurland 2031; Lisle 2017; Mathigon 2012; MCCH 2017; McGill 2007; NPR Staff 2014; O'Connor and Robertson 2002; Paulos 2014; Ransford 2017a, 2017b; Ransford and Eyghen 2018; Riley 2020; Rossis 2021; Scoles 2016; Tassone 2021; Tbakhi and Amr 2007; Webb 2015; Wikipedia (2021 – Cistercian Numerals; Georg Cantor, History of Calculus, Infinity, Zeno's Paradoxes); Zarepour 2020; Zhu 2022.

Section 5.6: Aafiya 2015; Abdelhamid 2021; Aron 2016; Aslan 2022; Bellos 2015; Bollobas 2016; Budd 2020; Devlin 1997, 2011; Dvorsky 2017; Forsey 2021; Gouvea 2014; Hannon 2010; HASD 2014; Kanigel 1991; Landau 2020; Lisle 2009; Livio 2003; MCCH 2017; Russo 2019; RW 2021; Sameer 2017; University of Pennsylvania 2005; Wilkerson 2015; Wilson 2002; Wilson 2021; Wikipedia (2021 – Aryabhata, Golden Ratio, Srinivasa Ramanujan).

Section 5.7: Dudley 1997; Esposito 2021; Harrison 2019; Gabriel 2017; Jayaram 2021; Jnana 2021; Schimmel 1993; Sujato 2010; Wikipedia (2021 – Biblical Numerology, Numerology, Numerology and the Church Fathers).

Section 5.8: Calvo 2021; Devlin 1998; Goldstein 2006; Hodges 2008; Monk 2018; Nagel and Newman 1958; Parc 2014; SA 2006; SOM 2021b; Wikipedia (2021 – Non-Euclidean Geometry).

Section 5.9: Archuman 2017; AW 2021; Bird 2021; Bogomolny 2021; Davis 1998; DT 2021; Kilanowski 2014; Look 2020; Manu 2018; McHargue 2015; McIntyre 2017; Monk 2018; Murphy 2021; Murray 2021; Nagasawa 2017; Otis 2020b; Parc 2014; Smith 2011; Strickland 2016; Suri and Bal 2007; Tent 2021; Vaughn 2021; Vestal 2007; Wikipedia (2021 – Carl Friedrich Gauss, Kurt Friedrich Godel, Leibniz, Non-Euclidean Geometry, Rene Descartes, Trademark Argument); Wilson 2014.

Section 5.10: Abeka 2017; Angier 1997; Bradley and Howell 2011; Davis 1998; Downey 2017; EBC 2021; GU 2021; Howell and Bradley 2001; Kelly 2021a, 2021b; McIntyre

2017; Nickel 2012; Poythress 2015; Stirrat and Cornwell 2013; Wikipedia (2021 -- Association of Christians in the Mathematical Sciences, Sir George Stokes, 1st Baronet).

Section 5.11: Akpan 2013; Als-Nielsen et al; Altman 1980; AMS 2019; Angell (2005); Bekelman, Li and Gross 2003; Benner 2020; Black 2000; Brenner 2002; Bret 2012; Bruning 2005; Bultheel 2015; Chandra and Holt 1999; Chiodo and Bursill 2019; Editorial 2019; Ernest 2020; Faerber and Kreling 2014; Ferguson 2012; Finkbeiner 2013; Fraenkel 2017; Gerardia, Goette and Meier 2013; Ghaemi, Shirzadi and Filkowski 2008; Gossard 2018; GUM 2017; Gunderman 2019; Halpern 2020; Hardy 1940; Harris 2015; Harris 2020; Harvard Health 2017; Harvey 2020; Haynes 2020; Hill 2014; Hirji 2008; Huckle 2021; Jalali et al 2020; Kalla 2010; Kassirer 2005; Kerekovska and Galunska 2015; Kolodny 2020; KW 2009; Lamb 2017; Lambert 2011; Leinster 2014a, 2014b; Lohr 2008; Magnello 2018; McCleary 2018; McNeill 2019; MT Staff 2019; Munoz 2020; Orhan 2021; Patterson 2009; Plackett 2013; Rehmeyer 2008; Ruiz 2010; Schaefer 2021; Schoenfeld 2001; Segal 1980, 2003, 2014; Sheller 2021; Siegmund-Schultze 2004; Spencer 2009; Spiegelhalter and Pearson 2009; Stewart 2012; TPM 2018; Van Dam 2009; Wikipedia (2021 – Crimean War, Emil J Gumbel, Financial Crisis of 2007–2008, Florence Nightingale, Hellmuth Kneser, IBM and the Holocaust, Misuse of Statistics, National Socialist German Lecturers League); Williams 2021.

Section 5.12: Alvarez Amado and London 2021; Green 2012; Kelly 2021a; Mason 2002; McKenzie 2020; Oyedele 2016; Saka 2021; Wikipedia (2021 – Blaise Pascal, Jafar al-Ṣadiq, Pascal's Wager, Precautionary Principle).

Section 5.13: Antonio 2012; Davis 1998; Davis and Hersh 1999; Sullivan and Zutavern 2017; Tu 2013; Wikipedia (2020 – Pythagorean Theorem).

Section 5.14: Calgar 2020; Davis 2018; Feron 2014; Hartnett 2017; Holt 2008; Kelly 2021b; Korner 2007; Larson and Witham 1998; Lennox 2020; Su 2012; UNHCR 2021; Vasak 2020.

6. RUMINATIONS

Section 6.1: http://math.furman.edu/~mwoodard/mqs, www.quotemaster.org/math+and+religion

Section 6.2: Agar 2008; Anonymous 1913; Burton 2022; ECLA 2008; Ellis-Petersen 2022; Esmaeilzadeh et al. 2022; Farley 2021; Gerrard 1914; GQ 2022a; Guest 2022; Harden 2021; Hathout 2006; Kramer 2018; Leon 2013; Mason 2022; Nelson 1988; NHS 2022; Pandey 2021; Poole 2022; Rahman 2022; Ratcliffe 2022; Reporter 2016; Rutherford 2022; Thangaraj 2017; Thomas 2017; Tong 2022a, 2022b; Wikipedia (2022 - New Eugenics).

Section 6.3: Al-Jazeera 2021; BBC 2022; Bricker 2021; CCSWK 2022; CE 2022; CEA 2022; Chandler 2021; Chandra 2021; CUP 2022; Dalai Lama and Alt 2020; Fang and Lerner 2019; Fici 2018; Foley 2020; Gahlau 2022; Graves-Fitzsimmons and Siddiqi 2021; GQ 2022b; Guru-Murthy 2020; Hanh 2022; Huxter 2022; Jenkins 2022; JW 2022; Kimeu 2022; Loy 2015; NASA 2022; OES 2015; Ozdemir 2020; PRC 2015; Radwan 2022; Rao 2021; Roy 2018; Sengupta 2019; Shraiky 2021; Snyder 2022; UN 2022; Vedachalam 2012; Wikipedia (2022 - Jehovah's Witnesses, The Islamic Declaration on Global Climate Change, Saudi Arabia); Zafar 2021.

Section 6.4: Armstrong 2022; Crawford 2019; Falconer 2021; Fromm 1961; Ghosh 2021; Gudynas 2018; Haupt 2022; Kashyap 2022; Lansdowne 2017; Moore and Tucker 2015; NGL 2023; Raihani 2021; Thakur 2021; Turney 2021; Wagner and Briggs 2016; Wikipedia (2023 - Marx's Theory of Alienation).

Section 6.5: EWT 2016; Goldman 2004; Hamer 2004; KR Editor 2010; Landless and Charles-Marcel 2021; Lewis 2022; Mohler 2004; PW Editor 2004; Silveira 2008; Wikipedia (2023 – God Gene).

7. FINALE

Section 7.1: Azarian 2022; Barrett 2021; Bragazzi et al. 2018; Cheong 2020; Cheong 2020; Costandi 2022a, 2022b; Gaw 2019; Heyes et al. 2020; Koch et al 2016; Pretorious 2016; Smith 2022; Tarhan 2022.

Section 7.2: Bradshaw and Ellison 2008; DeAngelis 2004; Hamer 2005; Hill 2022; Ironson and Ahmad 2022; McKee 2005; Robson 2022a, 2022b; UMC 2012; Wikipedia (2022 -- God Gene); Zyga 2011.

Section 7.3: ADL 2022; Ahmed 2013; Bazilchuk 2007; BBC 2022; Curry and Gushee 2010; DC 2016; Escobar 2019; FORE 2022; Gahlau 2022; Goldenberg 2013, 2016; Graves-Fitzsimmons and Siddiqi 2021; Harvey 2021; Hopkin 2015; Jenkins et al. 2018; Kejriwal and Vora 2019; Kilvert 2022; Lima 2019; Mitra 1993; Nuccitelli 2018; Petruzzello 2022; Richardson 2010; Rocha 2013; Sandell and Blakemore 2007; Taylor 2016; The Vatican 2019; USCCB 1989; Wikipedia (2022 - Amazon Rain Forest, Bishnoi, Chico Mendez, Chipko Movement, Gaura Devi, James Edward Hansen, Sunderlal Bahuguna, The Islamic Declaration on Global Climate Change).

Section 7.4: Altmann and Bowen 2021; Asher 2012; Bailey 2011; Berman 2018; Coulson 2015, 2018; Dandia 2020; Ducharme 2018; Ecklund 2010; Ecklund et al. 2016; Ecklund and Scheitle 2017; Ecklund and Johnson 2019; Fox 2022; Funk and Masci 2015; Hossenfelder 2022; Hough 2006; Kalthoff 2015; Kershaw 2012; Larson and Witham 1998; Marty 2002; McGrath 2017; Michael 2011; PC 2015; Resentiel 2009; Robinson 2022; Rutherford 2021; SMF 2019; Spencer 2009; Yang 2021; Wikipedia (2022 - Charles Darwin, Jim Al-Khalili, Michio Kaku, Stephen Hawking).

Section 7.5: RPS 2022; Wikipedia (2022 - Dietrich Bonhoeffer).

CREDITS

The sources for symbols are given in the order of appearance in the text.

FRONT MATTER

Endless Knot
Jarvisa: https://en.wikipedia.org/wiki/File:Endlessknot.svg#file
Endless Knot Symbol
By Dontpanic, https://commons.wikimedia.org/w/index.php?curid=1452494

1. INTRODUCTION

2. EMERGENCE

Neuron
Bruce Blaus, https://commons.wikimedia.org/w/index.php?curid=28761830

3. EUGENICS

Ranking of Occupational Classes – Francis Galton
https://en.wikipedia.org/wiki/File:Galton_class_eugenics.jpg
Eugenics Logo
https://psychology.wikia.org/wiki/Eugenics
Eugenics Poster, USA, 1926.
https://en.wikipedia.org/wiki/Eugenics_in_the_United_States
Better Babies Contest, Indiana, 1927
Indiana State Archives, Helfand J (2020).
Eugenics Certificate
http://sociologylegacy.pbworks.com/w/page/104997441/EUGENICS
Nazi Swastika
Wikipedia (2020 – Swastika)
Collection Bus for Killing Patients
Wikipedia (2020 – Nazi Eugenics)
Baltimore Branch NAACP Baby Contest Winners, 1946

https://www.loc.gov/item/2003674030/
WEB Du Bois, 1919
Nuriddin (2017).
M Sanger and MK Gandhi, 1935.
Pokorski (2013).

4. SCIENCE

Official Emblem of the Unification Church
Wikipedia (2021 -- Unification Movement)

5. MATHEMATICS

Ishango Bone
https://en.wikipedia.org/wiki/Ishango_bone
Quipu
https://en.wikipedia.org/wiki/Quipu
Right Triangle
Wikipedia (2021 – Right Triangle)
Pythagoras of Samoa
https://en.wikipedia.org/wiki/Pythagoras
First Six Triangular Numbers
https://en.wikipedia.org/wiki/File:First_six_triangular_numbers.svg
Unit Right Angled Triangle
Wikipedia (2021 – Right Triangle)
Pythagoras Teaching Women
https://en.wikipedia.org/wiki/Pythagoras
A *Sangaku* Problem
https://math.stackexchange.com/questions/4480/sangaku-a-geometrical-puzzle
Penrose Tile
Inductiveload - Own work, Public Domain,
https://commons.wikimedia .org/w/index.php?curid=5839079
Infinity
Wikipedia (2021 – Infinity)
Official Seal of the Jesuits
Moranski: https://commons.wikimedia.org/w/index.php?curid=5596804
Wallis Formula for Pi
Wikipedia (2022 – Pi)
Fibonacci Spiral
Jahobr - CC0, https://commons.wikimedia.org/w/index.php?curid=58460223
Nautilus Shell
Chris 73, Wikimedia Commons, CC BY-SA 3.0,
https://commons.wikimedia.org/w/index.php?curid=19711
Srinivasa Ramanujan
https://en.wikipedia.org/wiki/Srinivasa_Ramanujan
Cistercian Numerals
Wikipedia (2022 - Cistercian Numerals)
Three Geometries

Wikipedia (2021 – Non-Euclidean Geometry)
Godel's Proof
Source: AW (2021).
Emmy Noether
https://commons.wikimedia.org/wiki/File:Noether.jpg
Florence Nightingale in a Field Hospital
https://en.wikipedia.org/wiki/Florence_Nightingale
An Elegant Proof
https://en.wikipedia.org/wiki/Pythagorean_theorem

6. RUMINATIONS

Industrial Greenhouse
Goldlocki, https://commons.wikimedia.org/w/index.php?curid=1661760
The Saudi Arabian National Emblem
Wikipedia (2022 – Saudi Arabia)

7. FINALE

Walking Tree - Amazon Rain Forest
R Hardy - https://commons.wikimedia.org/w/index.php?curid=78906825
Green Anacoda - Amazon Rain Forest
https://commons.wikimedia.org/w/index.php?curid=380702
Chico Mendez
Miranda Smith, https://commons.wikimedia.org/w/index.php?curid=30268371
James Hansen Under Arrest
Wikipedia (2022 – James Hansen)
Khejri Tree (*Prosopis cineraria*)
LRBurdak: https://commons.wikimedia.org/w/index.php?curid=3945050
Chipko Tree Huggers
https://en.wikipedia.org/wiki/Chipko_movement#/media/
File:Big_chipko_movement_1522047126.jpg
Stephen Hawking
By NASA - https://commons.wikimedia.org/w/index.php?curid=1657641

REFERENCES

A scientist reads many books in his lifetime,
and knows he still has a lot to learn.
A religious man barely reads one book,
and thinks he knows everything.
Anonymous

Note: Wikipedia sources are identified by year of access and subject title.

1. INTRODUCTION

Aga Khan III (1954) *The Memoirs of the Aga Khan: World Enough and Time*, Cassell Publishers, London.

Boyett J (2016) *12 Major World Religions: The Beliefs, Rituals, and Traditions of Humanity's Most Influential Faiths*, Zephyros Press, Berkeley.

DK Publisher (2013) *The Religions Book: Big Ideas Simply Explained*, DK Press (Penguin Random House), New York.

Gandhi MK (1929, 1983) *Autobiography: The Story of My Experiments with Truth*, Dover Publications, New York.

HH (2022) Rosalind Franklin (1920-1958), *Humanist Heritage*, 4 May 2022, https://heritage.humanists.uk/rosalind-franklin/

Hirji KF (2022) *Religion, Politics and Society: A Progressive Primer*, Zand Graphics, Nairobi and Daraja Press, Montreal.

Trefil J and Hazen RM (1998) *The Sciences: An Integrated Approach* (second edition), John Wiley & Sons, New York.

Wikipedia (2022 - Albert Einstein, Endless Knot, Louis Pasteur, Michael Faraday, Rosalind Franklin).

2. EMERGENCE

Aaen-Stockdale C (2012) Neuroscience for the soul, *The Psychologist*, July 2012, 25:520—523, https://thepsychologist.bps.org.uk/volume-25/edition -7/neuroscience-soul

Ahlstrom D (2001) Free will versus genetic destiny, *The Irish Times*, 5 April 2001, https://www.irishtimes.com/news/free-will-versus-genetic-destiny-1.298180

Alimohammadi N et al. (2013) Nursing in Islamic thought: Reflection on application nursing metaparadigm concept: A philosophical inquiry, *Iranian Journal of Nursing and Midwifery Research*, 18(4):272—279.

Amaro AA (2015) Holistic mindfulness, *Mindfulness*, 6:63–73, https://doi.org/10.1007/s12671-014-0382-3

Anderson R (2014) Pharmaceutical industry gets high on fat profits, *BBC News*, 6 November 2014, www.bbc.com/news/business-28212223

Angell M (2005) *The Truth About the Drug Companies: How They Deceive Us and What to Do About It*, Random House, New York.

Ann K (2019) Choice, free will, and nicotine addiction, *Medium*, 30 July 2019, https://medium.datadriveninvestor.com/free-will-and-nicotine-addiction

Astbury N (2001) Alternative eye care, *British Journal of Ophthalmology*, 85:767-768.

Awaad R et al. (2021) Holistic healing: Islam's legacy of mental health, *Yaqeen Institute*, 27 May 2021, https://yaqeeninstitute.org/read/paper/holistic-healing-islams-legacy-of-mental-health

Ball P (2021) Why free will is beyond physics, *Physics World*, 6 January 2021, https://physicsworld.com/a/why-free-will-is-beyond-physics/

Ball P (2022) *The Book of Minds: How to Understand Ourselves and Other Beings: From Animals to AI and Aliens*, Picador, New York.

Bang C (2012) How bees decide what to be, *John Hopkins Medicine*, 17 September 2012, www.hopkinsmedicine.org/news/media/releases/how_bees_decide_what_to_be_

Bar-Yam Y (1999) *Dynamics of Complex Systems*, Westview Press, NY.

Barni M and Mahdany D (2017) Al Ghazali's thoughts on Islamic education curriculum, *Dinamika Ilmu*, 17(2):251—260, https://eric.ed.gov/?id=EJ1169434.

Barret LF (2021) *Seven and a Half Lessons about the Brain*, Mariner Books, Boston.

Bastian B (2018) Moral vitalism, In K Gray & J Graham (editors), *Atlas of Moral Psychology*, 303—309, The Guilford Press, New York.

Bastian B et al. (2019) Explaining illness with evil: Pathogen prevalence fosters moral vitalism, *Proceedings of the Royal Society*, Series B, 286: 20191576, http://dx.doi.org/10.1098/rspb.2019.1576

Baumeister RF (2013) Do you really have free will? Of course. Here's how it evolved, *Slate*, 25 September 2013, https://slate.com/technology/2013/09/free-will-debate-what-does-free-will-mean-and-how-did-it-evolve.html

Bhattacherjee A (2012) *Social Science Research: Principles, Methods, and Practices*, Global Text Project, https://open.umn.edu/opentextbooks/textbooks/79

BS (2021) Honeybee genome, *Bee Spotter*, 26 March 2021, https://beespotter.org/topics/genome/

BT (2019) Alan Watts: What is the self? *Big Think*, 4 April 2019, https://bigthink.com/thinking/alan-watts-self/

Burlew R (2022) The amazing genetic diversity in the best honey bee colonies, *Honey Bee Suite*, November 2022, www.honeybeesuite.com/genetic-diversity-within-a-honey-bee-colony/

Burton MJ et al. (2021) The Lancet Global Health Commission on Global Eye Health: Vision beyond 2020, *Lancet Global Health*, 9:e489–551.

Carey N (2012) *The Epigenetic Revolution*, Icon Books, London.

Cave S (2016) There's no such thing as free will, *The Atlantic*, June 2016, www.theatlantic.com/magazine/archive/2016/06/theres-no-such-thing-as-free-will/480750/

Chan CW (2016) *Guard Your Heart: Biblical Wisdom for Heart Health*, CreateSpace, New York.

Clarke R (2017) What can a Marxist approach tell us about science? *Culture Matters*, 28 November 2017, www.culturematters.org.uk/index.php/culture/science/item/2676-what-can-a-marxist-approach-tell-us-about-science

Clark T et al. (2018) *The Truth is the Whole: Essays in Honor of Richard Levins*, The Pumping Station, Arlington, MA.

Clayton P and Davies P (editors) (2006) *The Re-Emergence of Emergence: The Emergentist Hypothesis from Science to Religion*, Oxford University Press, Oxford.

Coyne CJ (2013) *Doing Bad by Doing Good: Why Humanitarian Action Fails*, Stanford Economics and Finance, Stanford.

Cozzarelli T (2020) Class reductionism is real, and it's coming from the Jacobin wing of the DSA, *Left Voices*, 16 June 2020, www.leftvoice.org/class-reductionism-is-real-and-its-coming-from-the-jacobin-wing-of-the-dsa/

Critchlow H (2020) How much do our genes restrict free will? *The Conversation*, 13 October 2020, https://theconversation.com/how-much-do-our-genes-restrict-free-will-134330

Davies L (2022) Book review: *Sedated: How Modern Capitalism Created Our Mental Health Crisis* by James Davis, *Counter Fire*, 14 April 2022, www.counterfire.org/articles/book-reviews/23140-sedated-how-modern-capitalism-created-our-mental-health-crisis-book-review

Davis J (2022) *Sedated: How Modern Capitalism Created our Mental Health Crisis*, Atlantic Books, New York.

Dhonden Y (2000) *Healing from the Source: The Science and Lore of Tibetan Medicine*, Snow Lion, New York.

Draper B and Uhl A (2019) How to (not) raise a reductionist: Reassessing the paradigms of child rearing for an age of flourishing, *Journal of Design and Science*, 1 March 2019, https://jods.mitpress.mit.edu/pub/sj6ivvh5/release/1

Editorial (2012) Can we live without free will? *New Scientist*, 8 August 2012, www.newscientist.com/article/mg21528772-300-can-we-live-without-free-will/

Editors (2019) Hinduism, *History.com*, 30 September 2019, www.history.com/topics/religion/Hinduism

Edwards S (2015) Love and the brain, *HMS News*, Spring 2015, https://hms.harvard.edu/news-events/publications-archive/brain/love-brain

Egnor M and Gallagher A (2022a) What do the world's 1.2 billion Hindus think about the mind? *Mind Matters*, 14 March 2022, https://mindmatters.ai/2022/03/what-do-the-worlds-1-2-billion-hindus-think-about-the-mind/

Egnor M and Gallagher A (2022b) What do Hindus think about the Big Bang? The cyclic Universe? *Mind Matters*, 28 March 2022, https://mindmatters.ai/2022/03/what-do-hindus-think-about-the-big-bang-the-cyclic-Universe/

Ellis G (2020) From chaos to free will, *Aeon*, 9 June 2020, https://aeon.co/essays/heres-why-so-many-physicists-are-wrong-about-free-will

Evans DJ et al. (2020) *Biblical Holism and Agriculture: Cultivating Our Roots*, William Carey Library, New York.

FH (2020) Biblical holism, *Food for the Hungry*, 23 May 2013, www.fh.org/blog/Biblical-holism/

Frank A (2021a) Reductionism vs. emergence: Are you "nothing but" your atoms? *Big Think*, 29 April 2021, https://bigthink.com/13-8/reductionism-vs-emergence-science-philosophy/

Frank A (2021b) Why condensed matter physicists reject reductionism, *Big Think*, 1 July 2021, https://bigthink.com/13-8/condensed-matter-physicists -reject-reductionism/

Gagnon C (1981) For a scientific vision of the world: Determinism or free will? *Proletarian Unity*, 5(2), www.marxists.org/history/erol/ca.secondwave/is-free-will.htm

Garnham S (2018) Against reductionism: Marxism and oppression, *Marxist Left Review*, Winter 2018, https://marxistleftreview.org/articles/against-reductionism-marxism-and-oppression/

Geiser M (2022) "More is different": Why reductionism fails at higher levels of complexity, *Big Think*, 9 March 2022, https://bigthink.com/13-8/reductionism-fails-complexity/

GQ (2022a) What is reductionism? *Got Questions Ministries*, 20 May 2022, www.gotquestions.org/reductionism.html

GQ (2022b) What does the Bible have to say about holistic medicine? *Got Questions Ministries*, 23 May 2022, www.gotquestions.org/holistic-medicine.html

Groopman J (2005) *The Anatomy of Hope: How People Prevail in the Face of Illness*, Random House, New York.

Guttridge N (2012) Job swapping makes its mark on honeybee DNA, *Nature*, 16 September 2012, www.nature.com/articles/nature.2012.11418

Hamdy SF (2009) Islam, fatalism, and medical intervention: lessons from Egypt on the cultivation of forbearance (sabr) and reliance on God (tawakkul), *Anthropology Quarterly*, 82:173–196, doi:10.1353/anq.0.0053

Handelman D (2015) Holism, religion and geopolitics, *E-International Relations*, 31 October 2015, www.e-ir.info/2015/10/31/holism-religion-and-geopolitics/

Harden KP (2021) *The Genetic Lottery: Why DNA Matters for Social Equality*, Princeton University Press, Princeton.

Hari K (2022) Engels, reductionism and epigenetics: The Lysenko debate, *Marxism & Sciences*, 1(1):157–191.

Harper A (2019) Holism and reductionism in psychology, *Owlcation*, 18 July 2019, https://owlcation.com/social-sciences/Holism-and-Reductionism-in -Psychology

Harris S (2012*) Free Will*, Free Press, New York.

Harvey D (2007) *A Brief History of Neoliberalism*, Oxford University Press, Oxford.

Hawramani I (2018) On Islam's view of psychology and scientific reductionism, *Imam Hawarami Website*, 6 December 2018, https://hawramani.com/on-islams-view-of-psychology-and-scientific-reductionism/

Helminski K (2022) Holistic Islam, *The Threshold Society*, 26 May 2022, https://sufism.org /library/articles/holistic-islam

Helminski K (2017) *Holistic Islam: Sufism, Transformation, and the Needs of Our Time*, White Cloud Press, Ashland, OR.

Hercher L (2018) Designer babies aren't futuristic. They're already here, *MIT Technology Review*, 22 October 2018, www.technologyreview.com/2018/10/22/139478/are-we-designing-inequality-into-our-genes/

Heuertz CL (2009) Filling in the holes of Holism, *Christianity Today*, 22 September 2022, www.christianitytoday.com/ct/2009/septemberweb-only/response3.html

Holt N (2017) Is holistic holy? *The Salvation Army*, 1 June 2017, www.salvationarmy.org.nz/our-community/faith-in-life/soul-food/is-holistic-holy

Horgan J (2022) Denying free will is physics reductionism at its absolute worst, *Big Think*, 3 May 2013, https://bigthink.com/articles/denying-free-will-is-physics-reductionism-at-its-absolute-worst/

HSS UK (2022) Sangh means to connect, *HSS UK*, 21 May 2022, https://hssuk.org/hss-uk-perspectives-sangh-means-to-connect/

Hoekstra HE and Robinson GE (2022) Behavioral genetics and genomics: Mendel's peas, mice, and bees, *PNAS*, 18 July 2022, https://doi.org/10.1073/pnas.2122154119

Hurd P et al. (2018) Epigenetic patterns determine if honeybee larvae become queens or workers, *Science Daily*, 22 August 2018, www.sciencedaily.com/releases/2018/08/180822130958.htm

IB (2018) Islam is a holistic approach to worship, *Islamic Bridge*, July 2018, https://islamicbridge.com/2018/07/islam-is-a-holistic-approach-to-worship/

Jaber D et al. (2021) Use of complementary and alternative therapies by patients with eye diseases: A hospital-based cross-sectional study from Palestine, *BMC Complementary Medical Therapies*, 21(1):3, doi:10.1186/s12906-020-03188-9

James SP (2007) Against holism: Rethinking Buddhist environmental ethics, *Environmental Values*, 16(4):447–61, doi:10.3197/096327107X243231.

Kangal K (2020) Engels's emergentist dialectics, *Monthly Review*, 1 November 2020, https://monthlyreview.org/2020/11/01/engelss-emergentist-dialectics/

Klug W et al. (2019) *Concepts of Genetics*, Pearson, New York.

Koch C (2014) Is consciousness universal? *Scientific American*, 1 January 2014, www.scientificamerican.com/article/is-consciousness-universal/

Kreeft P (2008) A philosophical refutation of reductionism, *Catholic Education Resource Center*, 2008, www.catholiceducation.org/en/religion-and-philosophy/apologetics/a-philosophical-refutation-of-reductionism.html.

Laughlin R (2005) *A Different Universe: Reinventing Physics from the Bottom Down*, Basic Books, New York.

Levine A et al. (1987) Marxism and methodological individualism, *New Left Review*, 1(162), https://newleftreview.org/issues/i162/articles/andrew-levine-elliott-sober-erik-olin-wright-marxism-and-methodological-individualism.pdf

List C, Caruso G and Clark C (2019) Free will: Real or illusion: A debate, *The Philosopher*, Autumn 2019, www.thephilosopher1923.org

Llewellin O (2022) The science of belief: Against reductionist approaches to religion, *Varsity*, 20 May 2022, www.varsity.co.uk/science/19653

Mandal A (2020) What is Genetics? *AZO Life Sciences*, 28 October 2020, www.azolifesciences.com/article/What-is-Genetics.aspx

Marable M (2015) *How Capitalism Underdeveloped Black America*, Haymarket Books, San Francisco.

Marc I (2018) Understanding Islam – Reductionism, reality, and intention, *Muslim Center*, 2 December 2018, https://muslim.center/understanding-islam-reductionism-reality-and-intention/

Meloni M (2016) If we're not careful, epigenetics may bring back eugenic thinking, *The Conversation*, 15 March 2016, https://theconversation.com/if-were-not-careful-epigenetics-may-bring-back-eugenic-thinking-56169

Miksha R (2016) Bumble bees and free will, *Bad Beekeeping Blog*, 8 September 2016, https://badbeekeepingblog.com/2016/09/08/bumble-bees-and-free-will/

Milanovic B (2021) *Capitalism, Alone: The Future of the System That Rules the World*, Harvard University Press, Cambridge, MA.

Miri SJ (2022) *Alternative Sociology: Probing into the Sociological Thought of Allama MT Jafari*, Al-Islam, 23 May 2022, https://www.al-islam.org /alternative-sociology-probing-sociological-thought-allama-m-t-jafari-seyed-javad-miri/chapter-3

Monbiot G (2016) Neoliberalism – the ideology at the root of all our problems, *The Guardian* (UK), 15 April 2016, www.theguardian.com/books/2016/apr/15/neoliberalism-ideology-problem-george-monbiot

Mouzelis N (1980) Reductionism in Marxist theory, *Telos,* 21 September 1980:173—185.

Mukherjee S (2017) *The Gene: An Intimate History*, Scribner, New York.

Murray D (2017) A plea for holistic Christianity, *Crossway*, 21 March 2017, www.crossway.org/articles/a-plea-for-holistic-christianity/

Nanda M (2016) Hindutva's science envy, *The Hindu*, 16 September 2016, https://frontline.thehindu.com/science-and-technology/hindutvas-science-envy/article9049883.ece

Narain V (2014) Determinism, free will, and moral responsibility, *The Humanist*, 21 October 2014, https://thehumanist.com/magazine/november-december-2014/philosophically-speaking/determinism-free-will-and-moral-responsibility/

Nintil (2015) The follies of holism: a reductionist critique, *Nintil*, 4 October 2015, https://nintil.com/the-follies-of-holism-a-reductionist-critique

NWE Contributors (2018) Holism, *New World Encyclopedia*, 12 January 2018, www.newworldencyclopedia.org/entry/Holism

Oake BJ (2021) The relationship between holistic practice and 'Spiritual but not religious' identity in the UK, *Secularism and Nonreligion*, 10(1):9, http://doi.org/10.5334/snr.150

Oberlander E (2022) A new theory in physics claims to solve the mystery of consciousness, *Neuroscience News*, 11 August 2022, https://neurosciencenews.com/physics-consciousness-21222/

O'Grady J (2022) What exactly do we mean by the mind? *The Spectator*, 6 August 2022, www.spectator.co.uk/article/what-exactly-do-we-mean-by-the-mind

Palmer DA (2021) Vitalism, *Medium*, 8 November 2021, https://medium.com/re-assembling-reality/vitalism-9cedd397704c

Palmer J (2010) Free will similar in animals, humans - but not so free, *BBC News*, 16 December 2010, www.bbc.com/news/science-environment-11998687

Patnaik U and Patnaik P (2019) Neoliberal capitalism at a dead end, *Monthly Review*, 1 July 2019, https://monthlyreview.org/2019/07/01/neoliberal-capitalism-at-a-dead-end/

PB (2023) The role of the drone bee, *Perfect Bee*, 1 March 2023, https://www.perfectbee.com/learn-about-bees/the-life-of-bees/role-of-the-drone-bee

Pipkin B (2019) Why free will doesn't exist and what that means, *Medium*, 17 February 2019, https://medium.com/@bretpipkin/why-free-will-doesnt-exist-and-what-that-means-a0bb7fb16c7

Preston JL et al. (2013) Neuroscience and the soul: Competing explanations for the human experience, *Cognition, 127(1):*31—37, https://doi.org/10.1016/j.cognition.2012.12.003.

Promtha S et al. (2021) Heisenberg's Uncertainty Principle in Buddhist philosophical perspective, *Social Science Research Network*, 3 March 2021, http://dx.doi.org/10.2139/ssrn.3859162

Rae A (2013) *Reductionism: A Beginner's Guide*, One World Publications, New York.

Rios E (2007) De Waal sides with Darwin: Morality is instinctual, evolved, *Emory Report*, 16 April 2007, www.emory.edu/EMORY_REPORT/erarchive/2007/April/DeWaal.htm

Rothman S (2001) *Lessons from the Living Cell: The Limits of Reductionism*, McGraw-Hill, New York.

Schmidt JB (2017) *The Horrors of Holistic Medicine*, Westbow Press, New York.

Schoonover-Shoffner K (2013) Holistic or wholistic? *Journal of Christian Nursing*, 30(3):133, doi:10.1097/CNJ.0b013e31829ab052

Sider R et al. (2019) Holistic Ministry defined, *Christians for Social Action*, 21 December 2019, https://christiansforsocialaction.org/resource/holistic-ministry-defined/

Siegel E (2022) Yes, the Universe really is 100% reductionist in nature, *Big Think*, 9 August 2022, https://bigthink.com/starts-with-a-bang/Universe-reductionist/

Skarbek R (2021) Are Buddhism and quantum physics the same story, with just a different name? *Medium*, 1 January 2021, https://medium.com/illumination/empowerment-coaching-are-buddhism-and-quantum-physics-the-same-story-789c4930fdfa

Slingerland E (2018) *Mind and Body in Early China: Beyond Orientalism and the Myth of Holism*, Oxford University Press, Oxford.

Spizziri LJ (1990) Too much eye surgery? *Archives of Ophthalmology*, 108(10):1379, doi:10.1001/archopht.1990.01070120025011

Stiglitz J (2019) Neoliberalism must be pronounced dead and buried. Where next? *The Guardian* (UK), 30 May 2019, www.theguardian.com/business/2019/may/30/neoliberalism-must-be-pronounced-dead-and-buried-where-next

Taylor J (2010) How emphatic evangelicalism becomes reductionist evangelicalism, *The Gospel Coalition*, 27 August 2010, www.thegospelcoalition.org/blogs/justin-taylor/how-emphatic-evangelicalism-becomes-reductionist-evangelicalism/

Thier H (2020) *A People's Guide to Capitalism: An Introduction to Marxist Economics*, Haymarket Books, San Francisco.

Thompson P (2011a) Karl Marx, part 1: Religion, the wrong answer to the right question, *The Guardian* (UK), 4 April 2011, www.theguardian.com/commentisfree/belief/2011/apr/04/karl-marx-religion

Thompson P (2011b) Karl Marx, part 2: How Marxism came to dominate socialist thinking, *The Guardian* (UK), 11 April 2011, www.theguardian.com/commentisfree/belief/2011/apr/11/marx-engels-science-marxism

Thompson P (2011c) Karl Marx, part 3: Men make their own history, *The Guardian* (UK), 18 April 2011, www.theguardian.com/commentisfree/belief/2011/apr/18/karl-marx-men-make-history

Thompson P (2011d) Karl Marx, part 4: 'Workers of the world, unite!' *The Guardian* (UK), 25 April 2011, www.theguardian.com/commentisfree/belief/2011/apr/25/karl-marx-communist-manifesto

Thompson p (2011e) Karl Marx, part 5: The problem of power, The Guardian (UK), 2 May 2011, www.theguardian.com/commentisfree/2011/may/02/karl-marx-power-dictatorship-proletariat

Thompson P (2011f) Karl Marx, part 6: The economics of power, *The Guardian* (UK), 9 May 2011, www.theguardian.com/commentisfree/belief/2011/may/09/karl-marx-part-6-economics

Thompson P (2011g) Karl Marx, part 7: The psychology of alienation, *The Guardian* (UK), 16 May 2011, www.theguardian.com/commentisfree/belief/2011/may/16/karl-marx-psychology-alienation

Thompson P (2011h) Karl Marx, part 8: Modernity and the privatization of hope, *The Guardian* (UK), 23 May 2011, www.theguardian.com/commentisfree/belief/2011/may/23/karl-marx-privatisation-of-hope

Tiessen T (2013) Is there a hole in our holism? *Thoughts Theological*, 6 December 2013, www.thoughtstheological.com/is-there-a-hole-in-our-holism/

Timpe K (2021) Free will, *Internet Encyclopedia of Philosophy*, 14 February 2021, https://iep.utm.edu/freewill/

Trefil J and Hazen RM (1998) *The Sciences: An Integrated Approach* (second edition), John Wiley & Sons, New York.

Tse PU (2015) *The Neural Basis of Free Will: Criterial Causation*, The MIT Press, Cambridge, MA.

TSL Editors (2022) Holism, *The Spiritual Life*, 16 May 2022, https://slife.org/holism/

van Bohemen S (2010) *Divergent Religious Conceptions of Nature: Dualism and Holism: A Study on Religion and Environmental Concern in the Netherlands*, Erasmus Universiteit Rotterdam, Rotterdam, 30 June 2010.

Verschuuren GM (2017) *The Holism-Reductionism Debate in Physics, Genetics, Biology, Neuroscience, Ecology, and Sociology*, CreateSpace, New York.

Voosholz J and Gabriel M (editors) (2021) *Top-Down Causation and Emergence*, Springer Verlag, New York.

Wagner R and Briggs A (2016) *The Penultimate Curiosity*, Oxford University Press, Oxford.

Wan PY (2011) *Reframing the Social: Emergentist Systemism and Social Theory*, Ashgate Publishing, New York.

Wattles J (2002) Religion and science: The scientific study of religion, *Kent State University Distributed Learning*, www.wabashcenter.wabash.edu/syllabi/w/wattles/science.htm.

Weber B (2019) Human nature is created, *Medium*, 13 December 2019, https://medium.com/the-philosophers-stone/human-nature-is-created-c96fd1a701af

Weinrib L (2019) *Healing the Soul: An Intuitive MD's Prescription for Health and Wholeness*, Author House, New York.

Wikipedia (2022 – Emergence, Emergentism, Free Will, Holism, Reductionism, Scientific Method, Vitalism).

Wintjen H (2022) Erwin Schrodinger: There is only one mind, *Hendrik Wintjen Blog*, 13 August 2022, www.hendrik-wintjen.info/consciousness/erwin-schroedinger-one-mind/

Yablon A (2020) We're all mad as hell, thanks to late capitalism, *Medium*, 22 October 2020, https://gen.medium.com/were-all-mad-as-hell-thanks-to-late-capitalism

3. EUGENICS

Adam AM (2003) Human genetics in the holy Quran and Sunna, *Journal of the Royal College of Physicians*, 33:44--45.

Adams MB (1989) The politics of human heredity in the USSR, 1920—1940, *Genome*, 31(2):879—884, doi:10.1139/g89-155.

Adams R (2021) University College London apologizes for role in promoting eugenics, *The Guardian* (UK), 7 January 2021, www.theguardian.com/education/2021/jan/07/university-college-london-apologises-for-role-in-promoting-eugenics

AES (1926) A Eugenics catechism, *American Eugenics Society*, www.uvm.edu/~eugenics/primarydocs/oraesec000026.xml

Agarwal P (2020) Malthusian theory of population, *Intelligent Economist*, 21 November 2020, www.intelligenteconomist.com/malthusian-theory/

Ali W (2007) On Islamo-Fascism and other vacuous epithets, *Counter Punch*, 20 November 2007, https://www.counterpunch.org/2007/11/20/on-islamo-fascism-and-other-vacuous-epithets/

Allen A (2018) Toby Young: What is 'progressive eugenics' and what does it have to do with meritocracy? *The Conversation*, 5 January 2018, https://theconversation.com/toby-young-what-is-progressive-eugenics-and-what-does-it-have-to-do-with-meritocracy-89671

Allen AT (2000) Feminism and eugenics in Germany and Britain, 1900-1940: A comparative perspective, *German Studies Review*, 23(3):477—505, https://doi.org/10.2307/1432830

Alomia H (2010) Fatal flirting: The Nazi state and the Seventh-day Adventist church, *Journal of Adventist Mission Studies*, 6(1):3—14.

AP (2020) Black Catholics: Words not enough as church decries racism, *VOA*, 22 June 2020, www.voanews.com/usa/race-america/black-catholics-words-not-enough-church-decries-racism

Appleman LI (2018) Deviancy, dependency, and disability: The forgotten history of eugenics and mass incarceration, *Duke Law Journal*, 68(3), December 2018.

Angus I (2009) Marx and Engels...and Darwin? *International Socialist Review*, No. 65, 2009, https://isreview.org/issue/65/marx-and-engelsand-darwin

Baker GJ (2014) Christianity and eugenics: The place of religion in the British Eugenics Education Society and the American Eugenics Society, c.1907–1940, *Social History of Medicine*, 27(2):281—302.

Baker R (2015) Top ten unlikely and surprising eugenicists, *Flash Back*, 18 March 2015, https://flashbak.com/top-ten-unlikely-and-surprising-eugenicists-32300/

Ball N (2021) Galton, Sir Francis, *Eugenics Archive*, 6 January 2021, https://eugenicsarchive.ca/discover/tree/518c1ed54d7d6e0000000002

Ball P (2011) Unnatural: *The Heretical Idea of Making People*, Vintage, New York.

Barnes EW (1949) The mixing of races and social decay, *Eugenics Review*, 61(1):11—16.

Bashford A and Levine P (2010) *The Oxford Handbook of the History of Eugenics*, Oxford University Press, Oxford.

Bayton DC (2016) *Defectives in the Land: Disability and Immigration in the Age of Eugenics*, University of Chicago Press, Chicago.

BBC (2021) Life in Nazi Germany, 1933--1939, *BBC News*, 16 January 2021, www.bbc.co.uk/bitesize/guides/zpq9p39/revision/4

Berenbaum M (2020) T4 Program, *Encyclopedia Britannica*, 30 November 2020, www.britannica.com/event/T4-Program

Berg H (2013) Nation of Islam, *Oxford Bibliographies*, 26 February 2013, www.oxfordbibliographies.com/view/document/obo-9780195390155/obo-97801953 90155-0130.xml

Bergman J (2001) Darwin's influence on ruthless laissez faire capitalism, *Acts & Facts*, 30(3), https://www.icr.org/article/darwins-influence-ruthless-laissez-faire-capitalis/

Bergman J (2011) Eugenics leader AE Wiggam was a disciple of Darwin, *Creation Matters*, 16(5), September/October 2011.

Berzin A (2020) The Nazi connection with Shambhala and Tibet, *Study Buddhism*, 31 December 2020, https://studybuddhism.com/en/advanced-studies/history-culture/shambhala/the-nazi-connection-with-shambhala-and-tibet

Bianculli D (2017) The Supreme Court ruling that led to 70,000 forced sterilizations, *NPR*, 24 March 2017, www.npr.org/2017/03/24/521360544/the-supreme-court-ruling-that-led-to-70-000-forced-sterilizations

Billinger M (2014) Degeneracy, *Eugenics Archive*, 28 April 2014, http://eugenicsarchive.ca/discover/tree/535eeb0d7095aa0000000218

Bjorkman M and Widmalm S (2011) Selling eugenics: The case of Sweden, *The Royal Society Journal of the History of Science*, 18 August 2010, https://royalsociety publishing.org/doi/10.1098/rsnr.2010.0009

Black E (2003) Eugenics and the Nazis -- the California connection, *SF Gate*, 9 November 2003, www.sfgate.com/opinion/article/Eugenics-and-the-Nazis-the-California-25497 71.php

Black E (2004) Hitler's debt to America, *The Guardian* (UK), 6 February 2004, www.theguardian.com/uk/2004/feb/06/race.usa

Black E (2012) *War against the Weak: Eugenics and America's Campaign to Create a Master Race*, Dialog Press, New York.

Blacker CP (1951) JBS Haldane on eugenics, *The Eugenics Review*, 146—151, 54(3), www.ncbi.nlm.nih.gov/pmc/articles/PMC2973346/

Blaich R (1993) Selling Nazi Germany abroad: The case of Hulda Jost, *Journal of Church and State*, 35(4):807—830, www.jstor.org/stable/23920858.

Blaich R (1996) Health reform and race hygiene: Adventists and the biomedical vision of the Third Reich, *Church History*, 65(3):425—440, www.jstor.org/stable/3169939.

Blakemore E (2019) Why the Nazis were obsessed with twins, *History.com*, 8 July 2019, www.history.com/news/nazi-twin-experiments-mengele-eugenics

Brezis ES and Young W (2003) The new views on demographic transition: A reassessment of Malthus's and Marx's approach to population, *European Journal of History of Economic Thought*, 10(1):25–45.

Brignell V (2010a) The eugenics movement Britain wants to forget, *The New Statesman*, 9 December 2010, www.newstatesman.com/society/2010/12/british-eugenics-disabled

Brignell V (2010b) When America believed in eugenics, *The New Statesman*, 10 December 2010, www.newstatesman.com/society/2010/12/disabled-america-immigration

Brian K (2016) Book Review: *An Image of God: The Catholic Struggle with Eugenics* by Sharon M Leon, *Canadian Bulletin of Medical History*, 33(1):228—229, https://muse.jhu.edu/article/631736/pdf

Brown D (2007) *Bury My Heart at Wounded Knee: An Indian History of the American West*, Picador, New York.

Bruce P (2016) Eugenics – Journey to the dark side at the dawn of statistics, *Statistics.com*, April 2016, www.kdnuggets.com/2016/04/eugenics-journey-dark-side-statistics.html

Burleigh M (1994) Between enthusiasm, compliance and protest: The churches, eugenics and the Nazi 'euthanasia' Program, *Contemporary European History*, 3(3):253—263, www.jstor.org/stable/20081526

Cain J (2020) Karl Pearson praised Hitler and Nazi race hygiene, *Professor Joe Cain Blog*, 26 November 2020, https://profjoecain.net/karl-pearson-praised-hitler-nazi-race-hygiene/

Cama A (2012) Albanians saved Jews from deportation in WWII, *DW*, 27 December 2012, www.dw.com/en/albanians-saved-jews-from-deportation-in-wwii/a-16481404

Carrel A (1939, 2010) *Man the Unknown*, Wilco International Library, USA.

Carter J (2017) 9 things you should know about eugenics, *The Gospel Coalition*, 25 July 2017, www.thegospelcoalition.org/article/9-things-you-should-know-about-eugenics/

Cavanaugh-O'Keefe J (2020) Introduction to eugenics, *Catholic Culture*, 22 November 20202, www.catholicculture.org/culture/library/view.cfm? recnum=302

Chaney DK (2016) Christian eugenics, *Religion Refuted*, 1 August 2016, https://religionrefuted.com/christian-eugenics/

Chase J (2017) Remembering the 'forgotten victims' of Nazi 'euthanasia' murders, *DW*, 26 January 2017, www.dw.com/en/remembering-the-forgotten-victims-of-nazi-euthanasia-murders/a-37286088

Chaudhury A (2018) Why white supremacists and Hindu nationalists are so alike, *Al-Jazeera News*, 13 December 2018, www.aljazeera.com/opinions/2018/12/13/why-white-supremacists-and-hindu-nationalists-are-so-alike

Chazan G (2018) Book review: *Hostile Takeover: How Islam Hinders Progress and Threatens Society*, by Thilo Sarrazin, *Financial Times*, 10 September 2018, www.ft.com/content/bfdeaea8-b1cd-11e8-8d14-6f049d06439c

Chesterton GK (1922, 2008) *Eugenics and Other Evils*, Cassell and Company, Limited, London, www.gutenberg.org/files/25308/25308-h/25308-h.htm

Chowdhury AR (2018) Vinayak Damodar Savarkar: He admired Hitler and other lesser-known facts about him, *The Indian Express*, 28 January 2018, https://indianexpress.com/article/research/vinayak-damodar-savarkar-135th-birth-anniversary-he-admired-hitler-and-other-lesser-known-facts-about-him-5194470/

Churchill W (2001) *A Little Matter of Genocide: Holocaust and Denial in the Americas 1492 to the Present*, City Lights Publishers, San Francisco.

Citizendium (2020) Eugenics, *Citizen Dium*, 20 November 2020, https://en.citizendium.org/wiki?title=Eugenics&oldid=100511461

Clark ES (2017) Religion and race in America, *Oxford Research Encyclopedia of American History*, February 2017, doi:10.1093/acrefore/9780199329175.013.322.

Clayton A (2020) How eugenics shaped statistics, *Nautilus*, 28 October 2020, http://nautil.us/issue/92/frontiers/how-eugenics-shaped-statistics

Cochran DC (2020) How Hitler found his blueprint for a German empire by looking to the American West, *Waging Nonviolence*, 7 October 2020, https://wagingnonviolence.org/2020/10/hitler-found-blueprint-german-empire-in-the-american-west/

Cohen AS (2016) Harvard's eugenics era, *Harvard Magazine*, March 2016, https://harvardmagazine.com/2016/03/harvards-eugenics-era

Cohen A (2017) *Imbeciles: The Supreme Court, American Eugenics, and the Sterilization of Carrie Buck*, Penguin Books, New York.

Dandia A (2020) Can atheists make their case without devolving into bigotry? Al-Jazeera News, 21 February 2020, www.aljazeera.com/indepth/opinion/atheists-case-devolving-bigotry-200220114842749.html

Darwin C (1871) *The Descent of Man, and Selection in Relation to Sex*, MacMillan, London.

Crew FAE et al. (1939) *Eugenics Manifesto*, Nature, 16 November 1939.

Delzell DAP and Poliak CD (2013) Karl Pearson and eugenics: Personal opinions and scientific rigor, *Science and Engineering Ethics*, 19:1057—1070.

deJong-Lambert W (2017) HJ Muller and JBS Haldane: Eugenics and Lysenkoism, In: deJong-Lambert W and Krementsov N (editors) *The Lysenko Controversy as a Global Phenomenon*, Volume 2, *Palgrave Studies in the History of Science and Technology*, Palgrave Macmillan, https://doi.org/10.1007/978-3-319-39179-3_4

Devolder K (2015) US complicity and Japan's wartime medical atrocities: Time for a response, *American Journal of Bioethics*, 15(6):40—9, doi:10.1080/15265161.2015.1028659.

Dhingra N (2018) Acts and eugenics, *The Living World*, 1 April 2018, https://livingchurch.org/covenant/2018/04/01/acts-and-eugenics/

Donahue J (2016) The supreme court, American eugenics, and the sterilization of Carrie Buck, *WAMC Public Radio*, 25 March 2016, www.wamc.org/post/supreme-court-american-eugenics-and-sterilization-carrie-buck

Draper AT (2019) Book Review: *Terence Keel, Divine Variations: How Christian Thought Became Racial Science, Studies in Christian Ethics*, 33(2):279—283.

Durst D (2017) *Eugenics and Protestant Social Reform: Hereditary Science and Religion in America, 1860–1940*, Wipf and Stock Publishers, Eugene, OR.

Duster T (2003) *Backdoor to Eugenics* (second edition), Routledge, New York.

EA Editors (2021) Religion, *Eugenics Archive*, 6 January 2021, www.eugenicsarchive.org/eugenics/religion

Ehle J (1998) *The Trail of Tears: The Rise and Fall of the Cherokee Nation*, Anchor Books, New York.

Eig J (2017) The real reason Muhammad Ali converted to Islam, *The Washington Post*, 26 October 2017, www.washingtonpost.com/news/acts-of-faith/wp/2017/10/26/the-real-reason-muhammad-ali-converted-to-islam/

Encyclopedia (2020) Eugenics and religious law: IV. Hinduism and Buddhism, *Encyclopedia.com*, 28 November 2020, www.encyclopedia.com/science/encyclopedias-almanacs-transcripts-and-maps/eugenics-and-religious-law-iv-hinduism-and-buddhism

English D (2004) *Unnatural Selections: Eugenics in American Modernism and the Harlem Renaissance*, University of North Carolina Press, Chapel Hill.

English DK (2016) Eugenics, *Oxford Bibliographies Online*, 28 June 2016, www.oxfordbibliographies.com/view/document/obo-9780190280024/obo-9780190280024-0029.xml

Epstein R (1999) Genetic engineering: A Buddhist assessment, *Tricycle*, Winter 1999, https://tricycle.org/magazine/genetic-engineering-buddhist-assessment/

Evans G (2018) The unwelcome revival of 'race science', *The Guardian* (UK), 2 March 2018, www.theguardian.com/news/2018/mar/02/the-unwelcome-revival-of-race-science

Farley A (2019) Book Review: *The Revival of Raced-Based Medicine: Eugenics, Religion, and the Black Experience, Marginalia*, 18 October 2019, https://marginalia.lareviewofbooks.org/revival-of-raced-based-medicine-eugenics-religion-and-the-black-experience/

Fiedler M (2013) Catholics and eugenics: A little-known history, *National Catholic Reporter*, 9 September 2013, www.ncronline.org/blogs/ncr-today/catholics-and-eugenics-little-known-history

Fitzpatrick M and Moses AD (2018) Nazism, socialism and the falsification of history, *ABC*, 28 August 2018, www.abc.net.au/religion/nazism-socialism-and-the-falsification-of-history/10214302

Fleury B (2015) *The Negro Project: Margaret Sanger's Diabolical, Duplicitous, Dangerous, Disastrous and Deadly Plan for Black America*, Dorrance Publishing, New York.

Follet C (2019) The cruel truth about population control, *Cato Institute*, 13 June 2019, www.cato.org/publications/commentary/cruel-truth-about-population-control

Foster JB (1998) Malthus' Essay on Population at age 200, *Monthly Review*, 1 December 1998, https://monthlyreview.org/1998/12/01/malthus-essay-on-population-at-age-200

Frazier I (2019) When WEB Du Bois made a laughingstock of a white supremacist, *The New Yorker*, 26 August 2019, www.newyorker.com/magazine/2019/08/26/when-w-e-b-du-bois-made-a-laughingstock-of-a-white-supremacist

Freedland J (2012) Eugenics: the skeleton that rattles loudest in the left's closet, *The Guardian* (UK), 17 February 2012, www.theguardian.com/commentisfree/2012/feb/17/eugenics-skeleton-rattles-loudest-closet-left

Freedland J (2019) Eugenics and the master race of the left – archive, 1997, *The Guardian (UK)*, 1 May 2019, www.theguardian.com/politics/from-the-archive-blog/2019/may/01/eugenics-founding-fathers-british-socialism-archive-1997

Friedman IR (2020) The other victims of the Nazis, *Social Studies*, 10 December 2020, www.socialstudies.org/sites/default/files/publications/se/5906/590606.html

Galton DJ (1998) Greek theories in eugenics, *Journal of Medical Ethics*, 24:263—267.

Galton F (1873) Africa for the Chinese, *The Times*, 5 June 1873, http://galton.org/letters/africa-for-chinese/AfricaForTheChinese.htm

Garver KL and Garver BL (1992) Eugenics, euthanasia and genocide, *Linacre Quarterly*, 59(3):24—51, August 1992.

Geiderman JM (2002) Ethics seminars: Physician complicity in the Holocaust: Historical review and reflections on emergency medicine in the 21st Century, Part I, *Academic Emergency Medicine*, 9:223–231, https://onlinelibrary.wiley.com/doi/pdf/10.1197/aemj.9.3.223.

Gerais R (2017) Better babies contests in the United States (1908–1916*), Embryo Project Encyclopedia*, 19 July 2017, http://embryo.asu.edu/handle/10776/12566

Gerhart GM (1996) Books review: *Season of Blood: A Rwandan Journey* and *Exterminate All the Brutes: One Man's Odyssey into the Heart of Darkness and the Origins of European Genocide*, *Foreign Affairs*, December 1996, www.foreignaffairs.com/reviews/capsule-review/1996-11-01/season-blood-rwandan-journey-exterminate-all-brutes-one-mans

Gershoni I (2012) Why the Muslims must fight against Nazi Germany: Muḥammad Najati Ṣidqi's plea, *Brill*, 52(3/4):471—498, www.jstor.org/stable/41722008

Gibbons SR (2020) Women's suffrage, *Eugenics Archive*, 26 December 2020, https://eugenicsarchive.ca/discover/connections/535eeeaf7095aa0000000264

Gordon J (2012) *Invisible War: The United States and the Iraq Sanctions*, Harvard University Press, Cambridge.

Gray J (2015) Evangelical atheism and its discontents, *ABC*, 16 June 2015, www.abc.net.au/religion/evangelical-atheism-and-its-discontents/10098184

Gould SJ (1981) *The Mis-Measure of Man*, WW Norton, New York.

GQ (2021a) Does the Bible support eugenics? *Got Questions Ministries*, 5 January 2021, www.gotquestions.org/eugenics-Bible.html

Grue L (2010) Eugenics and euthanasia – Then and now, *Scandinavian Journal of Disability Research*, 12:33–45, http://doi.org/10.1080

Haldane JBS (1934) Human biology and politics, *The British Science Guild*, 28 November 1934, www.marxists.org/archive/haldane/works/1930s/biology.htm

HE (2020) The Nazi rise to power, *Holocaust Explained*, 17 December 2020, www.theholocaustexplained.org/the-nazi-rise-to-power/

Helfand J (2020) Darwin, expression, and the lasting legacy of eugenics, *The MIT Press Reader*, 3 August 2020, https://thereader.mitpress.mit.edu/evolution-expression-and-the-lasting-legacy-of-eugenics/

Hemmer N (2017) 'Scientific racism' is on the rise on the right. But it's been lurking there for years, *Vox*, 28 March 2017, www.vox.com/the-big-idea/2017/3/28/15078400/scientific-racism-murray-alt-right-black-muslim-culture-trump

Hercher L (2018) Designer babies aren't futuristic. They're already here, *MIT Technology Review*, 22 October 2018, www.technologyreview.com/2018/10/22/139478/are-we-designing-inequality-into-our-genes/

Hirji KF (2019) El-Shifa: A forgotten war crime, *Pambazuka News*, 7 August 2019, www.pambazuka.org.

History Editors (2020) Women's history milestones: A timeline, *History*, 18 November 2020, www.history.com/topics/womens-history/womens-history-us-timeline

Hogben L (1948) Book review: *Principles of Medical Statistics* by AB Hill, *British Journal of Social Medicine*, 11:43.

Horvath A (2020a) The roots of the concentration camp were American and British eugenicists, not Nazis, *Eugenics US*, 15 June 2020, http://eugenics.us/the-roots-of-the-concentration-camp-were-american-and-british-eugenicists-not-nazis/77.htm

Horvath A (2020b) Carrel: Eugenics asks for the sacrifice of many individuals, *Eugenics US*, 15 June 2020, http://eugenics.us/carrel-eugenics-asks-for-the-sacrifice-of-many-individuals/589.htm

Hughes PW (2020) A passion for promoting the public good guided geneticist JBS Haldane's scholarship, *Science Magazine*, 12 August 2020, https://blogs.sciencemag.org/books/2020/08/12/a-dominant-character/

Ings S (2016) Eugenic America: How to exclude almost everyone, *New Scientist*, 22 March 2016, www.newscientist.com/article/mg22930663-000-how-to-exclude-everyone/

Jeffries S (2012) Sven Lindqvist: A life in writing, *The Guardian* (UK), 22 June 2012, www.theguardian.com/books/2012/jun/22/sven-lindqvist-life-in-writing

Johnson I (2011) *A Mosque in Munich*, Mariner Books, New York.

JVL (2020) Nazi medical experiments: Background & overview, *Jewish Virtual Library*, 16 December 2020, www.jewishvirtuallibrary.org/background-and-overview-of-nazi-medical-experiments

Kaminski M (2010) The German connection: How the Muslim Brotherhood found a haven in Europe, *Wall Street Journal*, 6 May 2010, www.wsj.com/articles/SB10001424052748703961104575226372646226094

Katznelson I (2017) What America taught the Nazis, *The Atlantic*, November 2017, www.theatlantic.com/magazine/archive/2017/11/what-america-taught-the-nazis/540630/

Keel T (2019) *Divine Variations: How Christian Thought Became Racial Science*, Stanford University Press, Stanford.

Kellner MA (2005) German, Austrian churches apologize for Holocaust actions, *Adventist News*, 16 August 2005, https://adventist.news/en/news/europe-german-austrian-churches-apologize-for-holocaust-actions

Kernan S (2020) The results of the Nazi IQ tests, *Medium*, 18 July 2020, https://medium.com/history-of-yesterday/the-results-of-the-nazi-iq-tests-c3a5e442f37c

Kevles D (1985) *In the Name of Eugenics: Genetics and the Uses of Human Heredity*, Knopf, New York.

Klein D (2011) Hitler, Nazis, socialism, and rightwing propaganda, *CSUN*, January 2011, www.csun.edu/~vcmth00m/NazismSocialism.html

Krementsov N (2011) From 'Beastly Philosophy' to medical genetics: Eugenics in Russia and the Soviet Union, *Annals of Science*, 68(1):61--92, doi:10.1080/00033790.2010.527162

Krementsov N (2018) *With and Without Galton: Vasilii Florinskii and the Fate of Eugenics in Russia*, Open Book Publishers, Cambridge, https://books.Openedition.org/obp/ 7405?

Krome-Lukens AL (2009) *A Great Blessing to Defective Humanity: Women and the Eugenics Movement in North Carolina, 1910-1940*, Master of Arts Thesis, Department of History, University of North Carolina, Chapel Hill.

Kuhl S (1994) *The Nazi Connection; Eugenics, American Racism, and German National Socialism*, Oxford University Press, New York.

Kurbegovic E (2013) Catholic Church issues Casti Connubii, *Eugenics Archive*, 13 September 2013, https://eugenicsarchive.ca/discover/tree/52329f2f5c2ec5000000000f

Kurbegovic E (2014a) Education, *Eugenics Archive*, 14 November 2014, https://eugenicsarchive.ca/discover/tree/54668bd62432860000000001

Kurbegovic E (2014b) Race betterment, *Eugenics Archive*, 29 April 2014, http://eugenicsarchive.ca/discover/connections/535eedae7095aa000000024f

Levine Y (2019) America sponsors far-right holocaust revisionist exhibit in Kiev (Part I), *MR Online*, 24 October 2019, https://mronline.org/2019/10/24/america-sponsors-far-right-holocaust-revisionist-exhibit-in-kiev-part-i/

Lienhard JH (2021) Darwin and racism, engines of our ingenuity, *University of Huston*, 19 January 2021, www.uh.edu/engines/epi617.htm

Lombardo P (2016) Excerpt: The banality of eugenics, *UN Dark*, 4 April 2016, https://undark.org/2016/04/04/the-banality-of-eugenics-tuskegee/

Lombardo P (2019) Eugenics and public health: Historical connections and ethical implications, *The Oxford Handbook of Public Health Ethics*, September 2019, doi:10.1093/oxfordhb/9780190245191.013.56

Lombardo PA (editor) (2011) *A Century of Eugenics in America: From the Indiana Experiment to the Human Genome Era*, Indiana University Press, Bloomington.

Lombardo P (2020) Eugenic sterilization laws, *Eugenics Archive*, 13 December 2020, www.eugenicsarchive.org/html/eugenics/essay8text.html

LT (2020) Pope Pius XI never condemned eugenics, *Medium*, 18 March 2020, https://medium.com/@LearningTools/pope-pius-xi-never-condemned -eugenics-e4b0c00751fa

Lucassen LL (2010) A brave new world: The left, social engineering, and eugenics in twentieth-century Europe, *IRSH*, 55:265–296 doi:10.1017/S0020859010000209

Luff J (2020) Book Review: *Little 'Red Scares': Anti-Communism and Political Repression in the United States, 1921-1946*, *Reviews in History*, 21 December 2020, doi:10.14296/RiH/2014/1730

Lyster C (2014), Gender, *Eugenics Archive*, 29 April 2014, https://eugenicsarchive.ca/discover/connections/535eec037095aa0000000229

MacKinnon H (2021) The history of eugenics in Quebec and at McGill, *McGill Tribune*, 10 January 2021, www.mcgilltribune.com/history-of-eugenics-mcgill-quebec/

Maisey R (2020) When science met socialism, *Tribune Magazine*, 22 December 2020, https://tribunemag.co.uk/2020/12/when-science-met-socialism/

Majfud J (2020) Hitler's ideologues: The US racism that bore fruit in Mein Kampf, *Common Dreams*, 5 July 2020, www.commondreams.org/views/2020/07/05/hitlers-ideologues-us-racism-bore-fruit-mein-kampf

Malik K (2019) The spirit of eugenics is still with us, as immigrants know to their cost, *The Guardian* (UK), 6 October 2019, www.theguardian.com/commentisfree/2019/oct/06/spirit-of-eugenics-is-still-with-us-as-immigrants-know-to-their-cost

Mandal A (2020) What is Genetics? *AZO Life Sciences*, 28 October 2020, www.azolifesciences.com/article/What-is-Genetics.aspx

McClay WM (2010) Chesterton's warning, *The American Interest*, 6(1), 1 September 2010, www.the-american-interest.com/2010/09/01/chestertons-warning/

Media PA (2020) London University removes names of Francis Galton and Karl Pearson from two lecture theatres and a building, *The Guardian* (UK), 19 June 2020, www.theguardian.com/education/2020/jun/19/ucl-renames-three-facilities-that-honoured-prominent-eugenicists

Meloni M (2016) If we're not careful, epigenetics may bring back eugenic thinking, *The Conversation*, 15 March 2016, https://theconversation.com/if-were-not-careful-epigenetics-may-bring-back-eugenic-thinking-56169

Mena (2017) Eugenics and feminism: A brief history, *Dangerous Women*, 28 February 2017, http://dangerouswomenproject.org/2017/02/28/eugenics-and-feminism/

Merricks PT (2012) 'God and the gene': EW Barnes on eugenics and religion, *Politics, Religion & Ideology*, 13(3):353--374, doi:10.1080/21567689.2012.698978

Miller RJ (2020) Nazi Germany's race laws, the United States, and American Indians, *Social Science Research Network*, 19 February 2020, http://dx.doi.org/10.2139 /ssrn.3541009

Milmo C (2011) Fury at DNA pioneer's theory: Africans are less intelligent than Westerners, *The Independent*, 18 September 2011, www.independent.co.uk/news/science/fury-dna-pioneer-s-theory-africans-are-less-intelligent-westerners-394898.html

Motadel D (2017) *Islam and Nazi Germany's War*, Belknap Press, Cambridge.

Moya-Smith S (2017) Hitler said to have been inspired by US Indian reservation system, *Indian Country Today*, 28 August 2017, https://indiancountrytoday.com/archive/hitler-said-to-have-been-inspired-by-us-indian-reservation-system

MPS (2020) History of the Kaiser Wilhelm Society under National Socialism, *Max Planck Society*, 14 December 2020, www.mpg.de/history/kws-under-national-socialism

Murphy JT (1935, 2008) Fascism! The socialist answer, *Marxists Internet Archive*, 2008, www.marxists.org/archive/murphy-jt/1935/x01/fascism.htm

Naicker L (2010) The role of eugenics and religion in the construction of race in South Africa, *Department of Christian Spirituality, Church History and Missiology*, 2010, University of South Africa, Pretoria, South Africa.

Nelson RH (2015) Prohibition and eugenics: Implicit religions that failed, *Social Science Research Network*, June 2015, https://ssrn.com/abstract=2996213

Nock AJ (1941) The Jewish problem in America, *Atlantic Monthly*, June 1941, https://en.wikiquote.org/wiki/Adolf_Hitler#1910s.

Nuriddin A (2017) The Black politics of eugenics, *Nursing Clio*, 1 June 2017, https://nursingclio.org/2017/06/01/the-black-politics-of-eugenics/

NYT (1975) Lancelot Hogben dead; Popularizer of science, *The New York Times*, 23 August 1975, www.nytimes.com/1975/08/23/archives/lancelot-hogben-dead-popularizer-of-science.html

O'Brien E (2020) A battle for the soul of the nation: Eugenics and religion in post-revolutionary Mexico, 1925–1935, *Berkeley Center for Religion, Peace and World Affairs*, 3 January 2020, https://berkleycenter.georgetown.edu/responses/a-battle-for-the-soul-of-the-nation-eugenics-and-religion-in-post-revolutionary-mexico-1925-1935

OC (2016) Umberto Eco makes a list of the 14 common features of fascism, *Open Culture*, November 2016, www.openculture.com/2016/11/umberto-eco-makes-a-list-of-the-14-common-features-of-fascism.html

OR (1924) Book Review: *The New Decalogue of Science* by Albert Edward Wiggam, *The Harvard Crimson*, 20 March 1924, www.thecrimson.com/article/1924/3/20/crimson-bookshelf-ponly-occasionally-is-a/

Oreskes N (2019) *Why Trust Science?* Princeton University Press, Princeton.

Oveyssi N (2015) Dangerous love: 'Positive' eugenics, mass media, and the scientific woman, 1900–1945, *Berkeley Undergraduate Journal*, 28(2), https://escholarship.org/uc/item/0hh4p12n

Pearson K (1901) *National Life from the Standpoint of Science*, Adam & Charles Black, London.

Pearson K (1912, 2010) *Darwinism, Medical Progress and Eugenics: The Cavendish Lecture, 1912, an Address to the Medical Profession*, Nabu Press, UK.

Pearson K (1930) *The Life, Letters and Labors of Francis Galton*, Cambridge University Press, Cambridge, https://galton.org/pearson/

Pernick NS (1997) Eugenics and public health in American history, *American Journal of Public Health*, 87:1767—1772.

Pernick MS (2002) Taking Better Baby Contests seriously, *American Journal of Public Health*, 92:707—708, doi:10.2105/ajph.92.5.707

Phillips M (2018) *Eugenics and Protestant Social Reform: Hereditary Science and Religion in America, 1860–1940* by Dennis L. Durst (review), *Journal of Southern History*, 84(3):750—751, doi:10.1353/soh.2018.0201

Polya G (2012) Genocidal racist Charles Dickens (1812-1870), Indian Holocaust and UK-US Muslim genocide, *Counter Currents*, 10 February 2012, https://countercurrents.org/polya100212.htm

Pontecorvo G and Pallardy R (2020) Hermann Joseph Muller, *Encyclopedia Britannica*, 25 November 2020, www.britannica.com/biography/Hermann-Joseph-Muller

Quartey M (2019) The immorality of silence: Adventist leadership in times of conflict, *Spectrum Magazine*, 21 November 2019, https://spectrummagazine.org/views/2019/immorality-silence-adventist-leadership-times-conflict

Ray M (2020) Were the Nazis socialists? *Encyclopedia Britannica*, 16 December 2020, www.britannica.com/story/were-the-nazis-socialists

Redmond GP (2017) Karma and eugenics, *Barnard Health*, 11 May 2017, www.barnardhealth.us/medical-ethics/karma-and-eugenics.html

Reggiani AH (2006) *God's Eugenicist: Alexis Carrel and the Sociobiology of Decline*, Berghahn Books, Ney York.

Rejon MR (2018) Two clashing giants: Marxism and Darwinism, *Open Mind*, 13 August 2018, www.bbvaopenmind.com/en/science/leading-figures/two-clashing-giants-marxism-and-darwinism/

Richardson S (2019) Insidious inspiration: How Jim Crow inspired the Nazis, *History Net*, August 2019, www.historynet.com/insidious-inspiration-how-jim-crow-inspired-the-nazis.htm

Rogers K (2021) Francis Galton, *Encyclopedia Britannica*, 6 January 2021, www.britannica.com/biography/Francis-Galton

Root D (2016) Progressives and eugenics: The case of Justice Brandeis, *Reason.com*, 12 April 2016, https://reason.com/2016/04/12/progressives-and-eugenics-the-case-of-ju/

Rosen C (2020) *Preaching Eugenics: Religious Leaders and the American Eugenics Movement*, Oxford University Press, 28 December 2020, https://oxford.universitypressscholarship.com/view/10.1093/019515679X.001.0001/acprof-9780195156799

Rosenthal K (2019) Disability and the Russian revolution, *International Socialist Review*, 2 May 2019, https://isreview.org/issue/102/disability-and-russian-revolution

Rudling PA (2014) Eugenics and racial biology in Sweden and the USSR: Contacts across the Baltic Sea, *Canadian Bulletin of Medical History*, 31(1).

Rudling P (2015) Russia (former USSR), *The Eugenics Archive*, 24 February 2015, https://eugenicsarchive.ca/discover/tree/54ece589642e09bce5000001

Sachedina A (1995) Eugenics and religious law: III. Islam, *Encyclopedia*, 1995, www.encyclopedia.com/science/encyclopedias-almanacs-transcripts-and-maps/eugenics-and-religious-law-iii-islam

Samaan AE (2020) *From a 'Race of Masters' to a 'Master Race': 1948 to 1848*, Library Without Walls, New York.

Sanger M (2011) *The Pivot of Civilization and A Plan for Peace*, Suzeteo Enterprises, New York.

Schroder C (2003) Seventh Day Adventists, *UCSB Oral History Project*, 1 January 2003, http://holocaust.projects.history.ucsb.edu/Research/Proseminar/corrieschroder.htm

Selden S (2021) Eugenics popularization, *Eugenics Archive*, 6 January 2021, www.eugenicsarchive.org/html/eugenics/essay6text.html

Sharma MS (2018) Why Hitler is not a dirty word in India, *The Times of India*, 29 April 2018, http://timesofindia.indiatimes.com/articleshow/63955029.cms

Shtrauchler N (2017) How Nazis courted the Islamic world during WWII, *DW*, 13 November 2017, www.dw.com/en/how-nazis-courted-the-islamic-world-during-wwii/ a-41358387

Shudy D (2012) Eugenics and Christianity, *Urbana Theological Seminary*, 21 June 2012, www.urbanatheologicalseminary.org/eugenics-and-christianity/

Singleton MM (2014) The 'science' of eugenics: America's moral detour, *Journal of American Physicians and Surgeons*, 19(4):22--125.

Slorach R (2020) From eugenics to scientific racism, *International Socialism*, 10 January 2020, http://isj.org.uk/from-eugenics-to-scientific-racism/

Smith WJ (2004) The role of religion in the rise of eugenics, *Discovery Institute*, 26 July 2004, www.discovery.org/a/2135/

Spenkuch JL and Tillmann P (2017) Elite influence? Religion and the electoral success of the Nazis, Research Paper, *Kellogg School of Management*, Northwestern University, March 2017.

Stahnisch D (2013) Twin studies, *Eugenics Archive*, 13 September 2020, http://eugenicsarchive.ca/discover/tree/5233682e5c2ec5000000003e

Stern AM (2014) Book Review: *Preaching Eugenics: Religious Leaders and the American Eugenics Movement*, *H-Indiana*, 23 January 2014, https://networks.h-net.org/node/7579/reviews/8470/stern-rosen-preaching-eugenics-religious-leaders-and-american-eugenics

Tabery J (2011) Commentary: Hogben vs the tyranny of averages, *International Journal of Epidemiology*, 40(6):1454–1458, https://doi.org/10.1093/ije/dyr027

Thangaraj S (2017) Many Hindus saw themselves as Aryans and backed Nazis. Does that explain their support for Donald Trump? *Quartz India*, 3 February 2017, https://qz.com/india/901244/many-hindus-saw-themselves-as-aryans-and-backed-nazis-does-that-explain-hindutvas-support-for-donald-trump/

TMSPP (2009) Giving it the Old School Try--Margaret Sanger and the Taft Plan for Human Betterment, *The Margaret Sanger Papers Project Newsletter No. 51*, Spring 2009, New York.

Tonn J (2017) White feminism and eugenics: The case of Gertrude Davenport, *Lady Science*, 14 December 2017, https://thenewinquiry.com/blog/white-feminism-and-eugenics-the-case-of-gertrude-davenport/

Toth M (2005) Book review: *Playing God*, *Claremont Review of Books*, 5(3), Summer 2005, https://claremontreviewofbooks.com/playing-god/

Trotsky L (1935) If America should go communist, *Liberty*, 23 March 1935, www.marxists.org/archive/trotsky/1934/08/ame.htm

UMC (2016) Book of Resolutions: Repentance for support of eugenics, *The Book of Resolutions of The United Methodist Church*, 2016, www.umc.org/en/content/book-of-resolutions-repentance-for-support-of-eugenics

USHMM (2020) Nuremberg Race Laws, *Holocaust Encyclopedia*, https://encyclopedia.ushmm.org/content/en/article/introduction-to-the-holocaust, 12 December 2020.

USHMM (2021) Nazi persecution of Jehovah's Witnesses, *US Holocaust Memorial Museum*, 14 January 2021, https://encyclopedia.ushmm.org/content/en/article/nazi-persecution-of-jehovahs-witnesses

Victoria B (2013) DT Suzuki, Zen and the Nazis, *The Asia Pacific Journal*, 11(3), 21 October 2013, https://apjjf.org/2013/11/43/Brian-Victoria/4019/article.html

Victoria B (2015) Sawaki Kodo, Zen and wartime Japan: Final pieces of the puzzle, *The Asia Pacific Journal*, 13(3), 4 May 2015.

Victoria B and Muneo N (2014) "War is a Crime": Takenaka Shogen and Buddhist resistance, *The Asia Pacific Journal*, 12(4), 4 August 2014.

von Lupke-Schwarz M (2013) Remembering the victims of Nazi eugenics, *DW*, 14 July 2013, www.dw.com/en/remembering-the-victims-of-nazi-eugenics/a-16945569

von Sponeck HC (2006) *A Different Kind of War: The UN Sanctions Regime in Iraq*, Berghahn Books, New York.

Wagenmakers EJ (2018) Karl Pearson's worst quotation? *Bayesian Spectacles*, 3 May 2018, www.bayesianspectacles.org/karl-pearsons-worst-quotation/

Walker A (2015) The twins of Auschwitz, *BBC News*, 28 January 2015, www.bbc.com/news/magazine-30933718

Waller JC (2012) Commentary: The birth of the twin study—a commentary on Francis Galton's 'The History of Twins,' *International Journal of Epidemiology*, 41(4)913--917, https://doi.org/10.1093/ije/dys100

Wark W (2010) Review: *A Mosque in Munich*, by Ian Johnson, *The Globe and Mail*, 20 August 2020, www.theglobeandmail.com/arts/books-and-media/review-a-mosque-in-munich-by-ian-johnson/article4324972/

Weber B (2019) Human nature is created, *Medium*, 13 December 2019, https://medium.com/the-philosophers-stone/human-nature-is-created-c96fd1a701af

Weindling P et al. (2016) The victims of unethical human experiments and coerced research under National Socialism, *Endeavour*, 40(1):1—6, doi:10.1016/j.endeavour.2015.10.005.

Weisman S (1970) Marx and Engels on the population bomb, *Eco-Catastrophe*, Canfield Press, https://web.archive.org/web/20000521124318

Whitman JQ (2017) *Hitler's American Model: The United States and the Making of Nazi Race Law*, Princeton University Press, Princeton.

Wiggam AE (1924) *The New Decalogue of Science*, The Bobbs-Merrill Co., Indianapolis.

Wikipedia (2022 -- Adolf Hitler, Alexander Sergeevich Serebrovsky, Alexis Carrel, Akton T4, Anti-communist Mass Killings, Anti-miscegenation Laws in the United States, Armenian

Genocide; Augustus Hopkins Strong, Auschwitz Concentration Camp, Catholic Church and Nazi Germany, Charles B Davenport, Compulsory Sterilization, Confessing Church, Daisetsu Teitaro Suzuki, David Starr Jordan, David Herbert Lawrence, Emanuel Hirsch, Ernst Rudin, Eugenics, Eugenics Manifesto, Extermination Camps, Feminist Eugenics, Franz Boas, George Bernard Shaw, Gertrude Davenport, Gilbert Keith Chesterton, Harry H Laughlin, Health Care in Russia, Hermann Muller, Hindu Mahasabha, Introduction to Genetics, Irreligion in the United Kingdom, Irving Fisher, Jack Lindsay, JBS Haldane, Jehovah's Witnesses, John Desmond Bernal, John Harvey Kellogg, Joseph Mengele, Kurds in Turkey, Lancelot Thomas Hogben, Law for the Prevention of Genetically Diseased Offspring, List of Prisoners of Dachau, Lothrop Stoddard, Margaret Sanger, Mary de Garmo, Matthew (3:10, 7:17, 7:18), Nazi Concentration Camp, Nazi Eugenics, Nazi Human Experimentation, Oliver Wendell Holmes, Jr., Oliver Wendell Holmes, Sr., Paul Althaus, Racism in the Work of Charles Dickens, Rashtriya Swayamsevak Sangh, Relations between Nazi Germany and the Arab World, Religion in America, Religion in Germany, Religious Views of Adolf Hitler, Takenaka Shogen, The Herero and Nama Genocide, The Holocaust, The Galton Society, The Nuremberg Laws, Theodore Lothrop Stoddard, WEB Du Bois, Women's Rights).

Wilkes J (2020) Why did Hitler choose the swastika, and how did a Sanskrit symbol become a Nazi emblem? *History Extra*, 1 June 2020, www.historyextra.com/period/second-world-war/how-why-sanskrit-symbol-become-nazi-swastika-svastika/

Williams C (2010) Are there too many people? *International Socialist Review*, Issue 68, https://isreview.org/issue/68/are-there-too-many-people

Wolff RD (2020) The looming specter of fascism in capitalist states, *Asia Times*, 15 October 2020, https://asiatimes.com/2020/10/the-looming-specter-of-fascism-in-capitalist-states/

4. SCIENCE

Aaen-Stockdale C (2012) Neuroscience for the soul, *The Psychologist*, 25(7), July 2012, www.thepsychologist.org.uk

ACLU (2021) Race and the death penalty, *ACLU*, 15 March 2021, www.aclu.org/other/race-and-death-penalty

ACS (2020) Why people start smoking and why it's hard to stop, *American Cancer Society*, 12 November 2020, www.cancer.org/healthy/stay-away-from-tobacco/why-people-start-using-tobacco.html

Aczel AD (2014) Why science does not disprove God, *Time*, 27 April 2017, https://time.com/77676/why-science-does-not-disprove-god/

Afsaruddin A (2005) Free will and predestination: Islamic concepts, *Encyclopedia*, 2005, www.encyclopedia.com/environment/encyclopedias-almanacs-transcripts-and-maps/free-will-and-predestination-islamic-concepts

Ahlstrom D (2001) Free will versus genetic destiny, *The Irish Times*, 5 April 2001, https://www.irishtimes.com/news/free-will-versus-genetic-destiny-1.298180

Al-Jazeera (2021) World on the verge of climate crisis 'abyss', warns UN, *Al-Jazeera News*, 19 April 2021, www.aljazeera.com/news/2021/4/19/world-on-the-verge-of-climate-crisis-abyss-warns-un

Al-Jubouri I (2004) Human acts in Islamic philosophy, *Philosophy Now*, https://philosophynow.org/issues/47/Human_Acts_in_Islamic_Philosophy

Al-Khalili J (2009) The 'first true scientist', *BBC News*, 4 January 2009, http://news.bbc.co.uk/2/hi/7810846.stm

Alpert M (2019) Can science rule out God? *Scientific American*, 23 December 2019, https://blogs.scientificamerican.com/observations/can-science-rule-out-god/

Ames WL (2003) Emptiness and quantum theory, In AB Wallace (editor) (2003) *Buddhism & Science: Breaking New Ground*, Columbia University Press, New York.

Ananthaswamy A (2012) Brain might not stand in the way of free will, *New Scientist*, 6 August 2012, www.newscientist.com/article/dn22144-brain-might-not-stand-in-the-way-of-free-will/

Anderson D (2008) Book review - John Byl: *The Divine Challenge on Matter, Mind, Math and Meaning, More Than Words*, 12 March 2008, http://mothwo.blogspot.com/2008/03/book-review-john-byl-divine-challenge.html

Anderson E (2020) The idea of God has one major flaw, *Medium*, 12 December 2020, https://ella-alderson.medium.com/the-idea-of-god-has-one-major-flaw-ab7de18fab88

Angell M (2005) *The Truth About the Drug Companies: How They Deceive Us and What to Do About It*, Random House, New York.

Ann K (2019) Choice, free will, and nicotine addiction, *Medium*, 30 July 2019, https://medium.datadriveninvestor.com/free-will-and-nicotine-addiction-8ba154b43924

Anne L (2016) Some problems with free will in evangelical Christianity, *Patheos*, 7 June 2016, www.patheos.com/blogs/lovejoyfeminism/2016/06/some-problems-with-free-will-in-evangelical-christianity.html

Anonymous (2003) DNA leaders call religion to account, *The Sidney Morning Herald*, 22 March 2003, www.smh.com.au/world/dna-leaders-call-religion-to-account-20030322-gdggx3.html

Anonymous (2020) Are we free? *Medium*, 6 June 2019, https://medium.com/swlh/are-we-free-a1c19d268b20

ANU (2021) Takdir and Ikhtiar: The problem of human destiny, *Australian National University*, 18 February 2021, http://press-files.anu.edu.au/downloads/press/p25911/html/ch03s05.html

Arnold C (2020) Clues to COVID-19 severity may lie in our genes, *Hopkins Bloomberg Public Health Magazine*, 2020 Special Issue, https://magazine.jhsph.edu/2020/clues-covid-19-severity-may-lie-our-genes

Asher R (2012) Can a scientist be religious? *University of Cambridge*, 11 March 2012, www.cam.ac.uk/research/discussion/can-a-scientist-be-religious

Assad W (2021) Religion and science — Does it work? *Medium*, 23 February 2021, https://willassad.medium.com/religion-and-science-does-it-work

Augustyn A (2022) False consciousness, *Encyclopedia Britannica*, 6 October 2022, www.britannica.com/topic/false-consciousness.

Austin JH (1998) *Zen and the Brain: Toward an Understanding of Meditation and Consciousness*, MIT Press, Cambridge, MA.

Baird J and Gleeson H (2017) 'Submit to your husbands': Women told to endure domestic violence in the name of God, *ABC News*, 18 July 2017, www.abc.net.au/news/2017-07-18/domestic-violence-church-submit-to-husbands/8652028?nw=0

Bais S (2010) *In Praise of Science: Curiosity, Understanding, and Progress*, The MIT Press, Cambridge, MA.

Ball P (2021) Why free will is beyond physics, *Physics World*, 6 January 2021, https://physicsworld.com/a/why-free-will-is-beyond-physics/

Barash D (2014) Is Buddhism the most science-friendly religion? *Scientific American*, 11 February 2014, https://blogs.scientificamerican.com/guest-blog/is-buddhism-the-most-science-friendly-religion/

Bardon A (2019) Can a sacred science that unites mysticism/spirituality and materialistic science be developed? *Medium*, 18 July 2019, https://medium.com/@angelabardon64/can-a-sacred-science-that-unites-mysticism-spirituality-and-materialistic-science-be-developed

Baars BJ (2003) The global brainweb: An update on global workspace theory, *Science and Consciousness Review*, October 2003.

Bar-Yam Y (1999) *Dynamics of Complex Systems*, Westview Press, NY.

Baumeister RF (2013) Do you really have free will? Of course. Here's how it evolved, *Slate*, 25 September 2013, https://slate.com/technology/2013/09/free-will-debate-what-does-free-will-mean-and-how-did-it-evolve.html

BBC (2022) Religion and science, *BBC News*, 29 September 2022, www.bbc.co.uk bitesize/guides/z2qxvcw/revision

Beaumont M (2019) Christian defense of free will in debate with Muslims in the early Islamic period, *Transformation*, 36(3):149—163, doi:10.1177/0265378819852269

Bentley W (2018) Balancing the transcendence and immanence of God in Nürnberger's theology, *Verbum Eccles*, 39(1), 2018, http://dx.doi.org/10.4102/ve.v39i1.1917

Berard T (2022) Do atheists believe in free will? *Medium*, 10 April 2022, https://tonyberard.medium.com/do-atheists-believe-in-free-will-d06c6c4 b9abc

Betuel E (2020) Brain study reveals how much of math ability is genetic, *Inverse*, 22 October 2020, www.inverse.com/innovation/5-ways-roborock-will-make-you-feel-like-youre-living-in-the-future

Bhat AR (2021) Free will and determinism: An overview of Muslim scholars' perspective, *Shah-i-Hamadan Institute of Islamic Studies*, University of Kashmir, Srinagar.

Bitesize (2021) Predestination, free will and judgement, *BBC News*, 18 February 2021, www.bbc.co.uk/bitesize/guides/zkdkw6f/revision/2

Blackford R (2016) Against accommodationism: How science undermines religion, *The Conversation*, 1 January 2016, https://theconversation.com/against-accommodationism-how-science-undermines-religion-52660

BL Editors (2019) Are gaps in scientific knowledge evidence for God? *Bio Logos*, 19 January 2019, https://biologos.org/common-questions/are-gaps-in-scientific-knowledge-evidence-for-god/

Bloom H (2016) *The God Problem: How a Godless Cosmos Creates,* Prometheus, New York.

Brass M, Furstenberg A and Mele AR (2019) Why neuroscience does not disprove free will, *Neuroscience & Biobehavioral Reviews*, 102:251--263, https://doi.org/10.1016/j.neubiorev.2019.04.024.

Brice M (2021) Absolutist theology vs. secularism, *Medium*, 2 January 2021, https://mansabrice.medium.com/absolutist-theology-vs-secularism

Brown EN and Behrmann M (2017) Controversy in statistical analysis of functional magnetic resonance imaging data, *PNAS*, 114(7), 25 April 2017, www.pnas.org/cgi/doi/10.1073/pnas.1705513114

BS (2021) Honeybee genome, *Bee Spotter*, 26 March 2021, https://beespotter.org/topics/genome/

Budson AE et al. (2022) Consciousness as a memory system, *Cognitive and Behavioral Neurology*, 10(1097), doi:10.1097/WNN.0000000000000319

Cain B (2021) Why we're all spellbound by myths, *Medium*, 22 February 2021, https://medium.com/interfaith-now/why-were-all-spellbound-by-myths

Capps D (1992) Religion and child abuse: Perfect together, *Journal for the Scientific Study of Religion*, 31(1), 1–14. https://doi.org/10.2307/1386828

Capra F (2010, 1975) *The Tao of Physics: An Exploration of the Parallels Between Modern Physics and Eastern Mysticism*, Shambala Publications, Boulder, CO.

Card D (2017) Book review: *Elbow Room*, by Daniel Dennett, *Medium*, 31 March 2017, https://dallascard.medium.com/elbow-room-by-daniel-dennett

Carey N (2012) *The Epigenetic Revolution*, Icon Books, London.

Carrier R (2018) Dennett vs. Harris on free will, *Richard Carrier Blog*, 31 May 2018, www.richardcarrier.info/archives/13814

Caruso GD (2016) Book review: *Restorative Free Will: Back to the Biological Base* by BN Waller, *Notre Dame Philosophical Reviews*, 16 January 2016, https://ndpr.nd.edu/reviews/restorative-free-will-back-to-the-biological-base/

Caruso GD (2020) Buddhism, free will, and punishment: Taking Buddhist ethics seriously, *Zygon*, 55(2), www.zygonjournal.org

Castillou G (2022) AI reveals links between individual differences within brain anatomy and those within autism spectrum disorder symptoms, *Neuroscience News*, 11 May 2022, https://neurosciencenews.com/cord7-gene-iq-20550/

Cave S (2016) There's no such thing as free will, *The Atlantic*, June 2016, www.theatlantic.com/magazine/archive/2016/06/theres-no-such-thing-as-free-will/480750/

CBC (2020) Buddhism and science: A doomed romance? *CBC Radio*, 2 October 2020, www.cbc.ca/radio/ideas/buddhism-and-science-a-doomed-romance-1.5745107

Chabris CF et al. (2012) Most reported genetic associations with general intelligence are probably false positives, *Psychological Science*, 23(11):1314—23, doi:10.1177/0956797611435528.

Chalabi M (2015) Are prisoners less likely to be atheists? *Five Thirty* Eight, 12 March 2015, https://fivethirtyeight.com/features/are-prisoners-less-likely-to-be-atheists/

Cherry K (2020) BF Skinner biography, *Very Well Mind*, 27 April 2020, www.verywellmind.com/b-f-skinner-biography-1904-1990-2795543

Chilton BG (2018) 4 views of free will, *Christian Post*, 1 December 2018, www.christianpost.com/voices/4-views-of-free-will.html

Choate A (2013) Free will vs fate in Hinduism, *Patheos*, 8 August 2013, www.patheos.com/blogs/whitehindu/2013/08/free-will-vs-fate-in-hinduism/

Choy et al. (2022) Larger striatal volume is associated with increased adult psychopathy, *Journal of Psychiatric Research*, 6 March 2022, doi:10.1016/j.jpsychires.2022.03.006

Clayton M (2018) The song remains the same: A review of Harris' *Free Will*, *Perspectives in Behavioral Science*, 41:653–656, https://doi.org/10.1007/s40614-018-0140-2

Collins F (2007) *The Language of God: A Scientist Presents Evidence for Belief*, Free Press, New York.

Colmez C and Schneps L (2019) The odds of innocence, *Medium*, 13 June 2019, https://medium.com/nautilus-magazine/the-odds-of-innocence-9597094158ef

Comfort N (2018) Genetic determinism rides again, *Nature*, 25 September 2018, www.nature.com/articles/d41586-018-06784-5

Conan N et al. (2010) Can genes and brain abnormalities create killers? *NPR*, 6 July 2921, www.npr.org/templates/story/story.php?storyId=128339306

Corbyn A (2022) Morgan Levine: 'Only 10-30% of our lifespan is estimated to be due to genetics', *The Guardian* (UK), 7 May 2022, www.theguardian .com/science/2022/may/07/morgan-levine-only-10-30-of-our-lifespan-is-estimated-to-be-due-to-genetics

Cowie S (2019) Brazil's flat Earthers to get their day in the sun, *The Guardian* (UK), 6 November 2019, www.theguardian.com/world/2019/nov/06/brazil-flat-Earth-conference-terra-plana

Craig WL (1999) The ultimate question of origins: God and the beginning of the Universe, *Astrophysics and Space Science*, 269-270:723—740, www.reasonablefaith.org/writings/scholarly-writings/the-existence-of-god/the-ultimate-question-of-origins-god-and-the-beginning-of-the-Universe

Craig WL (2000) *The Kalam Cosmological Argument*, Wipf and Stock, Eugene, Oregon.

Critchlow H (2020) How much do our genes restrict free will? *The Conversation*, 13 October 2020, https://theconversation.com/how-much-do-our-genes-restrict-free-will-134330

Crossman A (2020) Understanding Karl Marx's class consciousness and false consciousness, *ThoughtCo*, 27 August 2020, thoughtco.com/class-consciousness-3026135.

Dalai Lama (2006) *The Universe in a Single Atom: The Convergence of Science and Spirituality*, Random House, New York.

Davis N (2020) Long-term offenders have different brain structure, study says, *The Guardian* (UK), 18 February 2020, www.theguardian.com/science/2020/feb/17/long-term-offenders-have-different-brain-structure-study-says

Dawkins R (1995) *River Out of Eden: A Darwinian View of Life*, Basic Books, New York.

Dawkins R (2006) Why there almost certainly is no God, *Huffington Post*, 23 October 2006, www.huffpost.com/entry/why-there-almost-certainl_b_32164

Dawkins R (2008) *The God Delusion*, Mariner Books, New York.

Dawkins R (2015) *The Blind Watchmaker: Why the Evidence of Evolution Reveals a Universe without Design*, WW Norton, New York.

DeAngelis CD (2000) Conflict of interest and the public trust (editorial), *Journal of the American Medical Association*, 284:2237—2238.

DeAngelis CD (2006) The influence of money on medical science (editorial), *Journal of the American Medical Association*, 296:E1--E3, doi:10.1001/jama.296.8.jed60051.

Denigris M (2015) On the Kalam Cosmological Argument, *The Winnower*, 8:e144157.70670, doi:10.15200/winn.144157.70670

Dennett D (2015) *Elbow Room: The Varieties of Free Will Worth Wanting*, Bradford Books, Cambridge, MA.

Dennett DC and Caruso G (2018) Can we be held morally responsible for our actions? Yes, says Daniel Dennett. No, says Gregg Caruso. Reader, you decide, *Aeon*, 4 October 2018, https://aeon.co/essays/on-free-will-daniel-dennett-and-gregg-caruso-go-head-to-head

DiGravio G (2022) A new explanation for consciousness, *Neuroscience News*, 3 October 2022, https://neurosciencenews.com/consciousness-theory-21571/

Diodati M (2020) On real numbers, free will and open future, *Medium*, 18 April 2020, https://medium.com/amazing-science/on-real-numbers-free-will-and-open-future-57c28f3d7a72

Dixon T (2008) *Science and Religion: A Very Short Introduction*, Oxford University Press, Oxford.

Donis P (2021) Review of *The Dancing Wu Li Masters*, *Peter Donis Net*, 20 March 2021, www.peterdonis.net/philosophy/philosophyreview2.html

Doorn MV (2019) The most important question, *Medium*, 9 May 2019, https://medium.com/@maartenvandoorn/the-most-important-question-about-life-ffbc15ff7b8f

Ducharme J (2018) Stephen Hawking was an atheist. Here's what he said about God, heaven and his own death, *Time*, 14 March 2018, https://time.com/5199149/stephen-hawking-death-god-atheist/

Duffy G (2013) Human consciousness was always going to require a god, *The Irish Times*, 29 May 2013, https://www.irishtimes.com/news/social-affairs/religion-and-beliefs/human-consciousness-was-always-going-to-require-a-god-1.1342282

Duigman B (2021) Existence of God, *Encyclopedia Britannica*, 3 February 2021, www.britannica.com/topic/existence-of-God

Earp BD et al. (2021) Racial justice requires ending the War on Drugs, *The American Journal of Bioethics*, doi:10.1080/15265161.2020.1861364

Ebstein RP et al. (2015) Association between the dopamine D4 receptor gene exon III variable number of tandem repeats and political attitudes in female Han Chinese, *Proceedings of the Royal Society*, Series B, 28220151360, http://dx.doi.org/10.1098/rspb.2015.1360

Ecklund EH (2010) *Religion vs. Science: What Scientists Really Think*, Oxford University Press, Oxford.

Ecklund EH et al. (2016) Religion among scientists in international context: A new study of scientists in eight regions. *Socius*, 2, https://doi.org/10.1177/2378023116664353

Ecklund EH and Scheitle CT (2017) *Religion vs. Science: What Religious People Really Think*, Oxford University Press, Oxford.

Ecklund EH and Johnson DR (2019) Scientists talk religion around the globe, *Biologos*, 9 December 2019, https://biologos.org/articles/scientists-talk-religion-around-the-globe

Editorial (2004) Depressing research, *The Lancet*, 363:1335.

Editorial (2012) Can we live without free will? *New Scientist*, 8 August 2012, www.newscientist.com/article/mg21528772-300-can-we-live-without-free-will/

Ellis G (2007) Case not proven, *Physics World*, 1 May 2007, https://physicsworld.com/a/case-not-proven/

Ellis G (2020) From chaos to free will, *Aeon*, 9 June 2020, https://aeon.co/essays/heres-why-so-many-physicists-are-wrong-about-free-will

Ellison CG, Trinitapoli JA, Anderson KL and Johnson BR (2007) Race/ethnicity, religious involvement, and domestic violence, *Violence Against Women*, 2007, 13(11):1094—112, doi:0.1177/1077801207308259.

Engels F (1925) *Dialectics of Nature*, Progress Publishers, Moscow.

EPFL (2020) Breathing may change your mind about free will: Do you think that you are clicking on that button when your mind decides to do so? Think again, *Science Daily*, 6 February 2020, www.sciencedaily.com/releases/2020/02/200206080449.htm

Esteves JA (2020) Science, religion not opposing sides in humanity's progress, cardinal says, *CRUX*, 3 September 2020, https://cruxnow.com/vatican/2020/09/science-religion-not-opposing-sides-in-humanitys-progress-cardinal-says/

Evans CS (2018) Moral arguments for the existence of God, in EN Zalta (editor), *The Stanford Encyclopedia of Philosophy*, Fall 2018 edition, https://plato.stanford.edu/archives/fall2018/entries/moral-arguments-god/

Evans G (2018) The unwelcome revival of 'race science,' *The Guardian* (UK), 2 March 2018, www.theguardian.com/news/2018/mar/02/the-unwelcome -revival-of-race-science

Fallon F (2021) Integrated information theory of consciousness, *Internet Encyclopedia of Philosophy*, 5 February 2021, https://iep.utm.edu/int-info/

Farnsworth B (2020) What is the subconscious mind? *I Motions*, 18 February 2020, https://imotions.com/blog/what-is-the-subconscious-mind/

Federman A (2010) What kind of free will did the Buddha teach? *Philosophy East and West*, 60(1), doi:10.1353/pew.0.0086

Finkelstein DR (2003) Emptiness and relativity, In AB Wallace (editor) (2003) *Buddhism & Science: Breaking New Ground*, Columbia University Press, New York.

FIW (2020) Socrates, reductionism, and proper levels of explanation, *Medium*, 21 December 2020, https://medium.com/socrates-cafe/socrates-reductionism-and-proper-levels-of-explanation-135872cb8a2

Frank A (2021a) Reductionism vs. emergence: Are you "nothing but" your atoms? *Big Think*, 29 April 2021, https://bigthink.com/13-8/reductionism-vs-emergence-science-philosophy/

Frank A (2021b) Why condensed matter physicists reject reductionism, *Big Think*, 1 July 2021, https://bigthink.com/13-8/condensed-matter-physicists -reject-reductionism/

Fraser G (2015) Christianity, when properly understood, is a religion of losers, *The Guardian* (UK), 3 April 2015, www.theguardian.com/commentisfree/belief/2015/apr/03/christianity-when-properly-understood-religion-losers

Fuchs E (2013) Why the South is more violent than the rest of America, *Business Insider*, 18 September 2013, www.businessinsider.com/south-has-more-violent-crime-fbi-statistics-show-2013-9?IR=T

Funk C and Masci D (2015) 5 facts about the interplay between religion and science, *Pew Forum*, 22 October 2015, www.pewresearch.org/fact-tank/2015/10/22/5-facts-about-the-interplay-between-religion-and-science/

Gagnon C (1981) For a scientific vision of the world: Determinism or free will? *Proletarian Unity*, 5(2), www.marxists.org/history/erol/ca.secondwave/is-free-will.htm

Gajilan AC (2007) Are humans hard-wired for faith? *CNN*, 5 April 2007, www.cnn.com/2007/HEALTH/04/04/neurotheology/index.html

Galadari A (2019) Psychology of mystical experience: Muhammad and Siddhartha, *Anthropology of Consciousness*, 30(2): 152–178, https://ssrn.com/abstract=3451189

Garrido M (2013) Vedic philosophy and quantum mechanics on the soul, 15 April 2013, *Huffington Post*, www.huffpost.com/entry/vedic-philosophy-and-quantum-mechanics-on-the-soul_b_3082572

Geiser M (2022) "More is different": Why reductionism fails at higher levels of complexity, *Big Think*, 9 March 2022, https://bigthink.com/13-8/reductionism-fails-complexity/

Genschow O, Rigoni D and Brass M (2017) Belief in free will affects causal attributions when judging others' behavior, *PNAS*, 29 August 2017, www.pnas.org/content/early/2017/08/29/1701916114

Gewertz K (2007) Albert Einstein, civil rights activist, *The Harvard Gazette*, 12 April 2007, https://news.harvard.edu/gazette/story/2007/04/albert-einstein-civil-rights-activist/

Ghose T (2012a) Animals are moral creatures, scientist argues, *Live Science*, 15 November 2012, www.livescience.com/24802-animals-have-morals-book.html

Ghose T (2012b) 5 animals with a moral compass, *Live Science*, 15 November 2012, www.livescience.com/24800-animals-emotions-morality.html

Gholipour B (2019a) Philosophers and neuroscientists join forces to see whether science can solve the mystery of free will, *Science Magazine*, 21 March 2019, www.sciencemag.org/news/2019/03/philosophers-and-neuroscientists-join-forces-see-whether-science-can-solve-mystery-free

Gholipour B (2019b) A famous argument against free will has been debunked, *The Atlantic*, 10 September 2019, www.theatlantic.com/health/archive/2019/09/free-will-bereitschaftspotential/597736/

Gier NF and Kjellberg P (2004) Buddhism and the freedom of the will: Pali and Mahayanist responses, In JK Campbell et al. (editors) *Freedom and Determinism: Topics in Contemporary Philosophy*, MIT Press, Cambridge, 277—304, www.webpages.uidaho.edu/ngier/budfree.htm.

Giesinger J (2010) Free will and education, *Journal of Philosophy of Education*, 44(4):515—528.

GL (2022) The religious consciousness and deism, *The Gifford Lectures*, No 6, 2022, www.giffordlectures.org/books/interpretation-religious-experience-vol-2/lecture-sixth-religious-consciousness-and-deism

Gleeson H and Baird J (2017) Exposing the darkness within: Domestic violence and Islam, *ABC News*, 24 April 2017, www.abc.net.au/news/2017-04-24/confronting-domestic-violence-in-islam/8458116?nw=0

Glyn I (2013) *Elegance in Science: The Beauty of Simplicity*, Oxford University Press, Oxford.

Goetz S (2014) Is Sam Harris right about free will? *The Table*, 26 May 2014, https://cct.biola.edu/sam-harris-free-will-book-review/

Gooding P (2018) The psychology of believing in free will, *The Conversation*, 2 July 2018, https://theconversation.com/the-psychology-of-believing-in-free-will-97193

Gould SJ (1997) Non-overlapping magisteria, *Natural History*, No. 106, www.blc.arizona.edu/courses/schaffer/449/Gould-Nonoverlapping-Magisteria.htm

Gould SJ (1999) *Rock of Ages: Science and Religion in the Fullness of Life*, Vintage, New York.

Graffin GW and Provine WB (2007) Evolution, religion and free will, *American Scientist*, 95(4):294, doi:10.1511/2007.66.294

Graham-Leigh E (2018) Marxism and human nature, *Monthly Review*, 13 July 2018, https://mronline.org/2018/07/13/marxism-and-human-nature/

Griffin A (2016) Free will could all be an illusion, scientists suggest after study shows choice may just be brain tricking itself, *The Independent*, 30 April 2016, https://www.independent.co.uk/news/science/free-will-could-all-be-illusion-scientists-suggest-after-study-shows-choice-could-just-be-brain-tricking-itself-a7008181.html

Gross T and Wright R (2017) Can Buddhist practices help us overcome the biological pull of dissatisfaction? *NPR*, 7 August 2017, www.npr.org/transcripts/541610511

Hall NF and Hall LKB (1986) Is the war between science and religion over? *American Humanist*, May/June 1986, https://americanhumanist.org/what-is-humanism/war-science-religion/

Harman O (2014) 'Chance and Necessity' revisited, *LA Review of Books*, 24 July 2014, https://lareviewofbooks.org/article/chance-necessity-revisited/

Harris S (2012) *Free Will*, Free Press, New York.

Harrison P (2018) Saint of science: The religious life of Isaac Newton, *LA Review of Books*, 2 February 2018, https://marginalia.lareviewofbooks.org/saint-science-religious-life-isaac-newton/

Heflick N (2010) Having your God-is-all-knowing cake and eating your free will too, *Psychology Today*, 8 March 2010, www.psychologytoday.com/us/blog/the-big-questions/201003/having-your-god-is-all-knowing-cake-and-eating-your-free-will-too

Hirji KF (2009) No short-cut in assessing trial quality: A case study, *Trials*, 10:1, www.trialsjournal.com.

Hirji KF and Premji Z (2011) Pre-referral rectal Artesunate in severe malaria: A flawed trial, *Trials*, 12:188, www.trialsjournal.com/content/12/1/188.

Hirschfeld E (2009) Is Marxism determinist? *Marxist Theory of Art*, 26 February 2009, http://marxist-theory-of-art.blogspot.com/2009/02/is-marxism-determinist.html

Hoel EP (2021) Agent above, atom below: How agents causally emerge from their underlying microphysics, *Department of Biological Sciences*, Columbia University, New York, https://fqxi.org/community/forum/topic/2873

Hogan M (2019) The best available story of human consciousness, Integrated Information Theory, *Medium*, 15 July 2019, https://medium.com/thn/the-best-available-story-of-human-consciousness-6bc2db103996

Horgan J (2022) Denying free will is physics reductionism at its absolute worst, *Big Think*, 3 May 2013, https://bigthink.com/articles/denying-free-will-is-physics-reductionism-at-its-absolute-worst/

Hummel LM and Woloschak GE (2016) Chance, necessity, love: An evolutionary theology of cancer, *Zygon*, 51(2):293—317, 5 May 2016, https://doi.org/10.1111/zygo.2257

Hut P (2003) Conclusion: Life as a laboratory, In AB Wallace (editor) (2003) *Buddhism & Science: Breaking New Ground*, Columbia University Press, New York.

IDR Labs (2021) Russell on Buddhism, *IDR Labs*, 2021, www.idrlabs.com/quotes/bertrand-russell.php

Impey C (2020) What Buddhism and science can teach each other – and us – about the Universe, *The Conversation*, 16 June 2020, https://theconversation.com/what-buddhism-and-science-can-teach-each-other-and-us-about-the-universe-134322

Inch (2014) Debate: Can religion and science co-exist? *Inch Magazine*, November 2014, www.ineos.com/inch-magazine/articles/issue-7/debate/

Ioannidis JP (2005) Why most published research findings are false, *PLoS Medicine*, 2(8):e124.

Jaeger M et al. (2022) Where do cultural tastes come from? Genes, environments, or experiences, *Sociological Science*, doi:10.15195/v9.a11

Jammer M (2002) *Einstein and Religion: Physics and Theology*, Princeton University Press, Princeton.

Johns M (1999) Book review: *Rock of Ages: Science and Religion in the Fullness of Life* by Stephen Jay Gould, *Emory Report*, 27 September 1999, http://www.emory.edu/EMORY_REPORT/erarchive/1999/September/erseptember.27/9_27_99bookreview.html

Johnson G (2005a) Book review: 'The Universe in a Single Atom': Reason and faith, *The New York Times*, 18 September 2005.

Johnson G (2005b) Agreeing only to disagree on God's place in science, *The New York Times*, 27 September 2005, www.nytimes.com/2005/09/27/science/agreeing-only-to-disagree-on-gods-place-in-science.html

Jones S and Leicht C (2016) Why science and religion aren't as opposed as you might think, *The Conversation*, 24 March 2016, https://theconversation.com/why-science-and-religion-arent-as-opposed-as-you-might-think-56641

Jones SE (2011) A brief survey of Sir Isaac Newton's views on religion, in MD Rhodes and JW Moody (2011) *Converging Paths to Truth*, Deseret Book, Salt Lake City, https://rsc.byu.edu/converging-paths-truth/brief-survey-sir-isaac-newtons-views-religion

JW (2021) What does the Bible say about free will? Is God in control? *Jehovah's Witnesses*, 20 February 2021, www.jw.org/en/bible-teachings /questions/free-will-in-the-bible/

Jyotiraditya (2018) What is quantum mechanics? What is theory of relativity? *Medium*, 6 November 2018, https://medium.com/predict/what-is-quantum-mechanics-what-is-theory-of-relativity-fdbe87eb9c79

Kandi S (2013) Characteristics of mystical experiences and impact of meditation, *International Journal of Social Sciences*, 2(2):141, doi:10.5958/j.2321-5771.2.2.007

Karean NJ (2010) What it means to be "Scientifically Proven", *Digital Bits Skeptics*, 14 March 2010, www.dbskeptic.com/2010/03/14/what-it-means-to-be-scientifically-proven/

Khadka S et al. (2014) Genetic association of impulsivity in young adults: a multivariate study, *Translational Psychiatry*, 4:e451, https://doi.org/10.1038/tp.2014.95

Khalil L (2022) "Predatory cult": The shadow of Unification Church over Abe's funeral, *The Interpreter*, 28 September 2022, www.lowyinstitute.org/the-interpreter/predatory-cult-shadow-unification-church-over-abe-s-funeral

King BJ (2013) Frans de Waal's bottom-up morality: we're not good because of God, *NPR*, www.npr.org/sections/13.7/2013/03/21/174830095/frans-de-waals-bottom-up-morality-were-not-good-because-of-god

Klinghoffer D (2012) Darwin critic wins the Templeton Prize; Congratulations to Dalai Lama, *Evolution News*, 29 March 2012, https://evolutionnews.org/2012/03/darwin_critic_w/

Knight D (2013) Computer scientists 'prove' God exists, *ABC News*, 27 October 2013, https://abcnews.go.com/Technology/computer-scientists-prove-god-exists/story?id=20678984

Koch C (2009) A 'complex' theory of consciousness, *Scientific American*, 1 July 2009, www.scientificamerican.com/article/a-theory-of-consciousness/

Koch C (2018) What is consciousness? *Scientific American*, 318(6):60—64, www.scientificamerican.com/article/what-is-consciousness

Koenig LB et al. (2005) Genetic and environmental influences on religiousness: Findings for retrospective and current religiousness ratings, *Journal of Personality*, 73(2), 471-488, https://doi.org/10.1111/j.1467-6494.2005.00316.x

KR (2001) Book review: *Why God Won't Go Away*, *Kirkus Reviews*, 3 April 2001, www.kirkusreviews.com/book-reviews/andrew-md-newberg/why-god-wont-go-away/

Kulkarni V (2020) What Erwin Schrödinger said about the Upanishads, *The Wire*, 5 September 2020, https://science.thewire.in/the-sciences/erwin-schrodinger-quantum-mechanics-philosophy-of-physics-upanishads/

Larsson J (2011) False consciousness revisited. On Rousseau, Marx and the positive side of negative education, *KAPET*, 7(1), https://www.diva-portal.org/smash/get/diva2:490563/FULLTEXT01.pdf

Laughlin RB (2005) *A Different Universe: Reinventing Physics from the Bottom Down*, Basic Books, NY.

Lavers C (2001) A hands-off God? *The Guardian* (UK), 3 February 2001, www.theguardian.com/books/2001/feb/03/scienceandnature.science

Lawless C (2021) Religion and spirituality aren't enemies, *Medium*, 12 January 2021, https://inpurpledurance.medium.com/religion-and-spirituality-arent-enemies-72b29c6bcb38

Leasure R (2019) Predestination and free will, *Grace Bible*, 29 April 2019, www.gracebible.com/Predestination-And-Free-Will

Lee L (2018) Why atheists are not as rational as some like to think, *The Conversation*, 27 September 2018, https://theconversation.com/why-atheists-are-not-as-rational-as-some-like-to-think-103563

Levine P (2021) Freedom of the will or freedom from the will? (comparing Harry Frankfurt and Buddhism), *Peter Levine Website*, 17 February 2021, https://peterlevine.ws/?p=23772

Lewis R (2018) Why we should not be impressed by eerie coincidences, *Psychology Today*, 27 October 2018, www.psychologytoday.com/us/blog/finding-purpose/201810/why-we-should-not-be-impressed-eerie-coincidences

Lewis R (2019) The varieties of mystical experience, *Psychology Today*, 18 December 2019, www.psychologytoday.com/us/blog/finding-purpose/201912/the-varieties-mystical-experience

Lewis R (2020) Can DNA predict who might be a mass murderer? *Genetics Literacy Project*, 31 January 2020, https://geneticliteracyproject.org/2020/01/31/can-dna-predict-who-might-be-a-mass-murderer/

Lewis R (2021) How can so many people believe such weird things? *Psychology Today*, 3 February 2021, www.psychologytoday.com/us/blog/finding-purpose/202102/how-can-so-many-people-believe-such-weird-things

Libet B et al. (1983) Time of conscious intention to act in relation to onset of cerebral activity (readiness-potential): The unconscious initiation of a freely voluntary act, *Brain*, 106(3):623—642, doi:10.1093/brain/106.3.623.

Libet B (2007) The neural time factor in conscious and unconscious events, In GR Bock and J Marsh (editors) *Experimental and Theoretical Studies of Consciousness*, Ciba Foundation Symposium 174, https://doi.org/10.1002/9780470514412.ch7

List C, Caruso G and Clark C (2019) Free will: Real or illusion: A debate, *The Philosopher*, Autumn 2019, www.thephilosopher1923.org

Liu J et al. (2009) Religion and science: conflict or harmony? *Pew Forum*, 4 May 2009, www.pewforum.org/2009/05/04/religion-and-science-conflict-or-harmony/

Lopez DS (2010) *Buddhism and Science: A Guide for the Perplexed*, University of Chicago Press, Chicago.

Lopez DS (2021) The scientific Buddha, *University of Michigan*, 2021, https://info-buddhism.com/Scientific_Buddha_Lopez.html

Lott T (2013) Zen freedom, *Aeon*, 28 March 2013, https://aeon.co/essays/do-i-have-free-will-in-zen-the-question-makes-no-sense

Lurie A and Lurie R (2020) Single gene disorders not so simple after all, *Medical Xpress*, 12 October 2020, https://medicalxpress.com/news/2020-10-gene-disorders-simple.html

Mansfield V (2003) Time and impermanence in Middle Way Buddhism and modern physics, In AB Wallace (editor) (2003) *Buddhism & Science: Breaking New Ground*, Columbia University Press, New York.

Marcuse H (1991) *One-Dimensional Man: Studies in the Ideology of Advanced Industrial Society*, Beacon Press, Boston.

Marx K (1852, 1937) *The Eighteenth Brumaire of Louis Bonaparte*, Progress Publishers, Moscow, www.marxists.org/archive/marx/works/1852/18th-brumaire

Marx K (1859, 1993) *A Contribution to the Critique of Political Economy*, Progress Publishers, Moscow, www.marxists.org/archive/marx/works/1859/critique-pol-economy/index.htm

Marx K and Engels F (1846, 1988) *The German Ideology*, Prometheus Books, New York.

Masci D (2009) Religion and Science in the United States, *Pew Research Center*, 5 November 2009, www.pewresearch.org/religion/2009/11/05/scientists-and-belief/

McLeod SA (2019) Freewill vs determinism, *Simply Psychology*, 11 April 2019, www.simplypsychology.org/freewill-determinism.html

McLuhan R (2019) *Randi's Prize: What Sceptics Say about the Paranormal, Why They Are Wrong, and Why It Matters*, White Crow Books, New York.

McKenna J (2014) Pope Francis: 'Evolution ... is not inconsistent with the notion of creation', *National Catholic Reporter*, 27 October 2014, www.ncronline.org/news/vatican/pope-francis-evolution-not-inconsistent-notion-creation

McLeod SA (2019) Freewill vs determinism, *Simply Psychology*, 11 April 2019, www.simplypsychology.org/freewill-determinism.html

McMaster G (2020) Researchers reveal patterns of sexual abuse in religious settings, *Folio*, 5 August 2020, www.ualberta.ca/folio/2020/08/researchers-reveal-patterns-of-sexual-abuse-in-religious-settings.html

Mehta P (2014) There's a gene for that, *Jacobin*, 12 January 2014, https://jacobinmag.com/2014/01/theres-a-gene-for-that

Menaker D (2012) Have it your way, *The New York Times*, 13 July 2012, www.nytimes.com/2012/07/15/books/review/free-will-by-sam-harris.html

Merali Z (2018) How cosmic is the cosmos? *Aeon*, 31 July 2018, https://aeon.co/essays/can-buddhist-philosophy-explain-what-came-before-the-big-bang

Mitchell K (2019) Epigenetics: What impact does it have on our psychology? *The Conversation*, 24 January 2019, https://theconversation.com/epigenetics-what-impact-does-it-have-on-our-psychology-109516

Monod J (1970) *Chance and Necessity: Essay on the Natural Philosophy of Modern Biology*, Vintage Books, New York.

Montanye JA (2019) Free will: Hail and farewell, *Essays in the Philosophy of Humanism*, 27(6):98—124.

Morales V (2009) Are flying saucers real? *Voice of America News*, 1 November 2009, www.voanews.com/archive/are-flying-saucers-real

Mowe S (2011) Buddhists and evolution, *Tricycle*, 24 March 2011, https://tricycle.org/article/buddhists-and-evolution/

Mukherjee S (2016) *The Gene: An Intimate History*, Scribner, New York.

Mullin E (2020) The end of deafness, *Medium*, 29 September 2020, https://futurehuman.medium.com/the-end-of-deafness-670f06df39cd

Musser G (2011) Free will and quantum clones: How your choices today affect the Universe at its origin, *Scientific American*, 19 September 2011, https://blogs.scientificamerican.com/observations/free-will-and-quantum-clones-how-your-choices-today-affect-the-Universe-at-its-origin/

Narain V (2014) Determinism, free will, and moral responsibility, *The Humanist*, 21 October 2014, https://thehumanist.com/magazine/november-december-2014/philosophically-speaking/determinism-free-will-and-moral-responsibility/

Nelson RH (2017a) Arguments why God (very probably) exists, *The Conversation*, 11 May 2017, https://theconversation.com/arguments-why-god-very-probably-exists-75451

Nelson RH (2017b) Existence of God: The rational arguments from mathematics to human consciousness, *The Independent*, 30 May 2017, www.independent.co.uk/life-style/existence-god-rational-arguments-mathematics-human-consciousness-a7739841.html

Neri DJ (2018) The human instinct: A conversation with Ken Miller, *Behavioral Scientist*, 30 April 2018, https://behavioralscientist.org/the-human-instinct-a-conversation-with-ken-miller/

Newberg A, D'Aquili E and Rause V (2002) *Why God Won't Go Away: Brain Science and the Biology of Belief*, Ballantine Books, New York.

Newberg A (2010) Neurotheology: This is your brain on religion, *NPR*, 15 December 2010, www.npr.org/2010/12/15/132078267/neurotheology-where-religion-and-science-collide

Nsar SH (2006) Spirituality and science: Convergence or divergence? www.worldwisdom.com

NTU (2022) Scientists have established a key biological difference between psychopaths and normal people, *SciTech Daily*, 1 June 2022, https://scitechdaily.com/scientists-have-established-a-key-biological-difference-between-psychopaths-and-normal-people

NYT (1971) Chance and necessity, *The New York Times*, 21 November 1971, www.nytimes.com/1971/11/21/archives/chance-and-necessity-an-essay-on-the-natural-philosophy-of-modern.html

Oberlander E (2022) A new theory in physics claims to solve the mystery of consciousness, *Neuroscience News*, 11 August 2022, https://neurosciencenews.com/physics-consciousness-21222/

O'Brien B (2020) Buddhist teachings on the self, *Learn Religions*, 26 August 2020, www.learnreligions.com/self-no-self-whats-a-self-450190

O'Brien B (2021) An examination of free will and Buddhism, *Learn Religions*, 8 February 2021, www.learnreligions.com/free-will-and-buddhism-449 602.

O'Leary D (2017) Science does not understand our consciousness of God, but not for the reasons we might think, *The City: Science and Faith*, 14 February 2017, https://hbu.edu/news-and-events/2017/02/14/science-not-understand-consciousness-god-not-reasons-might-think/

Omundson J (2020) Can religions be objectively disproved? *Medium*, 20 October 2020, https://medium.com/excommunications/can-religions-be-objectively-disproved-210c536418e9

Oreskes N (2019) *Why Trust Science?* Princeton University Press, Princeton.

Oreskes N and Conway EM (2010) *Merchants of Doubt: How a Handful of Scientists Obscured the Truth on Issues from Tobacco Smoke to Global Warming*, Bloomsbury Press, New York.

Orr HA (1999) Gould on God; Can religion and science be happily reconciled? *Boston Review*, October/November 1999, https://bostonreview.net/archives/BR24.5/orr.html

OT (2015) *Introduction to Psychology, Behaviorism and the Question of Free Will*, Open Textbooks, www.opentextbooks.org.hk/ditatopic/18736

Palmer J (2010) Free will similar in animals, humans - But not so free, *BBC News*, 16 December 2010, www.bbc.com/news/science-environment-11998687

Park HD et al. (2020) Breathing is coupled with voluntary action and the cortical readiness potential, *Nature Communications*, 11(1), doi:10.1038/s41467-019-13967-9

Parrott J (2017) Reconciling the divine decree and free will in Islam, *Yaqeen Institute*, 31 July 2017, https://yaqeeninstitute.org/justin-parrott/reconciling-the-divine-decree-and-free-will-in-islam

Paulson S (2006) The believer, *Salon*, 7 August 2006, www.salon.com/2006/08/07/collins_6/

PBS (2007) Emergence, *PBS NOVA*, June 2007, www.pbs.org/wgbh/nova/sciencenow/3410/03-ever-nf.html

PBS (2021) Interview: Richard Dawkins, *PBS*, 20 February 2021, www.pbs.org/faithandreason/transcript/dawk-body.html

Pecorino PA (2021) *Philosophy of Religion*, CUNY, New York, www.qcc.cuny.edu/SocialSciences/ppecorino/PHIL_of_RELIGION_TEXT

Petkov M (2019) Do you choose to read this post? Sam Harris's *Free Will*, https://www.martinpetkov.com/your-opportunity/do-you-choose-to-read-this-post-sam-harriss-free-will-book-summary

Pines D (2014) Emergence: A unifying theme for 21st century science, *Medium*, 31 October 2014, https://medium.com/sfi-30-foundations-frontiers/emergence-a-unifying-theme-for-21st-century-science-4324ac0f951e

Pipkin B (2019) Why free will doesn't exist and what that means, *Medium*, 17 February 2019, https://medium.com/@bretpipkin/why-free-will-doesnt-exist-and-what-that-means-a0bb7fb16c7

Pleasants N (2019) Free will, determinism and the "problem" of structure and agency in the social sciences, *Philosophy of the Social Sciences*, 49(1):3—30, doi:10.1177/0048393118814952

Plomin R (2019) *Blueprint: How DNA Makes Us Who We Are*, The MIT Press, Cambridge, MA.

PLOS (2020) How genetic variation gives rise to differences in mathematical ability, *Medical Xpress*, 22 October 2020, https://medicalxpress.com/news/2020-10-genetic-variation-differences-mathematical-ability.html

Pretorius M (2016). Is consciousness a product of the brain or/and a divine act of God? Concise insights from neuroscience and Christian theology, *HTS Theological Studies*, 72(4):1—7, https://dx.doi.org/10.4102/hts.v72i4.3472

Priest G (2018) Marxism and Buddhism: Not such strange bedfellows, *Journal of the American Philosophical Association*, 4(1):2—13, https://doi.org/10.1017/apa.2017.40

Psychology Wiki (2021 – Emergence), https://psychology.wikia.org/wiki/Emergence

Randi J (1982) *Flim-Flam! Psychics: ESP, Unicorns, and Other Delusions*, Prometheus Books, New York.

Randi J (1995) *An Encyclopedia of Claims, Frauds, and Hoaxes of the Occult and Supernatural*, St Martins Press, New York.

Rawlette AH (2020) Are coincidences signs from God? *Psychology Today*, 5 February 2020, www.psychologytoday.com/us/blog/mysteries-consciousness/202002/are-coincidences-signs-god

Reed C (2012) The Rev Sun Myung Moon obituary, *The Guardian* (UK), 2 September 2012, www.theguardian.com/world/2012/sep/02/rev-sun-myung-moon

Rennie J (2018) How complex wholes emerge from simple parts, *Quanta Magazine*, 20 December 2018, www.quantamagazine.org/emergence-how-complex-wholes-emerge-from-simple-parts-20181220/

Repetti R (2015) Buddhist meditation and the possibility of free will. *Science, Religion and Culture*, 2(2): 81-98, http://researcherslinks.com/ current-issues/Buddhist-Meditation-and-the-Possibility-of-Free-Will/9/5/138/html

Resnik DB and Vorhaus DB (2006) Genetic modification and genetic determinism, *Philosophy, Ethics and Humanities in Medicine,* 1(9), https://doi.org/10.1186/1747-5341-1-9

Rhodes R (1999) Book review: *Rocks of Ages: Science and Religion in the Fullness of Life* by Stephen Jay Gould, *Bottom Layer*, 24 May 1999, www.bottomlayer.com/bottom/rocks.html

Ricker GA (2007) The trouble with NOMA: Why science and religion come into conflict, *Godless in America*, 2007, www.godlessinamerica.com/noma.html

Ridley M (1998) *The Origins of Virtue: Human Instincts and the Evolution of Cooperation*, Penguin Books, New Yok.

Rios E (2007) De Waal sides with Darwin: Morality is instinctual, evolved, *Emory Report*, 16 April 2007, www.emory.edu/EMORY_REPORT/erarchive/2007/April/DeWaal.htm

Romero GE and Perez D (2012) New remarks on the cosmological argument, *International Journal for Philosophy of Religion*, 72:103—113, http://dx.doi.org/10.1007/s11153-012-9337-6

Rovelli C (2022) Consciousness is irrelevant to Quantum Mechanics, *IAI News*, 19 July 2022, https://iai.tv/articles/consciousness-is-irrelevant-to-quantum-mechanics-auid-2187

PRR (2021) Free will, *Psychology Research and Reference*, 17 February 2021, http://psychology.iresearchnet.com/social-psychology/decision-making/free-will/

Russell B (1997) *Religion and Science*, Oxford University Press, Oxford.

Saini A (2020) Eugenics refuses to die – and now Andrew Sabisky has put it back in the headlines, *The Guardian* (UK), 19 February 2020, www.theguardian.com/commentisfree/2020/feb/19/eugenics-andrew-sabisky-right-ideas-human-breeding

Salleh A (2018) Are religion and science always at odds? Here are three scientists that don't think so, *ABC Science*, 24 May 2018, www.abc.net.au/news/science/2018-05-24/three-scientists-talk-about-how-their-faith-fits-with-their-work/9543772

Savage JE et al. (2018) Genome-wide association meta-analysis in 269,867 individuals identifies new genetic and functional links to intelligence, *Nature Genetics*, 50: 912—919, https://doi.org/10.1038/s41588-018-0152-6

Sayadaw M (2019) The theory of karma, *Buddha Net*, 20 October 2019, www.buddhanet.net/e-learning/karma.htm

SC (2021) Our definition of science, *Science Council*, 2 February 2021, https://sciencecouncil.org/about-science/our-definition-of-science/

Scharping N (2018) Can we blame our genes for our decisions? *Discover Magazine*, 11 December 2018, www.discovermagazine.com/health/can-we-blame-our-genes-for-our-decisions

Schuler A (2016) Gould's NOMA – A thorough analysis, *Patheos*, 19 January 2016, www.patheos.com/blogs/tippling/2016/01/19/goulds-noma-a-thorough-analysis-part-1-updated/

Schwartz GE (2017) *Super Synchronicity: Where Science and Spirit Meet*, Waterfront Digital Press, New York.

Schwartz A (2021) Radical religious belief and child abuse, 'Spare the Rod and Spoil the Child', *Mental Help*, 10 March 2021, www.mentalhelp.net/blogs/radical-religious-belief-and-child-abuse-quot-spare-the-rod-and-spoil-the-child-quot/

Schwartz GE and Simon WL (2007) *The G.O.D. Experiments: How Science Is Discovering God In Everything, Including Us*, Atria Books, New York.

SD (2012) Genes influence criminal behavior, research suggests, *Science Daily*, 26 January 2012, www.sciencedaily.com/releases/2012/01/120125151841.htm

Seth A (2007) Models of consciousness, *Scholarpedia*, 2(1):1328, www.scholarpedia .org/article/Models_of_consciousness

Sengupta H (2016) Does quantum physics have anything to do with the Upanishads? *Medium*, 18 July 2016, https://medium.com/@hindolsengupta/does-quantum-physics-have-anything-to-do-with-the-upanishads-1d802c54b16d

Sennett R (1971) Beyond freedom and dignity, *The New York Times*, 24 October 1971, www.nytimes.com/1971/10/24/archives/beyond-freedom-and-dignity-by-b-f-skinner-225-pp-new-york-alfred-a.html

Siegel E (2014) 22 messages of hope (and science) for creationists, *Medium*, 6 February 2014, https://medium.com/starts-with-a-bang/22-messages-of-hope-and-science-for-creationists-8712e42fbb0d

Siegel E (2017) Can science prove the existence of God? *Forbes Magazine*, 20 January 2017, www.forbes.com/sites/startswithabang/2017/01/20/can-science-prove-the-existence-of-god/

Siegel E (2022) Yes, the Universe really is 100% reductionist in nature, *Big Think*, 9 August 2022, https://bigthink.com/starts-with-a-bang/Universe-reductionist/

Skatssoon J (2006) Magic mushrooms hit the God spot, *ABC Science Online*, 12 July 2006, www.abc.net.au/science/news/health/HealthRepublish_1682610.htm

Smilansky S (2000) *Free Will and Illusion*, Clarendon Press, New York.

Snodgrass E (2022) Mass weddings and cult accusations: Who are the 'Moonies' and what is the Unification Church? *Insider*, 25 July 2022, www.insider.com/who-are-the-moonies-and-what-is-the-unification-church-2022-7

Spirkin A (1983) *Dialectical Materialism*, Progress Publishers, Moscow.

Spirkin A (1990) *Fundamentals of Philosophy*, Progress Publishers, Moscow.

Sproul RC (2021) What is free will? *Ligonier Ministries*, 20 February 2021, www.ligonier.org/learn/series/chosen_by_god/what-is-free-will/

Stafford T (2013) Does non-belief in free will make us better or worse? *BBC News*, 25 September 2013, www.bbc.com/future/article/20130924-how-belief-in-free-will-shapes-us

Stankorb S (2021) What happens after Christian prophets admit they were wrong about Trump? *Medium*, 14 February 2021, https://gen.medium .com/with-trumps-defeat-christian-prophets-admit-i-was-wrong-967a13abb8a7

Stenger VJ (2006) Do our values come from God? The evidence says no, *Free Inquiry*, August/September 2006.

Stenger V (2007) *God: The Failed Hypothesis: How Science Shows That God Does Not Exist*, Prometheus, New York.

Stenger V (2011) *The Fallacy of Fine-Tuning: Why the Universe Is Not Designed for Us*, Prometheus, New York.

Stenger VJ, Lindsay JA and Boghossian P (2015) Physicists are philosophers, too, *Scientific American*, 8 May 2018, www.scientificamerican.com/article/physicists-are-philosophers-too/

Stevenson T (2019) Reality does not exist, *Medium*, 6 June 2019, https://medium.com/ @tom.stevenson78/reality-does-not-exist-343fe045286b

Strawson G (2003) Evolution explains it all for you, *The New York Times*, 2 March 2003, www.nytimes.com/2003/03/02/books/evolution-explains-it-all-for-you.html

Sullivan B (2019) Why we like what we like: A scientist's surprising findings, *National Geographic*, September 2019, www.nationalgeographic.com/science/article/why-we-like-what-we-like-a-scientists-surprising-findings

Suttie J (2013) Finding morality in animals, *Greater Goods Magazine*, 9 July 2013, https:// greatergood.berkeley.edu/article/item/morality_animals

Tan CY (2019) How does the cosmos create itself? *Medium*, 26 August 2019, https:// medium.com/@tcherry/how-does-the-cosmos-create-itself-2da0618e58ef

Tharoor I (2014) Pope Francis says evolution is real and God is no wizard, *The Washington Post*, 28 October 2014, www.washingtonpost.com/news/worldviews/wp/2014/10/28/ pope-francis-backs-theory-of-evolution-says-god-is-no-wizard/

The Onion (1999) Scientists discover gene responsible for eating whole goddamn bag of chips, *The Onion*, 1 September 1999, www.theonion.com/scientists-discover-gene-responsible-for-eating-whole-g-18195652 91

Tielbeek JJ et al (2017) Genome-wide association studies of a broad spectrum of antisocial behavior, *JAMA Psychiatry*, 74(12):1242–1250, doi:10.1001/jamapsychiatry.2017.3069

Timpe K (2021) Free will, *Internet Encyclopedia of Philosophy*, 14 February 2021, https:// iep.utm.edu/freewill/

Trefil J and Hazen RM (1998) *The Sciences: An Integrated Approach (second edition)*, John Wiley & Sons, New York.

Vintiadis E (2021) Emergence, *Internet Encyclopedia of Philosophy*, March 2021, https:// iep.utm.edu/emergenc/

Vithoulkas G and Muresanu DF (2014) Conscience and consciousness: a definition, *Journal of Medicine and Life*, 7(1):104–108.

Vonasch AJ et al. (2017) Ordinary people associate addiction with loss of free will, *Addictive Behaviors Reports*, 5:56--66, https://doi.org/10.1016/j.abrep.2017.01.002.

Wagner A (2009) *Paradoxical Life: Meaning, Matter, and the Power of Human Choice*, Yale University Press, New Haven.

Wallace AB (editor) (2003) *Buddhism & Science: Breaking New Ground*, Columbia University Press, New York.

Wallace AB (2003) Introduction, In Wallace AB (editor) (2003) *Buddhism & Science: Breaking New Ground*, Columbia University Press, New York.

Waller BN (2015) *Restorative Free Will: Back to the Biological Base*, Lexington Books, New York.

Ward K (2008) *The Big Questions in Science and Religion*, Templeton Press, West Conshohocken, PA.

Wertz J et al. (2018) Genetics and crime: Integrating new genomic discoveries into psychological research about antisocial behavior, *Psychological Science*, 29(5):791–803, doi:10.1177/0956797617744542

Westmoreland D (2019) Teaching evolution to Tibetan monks, *Scientific American*, 10 April 2019, https://blogs.scientificamerican.com/observations/teaching-evolution-to-tibetan-monks/

Whitehead AN (1925) Religion and science, *The Atlantic*, August 1925, www.theatlantic.com/magazine/archive/1925/08/religion-and-science/ 304220/

WHO (2020) Tobacco, *WHO*, 27 May 2020, www.who.int/news-room/fact-sheets/detail/tobacco

Wikipedia (2021 – Alexander Spirkin, Alfred North Whitehead, Anthropic Principle, Big Bang Theory, Buddhism and Evolution, Buddhism and Science, Buddhist Cosmology, Comparative Religion, Does God Exist?, Elbow Room, Endless Knot, Evolution and the Catholic Church, Existence of God, False Consciousness, Francis Collins, Francis Crick, Free Will, Free Will in Theology, James Watson, Kalam Cosmological Argument, List of New Religious Movements, Marx's Theory of Alienation, Mind, Neuroscience of Free Will, Neuroscience of Religion, Non-overlapping Magisteria, Pratityasamutpada, Predestination in Islam, Reality, Relationship Between Religion and Science, Religious Interpretations of the Big Bang Theory, Science, Scholarly Approaches to Mysticism, Stephen Jay Gould, The Mind–Body Problem, The Unanswered Questions, Thomas Aquinas, Unification Church, Unification Movement, Victor J Stenger).

Wills V (2021) And he ate Jim Crow, In: Brandon Hogan et al. (editors), *The Movement for Black Lives*, Oxford University Press, doi:10.1093/oso/9780197507773.003.0003

Wintour P (2013) Genetics outweighs teaching, Gove adviser tells his boss, *The Guardian* (UK), 11 October 2013, www.theguardian.com/politics/2013/oct/11/genetics-teaching-gove-adviser

Williams PS (2006) The big bad wolf, theism and the foundations of Intelligent Design, *Evangelical Philosophical Society*, www.epsociety.org/library/articles.asp?pid=53&ap=2

Wilson LT (2009), Definition of science, *Explorable.com*, 16 June 2009, https://explorable.com/definition-of-science

Wintjen H (2022) Erwin Schrodinger: There is only one mind, *Hendrik Wintjen Blog*, 13 August 2022, www.hendrik-wintjen.info/consciousness/erwin-schroedinger-one-mind/

Wolchover N (2017) A theory of reality as more than the sum of its parts, *Quanta Magazine*, 1 June 2017, www.quantamagazine.org/a-theory-of-reality-as-more-than-the-sum-of-its-parts-20170601/

Woods A (2001) What the human genome means for socialists, *Marxist*, 16 February 2021, www.marxist.com/human-genome-socialism160201.htm

Wright R (2018) *Why Buddhism is True: The Science and Philosophy of Meditation and Enlightenment*, Simon & Schuster, New York.

Youra S (2020) Religion and science don't have to be at odds, *Medium*, 16 October 2020, https://medium.com/climate-conscious/religion-and-science-dont-have-to-be-at-odds-10c6436dfcd1

Ziliak ST and McCloskey DN (2008) *The Cult of Statistical Significance: How the Standard Error Costs Us Jobs, Justice, and Lives*, University of Michigan Press, Michigan.

Zinn H (1980) *A People's History of the United States*, Harper & Row, New York.

Zinn H (1990) *The Politics of History* (second edition), University of Illinois Press, Illinois.

Zuckerman P (2015) Op-Ed: Think religion makes society less violent? Think again, *The Los Angeles Times*, 30 October 2015, www.latimes.com/opinion/op-ed/la-oe-1101-zuckerman-violence-secularism-20151101-story.html

Zukav G (1983) *The Dancing Wu Li Masters: An Overview of the New Physics*, William Morrow, New York.

Zwart H (2014) *Genetic Determinism*, Springer Reference, Berlin, www.springerreference.com/index/chapterdbid/398875

5. MATHEMATICS

Aafiya (2015) Golden Ratio and Islam: The miracle of Mecca, *Islam Hashtag*, 15 July 2015, https://islamhashtag.com/golden-ratio-and-islam/

Abbot D (2013) Is mathematics invented or discovered? *Huffington Post*, 10 September 2013, www.huffpost.com/entry/is-mathematics-invented-o_b_3895622

Abdelhamid A (2021) We approximate pi & Allah knows best, *About Islam*, 12 March 2021, https://aboutislam.net/muslim-issues/science-muslim-issues/approximate-pi-allah-knows-best/

Abeka (2017) The Christian approach to teaching elementary math, *Abeka Academy*, 10 August 2017, www.abeka.com/blog/the-christian-approach-to-teaching-elementary-math/

Akpan N (2013) Did poor math skills cause the 2008 financial crisis: Fed study argues yes, *Medical Daily*, 24 July 2013, www.medicaldaily.com/did-poor-math-skills-cause-2008-financial-crisis-fed-study-argues-yes-247075

Alexander A (2014) *Infinitesimal: How a Dangerous Mathematical Theory Shaped the Modern World*, Farrar, Straus and Giroux, New York.

Allen D (1997) Pythagoras and the Pythagoreans, *Department of Mathematics*, Texas A&M University, 6 February 1997, www.math.tamu.edu/~don.allen/history/pythag/pythag.html

Als-Nielsen B et al (2003) Association of funding and conclusions in randomized drug trials: a reflection of treatment effect or adverse events? *JAMA*, 290(7):921—8.

Altman DG (1980) Statistics and ethics in medical research: Misuse of statistics is unethical, *BMJ*, 281:1182—1184.

Alvarez E, Amado Y and London WM (2021) Precautionary principle, *Oxford Bibliographies Online*, 24 February 2021, doi:10.1093/obo/9780199 756797-0046

AMS (2019) Ethical guidelines of the American Mathematical Society, *AMS*, January 2019, www.ams.org/about-us/governance/policy-statements/sec-ethics

Angell M (2005) *The Truth About the Drug Companies: How They Deceive Us and What To Do About It*, Random House, New York.

Angier N (1997) Survey of scientists finds a stability of faith in God, *The New York Times*, 3 April 1997, www.nytimes.com/1997/04/03/us/survey-of-scientists-finds-a-stability-of-faith-in-god.html

Antonio D (2012) Mathematics, *Stack Exchange*, 14 August 2012. https://math.stackexchange.com/users/31254/donantonio

Archuman (2017) Can math prove what God is? Mathematical mysticism, the Liar Paradox, and Gödel's Incompleteness Theorem, *Humanity Plus*, 9 February 2017, https://humanityplus.wordpress.com/2017/02/09/mathematical-mysticism-god-and-godels-incompleteness-theorem/

Aron J (2016) *The Man Who Knew Infinity* fails to break the mathematical mold, *New Scientist*, 20 April 2016, www.newscientist.com/article/mg23030701-000-the-man-who-knew-infinity-fails-to-break-the-mathematical-mould/

Ascher M (2002) *Mathematics Elsewhere: An Exploration Across Cultures*, Princeton University Press, Princeton.

Aslan A (2022) Why is the Cacvotic number system of Inyupiks one of the best counting methods? *Medium*, 3 November 2022, https://olMayanaergi.medium.com/why-is-the-cacvotic-number-system-of-inyupiks-one-of-the-best-counting-methods-27d8e3ca083d

Atiyah M (1995) Creation v discovery, *Times Higher Education Supplement*, 19 September 1995, www.timeshighereducation.com/books/creation-v-discovery/161513.article

AW (2021) Godel's ontological proof, *Apologetics Wiki*, 1 June 2021, https://apologetics.fandom.com/wiki/Godels_ontological_proof

Beckmann P (1967) *A History of Pi*, St. Martin's Press, New York.

Bekelman J, Li Y and Gross C (2003) Scope and impact of financial conflicts of interest in biomedical research: A systematic review, *JAMA*, 289(4):454—65.

Bellos A (2015) He ate all the pi: Japanese man memorizes π to 111,700 digits, *The Guardian* (UK), 13 March 2015, www.theguardian.com/science/alexs-adventures-in-numberland/ 2015/mar/13/pi-day-2015-memory-memorisation-world-record-japanese-akira-haraguchi

Benner K (2020) Purdue Pharma pleads guilty to role in opioid crisis as part of deal with Justice Dept, *The New York Times*, 24 November 2020, www.nytimes.com/2020/11/24/us/ politics/purdue-pharma-opioids-guilty-settlement.html

Berggren JL (2021a) Mathematics in the Islamic world (8th–15th century), *Encyclopedia Britannica*, 5 May 2021, www.britannica.com/science/mathematics/Mathematics-in-the-Islamic-world-8th-15th-century

Berggren JL (2021b) Omar Khayyam, *Encyclopedia Britannica*, 11 May 2021, www.britannica.com/science/mathematics/Omar-Khayyam#ref65995

Berman R (2018) Michio Kaku believes in God, if not that God, *Big Think*, 15 February 2018, https://bigthink.com/culture-religion/michio-kaku-believes-in-god-if-not-that-god/

Bernhard A (2020) How modern mathematics emerged from a lost Islamic library, *BBC Future*, 7 December 2020, www.bbc.com/future/article/20201204-lost-islamic-library-maths

Bird A (2021) Leibniz: We live in the best of all possible worlds, *Medium*, 27 April 2021, www.cantorsparadise.com/leibniz-we-live-in-the-best-of-all-possible-worlds-9b0ee1018552

Black E (2000) *IBM and the Holocaust: The Strategic Alliance Between Nazi Germany and America's Most Powerful Corporation*, Crown Publisher, New York, https:// archive.nytimes.com/www.nytimes.com/books/first/b/black-ibm.html

Blanc JL (2021) Infinity in theology and mathematics, *Mathematics Department*, Dartmouth College, 15 May 2021, https://math.dartmouth.edu/~matc/Readers/HowManyAngels/ Blanc.html

Bogomolny A (2021) Review: *A Certain Ambiguity*, *Cut the Knot*, 27 May 2021, www.cut-the-knot.org/books/Reviews/CertainAmbiguity.shtml

Bollobas B (2016) The man who taught infinity: How GH Hardy tamed Srinivasa Ramanujan's genius, *The Conversation*, 22 April 2016, https://theconversation.com/the-man-who-taught-infinity-how-gh-hardy-tamed-srinivasa-ramanujans-genius-57585

Boltz CL (1983) Galileo Galilei, *New Scientist*, 7 April 1983, www.newscientist.com/people/ galileo-galilei/

Bourgoin J (1973) *Arabic Geometrical Pattern and Design*, Dover Books, New York.

Boyer CB and Merzbach UC (1991) *A History of Mathematics*, John Wiley, New York.

BP (2021) Pythagoreanism, *The Basics of Philosophy*, 30 April 2021, www.philosophybasics.com/movements_pythagoreanism.html

Bradley J and Howell R (2011) *Mathematics Through the Eyes of Faith*, Harper One, New York.

Brenner AD (2002) *Emil J. Gumbel: Weimar German Pacifist and Professor*, Brill, Amsterdam.

Bret TL (2012) Are mathematics responsible for the financial crisis? *Huffington Post*, 19 March 2012, www.huffingtonpost.co.uk/theo-le-bret/are-mathematics-responsib_b_1362937.html

Broug E (2019) *Islamic Geometric Patterns*, Thames & Hudson, London.

Bruning J (2005) Book review: *Mathematicians Under the Nazis*, *Notices of the AMS*, April 2005:435—438.

Budd C (2020) Myths of maths: The golden ratio, *Plus Magazine*, 23 February 2020, https:// plus.maths.org/content/myths-maths-golden-ratio

Bultheel A (2015) Mathematicians and their gods, *European Mathematical Society*, 8 September 2015, https://euro-math-soc.eu/review/mathematicians-and-their-gods

Caldwell CK (2021) Ishango bone, *Mathematics and Statistics*, University of Tennessee at Martin, https://primes.utm.edu/glossary/page.php?sort=IshangoBone

Calegari D (2008) Review: *A Certain Ambiguity*, *Notices of the AMS*, February 2008, 55(2):235—237.

Calgar ME (2020) Why does intellectuality weaken faith and sometimes foster it? *Humanities and Social Science Communications*, 7(88), https://doi.org/10.1057/s41599-020-00567-y

Calvo SC (2021) The shape of space: The beginning of non-Euclidean geometry, *Encyclopedia*, 29 May 2021, www.encyclopedia.com/science/encyclopedias-almanacs-transcripts-and-maps/shape-space-beginning-non-euclidean-geometry

Cameron P (2009) Mathematics and religion? *Cameron Counts*, 21 December 2009, https://cameroncounts.wordpress.com/2009/12/21/mathematics-and-religion/

Campbell M (2016) Does infinity exist in the real world? *New Scientist*, 3 March 2016, www.newscientist.com/article/2079495-explanimator-does-infinity-exist-in-the-real-world/

Castelvecchi D (2020) Stirring biopic of the first woman to win top maths prize, *Nature*, 8 June 2020, www.nature.com/articles/d41586-020-01681-2

Chandra A and Holt GA (1999) Pharmaceutical advertisements: How they deceive patients, *Journal of Business Ethics*, 18(4):359—366, www.jstor.org/stable/25074060

Chiodo M and Bursill H (2019) Teaching ethics in mathematics, *EMS Newsletter*, December 2019, 38—41.

Clegg B (2009) The dangerous ratio, *NRICH*, https://nrich.maths.org/2671

Euclid AU and Heath A (2017) *Euclid's Elements (The Thirteen Books)*, Digireads Publishing, Kansas, USA.

Cordero-Soto RJ (2019) The applicability of mathematics and the naturalist die, *ACMS 22nd Biennial Conference Proceedings*, 2019:14—23.

Critchlow K (1984) *Islamic Patterns: An Analytical and Cosmological Approach*, WW Norton, New York.

Csillag R (2008) Math + Religion = Trouble, *The Star*, 26 January 2018, www.thestar.com/news/2008/01/26/math_religion_trouble.html

CSM (1985) The infinite nature of God, *Christian Science Monitor*, 19 April 1985, www.csmonitor.com/1985/0419/mrb660.html

David K (2017) Ramanujan, Hardy & the God debate, *Colombo Telegraph*, 26 March 2017, www.colombotelegraph.com/index.php/ramanujan-hardy-the-god-debate/

Davis DJ (2018) '*The Great Rift*' review: From comity to culture war, *Wall Street Journal*, 11 July 2018, www.wsj.com/articles/the-great-rift-review-from-comity-to-culture-war-1531347117

Davis PJ (1998) A brief look at mathematics and theology (Part I and Part II), *Scripps College*, Claremont, California.

Davis PJ and Hersh R (1986) *Descartes' Dream: The World According to Mathematics*, Houghton Mifflin Company. Boston.

Davis PJ and Hersh R (1999) *The Mathematical Experience*, Marine Books, New York.

Devlin K (1997) *Mathematics: The Science of Patterns*, Scientific American Library, New York.

Devlin K (1998) *Mathematics: The New Golden Age*, Penguin Books, London.

Devlin K (2002) The mathematical legacy of Islam, *Devlin's Angle*, Mathematical Association of America, July-August 2002, www.maa.org/external_archive/devlin/devlin_0708_02.html

Devlin K (2011) *The Man of Numbers: Fibonacci's Arithmetic Revolution*, Walker & Company, New York.

Dewdney AL (1999) *A Mathematical Mystery Tour: Discovering the Truth and Beauty of the Cosmos*, John Wiley, New York.

Dijkgraaf R (2020) The two forms of mathematical beauty, *Quanta*, 16 June 2020, www.quantamagazine.org/how-is-math-beautiful-20200616/

Downey A (2017) College freshmen are less religious than ever, *Scientific American*, 25 May 2017, https://blogs.scientificamerican.com/observations/college-freshmen-are-less-religious-than-ever/

DT (2021) Religious views of Carl Friedrich Gauss, the great mathematician, *Data Torch*, 31 May 2021, www.datatorch.com/life/Religious_views_of_Gauss_Great_Mathematician

Dudley U (1997) *Numerology: Or, What Pythagoras Wrought*, American Mathematical Society, New York.

Dunham W (2007) Medieval Muslims made stunning math breakthrough, *Reuters*, 22 February 2007, www.reuters.com/article/us-architecture-patterns-idUSN2245118920070222

Dutta AK (2002) Mathematics in ancient India, *Resonance*, April 2002, 1—16.

Dvorsky G (2017) 14 interesting examples of the Golden Ratio in nature, *Mathnasium*, 24 April 2017, www.mathnasium.com/examples-of-the-golden-ratio-in-nature

EBC (2021) Applied mathematics major, *Emmaus Bible College*, 5 June 2021, www.emmaus.edu/bachelor-of-science-in-bibletheology-and-applied-mathematics

Editorial (2019) The Guardian view on ethics for mathematicians: an essential addition, *The Guardian* (UK), 18 August 2019, www.theguardian.com/commentisfree/2019/aug/18/the-guardian-view-on-ethics-for-mathematicians-an-essential-addition

Ernest P (2020) Mathematics, ethics and purism: An application of MacIntyre's virtue theory, *Synthese*, 5 November 2020, https://doi.org/10.1007/s11229-020-02928-1

Esposito JL (2021) Numerology, *Oxford Islamic Studies Online*, 7 June 2021, www.oxfordislamicstudies.com/article/opr/t243/e253

Faena (2021) Is this the mathematical formula for God? *Faena*, 15 May 2021, www.faena.com/aleph/is-this-the-mathematical-formula-for-god

Faerber AE and Kreling DH (2014) Content analysis of false and misleading claims in television advertising for prescription and nonprescription drugs, Journal of General Internal Medicine, 29:110--118, https://doi.org/10.1007/s11606-013-2604-0

Ferencik J (2017) The harmony of the world: The Pythagoreans, *Medium*, 26 July 2017, https://jakubferencik.medium.com/the-harmony-of-the-world-the-pythagoreans-e61df7bd026

Ferguson CH (2012) *Predator Nation: Corporate Criminals, Political Corruption, and the Hijacking of America*, Crown Business, New York.

Feron H (2014) Human rights and faith: a 'world-wide secular religion'? *Ethics & Global Politics*, 7(4): 181--200, doi:10.3402/egp.v7.26262

Finkbeiner A (2013) Mathematicians and computer scientists shrug over the NSA hacking, *Scientific American*, 8 October 2013, www.scientificamerican.com/article/mathematicians-and-computer-scientists-shrug-over-the-nsa-hacking/

Forsey C (2021) The designer's guide to the Golden Ratio, *Hub Spot*, 23 May 2021, https://blog.hubspot.com/marketing/golden-ratio

Fraenkel AA (2017) Hitler's math, *Tablet Magazine*, 8 February 2017, www.tabletmag.com/sections/arts-letters/articles/hitlers-math

Freiberger M (2008) Review: *Is God a Mathematician? Plus Magazine*, 1 December 2008, https://plus.maths.org/content/god-mathematician

Freiberger M and Thomas R (2013) Do infinities exist in nature? *Plus Magazine*, 26 September 2013, https://plus.maths.org/content/do-infinities-exist-nature-0

FSTC (2013) Muslim founders of mathematics, *Muslim Heritage*, 30 October 2013, https://muslimheritage.com/muslim-founders-mathematics/

Gabriel B (2017) The Catholic Church & numerology, *Classroom*, 29 September 2017, https://classroom.synonym.com/the-catholic-church-numerology-12086108.html

Gerdes P (1994) On mathematics in the history of mathematics of sub-Saharan Africa, *Historia Mathematica*, 21:345—376.

Gerdes P (2007) *Drawings from Angola: Living Mathematics*, Lulu.com.

Gerardia K, Goette L and Meier S (2013) Numerical ability predicts mortgage default, *PNAS*, 2013.

Ghaemi SN, Shirzadi AA and Filkowski M (2008) Publication bias and the pharmaceutical industry: The case of lamotrigine in bipolar disorder, *Medscape Journal of Medicine*, 10(9):211.

Glutsyuk A (2014) Book review: *Naming Infinity: A True Story of Religious Mysticism and Mathematical Creativity*, *Notices of the AMS*, 61(1), January 2014.

Goldman JG (2012) Animals that can count, *BBC Future*, 28 November 2012, www.bbc.com/future/article/20121128-animals-that-can-count

Goldstein R (2006) *Incompleteness: The Proof and Paradox of Kurt Godel*, WW Norton, New York.

Gossard B (2018) How Big Pharma misleads patients with relative numbers, *Medium*, 27 March 2018, https://medium.com/@BlakeGossard/deceptive-drug-digits-8174ff2647d8

Gouvea FQ (2014) Review: *The Man Who Knew Infinity: A Life of the Genius Ramanujan*, *Mathematical Association of America*, 27 May 2014, www.maa.org/press/maa-reviews/the-man-who-knew-infinity-a-life-of-the-genius-ramanujan

Grabiner JV (2014) Review: *Infinitesimal: How a Dangerous Mathematical Theory Shaped the Modern World*, *Mathematical Association of America*, 12 June 2014, www.maa.org/press/maa-reviews/infinitesimal-how-a-dangerous-mathematical-theory-shaped-the-modern-world

Green N (2012) How to bet on climate change, *The Guardian* (UK), 3 July 2012, www.theguardian.com/science/2012/jul/03/climate-change-pascal-wager

GU (2021) Why study mathematics at Gordon? *Department of Mathematics and Computer Science*, Gordon University, www.gordon.edu/math

GUM (2017) Using statistics ethically to combat 'a scientific credibility crisis', *Science Daily*, 19 February 2017, www.sciencedaily.com/releases/2017/02/170219165147.htm

Gunderman D (2019) How one German city developed – and then lost – generations of math geniuses, *The Conversation*, 14 January 2019, https://theconversation.com/how-one-german-city-developed-and-then-lost-generations-of-math-geniuses-106750

Halpern P (2020) Genius in exile: The rise and fall of Göttingen's Mathematical Institute, *Medium*, 28 August 2020, https://medium.com/@phalpern/genius-in-exile-the-rise-and-fall-of-gottingens-mathematical-institute

Hannon K (2010) Thoughts on pi, *Christian Perspective*, 12 March 2010, www.christianperspective.net/blog/thoughts-on-pi

Hardy GH (1940) *A Mathematician's Apology*, Cambridge University Press, Cambridge.

Harris M (2015) *Mathematics without Apologies: Portrait of a Problematic Vocation*, Princeton University Press, Princeton.

Harris M (2020) Can mathematics be antiracist? *MWA*, 12 July 2020, https://mathematicswithoutapologies.wordpress.com/category/ethics/

Harrison P (2018) Saint of science: The religious life of Isaac Newton, *LA Review of Books*, 2 February 2018, https://marginalia.lareviewofbooks.org/saint-science-religious-life-isaac-newton/

Harrison T (2019) Divine numerology, *Church Times*, 26 April 2019, www.churchtimes.co.uk/articles/2019/26-april/faith/faith-features/divine-numerology

Hartnett K (2017) To live your best life, do mathematics, *Quanta Magazine*, 2 February 2017, www.quantamagazine.org/math-and-the-best-life-an-interview-with-francis-su-20170202/

Harvard Health (2017) Do not get sold on drug advertising, *Harvard Health*, 14 February 2017, www.health.harvard.edu/drugs-and-medications/do-not-get-sold-on-drug-advertising

Harvey L (2020) Research fraud: A long-term problem exacerbated by the clamor for research grants, *Quality in Higher Education*, 26(3):243—261, doi:10.1080/13538322.2020.1820126

HASD (2014) Pi in India, *Hinduism and Sanatan Dharma*, 20 January 2014, https://pparihar.com/2014/01/20/pi-in-india/

Haynes S (2020) How Florence Nightingale paved the way for the heroic work of nurses today, *Time*, 12 May 2020, https://time.com/5835150/florence-nightingale-legacy-nurses/

HC Editors (2019) Galileo Galilei, *History.com*, 24 October July 2019, www.history.com/topics/inventions/galileo-galilei

Hidetoshi F and Rothman T (2008) *Sacred Mathematics, Japanese Temple Geometry*, Princeton University Press, Princeton.

Hill K (2014) Mathematicians urge colleagues to refuse to work for the NSA, *Forbes Magazine*, 5 June 2014, www.forbes.com/sites/kashmirhill/2014/06/05/mathematicians-urge-colleagues-to-refuse-to-work-for-the-nsa/

Hirji KF (2008) Numerosis and numeritis: Twin pathologies of contemporary statistics, *Data Critica: International Journal of Critical Statistics*, 1(2):3—15.

Hobart ME (2018) *The Great Rift: Literacy, Numeracy, and the Religion-Science Divide*, Harvard University Press, Cambridge, MA.

Hodges A (2008) In retrospect: Godel's proof, *Nature*, 454:829, https://doi.org/10.1038/454829a

Holt J (2008) Proof, *The New York Times*, 13 January 2008, www.nytimes.com/2008/01/13/books/review/Holt-t.html

Hom EJ (2015) What is mathematics? *Live Science*, www.livescience.com/38936-mathematics.html

Hosch WL (2016) Infinitesimal, *Encyclopedia Britannica*, 17 October 2016, www.britannica.com/science/infinitesimal

Hosking RJ (2017) Solving Sangaku: A traditional solution to a nineteenth century Japanese temple problem, *Journal for History of Mathematics*, 30(2): 53—69, http://dx.doi.org/10.14477/jhm.2017.30.2.053

Howell R (2005) Book review: *The Divine Challenge: On Matter, Mind, Math and Meaning* by J Byl, *Mathematics Association of America*, 15 August 2005, www.maa.org/tags/mathematics-and-religion

Howell RW and Bradley WJ (editors) (2001) *Mathematics in a Postmodern Age: A Christian Perspective*, WB Eerdmans-Lightning Source, New York.

Huckle T (2021) Mathematicians during the Third Reich and World War II, *Institut fur Informatik*, 13 June 2021, www5.in.tum.de/~huckle/mathwar.html

Huffman C (2019) Pythagoreanism, *The Stanford Encyclopedia of Philosophy*, Fall 2019, EN Zalta (editor), https://plato.stanford.edu/archives/fall2019/entries/pythagoreanism/

Humphrey J (2018) To infinity and beyond: The numbers game behind God's 'existence', *Irish Times*, 6 February 2018, www.irishtimes.com/culture/to-infinity-and-beyond-the-numbers-game-behind-god-s-existence-1.3376646

IM (2021) Famous Islamic mathematicians, *Islamic Mathematics*, 4 May 2021, http://islamicmaths.weebly.com/famous-mathematicians.html

Jalali MS et al (2020) The opioid crisis: A contextual, social-ecological framework, *Health Research and Policy Systems*, 18(87): https://doi.org/10.1186/s12961-020-00596-8

Jarus O (2013) Timbuktu: History of fabled center of learning, *Live Science*, 2013, www.livescience.com/26451-timbuktu.html

Jayaram V (2021) Hinduism and numerology, *Hindu Website*, 7 June 2021, www.hinduwebsite.com/hinduism/h_numerology.asp

Jnana (2021) Buddhism in the numbers, *Urban Dharma*, 7 June 2021, www.urbandharma.org/udharma7/numbers.html

Jones SE (2011) A brief survey of Sir Isaac Newton's views on religion, in MD Rhodes and JW Moody (2011) *Converging Paths to Truth*, Deseret Book, Salt Lake City, https://rsc.byu.edu/converging-paths-truth/brief-survey-sir-isaac-newtons-views-religion

Jongsma C (2005) Review: *Mathematics and the Divine*, *Mathematics Association of America*, 5 November 2005, www.maa.org/tags/mathematics-and-religion

Joseph GG (2000) *The Crest of the Peacock: Non-European Roots of Mathematics*, Princeton University Press, Princeton.

Kaku M (2022) Is God a mathematician? *Big Think*, 2 November 2022, https://bigthink.com/the-well/mathematics/

Kalla S (2010) Ethics in statistics, *Explorable*, 16 April 2010, https://explorable.com/ethics-in-statistics

Kanigel R (1991) *The Man Who Knew Infinity: A Life of the Genius Ramanujan*, Washington Square Press, New York.

Kaplan R (1999) *The Nothing That Is: A Natural History of Zero*, Oxford University Press, Oxford.

Kassirer JP (2005) *On the Take: How Medicine's Complicity With Big Business Can Endanger Your Health*, Oxford University Press, Oxford.

Katz B (2019) An existence proof: The mathematicians of the African diaspora website, *American Mathematical Society*, 31 January 2018, https://blogs.ams.org/inclusionexclusion/2019/01/31/mathematicians-of-the-african-diaspora/

Kaufman M (2009) The structure of everything, *Washington Post*, 8 February 2009, www.washingtonpost.com/wp-dyn/content/article/2009/02/05/AR2009020502876_pf.html

Kelly BT (2021a) Wagering eternity: Why we study Blaise Pascal, *Thomas Aquinas College*, 7 May 2021, www.thomasaquinas.edu/a-liberating-education/why-we-study/why-we-study-mathematics

Kelly BT (2021b) Why we study mathematics, *Thomas Aquinas College*, 7 May 2021, www.thomasaquinas.edu/a-liberating-education/why-we-study/why-we-study-mathematics

Kerekovska A and Galunska B (2015) Publication bias in clinical research sponsored by pharmaceutical industry, *Scripta Scientifica Pharmaceutica*, http://dx.doi.org/10.14748/ssp.v1i1.598

Kilanowski H (2014) Axioms of faith, *Dominicana*, 20 November 2014, www.dominicanajournal.org/axioms-of-faith/

Kolodny A (2020) How FDA failures contributed to the opioid crisis, *AMA Journal of Ethics*, 22(8):E743—750, doi:10.1001/amajethics.2020.743.

Korner K (2007) A question of truth, Nature, 449(6):27, September 2007.

Kuhn RL (2010) Is mathematics invented or discovered? *Closer to Truth*, 1 April 2010, www.closertotruth.com/series/mathematics-invented-or-discovered

Kurland B (2031) Mathematics, the handmaiden of theology: Augustine and Cantor on the infinity of God, *Catholic Stand*, 22 August 2013, https://catholicstand.com/mathematics-the-handmaiden-of-theology-augustine-and-cantor-on-the-infinity-of-god/

KW (2009) Why economists failed to predict the financial crisis, *Knowledge@Wharton*, The Wharton School, University of Pennsylvania, 13 May 2009, https://knowledge.wharton.upenn.edu/article/why-economists-failed-to-predict-the-financial-crisis/

Lakshmi R (2017) Indians are certain they invented the zero. But can they prove it? *The Washington Post*, 11 March 2017, www.washingtonpost.com/world/asia_pacific/indians-are-convinced-they-discovered-the-zero-can-they-prove-it/2017/03/10/

Lamb E (2017) A math lesson from Hitler's Germany, *Undark*, 1 February 2017, https://undark.org/2017/02/01/math-lesson-hitlers-germany/

Lamb R (2011a) How math works, *How Stuff Works*, 26 April 2011, https://science.howstuffworks.com/math-concepts/math.htm

Lamb R (2011b) Math: Human discovery or human invention? *How Stuff Works*, 26 April 2011, https://science.howstuffworks.com/math-concepts/math.htm

Lambert A (2011) The Crimean War, *BBC News*, 29 March 2011, www.bbc.co.uk/history/british/victorians/crimea_01.shtml

Landau E (2020) The Fibonacci sequence is everywhere—even the troubled stock market, *Smithsonian Magazine*, 25 March 2020, www.smithsonianmagazine.com

Larson E and Witham L (1998) Leading scientists still reject God, *Nature*, 394(313) https://doi.org/10.1038/28478

Lawrence S and McCartney M (2015) *Mathematicians and Their Gods: Interactions between Mathematics and Religious Beliefs*, Oxford University Press, Oxford.

Leinster T (2014a) Maths spying: The quandary of working for the spooks, *New Scientist*, 23 April 2014, www.newscientist.com/article/mg22229660-200-maths-spying-the-quandary-of-working-for-the-spooks/

Leinster T (2014b) Mathematician spies, *Slate*, 27 April 2014, https://slate.com/technology/2014/04/mathematicians-at-the-nsa-and-gchq-is-it-ethical-to-work-for-spy-agencies.html

Lennox J (2020) Why science and atheism don't mix, *Evolution News*, 17 July 2020, https://evolutionnews.org/2020/07/why-science-and-atheism-don't-mix/

Lisle J (2017) The God of infinities, *Acts and Facts*, 46(4), 31 March 2017, www.icr.org/article/god-infinities

Lisle J (2009) As easy as pi, *Answers in Genesis*, 8 June 2009, https://answersingenesis.org/contradictions-in-the-bible/as-easy-as-pi/

Livingstone J (2018) Did math kill God? *New Republic*, 27 April 2018, https://newrepublic.com/article/148150/math-kill-god

Livio M (2003) *The Golden Ratio: The Story of Phi, the World's Most Astonishing Number*, Crown Publisher, New York.

Livio M (2008) Unreasonable effectiveness, *Plus Magazine*, 1 December 2008, https://plus.maths.org/content/os/issue49/features/livio/index

Livio M (2009) *Is God a Mathematician?* Simon & Schuster, New York.

Livio M (2015) Math: Discovered, invented, or both? *PBS*, 13 April 2015, www.pbs.org/wgbh/nova/article/great-math-mystery/

Lohr S (2008) In modeling risk, the human factor was left out, *The New York Times*, 5 November 2008, www.nytimes.com/2008/11/05/business/05risk.html

Look BC (2020) Gottfried Wilhelm Leibniz, In EN Zalta (editor) *The Stanford Encyclopedia of Philosophy*, Spring 2020 Edition, https://plato.stanford.edu/archives/spr2020/entries/leibniz/

Magnello E (2018) Florence Nightingale: The compassionate statistician, *Plus Magazine*, 8 December 2010, https://plus.maths.org/content/florence-nightingale-compassionate-statistician

Maluf F (2012) Mathematics and Christian education, *Catholicism*, 27 February 2012, https://catholicism.org/mathematics-and-christian-education.html

Manu S (2018) Fourier transforms and the Buddhist doctrine of no-self, *Medium*, 18 September 2018, https://medium.com/@sasha.manu95/fourier-transforms-and-the-buddhist-doctrine-of-no-self-b07591cd1754

Martinez AA (2018) Was Giordano Bruno burned at the stake for believing in exoplanets? *Scientific American*, 19 March 2018, https://blogs.scientificamerican.com/observations/was-giordano-bruno-burned-at-the-stake-for-believing-in-exoplanets/

Masci D (2009) Religion and science in the United States, *Pew Research Center*, 5 November 2009, www.pewforum.org/2009/11/05/scientists-and-belief/

Mason NA (2002) Formulating the precautionary principle, *Environmental Ethics*, 24:263—274.

Mastin L (2020) Islamic mathematics, *Story of Mathematics*, 11 January 2020, www.storyofmathematics.com/islamic.html

Mathigon (2012) Infinity: World of mathematics, *Mathigon*, 15 May 2021, https://mathigon.org/world/Infinity

McCleary J (2018) Book review: *Mathematics without Apologies, Notices of the AMS*, 65(10): 1280—1283.

McIntyre D (2017) Ten mathematicians who recognized God's hand in their work, *ACMS 21st Biennial Conference Proceedings*, 2017:117—133.

McKenzie J (2020) Climate change is a 'Pascal's wager': So how will you act? *Physics World*, 11 March 2020, https://physicsworld.com/a/climate-change-is-a-pascals-wager-so-how-will-you-act/

McNeill L (2019) The woman who reshaped maths, *BBC Future*, 1 November 2019, www.bbc.com/future/article/20191031-hilda-geiringer-mathematician-who-fled-the-nazis

MT Staff (2019) FDA fails to cite big pharma for false marketing and advertisements, *Mass Torts*, 18 April 2019. www.masstortnexus.com/News/4376/FDA-Fails-to-Cite-Big-Pharma-for-False-Marketing-and-Advertisements

Munoz D (2020) Opinion: COVID-19, masks and moral mathematics, *ABC*, 30 December 2020, www.abc.net.au/religion/covid19-masks-and-moral-mathematics/12495150

MCCH (2017) What does pi teach us about the Trinity of God? *Medium*, 19 April 2017, https://medium.com/@mrmhaws/what-does-pi-teach-us-about-the-nature-of-god-d8fbcbe0eff

McGill (2007) Gregor Cantor, *McGill University*, 2007, www.cs.mcgill.ca/~rwest/wikispeedia/wpcd/wp/g/Georg_Cantor.htm

McHargue M (2015) Axioms about faith, *Mike McHargue*, 24 March 2015, https://mikemchargue.com/blog/2015/3/24/axioms-about-faith

McIntyre D (2017) Ten mathematicians who recognized God's hand in their work, *Proceedings of the Twenty First Conference of the Association of Christians in the Mathematical Sciences*, Charleston Southern University, 117—133.

Monk R (2018) Kurt Godel and the romance of logic, *Prospect*, 13 December 2018, www.prospectmagazine.co.uk/magazine/kurt-godel-and-the-romance-of-logic

Mulcare C (2013) The lost mathematicians: Numbers in the (not so) dark ages, *Plus Magazine*, 8 August 2013, https://plus.maths.org/content/lost-mathematicians-numbers-not-so-dark-early-middle-ages

Murphy PA (2021) Who says nature is mathematical? *Cantor's Paradise*, 27 February 2021, www.cantorsparadise.com/who-says-nature-is-mathematical-1abdc1330224

Murray B (2021) Do abstract objects explain why anything exists at all? (Axiological arguments for existence), *Medium*, 15 January 2021, https://thinkingdeeply.medium.com/do-abstract-objects-explain-why-anything-exists-at-all-axiological-arguments-for-existence

Nagel E and Newman JR (1958) *Godel's Proof*, New York University Press, 1958.

Nagasawa Y (2017) Review: *Maximal God: A New Defense of Perfect Being Theism*, *Oxford University Press Blog*, 8 November 2017, https://blog.oup.com/2017/11/definitive-proof-existence-god/

Nelson E (2021) Mathematics and faith, *Department of Mathematics*, Princeton University, www.math.princeton.edu/nelson/papers.html

Nelson RH (2017) Existence of God: The rational arguments from mathematics to human consciousness, *The Independent*, 17 May 2017, www.independent.co.uk/life-style/existence-of-god-rational-arguments-mathematics-human-consciousness-a7739841.html

Nickel J (2012) *Mathematics: Is God Silent?* Ross House Books, California.

Pedigo M (2019) The unreasonable effectiveness of mathematics in the natural sciences, *Medium*, 29 September 2019, www.cantorsparadise.com/the-unreasonable-effectiveness-of-mathematics-in-the-natural-sciences-25bd8dc6429f

Pedigo M (2020) Review: *"Is God A Mathematician"* by Mario Livio, *Medium*, 3 January 2020, www.cantorsparadise.com/is-god-a-mathematician-by-mario-livio-a7ae4beec5e2

Poythress VS (2015) *Redeeming Mathematics: A God-Centered Approach*, Crossway, New York.

Nieder A (2020) The remarkable ways animals understand, *BBC Future*, 8 September 2020, numberswww.bbc.com/future/article/20200907-the-remarkable-ways-animals-understand-numbers

NPR Staff (2014) Far from 'Infinitesimal': A mathematical paradox's role in history, *NPR*, 20 April 2014, www.npr.org/2014/04/20/303716795/far-from-infinitesimal-a-mathematical-paradoxs-role-in-history

O'Connor JJ and Robertson EF (2002) Infinity, *Maths History*, February 2002, https://mathshistory.st-andrews.ac.uk/HistTopics/Infinity/

O'Connor JJ and Robertson EF (2021) Christianity and mathematics, *History Topics*, www-groups.dcs.st-and.ac.uk/history/HistTopics/Heliocentric.html

Orhan MA (2021) Dynamic interactionism between research fraud and research culture: A commentary to Harvey's analysis, *Quality in Higher Education*, 27(1):134—146, www.tandfonline.com/doi/full/10.1080/13538322.2020.1820126

Otis W (2020a) The connection between mathematics and empirical sciences, *Medium*, 30 May 2020, https://medium.com/however-mathematics/the-connection-between-mathematics-and-empirical-sciences-ec0a6d8f4f5

Otis W (2020b) Leibniz's mathematical approach to God, *Medium*, 4 June 2020, https://medium.com/however-mathematics/leibnizs-mathematical-approach-to-god-aa6844353b27

Oyedele A (2016) Warren Buffett on global warming: 'This issue bears a similarity to Pascal's Wager on the existence of God,' *Business Insider*, 27 February 2016, www.businessinsider.com/warren-buffett-on-climate-change-2016-2?

Parc S (editor) (2014) *50 Visions of Mathematics*, Oxford University Press, Oxford.

Patterson S (2009) *The Quants: How a New Breed of Math Whizzes Conquered Wall Street and Nearly Destroyed It*, Crown Business, New York.

Paulos JA (2009) *Irreligion: A Mathematician Explains Why the Arguments for God Just Don't Add Up*, Hill & Wang, New York.

Paulos JA (2014) The 16th century's line of fire, *The New York Times*, 7 April 2014, www.nytimes.com/2014/04/08/science/infinitesimal-looks-at-an-historic-math-battle.html

Plackett B (2013) Study finds most drug commercials misleading, *Scientific American*, 28 September 2013, www.scientificamerican.com/article/study-finds-most-drug-commercials-misleading/

Pomeroy SB (2013) *Pythagorean Women: Their History and Writings*, John Hopkins University Press, Baltimore.

Ransford HC (2017a) *God and the Mathematics of Infinity: What Irreducible Mathematics Says about Godhood*, Gardner Books, New York.

Ransford HC (2017b) God and the mathematics of infinity, *Conscious Connection*, 2017, Issue 1, www.consciousconnectionmagazine.com/2017/01/god-and-the-mathematics-of-infinity/

Ransford C and Eyghen HV (2018) God and mathematics, *The Religious Studies Project*, 12 February 2018, www.religiousstudiesproject.com/podcast/god-and-mathematics/

Rehmeyer J (2008) Florence Nightingale: The passionate statistician, *Science News*, 26 November 2008, www.sciencenews.org/index/generic/activity/view/id/38937/title/Florence_Nightingale_The_passionate_statistician

Renyi A (1967) *A Socratic Dialogue on Mathematics*, Holden Day Publishers, San Francisco.

Riley J (2020) If infinity is real, only God exists, *Medium*, 5 August 2020, https://medium.com/ascending-luminosity/if-infinity-is-real-only-god-exists

Rossis NC (2021) Medieval Cistercian numbers, *Nicholas Rossis*, 17 January 2021, https://nicholasrossis.me/2021/01/17/medieval-cistercian-numbers/

Ruiz R (2010) Ten misleading drug ads, *Forbes Magazine*, 2 February 2021, www.forbes.com/2010/02/02/drug-advertising-lipitor-lifestyle-health-pharmaceuticals-safety.html

Russo M (2019) Spiral as a guide: a comparison between Emerson's Nature and Lateralus by Tool, *The Serendipity Periodical*, 15 September 2019, www.theserendipityperiodical.it/2019/09/15/spiral-as-a-guide-a-comparison-between-emersons-nature-and-lateralus-by-tool-2/

RW (2021) Biblical value of pi, *Religions Wiki*, 23 May 2021, https://religions.wiki/index.php/Biblical_value_of_pi

SA (2006) What is Gödel's proof? *Scientific American*, 19 January 2006, www.scientificamerican.com/article/what-is-goumldels-proof/

Saka P (2021) Pascal's wager about God, *The Internet Encyclopedia of Philosophy*, 20 June 2021, https://iep.utm.edu/pasc-wag/

Sameer A (2017) The so-called mathematical miracle of the Quran, *Abdullah Sameer Insider*, 10 February 2017, https://abdullahsameer.com/the-so-called-mathematical-miracle-of-the-quran/

Schaefer R (2021) Book review: *Mathematics Without Apologies: Portrait of a Problematic Vocation*, New York Journal of Books, 9 June 2021, www.nyjournalofbooks.com/book-review/mathematics-without-apologies

Schimmel A (1993) *The Mystery of Numbers*, Oxford University Press, Oxford.

Schoenfeld G (2001) The punch-card conspiracy, *The New York Times*, 18 March 2001, https://archive.nytimes.com/www.nytimes.com/books/01/03/18/reviews/010318.18schoent.html

Scoles S (2016) How thinking about infinity changes kids' brains on math, *Aeon*, 9 March 2016, https://aeon.co/ideas/how-thinking-about-infinity-changes-kids-brains-on-math

Segal SL (1980) Helmut Hasse in 1934, *Historia Mathematica*, 7(1):46—56, https://doi.org/10.1016/0315-0860(80)90063-4.

Segal SL (2014) *Mathematicians under the Nazis*, Princeton University Press, Princeton.

Seife C (2000) *Zero: The Biography of a Dangerous Idea*, Penguin Books, New York.

Siegmund-Schultze R (2004) Review: *Mathematicians under the Nazis*, MAA Reviews, 10 February 2004, www.maa.org/press/maa-reviews/mathematicians-under-the-nazis

Sheller S (2021) Inside the pharmaceutical industry – Deceptive marketing schemes, *The Insider Exclusive*, 11 June 2021, https://insiderexclusive.com/inside-the-pharmaceutical-industry-deceptive-marketing-schemes/

Smith WG (2011) Leibniz's argument for God's existence, *Thoughts En Route*, 21 November 2011, https://wallacegsmith.wordpress.com/2011/11/21/leibnizs-argument-for-gods-existence/

Solis S (2013) Copernicus and the Church: What the history books don't say, *The Christian Science Monitor*, 19 February 2013, www.csmonitor.com/Technology/2013/0219/Copernicus-and-the-Church-What-the-history-books-don-t-say

SOM (2021a) Indian mathematics & mathematicians, *Story of Mathematics*, 4 May 2021, www.storyofmathematics.com/indian.html

SOM (2021b) Bertrand Russell & Alfred North Whitehead – *Principia Mathematica* 1+1=2, *Story of Mathematics*, 28 May 2021, www.storyofmathematics.com/20th_russell.html

Spencer A (2009) *Tower of Thieves*, Brick Tower Press, New York.

Spiegelhalter D and Pearson M (2009) Understanding uncertainty: The many ways of spinning risk, *Plus Magazine*, 1 March 2009, https://plus.maths.org/content/os/issue50/risk/index

Stewart I (2012) The mathematical equation that caused the banks to crash, *The Guardian* (UK), 12 February 2012, www.theguardian.com/science/2012/feb/12/black-scholes-equation-credit-crunch

Stirrat M and Cornwell RE (2013) Eminent scientists reject the supernatural: A survey of the Fellows of the Royal Society, *Evolution Education Outreach*, 6(33), https://doi.org/10.1186/1936-6434-6-33

Strevens M (2020) *The Knowledge Machine: How Irrationality Created Modern Science*, WW Norton, New York.

Strickland L (2016) Answering the biggest question of all: Why is there something rather than nothing? *The Conversation*, 11 November 2016, https://theconversation.com/answering-the-biggest-question-of-all-why-is-there-something-rather-than-nothing-65865

Stucki DJ (2007) Review: *Equations from God: Pure Mathematics and Victorian Faith*, *Mathematics Association of America*, 12 June 2007, www.maa.org/tags/mathematics-and-religion

Su F (2012) *Mathematics for Human Flourishing*, Yale University Press, Yale.

Sujato (201) Buddhist numerology, *Sujato's Blog*, 15 April 2010, https://sujato.wordpress.com/2010/04/15/buddhist-numerology/

Sullivan J and Zutavern A (2017), *The Mathematical Corporation: Where Machine Intelligence and Human Ingenuity Achieve the Impossible*, Public Affairs, New York.

Suri G and Bal HS (2007) *A Certain Ambiguity: A Mathematical Novel*, Princeton University Press, Princeton.

Sutton D (2007) *Islamic Design: A Genius for Geometry*, Wooden Books, London.

Tassone BG (2021) Aristotle on the infinite, *Infinity on Line*, 16 March 2021, https://infinityonline.valzorex.com/aristotle.html

Tate J (2010) Kepler's law, *Universe Today*, 11 February 2010, www.Universetoday.com/55423/keplers-law/

Tbakhi A and Amr SS (2007) Ibn Al-Haytham: father of modern optics, *Annals of Saudi Medicine*, 27(6):464—467, doi:10.5144/0256-947.2007.464

Thomas R (2002) Book review: *The Queen's Conjuror*, *Plus Magazine*, 1 September 2002, https://plus.maths.org/content/queens-conjuror

Timon A (2018) When magic gave way to numbers, *Open Mind*, 30 April 2018, www.bbvaopenmind.com/en/science/mathematics/when-magic-gave-way-to-numbers/

Tent MBW (2021) *Gottfried Wilhelm Leibnitz: The Polymath Who Brought Us Calculus*, CRC Press, Boca Raton.

TPM (2018) Mathematics and ethics education should go hand in hand, *TPM*, 20 June 2018, www.techpoweredmath.com/math-ethics-education/

Tu LW (2013) Remembering Raoul Bott (1923–2005), *Notices of the AMS*, https://celebratio.org/Bott_R/article/727/

Tyson P (2011) Describing nature with math, *PBS*, 10 November 2011, www.pbs.org/wgbh/nova/article/describing-nature-math/

UNHCR (2021) *A Human Rights Based Approach to Data - Leaving No One Behind in the 2030 Agenda for Sustainable Development*, UN Commission on Human Rights, Geneva, www.ohchr.org/EN/Issues/Indicators/Pages/documents.aspx

University of Pennsylvania (2005) Why is the helix such a popular shape? Perhaps because they are nature's space savers, *Science Daily*, 2 March 2005, www.sciencedaily.com/releases/2005/02/050223135535.htm.

Van Dam A (2009) Duo writes about how health statistics can mislead, *Health Journalism*, 9 December 2009, https://healthjournalism.org/blog/2009/12/duo-writes-about-how-health-statistics-can-mislead/

Vasak M (2020) Why mathematics isn't much different from religion, *Cantor's Paradise*, 7 August 2020, www.cantorsparadise.com/why-mathematics-isnt-much-different-from-religion-419ecb652724

Vaughn J (2021) Concerning attempts to logically justify the existence of God, Manuscript, *North Georgia University*, Georgia.

Vestal DL (2007) Review: *A Certain Ambiguity: A Mathematical Novel*, *Mathematical Association of America*, 26 September 2007, www.maa.org/press/maa-reviews/a-certain-ambiguity-a-mathematical-novel

Ward M (2018) India's impressive concept about nothing, *BBC News*, 8 August 2018, www.bbc.com/travel/story/20180807-how-india-gave-us-the-zero

Webb SH (2015) Is God really infinite? *First Things*, 10 March 2015, www.firstthings.com/web-exclusives/2015/03/is-god-really-infinite

Wigner E (1960) The unreasonable effectiveness of mathematics in the natural sciences, *Communications on Pure and Applied Mathematics*, 13:1–14.

Wikipedia (2021 – Aryabhata, Association of Christians in the Mathematical Sciences, Biblical Numerology, Blaise Pascal, Carl Friedrich Gauss, Cistercian Numerals, Crimean War, Emil J Gumbel, Financial Crisis of 2007–2008, Florence Nightingale, Galileo Galilei, Georg Cantor, Giordano Bruno, Godel's Ontological Proof, Golden Ratio, GW Leibniz, Hellmuth Kneser, Hippasus, History of Calculus, Indian Mathematics, Infinity, Isaac Newton, IBM and the Holocaust, Jafar al-Ṣadiq, Johannes Kepler, Kurt Friedrich Godel, Maryam Mirzakhani, Mathematics, Michio Kaku, Misuse of Statistics, National Socialist German Lecturers League, Nicolaus Copernicus, Non-Euclidean Geometry, Numerology, Numerology and the Church Fathers, Pascal's Wager, Precautionary Principle, Pythagoras, Pythagorean Theorem, Pythagoreanism, Rene Descartes, Sangaku, Shulba Sutras, Sir George Stokes, 1st Baronet, Srinivasa Ramanujan, Trademark Argument, The Unreasonable Effectiveness of Mathematics in the Natural Sciences, Tycho Brahe, Zeno's Paradoxes).

Wilkerson J (2015) God & math: Thinking Christianly about math education, *God and Math*, 14 March 2015, https://godandmath.com/tag/pi-day/

Williams J (2021) Editorial, *Quality in Higher Education*. 27(1):1—3.

Willson F (2002) Shapes, numbers, patterns, and the Divine Proportion in God's creation, *Institute for Creation Research*, 1 December 2002, www.icr.org/article/shapes-numbers-patterns-divine-proportion-gods-cre/

Wilson P (2014) The philosophy of mathematics, In S Parc (editor) (2014), *50 Visions of Mathematics*: 176—179.

Wilson P (2017) Frugal nature: Euler and the calculus of variations, *Plus Magazine*, 1 September 2007, https://plus.maths.org

Wilson R (2021) GH Hardy's Oxford years, *Mathematical Institute*, University of Oxford, 2021.

Zarepour MS (2020) Avicenna on mathematical Infinity, *Archiv fur Geschichte der Philosophie*, 102(3): 379—425, https://doi.org/10.1515/agph-2017-0032

Zhu D (2022) Cistercian numerals, *Probably World*, 2 November 2022, https://probably.world/posts/cistercian-numerals/

6. RUMINATIONS

Agar N (2008) *Liberal Eugenics: In Defense of Human Enhancement*, Wiley-Blackwell, New York.

Al-Jazeera (2021) World on the verge of climate crisis 'abyss', warns UN, *Al-Jazeera News*, 19 April 2021, www.aljazeera.com/news/2021/4/19/world-on-the-verge-of-climate-crisis-abyss-warns-un

Anonymous (1913) "Pastors for Eugenics", *Teaching American History*, 6 June 1913, https://teachingamericanhistory.org/document/new-york-times-pastors-for-eugenics/

Armstrong K (2022) *Sacred Nature: Restoring Our Ancient Bond with the Natural World*, Alfred K Knopf, New York.

BBC (2022) What is climate change? A really simple guide, *BBC News*, 2 November 2022, www.bbc.com/news/science-environment-24021772

Bricker V (2021) What does the Bible say about climate change? *Christianity.com*, 9 September 2021, www.christianity.com/wiki/christian-life/what-does-the-bible-say-about-climate-change.html

Burton K (2022) Book review: *Control* by Adam Rutherford, *Geographical*, 13 May 2022, https://geographical.co.uk/book-reviews/control-by-adam-rutherford

CCSWK (2022) Catholic social teaching 101: Climate change, *Catholic Charities of Southwest Kansas*, 16 November 2022. https://catholiccharitiesswks.org/about-us/46-home/news/social-justice/819-catholic-social-teaching-101-climate-change

CE (2022) Green washing files: Aramco, *Client Earth*, 14 November 2022, www.clientEarth.org/projects/the-greenwashing-files/aramco/

CEA (2022) Climate change, *Christian Enquiry Agency*, 16 November 2022, https://christianity.org.uk/article/climate-change

Chandler MF (2021) Pastor column: What does the Bible say about climate change? *Daily Press*, 24 November 2021, www.vvdailypress.com/story/lifestyle/2021/11/24/pastor-column-what-does-bible-say-climate-change/8735774002/

Chandra A (2021) How Modi's Hindu nationalism impairs global fight against climate change, *Chicago Sun Times*, 29 July 2021, https://chicago.suntimes.com/2021/7/29/22600092/narendra-modi-hindu-nationalism-climate-change-india

Crawford N (2019) Pentagon fuel use, climate change, and the costs of war, *Watson Institute for International & Public Affairs*, Brown University, 13 November 2019.

CUP (2022) Buddhism and climate change, *Cornell University Press Blog*, 10 November 2022, www.cornellpress.cornell.edu/buddhism-and-climate-change/

Dalai Lama and Alt F (2020) *Our Only Home: A Climate Appeal to the World*, Hanover Square Press, New York.

ECLA (2008) *Genetics and Faith: Power, Choice, and Responsibility*, Evangelical Lutheran Church in America, Chicago.

Ellis-Petersen H (2022a) India faces deepening demographic divide as it prepares to overtake China as the world's most populous country, *The Guardian* (UK), 14 November 2022, www.theguardian.com/world/2022/nov/14/india-faces-deepening-demographic-divide-as-it-prepares-to-overtake-china-as-the-worlds-most-populous-country

Ellis-Petersen H (2022b) India's energy conundrum: Committed to renewables but still expanding coal, *The Guardian* (UK), 15 November 2022, www.theguardian.com/world/2022/nov/15/india-committed-to-clean-energy-but-continues-to-boost-coal-production

Esmaeilzadeh F et al. (2022) Major thalassemia, screening or treatment: An economic evaluation study in Iran, *International Journal of Health Policy and Management*, 11(7):1112--1119, doi:10.34172/ijhpm.2021.04

EWT (2016) Book review: *The God Gene* by Dean Hamer, *Eleanor Writes Things*, 22 April 2016, https://eleanorwritesthings.com/2016/04/22/book-review-the-god-gene/

Falconer D (2021) A dazzling synthesis, *Sydney Review of Books*, 2 December 2021, https://sydneyreviewofbooks.com/review/ghosh-nutmegs-curse/

Fang L and Lerner S (2019) Saudi Arabia denies its key role in climate change even as it prepares for the worst, *The Intercept*, 18 September 2019, https://theintercept.com/2019/09/18/saudi-arabia-aramco-oil-climate-change/

Farley AC (2021) The eugenics roots of evangelical family values, *Religion and Politics*, 12 May 2021, https://religionandpolitics.org/2021/05/12/the-eugenics-roots-of-evangelical-family-values/

Fici CL (2018) Exploring the dharmic way to think about climate change, *The Wire*, 4 December 2018, https://thewire.in/religion/exploring-the-dharmic-way-to-think-about-climate-change

Foley A (2020) Climate change and the Bible, *Answers in Genesis*, 20 February 2020, https://answersingenesis.org/environmental-science/climate-change/climate-change-and-the-bible/

Fromm E (1961) *Marx's Concept of Man*, Frederick Ungar, New York.

Gahlau KD (2022) What is climate change? *Australian Academy of Sciences*, 4 November 2022, www.science.org.au/learning/general-audience/science-climate-change/

Gerrard T (1914) The Church and eugenics, In *The Catholic Encyclopedia*, 7 May 2022, New York, www.newadvent.org/cathen/16038b.htm

Goldman M (2004) Review: *The God Gene: How Faith is Hardwired into Our Genes*, *Nature Genetics*, 36, 1241, https://doi.org/10.1038/ng1204-1241

Ghosh A (2021) *The Nutmeg's Curse: Parables for a Planet in Crisis*, The University of Chicago Press, Chicago.

GQ (2022a) Does the Bible support eugenics? *Got Questions*, 7 May 2022, www.gotquestions.org/eugenics-Bible.html

GQ (2022b) How should a Christian view climate change? *Got Questions*, 16 November 2022, www.gotquestions.org/climate-change.html

Graves-Fitzsimmons G and Siddiqi M (2021) Religious Americans demand climate action, *American Progress*, 21 July 2021, www.americanprogress.org/article/religious-americans-demand-climate-action/

Gudynas E (2018) Religion and cosmovisions within environmental conflicts and the challenge of ontological openings, In E Berry & R Albro (editors) (2018) *Church, Cosmovision and the Environment: Religion and Social Conflict in Contemporary Latin America*, Routledge, New York.

Guest K (2022) *Control* by Adam Rutherford review – A warning from history about eugenics, *The Guardian* (UK), 10 February 2022, www.theguardian.com/books/2022/feb/10/control-by-adam-rutherford-review-a-warning-from-history-about-eugenics

Guru-Murthy K (2020) 'Buddha would be green': Dalai Lama calls for urgent climate action, *The Guardian* (UK), 11 November 2020, www.theguardian.com/world/2020/nov/11/buddha-would-be-green-dalai-lama-calls-for-urgent-climate-action

Hamer D (2004) *The God Gene: How Faith is Hardwired into our Genes*, Double Day, New York.

Hanh TN (2022) *Zen and the Art of Saving the Planet*, Harper One, New York.

Harden KP (2021) *The Genetic Lottery: Why DNA Matters for Social Equality*, Princeton University Press, Princeton.

Hathout H (2006) An Islamic perspective on human genetic and reproductive technologies, *Eastern Mediterranean Health Journal*, 12 (2), S22-S28.

Haupt M (2022) What is emerging? *Medium*, 13 September 2022, https://medium.com/society4/part-2-what-is-emerging-fc452737e600

Huxter M (2022) The Buddha's path of freedom and climate change mitigation, *Insight Timer Blog*, 10 November 2022, https://insighttimer.com/blog/buddhist-perspective-on-climate-change/

IEF (2022) Hindu declarations on climate change, 2009-2015, *International Environment Forum*, 2022 https://iefworld.org/hindu-cc

Jenkins J (2022) Bible demands action on climate change, Evangelicals say in new report, *The Washington Post*, 30 August 2022, www.washingtonpost.com/religion/2022/08/30/evangelicals-climate-change-bible/

Jenkins W et al. (2018) Religion and climate change, *Annual Review of Environment and Resources*, 43:85-108, https://doi.org/10.1146/annurev-environ-102017-025855

JW (2022) Climate change and our future—What the Bible says, *Jehovah's Witnesses*, 16 November 2022, www.jw.org/en/library/series/more-topics/climate-change-global-warming-bible/

Kashyap S (2022) Book review: *The Nutmeg's Curse: Parables for a Planet in Crisis* by Amitav Ghosh, *LSE Review of Books*, 22 May 2022, https://blogs.lse.ac.uk/usappblog/2022/05/22/book-review-the-nutmegs-curse-parables-for-a-planet-in-crisis-by-amitav-ghosh/

Kimeu C (2022) 'You'll rarely find a climate denier in East Africa', *The Guardian* (UK), 17 November 2022, www.theguardian.com/environment/2022/nov/17/youll-rarely-find-a-climate-denier-in-east-africa

KR Editor (2010) Review: *The God Gene: How Faith is Hardwired into Our Genes*, *Kirkus Review*, 20 May 2010, www.kirkusreviews.com/book-reviews/dean-h-hamer/the-god-gene/

Kramer B (2018) Science, race, and the Bible: Coming to terms with a messy history, *Bio Logos*, 27 February 2018, https://biologos.org/articles/science-race-and-the-bible-coming-to-terms-with-a-messy-history

Landless PN and Charles-Marcel ZL (2021) Is there a "God gene"? Fact or fancy? *The Adventist Review*, 1 November 2021, https://adventistreview.org/house-call/2111-69/

Lansdowne O (2017) Book review: *The Penultimate Curiosity*, *Bethinking*, 2017, www.bethinking.org/does-science-disprove-god/penultimate-curiosity-review

Leon SM (2013) *An Image of God: The Catholic Struggle with Eugenics*, University of Chicago Press, Chicago.

Lewis R (2022) In search of a religiosity gene, *PLOS DNA Science*, 22 December 2022, https://dnascience.plos.org/2022/12/22/in-search-of-a-religiosity-gene/

Loy D (2015) Awakening in the age of climate change, *Tricycle*, Spring 2015, https://tricycle.org/magazine/awakening-age-climate-change/

Mason D (2022) Book review: *Control* by Adam Rutherford, *Bookpage*, 15 November 2022, www.bookpage.com/reviews/control-adam-rutherford-book-review/

Mohler RA (2004) Review: *The God Gene*—Bad science meets bad theology, *Albert Mohler*, 1 October 2004, https://albertmohler.com/2004/10/01/the-god-gene-bad-science-meets-bad-theology

Moore KD and Tucker ME (2015) A roaring force from one unknowable moment, *Orion*, 12 May 2015, https://orionmagazine.org/2015/05/a-roaring-force-from-one-unknowable-moment/

NASA (2022) What is the greenhouse effect? *NASA*, 12 November 2022, https://climate.nasa.gov/faq/19/what-is-the-greenhouse-effect

Nelson JR (1988) Genetics and theology: A complementarity? *Religion Online*, 20 April 1988, www.religion-online.org/article/genetics-and-theology-a-complementarity/

NGL (2023) Anthropocene, *National Geographic Library*, 14 February 2023, https://education.nationalgeographic.org/resource/anthropocene

NHS (2022) Overview – Thalassemia, 17 October 2022, *NHS* (UK), www.nhs.uk/conditions/thalassaemia/

OES (2015) *The Time to Act is Now: A Buddhist Declaration on Climate Change*, One Earth Sanga, 20 September 2015, https://oneearthsangha.org/articles/buddhist-declaration-on-climate-change/

Ozdemir I (2020) What does Islam say about climate change and climate action? *Al-Jazeera News*, 12 August 2020, www.aljazeera.com/opinions/2020/8/12/what-does-islam-say-about-climate-change-and-climate-action

Pandey G (2021) NFHS: Does India really have more women than men? *BBC News*, 27 November 2021, www.bbc.com/news/world-asia-india-59428011

Poole S (2022) *Control* by Adam Rutherford review: A fizzy history of eugenics – and what science still can't do, *The Telegraph*, 18 February 2022, www.telegraph.co.uk/books/what-to-read/control-adam-rutherford-review-fizzy-history-eugenics-science/

PRC (2015) Religion and views on climate and energy issues, *Pew Research Center*, 22 October 2015, www.pewresearch.org/science/2015/10/22/religion-and-views-on-climate-and-energy-issues/

PW Editor (2004) *The God Gene: How Faith Is Hardwired into Our Genes*, *Publisher's Weekly*, 7 December 2004, www.publishersweekly.com/978-0-385-50058-6

Radwan R (2022) How Saudi Arabia is translating its climate-change ambitions into action, *Arab News*, 13 November 2022, www.arabnews.com/node/2198686/business-economy

Rahman SA (2022) India ruling party MP arrested over prophet remarks amid protests, *The Guardian* (UK) 26 August 2022, www.theguardian.com/world/2022/aug/26/india-ruling-party-mp-t-raja-singh-arrested-over-prophet-remarks-amid-protests

Raihani N (2021) *The Social Instinct: How Cooperation Shaped the World*, Jonathan Cape, London.

Raina R (2022) 'In Delhi I can see the climate catastrophe unfolding before my eyes', *The Guardian* (UK), 15 November 2022, www.theguardian.com/environment/2022/nov/15/delhi-climate-catastrophe-oxford-britain-crisis

Rao NR (2021) Climate change and Hinduism, 20 April 2021, *Psychiatric Times*, www.psychiatrictimes.com/view/climate-change-and-hinduism

Ratcliffe R (2022) Five years after the crackdown, Myanmar's remaining Rohingya 'living like animals', *The Guardian* (UK), 25 August 2022, www.theguardian.com/global-development/2022/aug/25/five-years-after-the-crackdown-myanmars-remaining-rohingya-living-like-animals

Reporter (2016) The troubling history of eugenics, *The Hindu*, 21 October 2016, www.thehindu.com/news/cities/bangalore/The-troubling-history-of-eugenics/article16077159.ece

Roy A (2018) The rising tide: Perhaps, 2018 flood is only a gentle warning, *The Week*, 2 August 2018, https://climateandcapitalism.com/2018/08/23/arundhati-roy-the-deadly-flood-in-kerala-may-be-only-a-gentle-warning/

Rutherford A (2022) *Control: The Dark History and Troubling Present of Eugenics*, W&N, UK.

Sengupta H (2019) Narendra Modi and India's new climate change norms, *Observer Research Foundation*, 18 July 2019, www.orfonline.org/expert-speak/narendra-modi-and-indias-new-climate-change-norms-53154/

Shraiky R (2021) Does Islam tackle the crisis of climate change? *The Review of Religions*, 16 March 2021, www.reviewofreligions.org/29427/does-islam-tackle-the-crisis-of-climate-change/

Silveira LA (2008) Experimenting with spirituality: Analyzing *The God Gene* in a non-majors laboratory course, *CBE Life Science Education*, 7(1):132-45. doi:10.1187/cbe.07-05-0029.

Snyder S (2022) Water in crisis – India, *The Water Project*, 12 November 2022, https://thewaterproject.org/water-crisis/water-in-crisis-india

Thakur AK (2021) Book review: *The Nutmeg's Curse: Parables for a Planet In Crisis* by Amitav Ghosh, *Outlook India*, 7 November 2021, www.outlookindia.com/website/story/entertainment-news-book-review-the-nutmegs-curse-parables-for-a-planet-in-crisis-by-amitav-ghosh/400080

Thangaraj S (2017) Many Hindus saw themselves as Aryans and backed Nazis. Does that explain their support for Donald Trump? *Quartz India*, 3 February 2017, https://qz.com/india/901244/many-hindus-saw-themselves-as-aryans-and-backed-nazis-does-that-explain-hindutvas-support-for-donald-trump/

Thomas M (2017) Hindu nationalists are trying to create designer babies that are fair, strong, and smart, *Quartz India*, 9 May 2017, https://qz.com/india/979007/rss-hindu-nationalists-are-trying-to-create-designer-babies-that-are-fair-strong-and-smart/

Tong Y (2022a) India's sex ratio at birth begins to normalize, *Pew Research Center*, 23 August 2022, www.pewresearch.org/religion/2022/08/23/indias-sex-ratio-at-birth-begins-to-normalize/

Tong Y (2022b) Sex ratios around the world, *Pew Research Center*, 23 August 2022, www.pewresearch.org/religion/2022/08/23/sidebar-sex-ratios-around-the-world/

Turney J (2021) Book review: *The Social Instinct* by Nichola Raihani, *The Arts Desk*, 4 June 2021, https://theartsdesk.com/books/nichola-raihani-social-instinct-review-habits-co-operation

UN (2022) What is climate change? *United Nations*, 4 November 2022, www.un.org/en/climatechange/what-is-climate-change

Vedachalam S (2012) Water supply and sanitation in India: Meeting targets and beyond, *Global Water Forum*, 23 September 2012, https://globalwaterforum.org/2012/09/23/water-supply-and-sanitation-in-india-meeting-targets-and-beyond/

Wagner R and Briggs A (2016) *The Penultimate Curiosity*, Oxford University Press, Oxford.

Wikipedia (2022 - God Gene, Jehovah's Witnesses, Marx's Theory of Alienation, New Eugenics, The Islamic Declaration on Global Climate Change, Saudi Arabia).

Zafar M (2021) Climate lessons from the Quran and Hadith, *Islamic Relief*, 6 May 2021, www.islamic-relief.org.uk/6-climate-lessons-from-the-quran-and-hadith/

7. FINALE

ADL (2022) Hindu tree huggers: The Chipko movement, *Academy for Distant Learning*, 15 December 2022, https://adlonlinecourses.com/hindu-tree-huggers-the-chipko-movement/

Ahmed N (2013) James Hansen: Fossil fuel addiction could trigger runaway global warming, *The Guardian* (UK), 10 July 2013, www.theguardian.com/environment/Earth-insight/2013/jul/10/james-hansen-fossil-fuels-runaway-global-warming

Altmann SL and Bowen EJ (2021) Obituary: Charles Alfred Coulson, 1910-1974, *The Royal Society*, 7 November 2021, https://royalsocietypublishing.org/

Asher R (2012) Can a scientist be religious? *University of Cambridge*, 11 March 2012, www.cam.ac.uk/research/discussion/can-a-scientist-be-religious

Azarian B (2022) *The Romance of Reality*, BenBella Books, Dallas, TX.

Bailey R (2011) Religion going "extinct" in nine countries, *Reason*, 24 March 2011, https://reason.com/2011/03/24/religion-going-extinct-in-nine/

Barrett LF (2021) *7½ Lessons About the Brain*, Mariner Books, New York.

Bazilchuk N (2007) New voice for the environment, *Environmental Health Perspectives*, 115(4):A190, doi:10.1289/ehp.115-a190a.

BBC (2022) What is climate change? A really simple guide, *BBC News*, 2 November 2022, www.bbc.com/news/science-environment-24021772

Berman R (2018) Michio Kaku believes in God, if not that God, *Big Think*, 15 February 2018, https://bigthink.com/culture-religion/michio-kaku-believes-in-god-if-not-that-god/

Bradshaw M and Ellison CG (2008) Do genetic factors influence religious life? Findings from a behavior genetic analysis of twin siblings, *Journal for the Scientific Study of Religion*, 47(4), doi:10.1111/j.1468-5906.2008.00425.x

Bragazzi N et al. (2018) Neurotheology of Islam and higher consciousness states, *Cosmos and History: The Journal of Natural and Social Philosophy*, 14(2), 315–321.

Cheong R (2020) The wonders of self-awareness, *Christian Counseling Coalition*, 8 April 2020, www.biblicalcounselingcoalition.org/2020/04/08/the-wonders-of-self-awareness/

Chitwood K (2015) Human consciousness & religious reality, *Religious Studies Project*, 24 September 2015, www.religiousstudiesproject.com/response/human-consciousness-religious-reality/

Costandi M (2022a) *Body Am I: The New Science of Self-Consciousness*, The MIT Press, Cambridge, MA.

Costandi M (2022b) Is the body key to understanding consciousness? The Guardian (UK), 2 October 2022, www.theguardian.com/science/2022/oct/02/is-the-body-key-to-understanding-consciousness

Coulson CA (2018) *Science, Technology and the Christian*, Reprint, Forgotten Books, New York.

Coulson CA (1955) *Science and Christian Belief*, University of North Carolina Press, Chapel Hill.

Curry JA and Gushee D (2010) Understanding conservative religious resistance to climate science, *Judith Curry Blog*, 20 December 2010, https://judithcurry.com/2010/12/20/understanding-conservative-religious-resistance-to-climate-science/

Dandia A (2020) Can atheists make their case without devolving into bigotry? *Al-Jazeera News*, 21 February 2020, www.aljazeera.com/opinions/2020/2/21/can-atheists-make-their-case-without-devolving-into-bigotry

DC (2016) Bengaluru steel flyover: A 'Chipko' protest to save the trees, *Deccan Chronicle*, 28 October 2016, www.deccanchronicle.com/nation/current-affairs/281016/steel-flyover-a-chipko-protest-to-save-the-trees.html

DeAngelis T (2004) Are beliefs inherited? *Monitor on Psychology*, April 2004, 35(4), www.apa.org/monitor/apr04/beliefs

Ducharme J (2018) Stephen Hawking was an atheist. Here's what he said about God, heaven and his own death, *Time*, 14 March 2018, https://time.com/5199149/stephen-hawking-death-god-atheist/

Ecklund EH (2010) *Religion vs. Science: What Scientists Really Think*, Oxford University Press, Oxford.

Ecklund EH et al. (2016) Religion among scientists in international context: A new study of scientists in eight regions, *Socius: Sociological Research for a Dynamic World*, 2:1—9, 1 September 2016, doi:10.1177/2378023116664353

Ecklund EH and Scheitle CT (2017) *Religion vs. Science: What Religious People Really Think*, Oxford University Press, Oxford.

Ecklund EH and Johnson DR (2019) Scientists talk religion around the globe, *Biologos*, 9 December 2019, https://biologos.org/articles/scientists-talk-religion-around-the-globe

Escobar H (2019) Brazil's new president has scientists worried. Here's why, *Science*, 22 January 2019, www.science.org/content/article/brazil-s-new-president-has-scientists-worried-here-s-why

FORE (2022) Chipko Movement, *Yale Forum on Religion and Ecology*, 15 December 2022, https://fore.yale.edu/World-Religions/Hinduism/Engaged-Projects/Chipko-Movement

Fox K (2022) Physicist Sabine Hossenfelder: 'There are quite a few areas where physics blurs into religion', *The Guardian* (UK), 26 November 2022, www.theguardian.com/science/2022/nov/26/physicist-sabine-hossenfelder-there-are-quite-a-few-areas-where-physics-blurs-into-religion-multiverse

Funk C and Masci D (2015) 5 facts about the interplay between religion and science, *Pew Research Center*, 22 October 2015, www.pewresearch.org/fact-tank/2015/10/22/5-facts-about-the-interplay-between-religion-and-science/

Gahlau KD (2022) What is climate change? *Australian Academy of Sciences*, 4 November 2022, www.science.org.au/learning/general-audience/science-climate-change/

Gamble D (2021) Poll: The rapid decline of US evangelicals, *Medium*, 12 July 2021, https://medium.com/science-and-critical-thinking/poll-the-rapid-decline-of-us-evangelicals-a51e4d3ea6d2

Gaw AC (2019) Religious belief at the level of the brain: Neural correlates and influence of culture, *Journal of Nervous and Mental Disorders*, doi:10.1097/NMD.0000000000001016

Goldenberg S (2013) Canadian oil minister Joe Oliver condemns climatologist James Hansen, *The Guardian* (UK), 24 April 2013, www.theguardian.com/environment/2013/apr/24/canada-joe-oliver-attack-james-hansen

Goldenberg S (2016) Climate experts urge leading scientists' association: reject Exxon sponsorship, *The Guardian* (UK), 22 February 2016, www.theguardian.com/environment/2016/feb/22/climate-change-scientists-exxonmobil-sponsorship-american-geophysical-union

Graves-Fitzsimmons G and Siddiqi M (2021) Religious Americans demand climate action, *American Progress*, 21 July 2021, www.americanprogress.org/article/religious-americans-demand-climate-action/

Hamer DH (2005) *The God Gene: How Faith is Hardwired into our Genes*, Anchor Books, New York.

Harvey F (2021) Top climate scientist warns PM over 'contemptuous' Cumbria coal mine plan, *The Guardian* (UK), 4 February 2021, www.theguardian.com/science/2021/feb/04/top-climate-scientist-warns-pm-over-contemptuous-cumbria-coalmine-plan

Heyes C et al. (2020) Knowing ourselves together: The cultural origins of metacognition, *Trends in Cognitive Sciences*, 24(5):349--362, https://doi.org/10.1016/j.tics.2020.02.007.

Hill A (2022) 'The shaman asks my spirit guides to gently cleanse me', *The Guardian* (UK), 3 December 2022, www.theguardian.com/world/2022/dec/03/the-shaman-asks-my-spirit-guides-to-gently-cleanse-me

Hopkin M (2015) James Hansen: Emissions trading won't work, but my global 'carbon fee' will, *The Conversation*, 2 December 2015, https://theconversation.com/james-hansen-emissions-trading-wont-work-but-my-global-carbon-fee-will-51676

Hossenfelder S (2022) *Existential Physics*, Viking, New York.

Hough A (2006) Not a gap in sight: Fifty years of Charles Coulson's science and Christian belief, *Theology*, 109(847):21—27, doi:10.1177/0040571X0610900104

Ironson G and Ahmad SS (2022) Praying for people you know predicts survival over 17 years among people living with HIV in the US, *Journal of Religion and Health*, 61(5):4081--4095, doi:10.1007/s10943-022-01622-5

Jenkins W et al. (2018) Religion and climate change, *Annual Review of Environment and Resources*, 43:85-108, https://doi.org/10.1146/annurev-environ-102017-025855

Kalthoff M (2015) Taming the rebel child: Science & God in the thought of CA Coulson, *The Imaginative Conservative*, https://theimaginativeconservative.org/2015/07/taming-the-rebel-child-science-technology-and-the-christian-idea-of-god-in-the-thought-of-c-a-coulson.html

Kejriwal S and Vora R (2019) In conversation with Dr Vandana Shiva: Chipko taught me humility, *Feminism in India*, 15 October 2019, https://feminisminindia.com/2019/10/15/vandana-shiva-interview-chipko-movement/

Kershaw T (2012) The religion and political views of Michio Kaku, *Hollow Verse*, 1 May 2012, https://hollowverse.com/michio-kaku

Kilvert N (2022) 'Tree hugger' might not be the insult you think it is — the term has deep historical roots, *ABC News*, 7 August 2022, https://www.abc.net.au/news/science/2022-08-07/tree-hugger-bishnoi-chipko-defiance-deep-historical-roots/101247020

Koch C et al. (2016) Neural correlates of consciousness: Progress and problems, Nature *Review of Neuroscience*, 17:307–321, https://doi.org/10.1038/nrn.2016.22

Larson EJ and Witham L (1998) Leading scientists still reject God, *Nature*, 394, 23 July 1998, 313.

Lima EC (2019) Eyeing Amazon synod, Brazil accuses church of 'leftist agenda', *National Catholic Reporter*, 26 January 2019, www.ncronline.org/Earthbeat/eyeing-amazon-synod-brazil-accuses-church-leftist-agenda

Marty C (2002) Darwin on a godless creation: "It's like confessing to a murder", *Scientific American*, 12 February 2009, www.scientificamerican.com/article/charles-darwin-confessions/#

McGrath A (2017) Making sense of reality – A scientist's journey into theology, *The Faraday Institute*, 29 September 2017, www.faraday.cam.ac.uk/churches/church-resources/posts/making-sense-of-reality-a-scientists-journey-into-theology/

McKee M (2005) Genes contribute to religious inclination, *New Scientist*, 16 March 2005, www.newscientist.com/article/dn7147-genes-contribute-to-religious-inclination/

Michael G (2011) Michio Kaku's religion of physics, *World Futures Review*, doi:10.1177/194675671100300304

Milman O (2018) Ex-Nasa scientist: 30 years on, world is failing 'miserably' to address climate change, *The Guardian* (UK), 19 June 2019, www.theguardian.com/environment/2018/jun/19/james-hansen-nasa-scientist-climate-change-warning

Mitra A (1993) Chipko: An unfinished mission, *Down to Earth*, 30 April 1993, www.downtoEarth.org.in/coverage/chipko-an-unfinished-mission-30883

Nuccitelli D (2018) 30 years later, deniers are still lying about Hansen's amazing global warming prediction, *The Guardian* (UK), 25 June 2018, www.theguardian.com/environment/climate-consensus-97-per-cent/2018/jun/25/30-years-later-deniers-are-still-lying-about-hansens-amazing-global-warming-prediction

PC (2015) The atheist: Science explains the Universe. I don't need God as well, *Premier Christianity*, 18 November 2015, www.premierchristianity.com/home/the-atheist-science-explains-the-universe-i-dont-need-god-as-well/801.article

Petruzzello M (2022) Chipko movement, *Encyclopedia Britannica*, 15 December 2022, www.britannica.com/topic/Chipko-movement

Pretorious M (2016) Is consciousness a product of the brain or/and a divine act of God? Concise insights from neuroscience and Christian theology, *HTS Theological Studies*, 76(4), http://dx.doi.org/10.4102/hts.v72i4.3472

Resentiel T (2009) Religion and science: Conflict or harmony? *Pew Research Center*, 4 May 2009, www.pewresearch.org/2009/05/04/can-science-and-religion-coexist-in-harmony/

Richardson JH (2010) This man wants to convince you global warming is a hoax, *Esquire*, 30 March 2010, www.esquire.com/news-politics/a7078/marc-morano-0410/

Robinson M (2022) A theology of the present moment, *New York Review of Books*, 22 December 2022, www.nybooks.com/articles/2022/12/22/a-theology-of-the-present-moment-marilynne-robinson/

Robson D (2022a) Beyond beliefs: Does religious faith lead to a happier, healthier life? *The Guardian* (UK), 3 December 2022, www.theguardian.com/world/2022/dec/03/beyond-beliefs-religious-faith-happier-healthier-life

Robson D (2022b) *The Expectation Effect: How Your Mindset Can Transform Your Life*, Canongate, London.

Rocha J (2013) Brazil salutes Chico Mendes 25 years after his murder, *The Guardian* (UK), 20 December 2013, www.theguardian.com/world/2013/dec/20/brazil-salutes-chico-mendes-25-years-after-murder

Rutherford A (2021) How should we address Charles Darwin's complicated legacy? *The Guardian* (UK), 13 February 2021, www.theguardian.com/science/2021/feb/13/how-should-we-address-charles-darwin-complicated-legacy

Sandell C and Blakemore B (2007) Science + religion = new alliance to save the planet, *ABC News*, 17 January 2007, https://abcnews.go.com/Technology/GlobalWarming/story?id=2800808&page=1

SMF (2019) Charles Alfred Coulson and his faith, *Science Meets Faith*, 13 December 2019, https://sciencemeetsfaith.wordpress.com/2019/12/13/charles-alfred-coulson-and-his-faith/

Smith PD (2022) *Body Am I* by Moheb Costandi review – the new science of self-consciousness, *The Guardian* (UK), 7 December 2022, www.theguardian.com/books/2022/dec/07/body-am-i-by-moheb-costandi-review-the-new-science-of-self-consciousness

Spencer N (2009) Darwin's religious beliefs, *The Faraday Institute for Science and Religion*, 19 February 2009, www.faraday.cam.ac.uk/news/darwins-religious-beliefs/

Tarhan N (2022) Consciousness, *Questions on Islam*, 5 December 2022, https://questionsonislam.com/article/consciousness

Taylor KR (2016) Francisco Alves Mendes Filho (December 15, 1944 - December 22, 1988), *Sundry Thoughts*, 9 November 2016, https://neatnik2009.wordpress.com/2016/11/09/feast-of-chico-mendes-december-22/

The Vatican (2019) Amazonia: New Paths for the Church and for an Integral Ecology, *The Vatican*, 26 October 2019, http://secretariat.synod.va/content/sinodoamazonico/en/documents/final-document-of-the-amazon-synod.html

UMC (2012) Distinct 'God spot' in the brain does not exist, study shows, *Science Daily*, 19 April 2012, www.sciencedaily.com/releases/2012/04/120419091223.htm

USCCB (1989) Statement on the death of Francisco Mendes, *The United States Conference of Catholic Bishops*, 9 January 1989, www.usccb.org/resources/statement-death-francisco-mendes-january-9-1989

Wikipedia (2022 - Amazon Rain Forest, Bishnoi, Charles Darwin, Chico Mendez, Chipko Movement, Gaura Devi, God Gene, James Edward Hansen, Jim Al-Khalili, Michio Kaku, Stephen Hawking, Sunderlal Bahuguna, The Islamic Declaration on Global Climate Change).

Yang M (2021) More Americans are shifting away from religious affiliation, new study finds, The Guardian (UK), 15 December 2021, www.theguardian.com/us-news/2021/dec/15/us-religious-affiliation-study-results

Zyga L (2011) Model predicts 'religiosity gene' will dominate society, *Phys Org*, 28 January 2011, https://phys.org/news/2011-01-religiosity-gene-dominate-society.html

AUTHOR PROFILE

It pays to keep an open mind,
but not so open your brains fall out.
Carl Sagan

If you will pursue the truth,
you will often end up having
to change your mind.
Nancy Murphy

KARIM F HIRJI is an award winning retired Professor of Medical Statistics who has published many statistical and biomedical research papers together with articles on education, politics and other issues, eight nonfiction books and one novel.

1. Hirji KF (2005) *Exact Analysis of Discrete Data*, Chapman and Hall/CRC Press, Boca Raton and London.
2. Hirji KF (editor) (2011) *Cheche: Reminiscences of a Radical Magazine*, Mkuki na Nyota Publishers, Dar es Salaam.
3. Hirji KF (2012) *Statistics in the Media: Learning from Practice*, Media Council of Tanzania, Dar es Salaam.
4. Hirji KF (2014) *Growing Up with Tanzania: Memories, Musings and Maths*, Mkuki na Nyota Publishers, Dar es Salaam.
5. Hirji KF (2017) *The Enduring Relevance of Walter Rodney's How Europe Underdeveloped Africa*, Daraja Press, Montreal.
6. Hirji KF (2017) *The Banana Girls*, Zand Graphics, Nairobi.
7. Hirji KF (2018) *The Travails of a Tanzanian Teacher*, Daraja Press, Montreal.
8. Hirji KF (2019) *Under-Education in Africa: From Colonialism to Neoliberalism*, Daraja Press, Montreal.
9. Hirji KF (2022) *Religion, Politics and Society*, Zand Graphics, Nairobi and Daraja Press, Wakefield.

He may be contacted at kfhirji@aol.com.

www.ingramcontent.com/pod-product-compliance
Lightning Source LLC
Chambersburg PA
CBHW080548270326
41929CB00019B/3229